STUDENT'S SOLUTIONS MANUAL

DAVID DUBRISKE

University of Arkansas, Fort Smith

CALCULUS AND ITS APPLICATIONS

NINTH EDITION

Marvin L. Bittinger
Indiana University Purdue University Indianapolis

David J. Ellenbogen
Community College of Vermont

PEARSON

Addison
Wesley

Boston   San Francisco   New York
London   Toronto   Sydney   Tokyo   Singapore   Madrid
Mexico City   Munich   Paris   Cape Town   Hong Kong   Montreal

Reproduced by Pearson Addison-Wesley from electronic files supplied by the author.

Copyright © 2008 Pearson Education, Inc.
Publishing as Pearson Addison-Wesley, 75 Arlington Street, Boston, MA 02116.

ISBN-13: 978-0-321-45056-2
ISBN-10: 0-321-45056-6

3 4 5 6 BRR 10 09 08 07

Table of Contents

Chapter R
Functions, Graphs, and Models

Exercise Set R.1

1. Graph $y = x + 4$.

We choose some x-values and calculate the corresponding y-values to find some ordered pairs that are solutions of the equation. Then we plot the points and connect them with a smooth curve.

x	y	(x, y)
-2	2	$(-2, 2)$
0	4	$(0, 4)$
3	7	$(3, 7)$

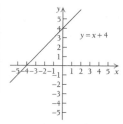

3. Graph $y = -3x$

We choose some x-values and calculate the corresponding y-values to find some ordered pairs that are solutions of the equation. Then we plot the points and connect them with a smooth curve.

x	y	(x, y)
-1	3	$(-1, 3)$
0	0	$(0, 0)$
2	-6	$(2, -6)$

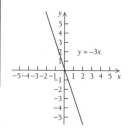

5. Graph $y = \dfrac{2}{3}x - 4$

We choose some x-values and calculate the corresponding y-values to find some ordered pairs that are solutions of the equation. Then we plot the points and connect them with a smooth curve.

x	y	(x, y)
-3	-6	$(-3, -6)$
0	-4	$(0, -4)$
3	-2	$(3, -2)$

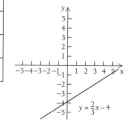

7. Graph $x + y = 5$

We solve for y first.
$$x + y = 5$$

$$y = 5 - x \qquad \text{subtract } x \text{ from both sides}$$

$$y = -x + 5 \qquad \text{cummutative property}$$

Next, we choose some x-values and calculate the corresponding y-values to find some ordered pairs that are solutions of the equation. Then we plot the points and connect them with a smooth curve.

x	y	(x, y)
-1	6	$(-1, 6)$
0	5	$(0, 5)$
2	3	$(2, 3)$

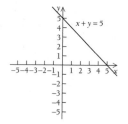

9. Graph $8y - 2x = 4$

We solve for y first.
$$8y - 2x = 4$$

$$8y = 2x + 4 \qquad \text{add } 2x \text{ to both sides}$$

$$y = \frac{2}{8}x + \frac{4}{8} \qquad \text{divide both sides by 8}$$

$$y = \frac{1}{4}x + \frac{1}{2}$$

Next, we choose some x-values and calculate the corresponding y-values to find some ordered pairs that are solutions of the equation. Then we plot the points and connect them with a smooth curve.

x	y	(x, y)
-2	0	$(-2, 0)$
2	1	$(2, 1)$
6	2	$(6, 2)$

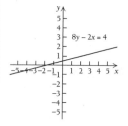

11. Graph $5x - 6y = 12$

We solve for y first.

$5x - 6y = 12$

$-6y = 12 - 5x$ 　　　subtract $5x$ from both sides

$y = \dfrac{1}{-6}(12 - 5x)$ 　　　divide both sides by -6

$y = -2 + \dfrac{5}{6}x$

$y = \dfrac{5}{6}x - 2$

We choose some x-values and calculate the corresponding y-values to find some ordered pairs that are solutions of the equation. Then we plot the points and connect them with a smooth curve.

x	y	(x, y)
-6	-7	$(-6, 7)$
0	-2	$(0, -2)$
6	3	$(6, 3)$

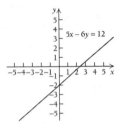

13. Graph $y = x^2 - 5$

We choose some x-values and calculate the corresponding y-values to find some ordered pairs that are solutions of the equation. Then we plot the points and connect them with a smooth curve.

x	y	(x, y)
-2	-1	$(-2, -1)$
-1	-4	$(-1, -4)$
0	-5	$(0, -5)$
1	-4	$(1, -4)$
2	-1	$(2, -1)$

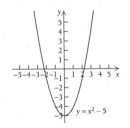

15. Graph $x = 2 - y^2$

Since x is expressed in terms of y we first choose values for y and then compute x. Then we plot the points that are found and connect them with a smooth curve.

x	y	(x, y)
-2	-2	$(-2, -2)$
1	-1	$(1, -1)$
2	0	$(2, 0)$
-1	1	$(-1, 1)$
-2	2	$(-2, 2)$

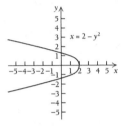

17. Graph $y = |x|$

We choose some x-values and calculate the corresponding y-values to find some ordered pairs that are solutions of the equation. Then we plot the points and connect them with a smooth curve.

x	y	(x, y)
-2	2	$(-2, 2)$
-1	1	$(-1, 1)$
0	0	$(0, 0)$
1	1	$(1, 1)$
2	2	$(2, 2)$

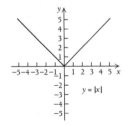

19. Graph $y = 7 - x^2$

We choose some x-values and calculate the corresponding y-values to find some ordered pairs that are solutions of the equation. Then we plot the points and connect them with a smooth curve.

x	y	(x,y)
-2	3	$(-2,3)$
-1	6	$(-1,6)$
0	7	$(0,7)$
1	6	$(1,6)$
2	3	$(2,3)$

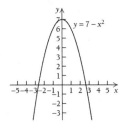

21. Graph $y + 1 = x^3$

First we solve for y.

$$y + 1 = x^3$$
$$y = x^3 - 1$$

Next, we choose some x-values and calculate the corresponding y-values to find some ordered pairs that are solutions of the equation. Then we plot the points and connect them with a smooth curve.

x	y	(x,y)
-2	-9	$(-2,-9)$
-1	-2	$(-1,-2)$
0	-1	$(0,-1)$
1	0	$(1,0)$
2	7	$(2,7)$

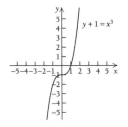

23. $R = -0.00582x + 15.3476$

We substitute 1954 in for x to get

$$R = -0.00582(1954) + 15.3476$$
$$= 3.97532$$

According to this model, the world record for the mile in 1954 is approximately 3.97532 minutes. To convert this to traditional minutes-seconds we multiply the decimal part by 60 seconds.

$$0.97532(60) = 58.5192$$

Therefore the world record for the mile in 1954 was 3:58.5.

Likewise, we substitute 2008 in for x to get

$$R = -0.00582(2008) + 15.3476$$
$$= 3.66104$$

According to the model, the world record for the mile in 2008 will be approximately 3.66104 minutes or 3:39.7.

25. $v(t) = 10.9t$

We substitute 2.34 in for t to get

$$v(2.34) = 10.9(2.34)$$
$$= 25.506$$
$$\approx 25.5$$

Durmont was traveling at 25.5 miles per hour when he reentered the halfpipe.

27. a) Locate 20 on the horizontal axis and go directly up to the graph. Then move left to the vertical axis and read the value there. We estimate the number of hearing-impaired Americans of age 20 is about 1.8 million.

Follow the same process for 40, 50, and 60 to determine the number of hearing-impaired Americans at each of those ages.

We estimate the number of hearing-impaired Americans of age 40 is about 3.7 million.

We estimate the number of hearing-impaired Americans of age 50 is about 4.4 million.

We estimate the number of hearing-impaired Americans of age 60 is about 4.5 million.

b) Locate 4 on the vertical axis and move horizontally across to the graph. There are two x-values that correspond to the y-value of 4. They are 44 and 67, so there are approximately 4 million Americans age 44 who are hearing-impaired and approximately 4 million Americans age 67 who are hearing-impaired.

c) The highest point on the graph appears to correspond to the *x*-value of 58. Therefore, age 58 appears to be the age at which the greatest number of Americans are hearing-impaired.

d) $\boxed{tw}$

29. a) $A = P(1+i)^t$

$A = 1000(1+0.06)^1$

$\quad = 1000(1.06)$

$\quad = 1060$

At the end of 1 year, $1060 is in the account.

b) $A = P\left(1+\dfrac{i}{n}\right)^{nt}$

$A = 1000\left(1+\dfrac{0.06}{2}\right)^{2\cdot 1}$

$\quad = 1000(1+0.03)^2$

$\quad = 1000(1.03)^2$

$A = 1000(1.0609)$

$\quad = 1060.90$

At the end of 1 year, $1060.90 is in the account.

c) $A = P\left(1+\dfrac{i}{n}\right)^{nt}$

$A = 1000\left(1+\dfrac{0.06}{4}\right)^{4\cdot 1}$

$\quad = 1000(1+0.015)^4$

$\quad = 1000(1.015)^4$

$\quad = 1000(1.061363551)$

$\quad = 1061.363551$

$\quad \approx 1061.36$

At the end of 1 year, approximately $1061.36 is in the account.

d) $A = P\left(1+\dfrac{i}{n}\right)^{nt}$

$A = 1000\left(1+\dfrac{0.06}{365}\right)^{365\cdot 1}$

$\quad = 1000(1+0.000164384)^{365}$

$\quad = 1000(1.000164384)^{365}$

$\quad = 1000(1.06183131)$

$\quad = 1061.83131$

$\quad \approx 1061.83$

At the end of 1 year, approximately $1061.83 is in the account.

e) There are $24 \cdot 365 = 8760$ hours in one year.

$A = P\left(1+\dfrac{i}{n}\right)^{nt}$

$A = 1000\left(1+\dfrac{0.06}{8760}\right)^{8760\cdot 1}$

$\quad = 1000(1+0.000006849)^{8760}$

$\quad = 1000(1.000006849)^{8760}$

$\quad = 1000(1.061836329)$

$\quad = 1061.836329$

$\quad \approx 1061.84$

At the end of 1 year, approximately $1061.84 is in the account.

31. a) $A = P(1+i)^t$

$A = 2000(1+0.04)^3$

$\quad = 2000(1.04)^3$

$\quad = 2249.728$

$\quad \approx 2249.73$

At the end of 3 years, approximately $2249.73 is in the account.

b) $A = P\left(1+\dfrac{i}{n}\right)^{nt}$

$A = 2000\left(1+\dfrac{0.04}{2}\right)^{2\cdot 3}$

$\quad = 2000(1.02)^6$

$\quad = 2000(1.126162419)$

$\quad = 2252.324839$

$\quad \approx 2252.32$

At the end of 3 years, $2252.32 is in the account.

c) $A = P\left(1+\dfrac{i}{n}\right)^{nt}$

$A = 2000\left(1+\dfrac{0.04}{4}\right)^{4\cdot 3}$

$\quad = 2000(1.01)^{12}$

$\quad = 2000(1.12682503)$

$\quad = 2253.65006$

$\quad \approx 2253.65$

At the end of 3 years, approximately $2253.65 is in the account.

d) $A = P\left(1+\dfrac{i}{n}\right)^{nt}$

$A = 2000\left(1+\dfrac{0.04}{365}\right)^{365\cdot 3}$

$\quad = 2000(1.000109589)^{1095}$

$\quad = 2000(1.127489438)$

$\quad = 2254.978877$

$\quad \approx 2254.98$

At the end of 3 years, approximately $2254.98 is in the account.

e) There are $24 \cdot 365 = 8760$ hours in one year.

$A = P\left(1+\dfrac{i}{n}\right)^{nt}$

$A = 2000\left(1+\dfrac{0.04}{8760}\right)^{8760\cdot 3}$

$\quad = 2000(1.000004566210046)^{26280}$

$\quad = 2000(1.127496543)$

$\quad = 2254.993083$

$\quad \approx 2254.99$

At the end of 3 years, approximately $2254.99 is in the account.

33. Using the formula:

$M = P\left[\dfrac{\dfrac{i}{12}\left(1+\dfrac{i}{12}\right)^{n}}{\left(1+\dfrac{i}{12}\right)^{n}-1}\right]$

We substitute 18,000 for P, 0.0975

$\left(9\dfrac{3}{4}\% = 0.0975\right)$ for i, and 36 $(3\cdot 12 = 36)$ for

n. Then we use a calculator to perform the computation.

$M = 18{,}000\left[\dfrac{\dfrac{0.0975}{12}\left(1+\dfrac{0.0975}{12}\right)^{36}}{\left(1+\dfrac{0.0975}{12}\right)^{36}-1}\right]$

$\quad \approx 578.70$

The monthly payment on the loan will be approximately $578.70.

35. $W = P\left[\dfrac{(1+i)^{n}-1}{i}\right]$

We substitute 3000 for P, 0.08 $(8\% = 0.08)$ in for i, and 18 for n.

$W = 3000\left[\dfrac{(1+0.08)^{18}-1}{0.08}\right]$

$\quad \approx 112{,}351$

Rounded to the nearest dollar, the annuity will be worth $112,351 after 18 years.

37. a) Locate 400,000 and 700,000 on the vertical axis and then think of horizontal lines extending across the graph from these points. The years for which the graph lies between these two lines are the years for which the population was common. Those time periods are $1800 - 1858$, $1862 - 1888$, $1900 - 1937$, $1957 - 1984$, and $1987 - 2000$.

b) Locate 700,000 on the vertical axis and then think of a horizontal line extending across the graph from this point. The years for which the graph lies above this line are the years for which the population was abundant. Those time periods are $1937 - 1957$ and $1984 - 1987$.

c) Locate 400,000 on the vertical axis and then think of a horizontal line extending across the graph from this point. The years for which the graph lies below this line are the years for which the population was scarce. Those time periods are $1858 - 1862$ and $1888 - 1900$.

39. Graph $y = x - 150$

We use the following window.

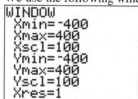

Next, we type the equation into the calculator.

```
Plot1 Plot2 Plot3
\Y1◼X-150
\Y2=
\Y3=
\Y4=
\Y5=
\Y6=
\Y7=
```

The resulting graph is:

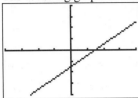

41. Graph $y = x^3 + 2x^2 - 4x - 13$

We use the following window:

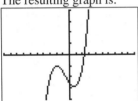

Next, we type the equation in to the calculator.

The resulting graph is:

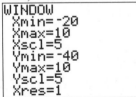

43. Graph $9.6x + 4.2y = -100$

First, we solve for y.

$9.6x + 4.2y = -100$

$4.2y = -100 - 9.6x$ subtract $9.6x$ from both sides

$$y = \frac{-9.6x - 100}{4.2}$$

Next, we set the window to be:

Next, we type the equation into the calculator.

The resulting graph is:

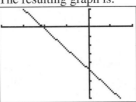

45. Graph $x = 4 + y^2$

First we solve for y.

$$x = 4 + y^2$$

$x - 4 = y^2$ subtracting 4 from both sides

$\pm\sqrt{x - 4} = y$ taking the square root of both sides

Next, we set the window to the standard window:

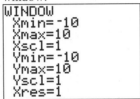

It is important to remember that we must graph both the positive and the negative root of the equation. Typing both functions into the calculator we have:

This resulting graph is:

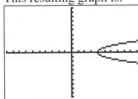

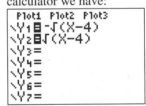

Exercise Set R.2

1. The correspondence is a function because each member of the domain corresponds to only one member of the range.

3. The correspondence is *not* a function because one member of the domain, 6, corresponds to two members of the range, -6 and -7.

5. The correspondence is a function because each member of the domain corresponds to only one member of the range, even though two members of the domain, Quarter Pounder ® and Big N' Tasty ® correspond to $2.29.

7. The correspondence is a function because each student has exactly one ID number.

9. The correspondence is a function because each student has exactly one shoe size.

11. The correspondence is a function because any number squared and then increased by 8, corresponds to exactly one number greater than or equal to 8.

13. The correspondence is a function because every male has exactly one biological father.

15. This correspondence is *not* a function, because it is reasonable to assume at least one avenue is intersected by more than one cross street.

17. The correspondence is a function because each shape has exactly one area.

19. a) $f(x) = 3x + 2$

$$f(4.1) = 3(4.1) + 2 = 14.3$$
$$f(4.01) = 3(4.01) + 2 = 14.03$$
$$f(4.001) = 3(4.001) + 2 = 14.003$$
$$f(4) = 3(4) + 2 = 14$$

x	4.1	4.01	4.001	4
$f(x)$	14.3	14.03	14.003	14

b) $f(x) = 3x + 2$

$$f(5) = 3(5) + 2 = 17$$
$$f(-1) = 3(-1) + 2 = -1$$
$$f(k) = 3(k) + 2 = 3k + 2$$
$$f(1 + t) = 3(1 + t) + 2 = 3 + 3t + 2 = 3t + 5$$
$$f(x + h) = 3(x + h) + 2 = 3x + 3h + 2$$

21. $g(x) = x^2 - 3$

$$g(-1) = (-1)^2 - 3 = 1 - 3 = -2$$
$$g(0) = (0)^2 - 3 = 0 - 3 = -3$$
$$g(1) = (1)^2 - 3 = 1 - 3 = -2$$
$$g(5) = (5)^2 - 3 = 25 - 3 = 22$$
$$g(u) = (u)^2 - 3 = u^2 - 3$$
$$g(a + h) = (a + h)^2 - 3 = a^2 + 2ah + h^2 - 3$$
$$\frac{g(a + h) - g(a)}{h} = \frac{(a + h)^2 - 3 - \left[(a)^2 - 3\right]}{h}$$
$$= \frac{a^2 + 2ah + h^2 - 3 - \left[a^2 - 3\right]}{h}$$
$$= \frac{2ah + h^2}{h}$$
$$= \frac{h(2a + h)}{h}$$
$$= 2a + h$$

23. $f(x) = \dfrac{1}{(x + 3)^2}$

a) $f(4) = \dfrac{1}{((4) + 3)^2} = \dfrac{1}{(7)^2} = \dfrac{1}{49}$

$f(-3) = \dfrac{1}{((-3) + 3)^2} = \dfrac{1}{(0)^2}$, Output is undefined.

$f(0) = \dfrac{1}{((0) + 3)^2} = \dfrac{1}{(3)^2} = \dfrac{1}{9}$

$f(a) = \dfrac{1}{((a) + 3)^2} = \dfrac{1}{(a + 3)^2}$

$f(t + 4) = \dfrac{1}{((t + 4) + 3)^2} = \dfrac{1}{(t + 7)^2}$

$f(x + h) = \dfrac{1}{((x + h) + 3)^2} = \dfrac{1}{(x + h + 3)^2}$

$$\frac{f(x+h)-f(x)}{h}$$

$$=\frac{\dfrac{1}{(x+h+3)^2}-\dfrac{1}{(x+3)^2}}{h}$$

$$=\frac{\dfrac{(x+3)^2}{(x+h+3)^2(x+3)^2}-\dfrac{(x+h+3)^2}{(x+h+3)^2(x+3)^2}}{h}$$

$$=\frac{x^2+6x+9-\left(x^2+2hx+6x+h^2+6h+9\right)}{h(x+h+3)^2(x+3)^2}$$

$$=\frac{-2hx-h^2-6h}{h(x+h+3)^2(x+3)^2}$$

$$=\frac{h(-2x-h-6)}{h(x+h+3)^2(x+3)^2}$$

$$=\frac{-2x-h-6}{(x+h+3)^2(x+3)^2},\quad h\neq 0$$

b) The function squares the input, then it adds six times the input, then it adds 9 and then it takes the reciprocal of the result.

25. Graph $f(x)=2x-5$

First, we choose some values for x and compute the values for $f(x)$, in order to form the ordered pairs that we will plot on the graph.

$$f(-1)=2(-1)-5=-7$$
$$f(0)=2(0)-5=-5$$
$$f(1)=2(1)-5=-3$$
$$f(2)=2(2)-5=-1$$

x	$f(x)$	$(x,f(x))$
-1	-7	$(-1,-7)$
0	-5	$(0,-5)$
1	-3	$(1,-3)$
2	-1	$(2,-1)$

Next we plot the input – output pairs from the table and, in this case, draw the line to complete the graph.

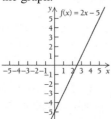

27. Graph $g(x)=-4x$

First, we choose some values for x and compute the values for $g(x)$, in order to form the ordered pairs that we will plot on the graph.

$$g(-1)=-4(-1)=4$$
$$g(0)=-4(0)=0$$
$$g(1)=-4(1)=-4$$

x	$g(x)$	$(x,g(x))$
-1	4	$(-1,4)$
0	0	$(0,0)$
1	-4	$(0,-4)$

Next we plot the input – output pairs from the table and, in this case, draw the line to complete the graph.

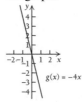

29. Graph $f(x)=x^2-2$

First, we choose some values for x and compute the values for $f(x)$, in order to form the ordered pairs that we will plot on the graph.

$$f(-2)=(-2)^2-2=2$$
$$f(-1)=(-1)^2-2=-1$$
$$f(0)=(0)^2-2=-2$$
$$f(1)=(1)^2-2=-1$$
$$f(2)=(2)^2-2=2$$

x	$f(x)$	$(x,f(x))$
-2	2	$(-2,2)$
-1	-1	$(-1,-1)$
0	-2	$(0,-2)$
1	-1	$(1,-1)$
2	2	$(2,2)$

Next we plot the input – output pairs from the table on the previous page and, in this case, draw the curve to complete the graph.

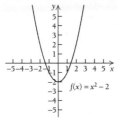

$f(x) = x^2 - 2$

31. Graph $f(x) = 6 - x^2$

First, we choose some values for x and compute the values for $f(x)$, in order to form the ordered pairs that we will plot on the graph.

$f(-2) = 6 - (-2)^2 = 2$

$f(-1) = 6 - (-1)^2 = 5$

$f(0) = 6 - (0)^2 = 6$

$f(1) = 6 - (1)^2 = 5$

$f(2) = 6 - (2)^2 = 2$

x	$f(x)$	$(x, f(x))$
-2	2	$(-2, 2)$
-1	5	$(-1, 5)$
0	6	$(0, 6)$
1	5	$(1, 5)$
2	2	$(2, 2)$

Next we plot the input – output pairs from the table and, in this case, draw the curve to complete the graph.

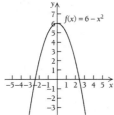

$f(x) = 6 - x^2$

33. Graph $g(x) = x^3$

First, we choose some values for x and compute the values for $g(x)$, in order to form the ordered pairs that we will plot on the graph.

$g(-2) = (-2)^3 = -8$

$g(-1) = (-1)^3 = -1$

$g(0) = (0)^3 = 0$

$g(1) = (1)^3 = 1$

$g(2) = (2)^3 = 8$

x	$f(x)$	$(x, f(x))$
-2	-8	$(-2, -8)$
-1	-1	$(-1, -1)$
0	0	$(0, 0)$
1	1	$(1, 1)$
2	8	$(2, 8)$

Next we plot the input – output pairs from the table and, in this case, draw the curve to complete the graph.

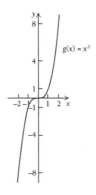

$g(x) = x^3$

35. The graph is a function, it is impossible to draw a vertical line that intersects the graph more than once.

37. The graph is a function, it is impossible to draw a vertical line that intersects the graph more than once.

39. The graph is not that of a function. A vertical line can intersect the graph more than once.

41. The graph is not that of a function. A vertical line can intersect the graph more than once.

43. The graph is a function, it is impossible to draw a vertical line that intersects the graph more than once.

45. The graph is a function, it is impossible to draw a vertical line that intersects the graph more than once.

47. Graph $x = y^2 - 2$

a) First, we choose some values for y (since x is expressed in terms of y) and compute the values for x, in order to form the ordered pairs that we will plot on the graph.

For $y = -2; x = (-2)^2 - 2 = 2$

For $y = -1; x = (-1)^2 - 2 = -1$

For $y = 0; x = (0)^2 - 2 = -2$

For $y = 1; x = (1)^2 - 2 = -1$

For $y = 2; x = (2)^2 - 2 = 2$

x	y	(x, y)
2	−2	(2, −2)
−1	−1	(−1, −1)
−2	0	(−2, 0)
−1	1	(−1, 1)
2	2	(2, 2)

Next we plot the input – output pairs from the table and, in this case, draw the curve to complete the graph.

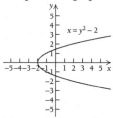

b) The graph is not that of a function. A vertical line can intersect the graph more than once.

49. $f(x) = x^2 - 3x$

$$\frac{f(x+h) - f(x)}{h}$$

$$= \frac{(x+h)^2 - 3(x+h) - \left[x^2 - 3x\right]}{h}$$

$$= \frac{x^2 + 2xh + h^2 - 3x - 3h - \left[x^2 - 3x\right]}{h}$$

$$= \frac{2xh + h^2 - 3h}{h} \quad \text{combining like terms}$$

$$= \frac{h(2x + h - 3)}{h} \quad \text{Factoring}$$

$$= 2x + h - 3, \quad h \neq 0$$

51. To find $f(-1)$ we need to locate which piece defines the function on the domain that contains $x = -1$. When $x = -1$, the function is defined by $f(x) = -2x + 1; \quad \text{for } x < 0$; therefore,

$$f(-1) = -2(-1) + 1 = 2 + 1 = 3.$$

To find $f(1)$ we need to locate which piece defines the function on the domain that contains $x = 1$. When $x = 1$, the function is defined by $f(x) = x^2 - 3; \quad \text{for } 0 < x < 4$; therefore,

$$f(1) = (1)^2 - 3 = 1 - 3 = -2.$$

53. To find $f(0)$ we need to locate which piece defines the function on the domain that contains $x = 0$.

When $x = 0$, the function is defined by $f(x) = 17; \quad \text{for } x = 0$; therefore,

$$f(0) = 17.$$

To find $f(10)$ we need to locate which piece defines the function on the domain that contains $x = 10$. When $x = 10$, the function is defined

by $f(x) = \frac{1}{2}x + 1; \quad \text{for } x \geq 4$; therefore,

$$f(10) = \frac{1}{2}(10) + 1 = 5 + 1 = 6.$$

55. Graph $f(x) = \begin{cases} 1 & \text{for } x < 0 \\ -1 & \text{for } x \geq 0 \end{cases}$

First, we graph $f(x) = 1$ for inputs less than 0. We note for any x-value less than 0, the graph is the horizontal line $y = 1$. Note that for

$$f(x) = 1$$

$$f(-3) = 1$$

$$f(-2) = 1$$

$$f(-1) = 1$$

The open circle indicates that $(0, 1)$ is not part of the graph.

Next, we graph $f(x) = -1$ for inputs greater than or equal to 0. We note for any x-value less than 0, the graph is the horizontal line $y = -1$.

Note that for $f(x) = -1$.

$$f(0) = -1$$

$$f(1) = -1$$

$$f(2) = -1$$

The solid dot indicates that $(0,-1)$ is part of the graph.

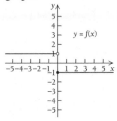

57. Graph $f(x) = \begin{cases} 6, & \text{for } x = -2 \\ x^2, & \text{for } x \neq -2 \end{cases}$

First, we graph $f(x) = 6$ for $x = -2$.

This graph consists of only one point, $(-2,6)$.

The solid dot indicates that $(-2,6)$ is part of the graph.

Next, we graph $f(x) = x^2$ for inputs $x \neq -2$.

Note that for $f(x) = x^2$

$f(-3) = (-3)^2 = 9$

$f(-1) = (-1)^2 = 1$

$f(0) = (0)^2 = 0$

$f(1) = (1)^2 = 1$

$f(2) = (2)^2 = 4$

x	$f(x)$	$(x, f(x))$
-3	9	$(-3,9)$
-1	1	$(-1,1)$
0	0	$(0,0)$
1	1	$(1,1)$
2	4	$(2,4)$

Since the input $x = -2$ is not defined on this part of the graph, the point $(-2,4)$ is not part of the graph. The open circle indicates that $(-2,4)$ is not part of the graph.

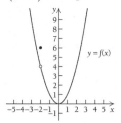

59. Graph $g(x) = \begin{cases} -x, & \text{for } x < 0 \\ 4, & \text{for } x = 0 \\ x+2, & \text{for } x > 0 \end{cases}$

First, we graph $g(x) = -x$ for inputs $x < 0$.

Creating the input – output table, we have:

x	$g(x)$	$(x, g(x))$
-3	3	$(-3,3)$
-2	2	$(-2,2)$
-1	1	$(-1,1)$

The open circle indicates that $(0,0)$ is not part of the graph. The graph is shown on the next page.

Next, we graph $g(x) = 4$ for $x = 0$. This part of the graph consists of a single point. The solid dot indicates that $(0,4)$ is part of the graph.

Next, we graph $g(x) = x + 2$ for inputs $x > 0$.

Creating the input – output table, we have:

x	$g(x)$	$(x, g(x))$
1	3	$(1,3)$
2	4	$(2,4)$
3	5	$(3,5)$

The open circle indicates that $(0,2)$ is not part of the graph. The graph is shown on the next page.

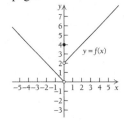

61. Graph $g(x) = \begin{cases} \frac{1}{2}x - 1, & \text{for } x < 2 \\ -4, & \text{for } x = 2 \\ x - 3, & \text{for } x > 2 \end{cases}$

First, we graph $g(x) = \frac{1}{2}x - 1$ for inputs $x < 2$.

Creating the input – output table, we have:

x	$g(x)$	$(x, g(x))$
-2	-2	$(-2,-2)$
0	-1	$(0,-1)$
1	$-\frac{1}{2}$	$\left(1, -\frac{1}{2}\right)$

The open circle on the graph indicates that $(2,0)$ is not part of the graph.

Next, we graph $g(x) = -4$ for $x = 2$. This part of the graph consists of a single point. The solid dot indicates that $(2,-4)$ is part of the graph.

Next, we graph $g(x) = x - 3$ for inputs $x > 2$. Creating the input – output table, we have:

x	$g(x)$	$(x, g(x))$
3	0	$(3,0)$
4	1	$(4,1)$
5	2	$(5,2)$

The open circle indicates that $(2,-1)$ is not part of the graph.

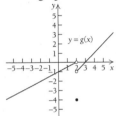

63. Graph $f(x) = \begin{cases} -7, & \text{for } x = 2 \\ x^2 - 3, & \text{for } x \neq 2 \end{cases}$

First, we graph $f(x) = -7$ for $x = 2$.

This graph consists of only one point, $(2,-7)$.

The solid dot indicates that $(2,-7)$ is part of the graph.

Next, we graph $f(x) = x^2 - 3$ for inputs $x \neq -2$. Note that for $f(x) = x^2 - 3$

$f(-3) = (-3)^2 - 3 = 6$

$f(-1) = (-1)^2 - 3 = -2$

$f(0) = (0)^2 - 3 = -3$

$f(1) = (1)^2 - 3 = -2$

$f(3) = (3)^3 - 3 = 6$

x	$f(x)$	$(x, f(x))$
-3	6	$(-3,6)$
-1	-2	$(-1,-2)$
0	-3	$(0,-3)$
1	-2	$(1,-2)$
3	6	$(3,6)$

Since the input $x = 2$ is not defined on this part of the graph, the point $(2,1)$ is not part of the graph. The open circle indicates that $(2,1)$ is not part of the graph.

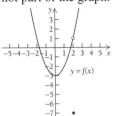

65. a) Locate 5 on the horizontal axis and move directly up to the graph. Then move across to the vertical axis and read the value there. We estimate that approximately 600 million CD's were sold in 2005.

b) Locate 710 (million) on the vertical axis and move horizontally across to the graph. The input that corresponds to 710 is approximately 1. Therefore, we estimate that in 2001 approximately 710 million CD's were sold.

67. a) Yes, each graph is a function. We can test this by applying the vertical line test. We see that every vertical line will intersect both graphs exactly once.

b) Locate 1990 on the horizontal axis and move directly up to the blue graph (gross receipts). Then move across to the vertical axis and read the value there. We estimate that the gross receipts of the federal government in 1990 were approximately $1050 billion.

c) Locate 800 (billion) on the vertical axis and move horizontally across to the red graph (gross outlays). Then move down to the horizontal axis and read the value of the input there. We estimate that gross federal outlays were $800 billion around the years 1982 or 1983.

69. $A(t) = P\left(1 + \dfrac{0.06}{4}\right)^{4t}$

We substitute 500 in for P and 2 in for t.

$A(t) = 500\left(1 + \dfrac{0.06}{4}\right)^{4 \cdot 2}$

$= 500(1.015)^8$

$= 500(1.126492587)$

$= 563.2462933$

≈ 563.25

The investment will be worth approximately $563.25 after 2 years.

71. $s = \sqrt{\dfrac{hw}{3600}}$

   a) We substitute 170 for h and 70 for w.

$s = \sqrt{\dfrac{(170)(70)}{3600}} \approx 1.818$

The patient's approximate surface area is $1.818 \, m^2$

   b) We substitute 170 for h and 100 for w.

$s = \sqrt{\dfrac{(170)(100)}{3600}} \approx 2.173$

The patient's approximate surface area is $2.173 \, m^2$

   c) We substitute 170 for h and 50 for w.

$s = \sqrt{\dfrac{(170)(50)}{3600}} \approx 1.537$

The patient's approximate surface area is $1.537 \, m^2$

73. a) Yes, the table represents a function. Each event is assigned exactly one scale of impact number.

   b) The inputs are the events; the outputs are the scale of impact numbers.

75. First we solve the equation for y.

$2y^2 + 3x = 4x + 5$

$2y^2 = x + 5$ subtract $3x$ from both sides

$y^2 = \dfrac{x + 5}{2}$ divide both sides by 2

$y = \pm\sqrt{\dfrac{x + 5}{2}}$ take the square root of both sides

We sketch a graph of the function.

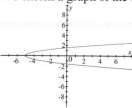

We can see that a vertical line will intersect the graph more than once; therefore, this is not a function.

77. First, we solve the equation for y.

$\left(3y^{3/2}\right)^2 = 72x$

$9y^3 = 72x$

$y^3 = 8x$

$y = \sqrt[3]{8x}$

$y = 2\sqrt[3]{x}$ $[y \geq 0]$

Note: since y must be positive to satisfy the original equation, we only graph the positive values of the function.

Next, we sketch a graph of the equation:

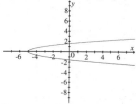

No vertical line meets the graph more than once. Thus, the equation represents a function.

79. $\boxed{tw}$ Yes, $x = 4$ is in the domain of f in Exercises 51-54. The function is defined for all values of x.

81. $f(x) = \dfrac{3}{x^2 - 4}$

We begin by setting up the table in the following manner:

Next, we will type in the equation into the graphing editor.

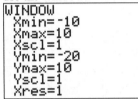

Now, we are able to look at the table:

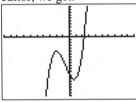

83. Each graph is shown below.

In order to graph $f(x) = x^3 + 2x^2 - 4x - 13$, we use the window:

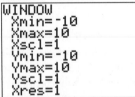

After entering the function into the graphing editor, we get:

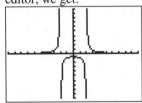

In order to graph $f(x) = \dfrac{3}{x^2 - 4}$, we use the standard window:

After entering the function into the graphing editor, we get:

In order to graph $f(x) = |x - 2| + |x + 1| - 5$, we use the standard window.

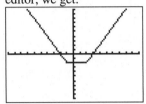

After entering the function into the graphing editor, we get:

Exercise Set R.3

1. $[-2, 4]$

3. $(0, 5)$

5. $[-9, -4)$

7. $[x, x + h]$

9. (p, ∞)

11. $[-2, 2]$

13. $[-4, -1)$

15. $(-\infty, -2]$

17. $(-2, 3]$

19. $(-\infty, 12.5)$

21. a) First, we locate 1 on the horizontal axis and then we look vertically to find the point on the graph for which 1 is the first coordinate. From that point, we look to the vertical axis to find the corresponding y-coordinate, 3. Thus, $f(1) = 3$.

   b) The domain is the set of all x-values of the points on the graph. The domain is $\{-3, -1, 1, 3, 5\}$.

   c) First, we locate 2 on the vertical axis and then we look horizontally to find any points on the graph for which 2 is the second coordinate. One such point exists, $(3, 2)$.

Thus the x-value for which $f(x) = 2$ is

$x = 3$.

   d) The range is the set of all y-values of the points on the graph. The range is $\{-2, 0, 2, 3, 4\}$.

23. a) First, we locate 1 on the horizontal axis and then we look vertically to find the point on the graph for which 1 is the first coordinate. From that point, we look to the vertical axis to find the corresponding y-coordinate, 4. Thus, $f(1) = 4$.

   b) The domain is the set of all x-values of the points on the graph. The domain is $\{-5, -3, 1, 2, 3, 4, 5\}$.

   c) First, we locate 2 on the vertical axis and then we look horizontally to find any points on the graph for which 2 is the second coordinate. Three such point exists, $(-5, 2); (-3, 2);$ and $(4, 2)$. Thus the x-values for which $f(x) = 2$ are $\{-5, -3, 4\}$.

   d) The range is the set of all y-values of the points on the graph. The range is $\{-3, 2, 4, 5\}$

25. a) First, we locate 1 on the horizontal axis and then we look vertically to find the point on the graph for which 1 is the first coordinate. From that point, we look to the vertical axis to find the corresponding y-coordinate, -1. Thus, $f(1) = -1$.

   b) The domain is the set of all x-values of the points on the graph. These extend from -2 to 4. Thus, the domain is $\{x \mid -2 \le x \le 4\}$, or in interval notation $[-2, 4]$.

   c) First, we locate 2 on the vertical axis and then we look horizontally to find any points on the graph for which 2 is the second coordinate. One such point exists, $(3, 2)$.

Thus the x-value for which $f(x) = 2$ is

$x = 3$.

   d) The range is the set of all y-values of the points on the graph. These extend from -3 to 3. Thus, the range is $\{y \mid -3 \le y \le 3\}$, or, in interval notation $[-3, 3]$.

27. a) First, we locate 1 on the horizontal axis and then we look vertically to find the point on the graph for which 1 is the first coordinate. From that point, we look to the vertical axis to find the corresponding y-coordinate, -2. Thus, $f(1) = -2$.

   b) The domain is the set of all x-values of the points on the graph. These extend from -4 to 2. Thus, the domain is $\{x \mid -4 \le x \le 2\}$, or, in interval notation $[-4, 2]$.

   c) First, we locate 2 on the vertical axis and then we look horizontally to find any points on the graph for which 2 is the second coordinate. One such point exists, $(-2, 2)$.

Thus the x-value for which $f(x) = 2$ are

$x = -2$.

   d) The range is the set of all y-values of the points on the graph. These extend from -3 to 3. Thus, the range is $\{y \mid -3 \le y \le 3\}$, or in interval notation $[-3, 3]$.

29. a) First, we locate 1 on the horizontal axis and then we look vertically to find the point on the graph for which 1 is the first coordinate. From that point, we look to the vertical axis to find the corresponding y-coordinate, 3. Thus, $f(1) = 3$.

   b) The domain is the set of all x-values of the points on the graph. These extend from -3 to 3. Thus, the domain is $\{x \mid -3 \le x \le 3\}$, or, in interval notation $[-3, 3]$.

   c) First, we locate 2 on the vertical axis and then we look horizontally to find any points on the graph for which 2 is the second coordinate. Two such point exists, $(-1.4, 2)$ and $(1.4, 2)$. Thus the x-values for which $f(x) = 2$ are $\{-1.4, 1.4\}$.

d) The range is the set of all y-values of the points on the graph. These extend from -5 to 4. Thus, the range is $\{y \mid -5 \le y \le 4\}$, or in interval notation $[-5, 4]$.

31. a) First, we locate 1 on the horizontal axis and then we look vertically to find the point on the graph for which 1 is the first coordinate. From that point, we look to the vertical axis to find the corresponding y-coordinate, 1. Thus, $f(1) = 1$.

b) The domain is the set of all x-values of the points on the graph. These extend from -5 to 5. However, the open circle at the point $(5, 2)$ indicates that 5 is not in the domain. Thus, the domain is $\{x \mid -5 \le x < 5\}$, or in interval notation $[-5, 5)$.

c) First, we locate 2 on the vertical axis and then we look horizontally to find any points on the graph for which 2 is the second coordinate. We notice all the points with x-values in the set $\{x \mid 3 \le x < 5\}$ Thus the x-values for which $f(x) = 2$ are $\{x \mid 3 \le x < 5\}$, or $[3, 5)$.

d) The range is the set of all y-values of the points on the graph. The range is $\{-2, -1, 0, 1, 2\}$.

33. $f(x) = \dfrac{6}{2 - x}$

Since the function value cannot be calculated when the denominator is equal to 0, we solve the following equation to find those real numbers that must be excluded from the domain of f.

$2 - x = 0$ setting the denominator equal to 0

$2 = x$ adding x to both sides

Thus, 2 is not in the domain of f, while all other real numbers are. The domain of f is $\{x \mid x \text{ is a real number and } x \ne 2\}$; or, in interval notation, $(-\infty, 2) \cup (2, \infty)$.

35. $f(x) = \sqrt{2x}$

Since the function value cannot be calculated when the radicand is negative, the domain is all real numbers for which $2x \ge 0$. We find them by solving the inequality.

$2x \ge 0$ setting the radicand ≥ 0

$x \ge 0$ dividing both sides by 2

The domain of f is $\{x \mid x \text{ is a real number and } x \ge 0\}$; or, in interval notation, $[0, \infty)$.

37. $f(x) = x^2 - 2x + 3$

We can calculate the function value for all values of x, so the domain is the set of all real numbers .

39. $f(x) = \dfrac{x - 2}{6x - 12}$

Since the function value cannot be calculated when the denominator is equal to 0, we solve the following equation to find those real numbers that must be excluded from the domain of f.

$6x - 12 = 0$ setting the denominator equal to 0

$6x = 12$ adding 12 to both sides

$x = 2$ dividing both sides by 6

Thus, 2 is not in the domain of f, while all other real numbers are. The domain of f is $\{x \mid x \text{ is a real number and } x \ne 2\}$; or, in interval notation, $(-\infty, 2) \cup (2, \infty)$

41. $f(x) = |x - 4|$

We can calculate the function value for all values of x, so the domain is the set of all real numbers .

43. $f(x) = \dfrac{3x - 1}{7 - 2x}$

Since the function value cannot be calculated when the denominator is equal to 0, we solve the equation on the following page to find those real numbers that must be excluded from the domain of f.

$7 - 2x = 0$ setting the denominator equal to 0

$7 = 2x$ adding $2x$ to both sides

$\dfrac{7}{2} = x$ dividing both sides by 2

Thus, $\dfrac{7}{2}$ is not in the domain of f, while all other real numbers are. The domain of f is $\left\{x \mid x \text{ is a real number and } x \neq \dfrac{7}{2}\right\}$; or, in interval notation, $\left(-\infty, \dfrac{7}{2}\right) \cup \left(\dfrac{7}{2}, \infty\right)$.

45. $g(x) = \sqrt{4 + 5x}$

Since the function value cannot be calculated when the radicand is negative, the domain is all real numbers for which $4 + 5x \geq 0$. We find them by solving the inequality.

$\begin{aligned} 4 + 5x &\geq 0 && \text{setting the radicand } \geq 0 \\ 5x &\geq -4 && \text{subtracting 4 from both sides} \\ x &\geq -\dfrac{4}{5} && \text{dividing both sides by 5} \end{aligned}$

The domain of g is $\left\{x \mid x \text{ is a real number and } x \geq -\dfrac{4}{5}\right\}$; or, in interval notation, $\left[-\dfrac{4}{5}, \infty\right)$.

47. $g(x) = x^2 - 2x + 1$

We can calculate the function value for all values of x, so the domain is the set of all real numbers .

49. $g(x) = \dfrac{2x}{x^2 - 25}$

Since the function value cannot be calculated when the denominator is equal to 0, we solve the following equation to find those real numbers that must be excluded from the domain of g.

$\begin{aligned} x^2 - 25 &= 0 && \text{setting the denominator equal to 0} \\ x^2 &= 25 && \text{adding 25 to both sides} \\ x &= \pm\sqrt{25} && \text{taking the square root or both sides} \\ x &= \pm 5 \end{aligned}$

Thus, -5 and 5 are not in the domain of g, while all other real numbers are. The domain of g is $\{x \mid x \text{ is a real number and } x \neq -5, \; x \neq 5\}$; or, in interval notation, $(-\infty, -5) \cup (-5, 5) \cup (5, \infty)$.

51. $g(x) = |x| + 1$

We can calculate the function value for all values of x, so the domain is the set of all real numbers .

53. $g(x) = \dfrac{2x - 6}{x^2 - 6x + 5}$

Since the function value cannot be calculated when the denominator is equal to 0, we solve the following equation to find those real numbers that must be excluded from the domain of f.

$\begin{aligned} x^2 - 6x + 5 &= 0 && \text{setting the denominator equal to 0} \\ (x - 5)(x - 1) &= 0 && \text{factoring the quadratic equation} \\ x - 5 = 0 \;\; &\text{or} \;\; x - 1 = 0 && \text{Using the principle of zero products} \\ x = 5 \;\; &\text{or} \quad\;\; x = 1 \end{aligned}$

Thus, 1 and 5 are not in the domain of g, while all other real numbers are. The domain of g is $\{x \mid x \text{ is a real number and } x \neq 1, x \neq 5\}$; or, in interval notation, $(-\infty, 1) \cup (1, 5) \cup (5, \infty)$.

55. The graph of f lies on or below the x-axis when $f(x) \leq 0$, so we scan the graph from left to right looking for the values of x for which the graph lies on or below the x axis. Those values extend from -1 to 2. So the set of x-values for which $f(x) \leq 0$ is $\{x \mid -1 \leq x \leq 2\}$, or, in interval notation, $[-1, 2]$

57. a) We use the compound interest formula from Theorem 2 in section R.1 and substitute 5000 for P, 2 for n and 0.08 (8%) for i. The equation for this function is:

$$A = P\left(1 + \dfrac{i}{n}\right)^{nt}$$

$$A = 5000\left(1 + \dfrac{0.08}{2}\right)^{2t}$$

$$A = 5000(1.04)^{2t}$$

b) The independent variable t is the time in years the principal has been invested in the account. It would not make sense to have time be a negative number in this case. Therefore, the domain is the set of all non-negative real numbers. $\{t \mid 0 \leq t < \infty\}$.

59. a) The graph extends from $x = 0$ to $x = 84.7$, so the domain, in interval notation, of the function N is $[0, 84.7]$

b) The graph extends from $N(x) = 0$ to $N(x) = 4.6$ million. Therefore, the range, in interval notation, of the function N is $[0, 4,600,000]$.

c) $\boxed{tw}$ Answers will vary. We would target the 50 year old to 60 year old age group, because that is the age group that has the most number of hearing-impaired Americans.

61. a) The graph extends from $t = 0$ to $t = 65$, so the domain, in interval notation, of the function L is $[0, 65]$.

b) The graph extends from $L(x) = 8$ to $L(x) = 75$, so the range, in interval notation, of the function L is $[8, 75]$.

63. $\boxed{tw}$ Answers may vary. Some possible answers are. The output -5 corresponds with the input 2. The point $(2, -5)$ is on the graph of the function. The value $x = 2$ is the solution to the equation $f(x) = -5$.

65. $\boxed{tw}$ Answers may vary. Consider the function $f(x) = \dfrac{1}{x-3}$. Here the number 3 is not in the domain of f because replacing x with 3 results in division by zero.

67. The range in interval notation for each function is:

Exercise 34: $(-\infty, 0) \cup (0, \infty)$

Exercise 36: $[0, \infty)$

Exercise 48: $(-\infty, \infty)$

Exercise 51: $[1, \infty)$

Exercise 54: $(-\infty, \infty)$

Exercise Set R.4

1. Graph $x = 3$.

The graph consists of all ordered pairs whose first coordinate is 3. This results in a vertical line whose x-intercept is the point $(3, 0)$

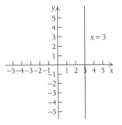

3. Graph $y = -2$.

The graph consists of all ordered pairs whose second coordinate is -2. This results in a horizontal line whose y-intercept is the point $(0, -2)$

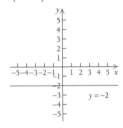

5. Graph $x = -4.5$.

The graph consists of all ordered pairs whose first coordinate is -4.5. This results in a vertical line whose x-intercept is the point $(-4.5, 0)$

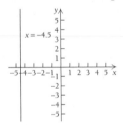

7. Graph $y = 3.75$.

The graph consists of all ordered pairs whose second coordinate is 3.75. This results in a horizontal line whose y-intercept is the point $(0, 3.75)$

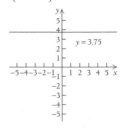

9. Graph $y = -2x$.

Using Theorem 4, The graph of y is the straight line through the origin $(0, 0)$ and the point $(1, -2)$. We plot these two points and connect them with a straight line.

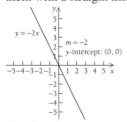

The function $y = -2x$ has slope -2, and y-intercept $(0, 0)$.

11. Graph $f(x) = 0.5x$.

Using Theorem 4, The graph of $f(x)$ is the straight line through the origin $(0, 0)$ and the point $(1, 0.5)$. We plot these two points and connect them with a straight line.

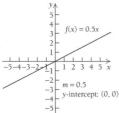

The function $f(x) = 0.5x$ has slope 0.5, and y-intercept $(0, 0)$.

13. Graph $y = 3x - 4$

First, we make a table of values. We choose any number for x and then determine y by substitution.

When $x = -1, y = 3(-1) - 4 = -7$.

When $x = 0, y = 3(0) - 4 = -4$.

When $x = 2, y = 3(2) - 4 = 2$.

We organize these values into an input – output table

x	y	(x, y)
-1	-7	$(-1, -7)$
0	-4	$(0, -4)$
2	2	$(2, 2)$

Next, we plot these ordered pairs and connect them with a straight line.

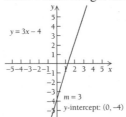

The function $y = 3x - 4$ has slope 3, and y-intercept $(0, -4)$.

15. Graph $g(x) = -x + 3$.

First, we make a table of values. We choose any number for x and then determine y by substation. Because the function is linear, we save ourselves some work by only plotting two points.

When $x = 0, g(0) = -(0) + 3 = 3$.

When $x = 1, g(1) = -(1) + 3 = 2$.

We organize these values into an input – output table

x	$g(x)$	$(x, g(x))$
0	3	$(0, 3)$
1	2	$(1, 2)$

Next, we plot these ordered pairs from the table and connect them with a straight line.

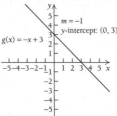

The function $g(x) = -x + 3$ has slope -1, and y-intercept $(0, 3)$.

17. Graph $y = 7$.

The graph consists of all ordered pairs whose second coordinate is 7. This results in a horizontal line, whose y-intercept is the point $(0, 7)$

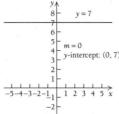

Since the graph is horizontal, the slope is 0 and the y-intercept is $(0, 7)$.

19. First, we solve the equation for y.
$$y - 3x = 6$$
$$y = 3x + 6 \qquad \text{adding } 3x \text{ to both sides}$$
The slope is 3.
The y-intercept is $(0, 6)$.

21. First, we solve the equation for y.
$$2x + y - 3 = 0$$
$$2x + y = 3 \qquad \text{adding 3 to both sides}$$
$$y = -2x + 3 \qquad \text{subtracting } 2x \text{ from both sides}$$
The slope is -2.
The y-intercept is $(0, 3)$.

23. First, we solve the equation for y.
$$2x + 2y + 8 = 0$$
$$2x + 2y = -8 \qquad \text{subtracting 8}$$
$$2y = -2x - 8 \qquad \text{subtracting } 2x$$
$$y = \frac{-2x - 8}{2} \qquad \text{dividing both sides by 2}$$
$$y = -x - 4 \qquad \text{simplifying}$$
The slope is -1.
The y-intercept is $(0, -4)$.

25. First, we solve the equation for y.
$$x = 3y + 7$$
$$3y + 7 = x \qquad \text{cummutative property of equality}$$
$$3y = x - 7 \qquad \text{subtracting 7}$$
$$y = \frac{x - 7}{3} \qquad \text{dividing by 3}$$
$$y = \frac{1}{3}x - \frac{7}{3} \qquad \text{simplifying}$$
The slope is $\frac{1}{3}$.
The y-intercept is $\left(0, -\frac{7}{3}\right)$.

27. $y - y_1 = m(x - x_1)$
$$y - (-3) = -5(x - (-2)) \qquad \text{Substituting}$$
$$y + 3 = -5(x + 2) \qquad \text{Simplifying}$$
$$y + 3 = -5x - 10$$
$$y = -5x - 13 \qquad \text{Subtracting 3}$$

29. $y - y_1 = m(x - x_1)$
$$y - (3) = -2(x - (2)) \qquad \text{Substituting}$$
$$y - 3 = -2x + 4$$
$$y = -2x + 7 \qquad \text{Adding 3}$$

31. $y - y_1 = m(x - x_1)$
$$y - (0) = 2(x - (3)) \qquad \text{Substituting}$$
$$y = 2x - 6$$

33. $y = mx + b$
$$y = \frac{1}{2}x + (-6) \qquad \text{Substituting}$$
$$y = \frac{1}{2}x - 6 \qquad \text{Simplifying}$$

35. $y - y_1 = m(x - x_1)$

$y - (3) = 0(x - (2))$ Substituting

$y - 3 = 0$ Simplifying

$y = 3$ Adding 3

37. $m = \dfrac{y_2 - y_1}{x_2 - x_1}$

Substituting, we have:

$m = \dfrac{1 - (-3)}{-2 - 5}$

$= \dfrac{1 + 3}{-2 - 5}$

$= \dfrac{4}{-7}$

$= -\dfrac{4}{7}$

Note: it does not matter which point is taken first, as long as we subtract coordinates in the same order. We could also find the slope as follows.

$m = \dfrac{(-3) - 1}{5 - (-2)}$

$= \dfrac{-3 - 1}{5 + 2}$

$= \dfrac{-4}{7}$

$= -\dfrac{4}{7}$

39. $m = \dfrac{y_2 - y_1}{x_2 - x_1}$

Substituting, we have:

$m = \dfrac{-4 - (-3)}{-1 - 2}$

$= \dfrac{-4 + 3}{-1 - 2}$

$= \dfrac{-1}{-3}$

$= \dfrac{1}{3}$

41. $m = \dfrac{y_2 - y_1}{x_2 - x_1}$

Substituting, we have:

$m = \dfrac{-9 - (-7)}{3 - 3}$

$= \dfrac{-9 + 7}{3 - 3}$

$= \dfrac{-2}{0}$

Since we cannot divide by 0, the slope is undefined.

43. $m = \dfrac{y_2 - y_1}{x_2 - x_1}$

Substituting, we have:

$m = \dfrac{\dfrac{2}{5} - (-3)}{\dfrac{1}{2} - \dfrac{4}{5}}$

$= \dfrac{\dfrac{2}{5} - \left(-\dfrac{15}{5}\right)}{\dfrac{5}{10} - \dfrac{8}{10}}$ finding a common denominator

$= \dfrac{\dfrac{17}{5}}{-\dfrac{3}{10}}$ adding fractions

$= \dfrac{17}{5} \cdot \left(-\dfrac{10}{3}\right)$ Multiplying by the reciprical

$= -\dfrac{34}{3}$

45. $m = \dfrac{y_2 - y_1}{x_2 - x_1}$

Substituting, we have:

$m = \dfrac{3 - (3)}{-1 - 2}$

$= \dfrac{0}{-3}$

$= 0$

47. $m = \dfrac{y_2 - y_1}{x_2 - x_1}$

Substituting, we have:

$m = \dfrac{3(x+h)-(3x)}{(x+h)-x}$

$= \dfrac{3x+3h-3x}{x+h-x}$

$= \dfrac{3h}{h}$

$= 3$

49. $m = \dfrac{y_2 - y_1}{x_2 - x_1}$

Substituting, we have:

$m = \dfrac{[2(x+h)+3]-(2x+3)}{(x+h)-x}$

$= \dfrac{2x+2h+3-(2x+3)}{x+h-x}$

$= \dfrac{2h}{h}$

$= 2$

51. From Exercise 37, we know that the slope is $-\dfrac{4}{7}$. Using the point $(5,-3)$, we substitute into the point-slope equation.

$y-(-3) = -\dfrac{4}{7}(x-5)$

$y+3 = -\dfrac{4}{7}x+\dfrac{20}{7}$

$y = -\dfrac{4}{7}x+\dfrac{20}{7}-3$

$y = -\dfrac{4}{7}x-\dfrac{1}{7}$

Note: We could have found the equation of the line using the point $(-2,1)$.

$y-(1) = -\dfrac{4}{7}(x-(-2))$

$y-1 = -\dfrac{4}{7}x-\dfrac{8}{7}$

$y = -\dfrac{4}{7}x-\dfrac{8}{7}+1$

$y = -\dfrac{4}{7}x-\dfrac{1}{7}$

53. From Exercise 39, we know that the slope is $\dfrac{1}{3}$. Using the point $(2,-3)$, we substitute into the point-slope equation.

$y-(-3) = \dfrac{1}{3}(x-2)$

$y+3 = \dfrac{1}{3}x-\dfrac{2}{3}$

$y = \dfrac{1}{3}x-\dfrac{2}{3}-3$

$y = \dfrac{1}{3}x-\dfrac{11}{3}$

Note: We could have found the equation of the line using the point $(-1,-4)$.

$y-(-4) = \dfrac{1}{3}(x-(-1))$

$y+4 = \dfrac{1}{3}x+\dfrac{1}{3}$

$y = \dfrac{1}{3}x+\dfrac{1}{3}-4$

$y = \dfrac{1}{3}x-\dfrac{11}{3}$

55. From Exercise 41, we know that the slope is undefined. The graph is a line which contains all ordered pairs whose first coordinate is 3. The equation of the line is $x = 3$

57. From Exercise 43, we know that the slope is $-\dfrac{34}{3}$. Using the point $\left(\dfrac{4}{5},-3\right)$, we substitute into the point-slope equation.

$y-(-3) = -\dfrac{34}{3}\left(x-\dfrac{4}{5}\right)$

$y+3 = -\dfrac{34}{3}x+\dfrac{136}{15}$

$y = -\dfrac{34}{3}x+\dfrac{136}{15}-3$

$y = -\dfrac{34}{3}x+\dfrac{91}{15}$

Note, in Exercise 57, we could have found the equation of the line using the point $\left(\dfrac{1}{2},\dfrac{2}{5}\right)$.

$$y-\left(\dfrac{2}{5}\right)=-\dfrac{34}{3}\left(x-\dfrac{1}{2}\right)$$

$$y-\dfrac{2}{5}=-\dfrac{34}{3}x+\dfrac{17}{3}$$

$$y=-\dfrac{34}{3}x+\dfrac{17}{3}+\dfrac{2}{5}$$

$$y=-\dfrac{34}{3}x+\dfrac{91}{15}$$

59. From Exercise 45, we know that the slope is 0. The line is horizontal, thus the equation is $y=3$.

61. $\text{Slope}=\dfrac{2\,ft}{5\,ft}=\dfrac{2}{5}=0.4$

The slope of the skateboard ramp is 0.4.

63. $\text{Slope}=\dfrac{43.33\,ft}{1238\,ft}=\dfrac{43.33}{1238}=\dfrac{7}{200}=0.035$

Expressing the slope as a percentage, we find the head of the river is 3.5%.

65. The average rate of change can be found using the coordinates of any two points on the line. We use the given coordinates $(2000,6438)$ and $(2005,10,880)$.

$$\text{Rate of Change}=\dfrac{\text{change in premiums}}{\text{corresponding change in time}}$$

$$=\dfrac{10,880-6438}{2005-2000}$$

$$=\dfrac{4442}{5}$$

$$=888.40$$

The average rate of change in annual premium for a family's health insurance was $888 per year.

67. The average rate of change can be found using the coordinates of any two points on the line. We use the given coordinates $(1995,1192)$ and $(2004,1670)$.

$$\text{Rate of Change}=\dfrac{1670-1192}{2004-1995}$$

$$=\dfrac{478}{9}\approx 53.11$$

The average rate of change of the tuition and fees at public two-year colleges is approximately $53 per year.

69. a) If R is directly proportional to T, then there is some positive constant m such that $R=mT$.
   To find m, substitute 12.51 for R, and 3 for T into the equation $R=mT$ and solve for m.
   $$12.51=m\cdot 3$$
   $$\dfrac{12.51}{3}=m$$
   $$4.17=m$$
   The variation constant $m=4.17$. The equation of variation is $R=4.17T$.

   b) Using the equation of variation found in part (a), we substitute 6 for T.
   $$R=4.17(6)$$
   $$=25.02$$
   The R-Factor for insulation that is 6 inches thick is 25.02.

71. a) Since the muscle weight M is directly proportional to the person's body weight W, there is a positive constant m such that $M=mW$.
   To find m, we substitute 80 for M and 200 for W into $M=mW$ solve for m.
   $$80=m\cdot 200$$
   $$\dfrac{80}{200}=m$$
   $$0.4=m$$
   The constant of variation $m=0.4$. The equation of variation is $M=0.4W$.

   b) The constant of variation $m=0.4$ is equivalent to 40%. Thus the equation of variation is $M=40\%W$. This implies that muscle weight, M, makes up 40% of total body weight, W.

c) Substituting 120 for W into the equation of variation we have:

$$M = 0.4(120)$$

$$= 48$$

A person weighing 120 pounds has 48 pounds of muscle weight.

73. a) If T, the amount of the toll, is directly proportional to the weight of the vehicle, w. Then there exists a constant m such that $T = mw$.

To find m, we substitute 2.70 for T and 3350 for w into $T = mw$ and solve for m.

$$2.70 = m \cdot 3350$$

$$\frac{2.70}{3350} = m$$

$$0.0008 \approx m$$

The constant of variation is $m = 0.0008$.
The equation of variation is $T = 0.0008w$.

b) Substituting 3700 for w into the equation of variation, we have:

$$T = 0.0008(3700)$$

$$= 2.96$$

The toll would be $2.96 for a Jeep Cherokee.

75. a) Total costs = Variables costs + Fixed Costs
To produce x calculators, it costs $20 dollars per calculator, that is the variable costs are $20x$. In addition to the variable costs the fixed costs are $100,000. The total cost is

$$C(x) = 20x + 100,000.$$

The graph is shown at the top of the next column.

b) Revenue = Price times Quantity.
The price of the calculator is $45, therefore, the revenue from selling x calculators is given by

$$R(x) = 45x.$$

The graph is shown at the top of the next column.

c) Profit = Revenue – Cost.
Using the Cost and Revenue functions found in part (a) and (b) we have:

$$P(x) = R(x) - C(x)$$

$$= 45x - (20x + 100,000)$$

$$= 25x - 100,000$$

The graph is shown at the top of the next column.

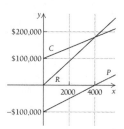

d) Substituting 150,000 for x into the profit equation, we have

$$P(150,000) = 25(150,000) - 100,000$$

$$= 3,650,000$$

The company would realize a $3,650,000 profit if they reach their expected sales of 150,000 calculators.

e) The break even point occurs when $P(x) = 0$. Therefore, we set the profit function equal to 0 and solve for x.

$$25x - 100,000 = 0$$

$$25x = 100,000$$

$$x = \frac{100,000}{25}$$

$$x = 4000$$

The company would need to sell 4000 calculators at $45 each in order to break even.

77. a) $V(t) = C - t\left(\dfrac{C - S}{N}\right)$

Substituting 5200 for C, 1100 for S, and 8 for N into the equation we have the straight line depreciation $V(t)$:

$$V(t) = 5200 - t\left(\frac{5200 - 1100}{8}\right)$$

$$= 5200 - t\left(\frac{4100}{8}\right)$$

$$= 5200 - t(512.5)$$

$$= 5200 - 512.5t$$

b) $V(t) = 5200 - 512.5t$

$V(0) = 5200 - 512.5(0) = 5200.00$

$V(1) = 5200 - 512.5(1) = 4687.50$

$V(2) = 5200 - 512.5(2) = 4175.00$

$V(3) = 5200 - 512.5(3) = 3662.50$

$V(4) = 5200 - 512.5(4) = 3150.00$

$V(7) = 5200 - 512.5(7) = 1612.50$

$V(8) = 5200 - 512.5(8) = 1100.00$

79. $V(t) = C - t\left(\dfrac{C - S}{N}\right)$

First, we figure the total cost of the improvements. 25,000 at $40 per square foot gives us a total cost of $C = (40)(25,000) = 1,000,000$.

After 39 years, the salvage value S is 0. Therefore, we substitute 1,000,000 for C, 39 for N, and 0 for S into the straight-line depreciations formula.

$V(t) = 1,000,000 - t\left(\dfrac{1,000,000 - 0}{39}\right)$

$V(t) = 1,000,000 - t(25,641.02564)$

$V(t) = 1,000,000 - 25,641.02564t$

In order to find the depreciated value of the improvements after 10 years, we substitute 10 for t.

$V(10) = 1,000,000 - 25,641.02564(10)$

$\quad = 743,589.7436$

$\quad \approx 743,589.74$

The depreciated value of the improvements after 10 years is approximately $743,590.

81. $B(t) = -700t + 3500$

a) The number -700 is the rate of change in the value of the photocopier per year. In other words, it tells us that the photocopier depreciates $700 per year.
The number 3500 is the initial value of the photocopier. In other words, the value of the photocopier at the time of purchase was $3500.

b) From the graph we see that $B(t) = 0$ when $t = 5$. However, since we have the equation, we set $B(t) = 0$ and solve for t to double check the graph.

$-700t + 3500 = 0$

$3500 = 700t$

$\dfrac{3500}{700} = t$

$5 = t$

It will take 5 years for the photocopier to depreciate completely.

c) $\boxed{tw}$

83. a) $D(r) = \dfrac{11r + 5}{10}$

$D(5) = \dfrac{11(5) + 5}{10} = 6$

When traveling 5 miles per hour, the cars reaction distance is 6 feet.

$D(10) = \dfrac{11(10) + 5}{10} = 11.5$

When traveling 10 miles per hour, the cars reaction distance is 11.5 feet.

$D(5) = \dfrac{11(20) + 5}{10} = 22.5$

When traveling 20 miles per hour, the cars reaction distance is 22.5 feet.

$D(50) = \dfrac{11(50) + 5}{10} = 55.5$

When traveling 50 miles per hour, the cars reaction distance is 55.5 feet.

$D(65) = \dfrac{11(65) + 5}{10} = 72$

When traveling 65 miles per hour, the cars reaction distance is 72 feet.

b) Plotting the points found in part (a) and connecting the points with a smooth curve we have:

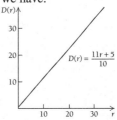

c) $\boxed{tw}$

85. a) We find the equation of the line that contains the points $(1999,1873)$ and $(2006,3116)$. First we find the slope of the line.

$$m = \frac{3116-1873}{2006-1999} = \frac{1243}{7} \approx 177.57 \approx 178$$

Next, we use the point-slope equation. We will use the point $(1999,1873)$.

$$y - y_1 = m(x - x_1)$$
$$y - 1873 = 178(x - 1999)$$
$$y - 1873 = 178x - 355,822$$
$$y = 178x - 353,949 \qquad (1)$$

Note: due to round off error, if we used the point $(2006,3116)$ we would have:

$$y - y_1 = m(x - x_1)$$
$$y - 3116 = 178(x - 2006)$$
$$y - 3116 = 178x - 357,068$$
$$y = 178x - 353,952 \qquad (2)$$

b) Using the first equation substituting 2000 for x we have:

$$y = 178(2000) - 353,949 = 2051$$

Approximately 2051 manatees were counted in January of 2000.
Alternatively, using the second equation and substituting 2000 for x we have:

$$y = 178(2000) - 353,952 = 2048$$

Approximately 2048 manatees were counted in January of 2000.

c) $\boxed{tw}$

87. a) We know that $N = P + 0.02P = 1.02P$.
Therefore the equation of variation is $N = 1.02P$.

b) Substituting 200,000 for P we have
$$N = 1.02(200,000) = 204,000.$$

The new population is 204,000 after a growth of 2%.

c)   Substituting 367,200 for N we have
$$367,200 = 1.02P$$
$$\frac{367,200}{1.02} = P$$
$$360,000 = P$$
The previous population was 360,000.

89. $\boxed{tw}$

91. a) Graph III is appropriate, because it shows the rate before January 1 is approximately $3000 per month, and the rate after January 1 is approximately $2000 per month.

b) Graph IV is appropriate, because it shows the rate before January 1 is approximately $3000 per month, and the rate after January 1 is approximately –$4000 per month.

c) Graph I is appropriate, because it shows the rate before January 1 is approximately $1000 per month, and the rate after January 1 is approximately $2000 per month.

d) Graph II is appropriate, because it shows the rate before January 1 is approximately $4000 per month, and the rate after January 1 is approximately –$2000 per month.

Exercise Set R.5

1.   Graph $y = \dfrac{1}{2}x^2$ and $y = -\dfrac{1}{2}x^2$

Starting with $y = \dfrac{1}{2}x^2$, we first find the vertex or the turning point. The x-coordinate of the vertex is

$$x = -\frac{b}{2a}$$
$$= -\frac{0}{2\left(\frac{1}{2}\right)}$$
$$= 0$$

Substituting 0 for x into the equation, we find the second coordinate of the vertex:

$$y = \frac{1}{2}(0)^2 = 0.$$

The vertex is $(0,0)$. The y-axis (The vertical line $x = 0$.) is the axis of symmetry. Next, we choose some x-values on each side of the vertex and compute the y-values.

When $x = 1, y = \dfrac{1}{2}(1)^2 = \dfrac{1}{2} \cdot 1 = \dfrac{1}{2}$

When $x = 2, y = \dfrac{1}{2}(2)^2 = \dfrac{1}{2} \cdot 4 = 2$

When $x = -1, y = \dfrac{1}{2}(-1)^2 = \dfrac{1}{2} \cdot 1 = \dfrac{1}{2}$

When $x = -2, y = \dfrac{1}{2}(-2)^2 = \dfrac{1}{2} \cdot 4 = 2$

We organize these values into an input-output table at the top of the next page.

x	y
0	0
1	$\frac{1}{2}$
2	2
−1	$\frac{1}{2}$
−2	2

We plot these points and connect them with a smooth curve on the axis below.

Next, we graph $y = -\frac{1}{2}x^2$. First, we find the vertex or the turning point. The x-coordinate of the vertex is

$$x = -\frac{b}{2a}$$

$$x = -\frac{0}{2\left(-\frac{1}{2}\right)} = 0$$

Substituting 0 for x into the equation, we find the second coordinate of the vertex:

$$y = -\frac{1}{2}(0)^2 = 0.$$

The vertex is $(0,0)$. The y-axis (The vertical line $x = 0$.) is the axis of symmetry. Next, we choose some x-values on each side of the vertex and compute the y-values.

When $x = 1, y = -\frac{1}{2}(1)^2 = -\frac{1}{2} \cdot 1 = -\frac{1}{2}$

When $x = 2, y = -\frac{1}{2}(2)^2 = -\frac{1}{2} \cdot 4 = -2$

When $x = -1, y = -\frac{1}{2}(-1)^2 = -\frac{1}{2} \cdot 1 = -\frac{1}{2}$

When $x = -2, y = -\frac{1}{2}(-2)^2 = -\frac{1}{2} \cdot 4 = -2$

x	y
0	0
1	$-\frac{1}{2}$
2	−2
−1	$-\frac{1}{2}$
−2	−2

We plot these points and connect them with a smooth curve on the axis below.

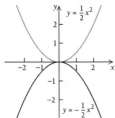

3. Graph $y = x^2$ and $y = x^2 - 1$

Starting with $y = x^2$, we first find the vertex or the turning point. The x-coordinate of the vertex is

$$x = -\frac{b}{2a} = -\frac{0}{2(1)} = 0$$

Substituting 0 for x into the equation, we find the second coordinate of the vertex:

$$y = (0)^2 = 0.$$

The vertex is $(0,0)$. The y-axis (The vertical line $x = 0$.) is the axis of symmetry. Next, we choose some x-values on each side of the vertex and compute the y-values.

When $x = 1, y = (1)^2 = 1$

When $x = 2, y = (2)^2 = 4$

When $x = -1, y = (-1)^2 = 1$

When $x = -2, y = (-2)^2 = 4$

x	y
0	0
1	1
2	4
−1	1
−2	4

We plot these points and connect them with a smooth curve on the axis below.

Next, we graph $y = x^2 - 1$. First, we find the vertex or the turning point. The x-coordinate of the vertex is

$$x = -\frac{b}{2a}$$

$$= -\frac{0}{2(1)}$$

$$= 0$$

Substituting 0 for x into the equation, we find the second coordinate of the vertex:

$$y = (0)^2 - 1 = -1.$$

The vertex is $(0,-1)$. The y-axis (The vertical line $x = 0$.) is the axis of symmetry. Next, we choose some x-values on each side of the vertex and compute the y-values.

When $x = 1, y = (1)^2 - 1 = 1 - 1 = 0$

When $x = 2, y = (2)^2 - 1 = 4 - 1 = 3$

When $x = -1, y = (-1)^2 - 1 = 1 - 1 = 0$

When $x = -2, y = (-2)^2 - 1 = 4 - 1 = 3$

x	y
0	−1
1	0
2	3
−1	0
−2	3

We plot these points and connect them with a smooth curve on the axis below.

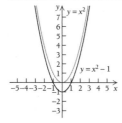

5. Graph $y = -2x^2$ and $y = -2x^2 + 1$

Starting with $y = -2x^2$, we first find the vertex or the turning point. The x-coordinate of the vertex is:

$$x = -\frac{b}{2a}$$

$$= -\frac{0}{2(-2)}$$

$$= 0$$

Substituting 0 for x into the equation, we find the second coordinate of the vertex:

$$y = -2(0)^2 = 0.$$

The vertex is $(0,0)$. The y-axis (The vertical line $x = 0$.) is the axis of symmetry. Next, we choose some x-values on each side of the vertex and compute the y-values.

When $x = 1, y = -2(1)^2 = -2 \cdot 1 = -2$

When $x = 2, y = -2(2)^2 = -2 \cdot 4 = -8$

When $x = -1, y = -2(-1)^2 = -2 \cdot 1 = -2$

When $x = -2, y = -2(-2)^2 = -2 \cdot 4 = -8$

x	y
0	0
1	−2
2	−8
−1	−2
−2	−8

We plot these points and connect them with a smooth curve on the next page.

Next, we graph $y = -2x^2 + 1$. First, we find the vertex or the turning point. The x-coordinate of the vertex is

$$x = -\frac{b}{2a}$$

$$= -\frac{0}{2(-2)}$$

$$= 0$$

Substituting 0 for x into the equation, we find the second coordinate of the vertex:

$$y = -2(0)^2 + 1 = 1.$$

The vertex is $(0,1)$. The y-axis (The vertical line $x = 0$.) is the axis of symmetry. Next, we choose some x-values on each side of the vertex and compute the y-values.

When $x = 1, y = -2(1)^2 + 1 = -2 + 1 = -1$

When $x = 2, y = -2(2)^2 + 1 = -8 + 1 = -7$

When $x = -1, y = -2(-1)^2 + 1 = -2 + 1 = -1$

When $x = -2, y = -2(-2)^2 + 1 = -8 + 1 = -7$

x	y
0	1
1	−1
2	−7
−1	−1
−2	−7

We plot these points and connect them with a smooth curve on the axis below.

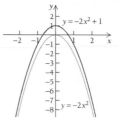

7.   Graph $y = |x|$ and $y = |x - 3|$

Starting with $y = |x|$, we choose some x-values and compute the y-values to make a table of points.

When $x = -2, y = |-2| = -(-2) = 2$

When $x = -1, y = |-1| = -(-1) = 1$

When $x = 0, y = |0| = 0$

When $x = 1, y = |1| = 1$

When $x = 2, y = |2| = 2$

x	y
-2	2
-1	1
0	0
1	1
2	2

We plot these points and connect them with a smooth curve on the axis at the top of the next column.

Next, we graph $y = |x - 3|$. We choose some x-values and compute the y-values to make a table of points.

When $x = -2, y = |-2 - 3| = |-5| = -(-5) = 5$

When $x = 0, y = |0 - 3| = |-3| = -(-3) = 3$

When $x = 3, y = |3 - 3| = |0| = 0$

When $x = 6, y = |6 - 3| = |3| = 3$

When $x = 8, y = |8 - 3| = |5| = 5$

x	y
-2	5
0	3
3	0
6	3
8	5

We plot these points and connect them with a smooth curve on the axis below.

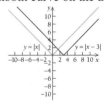

9.   Graph $y = x^3$ and $y = x^3 + 2$

Starting with $y = x^3$, we choose some x-values and compute the y-values to make a table of points.

When $x = -2, y = (-2)^3 = -8$

When $x = -1, y = (-1)^3 = -1$

When $x = 0, y = (0)^3 = 0$

When $x = 1, y = (1)^3 = 1$

When $x = 2, y = (2)^3 = 8$

x	y
-2	-8
-1	-1
0	0
1	1
2	8

We plot the points from the previous page and connect them with a smooth curve on the axis below.

Next, we graph $y = x^3 + 2$. We choose some x-values and compute the y-values to make a table of points.

When $x = -2, y = (-2)^3 + 2 = -8 + 2 = -6$

When $x = -1, y = (-1)^3 + 2 = -1 + 2 = 1$

When $x = 0, y = (0)^3 + 2 = 0 + 2 = 2$

When $x = 1, y = (1)^3 + 2 = 1 + 2 = 3$

When $x = 2, y = (2)^3 + 2 = 8 + 2 = 10$

x	y
-2	-6
-1	1
0	2
1	3
2	10

We plot these points and connect them with a smooth curve.

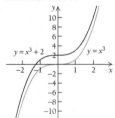

11. Graph $y = \sqrt{x}$ and $y = \sqrt{x-1}$

Starting with $y = \sqrt{x}$, we choose some x-values and compute the y-values to make a table of points. The domain of the function is the set of all nonnegative real numbers, so we choose x-values that are in the set $[0, \infty)$.

When $x = 0, y = \sqrt{0} = 0$

When $x = 1, y = \sqrt{1} = 1$

When $x = 4, y = \sqrt{4} = 2$

When $x = 9, y = \sqrt{9} = 3$

x	y
0	0
1	1
4	2
9	3

We plot these points and connect them with a smooth curve on the axis below.

Next, we graph $y = \sqrt{x-1}$. we choose some x-values and compute the y-values to make a table of points. The domain of the function is the set of all positive real numbers greater than or equal to 1, so we choose x-values that are in the set $[1, \infty)$.

When $x = 1, y = \sqrt{1-1} = \sqrt{0} = 0$

When $x = 2, y = \sqrt{2-1} = \sqrt{1} = 1$

When $x = 5, y = \sqrt{5-1} = \sqrt{4} = 2$

When $x = 10, y = \sqrt{10-1} = \sqrt{9} = 3$

x	y
1	0
2	1
5	2
10	3

We plot these points and connect them with a smooth curve on the axis below.

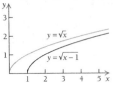

13. $y = x^2 + 4x - 7$

This function is of the form $y = ax^2 + bx + c,\ a \neq 0$, so its graph is a parabola.

We have $a = 1$ and $b = 4$, so the first coordinate of the vertex is

$$x = -\frac{b}{2a}$$
$$= -\frac{4}{2(1)} = -2.$$

Substituting -2 into the equation, we find the second coordinate of the vertex:

$$y = (-2)^2 + 4(-2) - 7$$
$$= 4 - 8 - 7$$
$$= -11.$$

The vertex is $(-2, -11)$.

15. $y = 2x^4 - 4x^2 - 3$

The function is not of the form $y = ax^2 + bx + c,\ a \neq 0$, so its graph is not a parabola.

17. Graph $y = x^2 - 4x + 3$

First, we should recognize that this function is a quadratic function. We find the vertex or the turning point. The x-coordinate of the vertex is

$$x = -\frac{b}{2a}$$
$$= -\frac{-4}{2(1)}$$
$$= 2$$

Substituting 2 for x into the equation, we find the second coordinate of the vertex:

$$y = (2)^2 - 4(2) + 3$$
$$= 4 - 8 + 3 = -1.$$

The vertex is $(2, -1)$. The vertical line $x = 2$ is the axis of symmetry. Next, we choose some x-values on each side of the vertex and compute the y-values. The computations are shown at the top of the next page.

When $x = -1, y = (-1)^2 - 4(-1) + 3 = 8$

When $x = 0, y = (0)^2 - 4(0) + 3 = 3$

When $x = 1, y = (1)^2 - 4(1) + 3 = 0$

When $x = 3, y = (3)^2 - 4(3) + 3 = 0$

When $x = 4, y = (4)^2 - 4(4) + 3 = 3$

When $x = 5, y = (5)^2 - 4(5) + 3 = 8$

x	y
-1	8
0	3
1	0
2	-1
3	0
4	3
5	8

We plot these points and connect them with a smooth curve on the axis below.

19. Graph $y = -x^2 + 2x - 1$

First, we should recognize that this function is a quadratic function. We find the vertex or the turning point. The x-coordinate of the vertex is

$$x = -\frac{b}{2a} = -\frac{2}{2(-1)} = 1$$

Substituting 1 for x into the equation, we find the second coordinate of the vertex:

$$y = -(1)^2 + 2(1) - 1$$
$$= -1 + 2 - 1 = 0.$$

The vertex is $(1, 0)$. The vertical line $x = 1$ is the axis of symmetry. Next, we choose some x-values on each side of the vertex and compute the y-values.

When $x = -1, y = -(-1)^2 + 2(-1) - 1 = -4$

When $x = 0, y = -(0)^2 + 2(0) - 1 = -1$

When $x = 2, y = -(2)^2 + 2(2) - 1 = -1$

When $x = 3, y = -(3)^2 + 2(3) - 1 = -4$

x	y
-1	-4
0	-1
1	0
2	-1
3	-4

We plot these points and connect them with a smooth curve on the axis below.

21. Graph $f(x) = 2x^2 - 6x + 1$

First, we should recognize that this function is a quadratic function. We find the vertex or the turning point. The x-coordinate of the vertex is

$$x = -\frac{b}{2a} = -\frac{-6}{2(2)} = \frac{3}{2}$$

Substituting $\frac{3}{2}$ for x into the equation, we find the second coordinate of the vertex:

$$f\left(\frac{3}{2}\right) = 2\left(\frac{3}{2}\right)^2 - 6\left(\frac{3}{2}\right) + 1$$

$$f\left(\frac{3}{2}\right) = 2\left(\frac{9}{4}\right) - 9 + 1 =$$

$$= \frac{9}{2} - 8 = -\frac{7}{2}.$$

The vertex is $\left(\frac{3}{2}, -\frac{7}{2}\right)$. The vertical line $x = \frac{3}{2}$ is the axis of symmetry.

Next, we choose some x-values on each side of the vertex and compute the y-values.

When $x = 0, y = 2(0)^2 - 6(0) + 1 = 1$

When $x = 1, y = 2(1)^2 - 6(1) + 1 = -3$

When $x = 2, y = 2(2)^2 - 6(2) + 1 = -3$

When $x = 3, y = 2(3)^2 - 6(3) + 1 = 1$

x	y
0	1
1	-3
$\frac{3}{2}$	$-\frac{7}{2}$
2	-3
3	1

We plot these points and connect them with a smooth curve on the axis below.

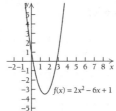

23. Graph $g(x) = -3x^2 - 4x + 5$

First, we should recognize that this function is a quadratic function. We find the vertex or the turning point. The x-coordinate of the vertex is

$$x = -\frac{b}{2a} = -\frac{-4}{2(-3)} = -\frac{2}{3}$$

Substituting $-\frac{2}{3}$ for x into the equation, we find the second coordinate of the vertex at the top of the next page.

$$f\left(-\frac{2}{3}\right) = -3\left(-\frac{2}{3}\right)^2 - 4\left(-\frac{2}{3}\right) + 5$$

$$= -3\left(\frac{4}{9}\right) + \frac{8}{3} + 5 =$$

$$= -\frac{4}{3} + \frac{8}{3} + 5 = \frac{19}{3}.$$

The vertex is $\left(-\frac{2}{3}, \frac{19}{3}\right)$. The vertical

line $x = -\frac{2}{3}$ is the axis of symmetry. Next, we choose some x-values on each side of the vertex and compute the y-values.

When $x = -2, y = -3(-2)^2 - 4(-2) + 5 = 1$

When $x = -1, y = -3(-1)^2 - 4(-1) + 5 = 6$

When $x = 0, y = -3(0)^2 - 4(0) + 5 = 5$

When $x = 1, y = -3(1)^2 - 4(1) + 5 = -2$

x	y
-2	1
-1	6
$-\frac{2}{3}$	$\frac{19}{3}$
0	5
1	-2

We plot these points and connect them with a smooth curve on the axis below.

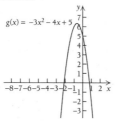

25. Graph $y = \frac{2}{x}$

First we determine the domain. The domain is all real numbers except for 0, since substituting 0 for x would result in division by 0. Now, to find y-values, we substitute any value for x other than 0, and compute the value for y.

When $x = -4, y = \frac{2}{-4} = -\frac{1}{2}$

When $x = -2, y = \frac{2}{-2} = -1$

When $x = -1, y = \frac{2}{-1} = -2$

When $x = 1, y = \frac{2}{1} = 2$

When $x = 2, y = \frac{2}{2} = 1$

When $x = 4, y = \frac{2}{4} = \frac{1}{2}$

We create an input-output table at the top of the next column.

x	y
-4	$-\frac{1}{2}$
-2	-1
-1	-2
1	2
2	1
4	$\frac{1}{2}$

We plot these points and connect them with a smooth curve on the axis below.

27. Graph $y = -\frac{2}{x}$

First we determine the domain. The domain is all real numbers except for 0, since substituting 0 for x would result in division by 0. Now, to find y-values, we substitute any value for x other than 0, and compute the value for y.

When $x = -4, y = -\frac{2}{-4} = \frac{1}{2}$

When $x = -2, y = -\frac{2}{-2} = 1$

When $x = -1, y = -\frac{2}{-1} = 2$

When $x = 1, y = -\frac{2}{1} = -2$

When $x = 2, y = -\frac{2}{2} = -1$

When $x = 4, y = -\frac{2}{4} = -\frac{1}{2}$

We create an input output table on the next page.

x	y
-4	$\frac{1}{2}$
-2	1
-1	2
1	-2
2	-1
4	$-\frac{1}{2}$

We plot these points and connect them with a smooth curve on the axis below.

29. Graph $y = \dfrac{1}{x^2}$

First we determine the domain. The domain is all real numbers except for 0, since substituting 0 for x would result in division by 0. Now, to find y-values, we substitute any value for x other than 0, and compute the value for y.

When $x = -2$, $y = \dfrac{1}{(-2)^2} = \dfrac{1}{4}$

When $x = -1$, $y = \dfrac{1}{(-1)^2} = 1$

When $x = -\dfrac{1}{2}$, $y = \dfrac{1}{\left(-\dfrac{1}{2}\right)^2} = \dfrac{1}{\dfrac{1}{4}} = 4$

When $x = \dfrac{1}{2}$, $y = \dfrac{1}{\left(\dfrac{1}{2}\right)^2} = \dfrac{1}{\dfrac{1}{4}} = 4$

When $x = 1$, $y = \dfrac{1}{(1)^2} = 1$

When $x = 2$, $y = \dfrac{1}{(2)^2} = 4$

We create an input-output table at the top of the next column.

x	y
-2	$\frac{1}{4}$
-1	1
$-\frac{1}{2}$	4
$\frac{1}{2}$	4
1	1
2	$\frac{1}{4}$

We plot these points and connect them with a smooth curve on the axis below.

31. Graph $y = \sqrt[3]{x}$

First we determine the domain of the function. Since the index of the radicand is odd (3) the domain is all real numbers. We are free to choose any number for x and compute the value for y.

When $x = -8$, $y = \sqrt[3]{-8} = -2$

When $x = -1$, $y = \sqrt[3]{-1} = -1$

When $x = 0$, $y = \sqrt[3]{0} = 0$

When $x = 1$, $y = \sqrt[3]{1} = 1$

When $x = 8$, $y = \sqrt[3]{8} = 2$

x	y
-8	-2
-1	-1
0	0
1	1
8	2

We plot these points and connect them with a smooth curve on the axis below.

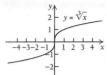

33. Graph $f(x) = \dfrac{x^2 + 5x + 6}{x + 3}$

First, we simplify the function, by factoring the numerator and removing a factor of 1 as follows:

$$f(x) = \frac{x^2 + 5x + 6}{x + 3}$$

$$= \frac{(x + 3)(x + 2)}{x + 3}$$

$$f(x) = \frac{x + 3}{x + 3} \cdot \frac{x + 2}{1}$$

$$= x + 2, \quad x \neq -3$$

The simplification assumes that x is not -3. The number -3 is not in the domain of the original function because it would result in division by zero. Thus we can express the function as follows:

$$y = f(x) = x + 2, \quad x \neq -3.$$

To find function values, we substitute any value for x other than -3 and calculate the y-values.

x	y
-5	-3
-4	-2
-2	0
-1	1
0	2
1	3
2	4

We plot these points and draw the graph. The open circle at the point $(-3, -1)$ indicates that it is not part of the graph.

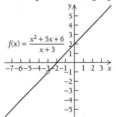

35. Graph $f(x) = \dfrac{x^2 - 1}{x - 1}$

First, we simplify the function, by factoring the numerator and removing a factor of 1 as follows:

$$f(x) = \frac{x^2 - 1}{x - 1}$$

$$= \frac{(x-1)(x+1)}{x-1}$$

$$f(x) = \frac{x-1}{x-1} \cdot \frac{x+1}{1}$$

$$= x+1, \quad x \neq 1$$

The simplification assumes that x is not 1. The number 1 is not in the domain of the original function because it would result in division by zero. Thus we can express the function as follows:

$$y = f(x) = x + 1, \quad x \neq 1.$$

To find function values, we substitute any value for x other than 1 and calculate the y-values.

x	y
-2	-1
-1	0
0	1
2	3
3	4
4	5

We plot these points and draw the graph. The open circle at the point $(1, 2)$ indicates that it is not part of the graph.

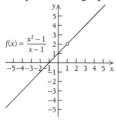

37. Solve $x^2 - 2x = 2$

First, we put the equation in standard form.

$x^2 - 2x - 2 = 0$ subtract 2 from both sides

The equation is in standard form with

$a = 1, b = -2, c = -2$

Next, we apply the quadratic formula.

$$x = \frac{-b \pm \sqrt{b^2 - 4ac}}{2a}$$

Substituting the values for a, b, and c, we get:

$$x = \frac{-(-2) \pm \sqrt{(-2)^2 - 4(1)(-2)}}{2(1)}$$

$$= \frac{2 \pm \sqrt{4 + 8}}{2} = \frac{2 \pm \sqrt{12}}{2}$$

$$= \frac{2 \pm 2\sqrt{3}}{2} \qquad \left(\sqrt{12} = \sqrt{4 \cdot 3} = 2\sqrt{3}\right)$$

$$= \frac{2\left(1 \pm \sqrt{3}\right)}{2 \cdot 1} = 1 \pm \sqrt{3}$$

The solutions are $1 + \sqrt{3}$ and $1 - \sqrt{3}$.

39. Solve $x^2 + 6x = 1$

First, we put the equation in standard form.

$x^2 + 6x - 1 = 0$ subtract 1 from both sides

The equation is in standard form with

$a = 1, b = 6, c = -1$

Next, we apply the quadratic formula.

$$x = \frac{-b \pm \sqrt{b^2 - 4ac}}{2a}$$

Substituting the values for a, b, and c, we get:

$$x = \frac{-(6) \pm \sqrt{(6)^2 - 4(1)(-1)}}{2(1)}$$

$$= \frac{-6 \pm \sqrt{36 + 4}}{2} = \frac{-6 \pm \sqrt{40}}{2}$$

$$= \frac{-6 \pm 2\sqrt{10}}{2} \qquad \left(\sqrt{40} = \sqrt{4 \cdot 10} = 2\sqrt{10}\right)$$

$$= \frac{2\left(-3 \pm \sqrt{10}\right)}{2 \cdot 1} = -3 \pm \sqrt{10}$$

The solutions are $-3 + \sqrt{10}$ and $-3 - \sqrt{10}$.

41. Solve $4x^2 = 4x + 1$

First, we put the equation in standard form.

$4x^2 - 4x - 1 = 0$ subtract $4x$ and 1 from both sides

The equation is in standard form with

$a = 4, b = -4, c = -1$

Next, we apply the quadratic formula.

$$x = \frac{-b \pm \sqrt{b^2 - 4ac}}{2a}$$

Substituting the values for *a*, *b*, and *c*, we get:

$$x = \frac{-(-4) \pm \sqrt{(-4)^2 - 4(4)(-1)}}{2(4)}$$

$$= \frac{4 \pm \sqrt{16 + 16}}{8}$$

$$= \frac{4 \pm \sqrt{32}}{8}$$

$$x = \frac{4 \pm 4\sqrt{2}}{8} \qquad \left(\sqrt{32} = \sqrt{16 \cdot 2} = 4\sqrt{2}\right)$$

$$= \frac{4\left(1 \pm \sqrt{2}\right)}{4 \cdot 2} = \frac{1 \pm \sqrt{2}}{2}$$

The solutions are $\dfrac{-1 + \sqrt{2}}{2}$ and $\dfrac{-1 - \sqrt{2}}{2}$.

43. Solve $3y^2 + 8y + 2 = 0$

The equation is in standard form with
$a = 3, b = 8, c = 2$

Next, we apply the quadratic formula.

$$y = \frac{-b \pm \sqrt{b^2 - 4ac}}{2a}$$

Substituting the values for *a*, *b*, and *c*, we get:

$$y = \frac{-(8) \pm \sqrt{(8)^2 - 4(3)(2)}}{2(3)}$$

$$= \frac{-8 \pm \sqrt{64 - 24}}{6} = \frac{-8 \pm \sqrt{40}}{6}$$

$$= \frac{-8 \pm 2\sqrt{10}}{6} \qquad \left(\sqrt{40} = \sqrt{4 \cdot 10} = 2\sqrt{10}\right)$$

$$= \frac{2\left(-4 \pm \sqrt{10}\right)}{2 \cdot 3} = \frac{-4 \pm \sqrt{10}}{3}$$

The solutions are $\dfrac{-4 + \sqrt{10}}{3}$ and $\dfrac{-4 - \sqrt{10}}{10}$.

45. Solve $x + 7 + \dfrac{9}{x} = 0$

Multiplying both sides by *x*, we get:

$$x \cdot \left(x + 7 + \frac{9}{x}\right) = 0 \cdot x$$

$$x^2 + 7x + 9 = 0.$$

This is a quadratic equation in standard form
with $a = 1, b = 7,$ and $c = 9$.

Next, we apply the quadratic formula.

$$x = \frac{-b \pm \sqrt{b^2 - 4ac}}{2a}$$

Substituting the values for *a*, *b*, and *c*, we get:

$$x = \frac{-(7) \pm \sqrt{(7)^2 - 4(1)(9)}}{2(1)}$$

$$= \frac{-7 \pm \sqrt{49 - 36}}{2} = \frac{-7 \pm \sqrt{13}}{2}$$

The solutions are $\dfrac{-7 + \sqrt{13}}{2}$ and $\dfrac{-7 - \sqrt{13}}{2}$.

47. $\sqrt{x^3} = x^{3/2} \qquad \left(\text{The index is 2; } \sqrt[n]{a^m} = a^{m/n}\right)$

49. $\sqrt[5]{a^3} = a^{3/5} \qquad \left(\sqrt[n]{a^m} = a^{m/n}\right)$

51. $\sqrt[7]{t} = t^{1/7}$

53. $\sqrt[4]{x^{12}} = x^{12/4} = x^3$

55. $\dfrac{1}{\sqrt{t^5}} = \dfrac{1}{t^{5/2}} \qquad \left(\sqrt[n]{a^m} = a^{m/n}\right)$

$\qquad = t^{-5/2} \qquad \left(\dfrac{1}{a^n} = a^{-n}\right)$

57. $\dfrac{1}{\sqrt{x^2 + 7}} = \dfrac{1}{\left(x^2 + 7\right)^{1/2}} \qquad \left(\sqrt[n]{a^m} = a^{m/n}\right)$

$\qquad = \left(x^2 + 7\right)^{-1/2} \qquad \left(\dfrac{1}{a^n} = a^{-n}\right)$

59. $x^{1/5} = \sqrt[5]{x^1} = \sqrt[5]{x} \qquad \left(a^{m/n} = \sqrt[n]{a^m}\right)$

61. $y^{2/3} = \sqrt[3]{y^2} \qquad \left(a^{m/n} = \sqrt[n]{a^m}\right)$

63. $t^{-2/5} = \dfrac{1}{t^{2/5}} \qquad \left(a^{-n} = \dfrac{1}{a^n}\right)$

$\qquad = \dfrac{1}{\sqrt[5]{t^2}} \qquad \left(a^{m/n} = \sqrt[n]{a^m}\right)$

65. $b^{-1/3} = \dfrac{1}{b^{1/3}} \qquad \left(a^{-n} = \dfrac{1}{a^n}\right)$

$\qquad = \dfrac{1}{\sqrt[3]{b}} \qquad \left(a^{m/n} = \sqrt[n]{a^m}\right)$

67. $e^{-17/6} = \dfrac{1}{e^{17/6}}$ $\qquad \left(a^{-n} = \dfrac{1}{a^n}\right)$

$\qquad = \dfrac{1}{\sqrt[6]{e^{17}}}$ $\qquad \left(a^{m/n} = \sqrt[n]{a^m}\right)$

69. $\left(x^2 - 3\right)^{-1/2} = \dfrac{1}{\left(x^2 - 3\right)^{1/2}}$ $\qquad \left(a^{-n} = \dfrac{1}{a^n}\right)$

$\qquad = \dfrac{1}{\sqrt{x^2 - 3}}$ $\qquad \left(a^{m/n} = \sqrt[n]{a^m}\right)$

71. $\dfrac{1}{t^{2/3}} = \dfrac{1}{\sqrt[3]{t^2}}$ $\qquad \left(a^{m/n} = \sqrt[n]{a^m}\right)$

73. $9^{3/2}$

$\qquad = \left(9^{1/2}\right)^3$ $\qquad \left(\tfrac{3}{2} = \tfrac{1}{2} \cdot 3; \; a^{m \cdot n} = \left(a^m\right)^n\right)$

$\qquad = \left(\sqrt{9}\right)^3$ $\qquad \left(a^{1/n} = \sqrt[n]{a}\right)$

$\qquad = (3)^3 = 27$

75. $64^{2/3}$

$\qquad = \left(64^{1/3}\right)^2$ $\qquad \left(\tfrac{2}{3} = \tfrac{1}{3} \cdot 2; \; a^{m \cdot n} = \left(a^m\right)^n\right)$

$\qquad = \left(\sqrt[3]{64}\right)^2$ $\qquad \left(a^{1/n} = \sqrt[n]{a}\right)$

$\qquad = (4)^2$ $\qquad \left(\sqrt[3]{64} = 4\right)$

$\qquad = 16$

77. $16^{3/4}$

$\qquad = \left(16^{1/4}\right)^3$ $\qquad \left(\tfrac{3}{4} = \tfrac{1}{4} \cdot 3; \; a^{m \cdot n} = \left(a^m\right)^n\right)$

$\qquad = \left(\sqrt[4]{16}\right)^3$ $\qquad \left(a^{1/n} = \sqrt[n]{a}\right)$

$\qquad = (2)^3$ $\qquad \left(\sqrt[4]{16} = 2\right)$

$\qquad = 8$

79. The domain of a rational function is restricted to those input values that do not result in division by 0. To determine the domain of

$$f(x) = \dfrac{x^2 - 25}{x - 5}$$

we set the denominator equal to zero and solve:

$x - 5 = 0$

$\qquad x = 5.$

Therefore, 5 is not in the domain. The domain of f consists of all real numbers except 5.

81. The domain of a rational function is restricted to those input values that do not result in division by 0. To determine the domain of

$$f(x) = \dfrac{x^3}{x^2 - 5x + 6}$$

we set the denominator equal to zero and solve:

$x^2 - 5x + 6 = 0$

$(x - 2)(x - 3) = 0 \qquad$ factoring

$x - 2 = 0 \;$ or $\; x - 3 = 0 \qquad$ Principle of Zero Products

$\qquad x = 2 \;$ or $\qquad x = 3.$

Therefore, 2 and 3 are not in the domain. The domain of f consists of all real numbers except 2 and 3.

83. The domain of the radical function

$\quad f(x) = \sqrt{5x + 4}$ is restricted to those input

values that result in the value of the radicand being greater than or equal to 0. In other words, the domain will be the set of real numbers that satisfy the inequality $5x + 4 \geq 0$.

To find the domain, we solve the inequality:

$5x + 4 \geq 0$

$\qquad 5x \geq -4$

$\qquad x \geq -\dfrac{4}{5}$

Therefore, the domain of f consists of all real

numbers greater than or equal to $-\dfrac{4}{5}$, or in

interval notation $\left[-\dfrac{4}{5}, \infty\right)$.

85. The domain of the radical function

$f(x) = \sqrt[4]{7-x}$ is restricted to those input

values that result in the value of the radicand
being greater than or equal to 0. In other words,
the domain will be the set of real numbers that
satisfy the inequality $7 - x \geq 0$.

To find the domain, we solve the inequality:

$7 - x \geq 0$

$\quad 7 \geq x$

Therefore, the domain of f consists of all real
numbers less than or equal to 7, or in interval
notation $(-\infty, 7]$.

87. We set the demand equation equal to the supply
equation and solve for x.

$1000 - 10x = 250 + 5x$

$\qquad 750 = 15x$

$\qquad \ 50 = x$

Thus, the equilibrium price is $50. To find the
equilibrium quantity, we substitute 50 for x into
either the demand equation or supply equation.
We use the demand equation.

$q = 1000 - 10(50)$

$q = 1000 - 500$

$q = 500$

The equilibrium quantity is 500 units. The
equilibrium point is $(50, 500)$.

89. We set the demand equation equal to the supply
equation and solve for x.

$\dfrac{5}{x} = \dfrac{x}{5}$

$25 = x^2 \qquad$ multiply both sides by $5x$

$\sqrt{25} = \sqrt{x^2} \qquad$ take the square root of both sides

$\pm 5 = x$

Since it is not appropriate to have a negative
price, the equilibrium price is 5 hundred dollars
or $500. To find the equilibrium quantity, we
substitute 5 for x into either the demand
equation or supply equation. We use the
demand equation.

$q = \dfrac{5}{(5)} = 1$

The equilibrium quantity is 1 thousand units or
1000 units. The equilibrium point is $(5, 1)$.

91. We set the demand equation equal to the supply
equation and solve for x.

$(x - 3)^2 = x^2 + 2x + 1$

$x^2 - 6x + 9 = x^2 + 2x + 1$

$\quad -6x + 9 = 2x + 1 \quad$ subtracting x^2 from both sides

$\qquad \quad 8 = 8x$

$\qquad \quad 1 = x$

The equilibrium price is $1.

To find the equilibrium quantity, we substitute 1
for x into either the demand equation or supply
equation. We use the demand equation.

$q = (1 - 3)^2$

$\quad = (-2)^2 = 4$

The equilibrium quantity is 4 hundred units or
400 units. The equilibrium point is $(1, 4)$.

93. We set the demand equation equal to the supply
equation and solve for x.

$5 - x = \sqrt{x + 7} \qquad 0 \leq x \leq 5$

$(5 - x)^2 = \left(\sqrt{x + 7}\right)^2 \quad$ Squaring both sides

$25 - 10x + x^2 = x + 7$

$18 - 11x + x^2 = 0$

$(9 - x)(2 - x) = 0 \qquad$ Factoring the quadratic

$9 - x = 0 \ $ or $\ 2 - x = 0 \quad$ Principle of Zero Products

$\quad 9 = x \ $ or $\qquad 2 = x$

Since 9 is not in the domain of the demand
function, the equilibrium price is 2 thousand
dollars or $2000. To find the equilibrium
quantity, we substitute 2 for x into either the
demand equation or supply equation. We use
the demand equation.

$q = 5 - (2) = 3$

The equilibrium quantity is 3 thousand units or
3000 units. The equilibrium point is $(2, 3)$.

95. If the price per share S is inversely proportional
to the prime rate R, then we have:

$S = \dfrac{k}{R}$

We find the constant of variation by substituting
86.89 for S and 0.0675 (6.75%) for R.

$86.89 = \dfrac{k}{0.0675}$

$5.865075 = k$

The equation of variation is $S = \dfrac{5.865075}{R}$.

If the prime rate rose to 7.50% we find S by substituting 0.075 in for R.

$$S = \frac{5.865075}{0.075}$$

$$= 78.201$$

$$\approx 78.20$$

The price per share would be approximately $78.20 if the assumption of inverse proportionality is correct.

97. a) $R(x) = 11.74x^{0.25}$

$$R(40,000) = 11.74(40,000)^{0.25}$$

$$= 11.74(14.14213562)$$

$$= 166.0286722$$

$$\approx 166$$

The maximum range will be approximately 166 miles when the peak power is 40,000 watts.

$$R(50,000) = 11.74(50,000)^{0.25}$$

$$= 11.74(14.95348781)$$

$$R(50,000) = 175.5539469$$

$$\approx 176$$

The maximum range will be approximately 176 miles when the peak power is 50,000 watts.

$$R(60,000) = 11.74(60,000)^{0.25}$$

$$= 11.74(15.6508458)$$

$$= 183.7409297$$

$$\approx 184$$

The maximum range will be approximately 184 miles when the peak power is 60,000 watts.

b) Plotting the points found in part (a) and connecting them with a smooth curve we see:

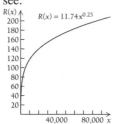

99. a) In 2005, $t = 2005 - 1970 = 35$.

$$P = 1000(35)^{5/4} + 14,000 \approx 99,130$$

In 2005, average pollution was approximately 99,130 particles per cubic centimeter.

In 2008, $t = 2008 - 1970 = 38$.

$$P = 1000(38)^{5/4} + 14,000 \approx 108,347$$

In 2008, average pollution will be approximately 108,347 particles per cubic centimeter.

In 2014, $t = 2014 - 1970 = 44$.

$$P = 1000(44)^{5/4} + 14,000 \approx 127,322$$

In 2005, average pollution will be approximately 127,322 particles per cubic centimeter.

b) Plot the points above and others, if necessary, and draw the graph.

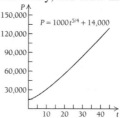

101. The number of cities N with a population greater than S is inversely proportional to S.

$$N = \frac{k}{S}$$

$$51 = \frac{k}{350,000} \quad \text{Substituting}$$

$$17,850,000 = k$$

The equation of variation is

$$N = \frac{17,850,000}{S}.$$

We find N when S is 500,000.

$$N = \frac{17,850,000}{500,000} = 35.7 \approx 36$$

Using the fact that there were 51 cities with a population greater than 350,000 and 36 cities with a population of 500,000 or greater, we estimate that there are $51 - 36 = 15$ cities with a population between 350,000 and 500,000.

To estimate the number of cities with a population between 300,000 and 600,000 we find N when S is 300,000 and when S is 600,000.

$$N = \frac{17,850,000}{300,000} = 59.5 \approx 60$$

$$N = \frac{17,850,000}{600,000} = 29.75 \approx 30$$

There are 60 cities with a population greater than 300,000 and 30 cities with a population greater than 600,000, so there are $60 - 30 = 30$ cities with a population between 300,000 and 600,000.

103. [tw]

105. $f(x) = 2x^3 - x^2 - 14x - 10$

Enter the function into your calculator.

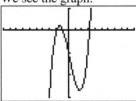

Using the window:

We see the graph:

Now using the ZERO feature on the calculator, we approximate the zeros. The zeros are -1.831, -0.856, 3.188.

107. $f(x) = x^4 + 4x^3 - 36x^2 - 160x + 300$

Enter the function into your calculator.

Using the window:

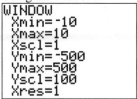

We see the graph:

Now using the ZERO feature on the calculator, we approximate the zeros. The zeros are 1.489 and 5.673.

109. $f(x) = |x+1| + |x-2| - 5$

Enter the function into your calculator.

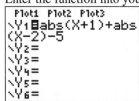

Using the standard window:

We see the graph:

Now using the ZERO feature on the calculator, we approximate the zeros. The zeros are –2 and 3.

111. $f(x) = |x+1| + |x-2| - 3$

Enter the function into your calculator.

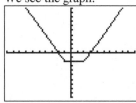

Using the standard window:

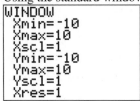

We see the graph:

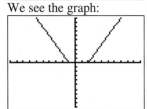

We see that the graph intersects the x-axis between $-1 \le x \le 2$. The zeros of this function are all real numbers in $[-1, 2]$.

113. We enter the demand and supply equations into the graphing editor on the calculator.

Using the window:

We see the graph:

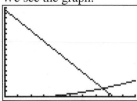

Using the intersect feature on the calculator we find the intersection to be:

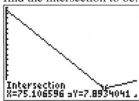

The equilibrium point for this market is $(75.11, 7.893)$. In other words, 7893 units will be sold at a price of $75.11.

Exercise Set R.6

1. The data is decreasing at a constant rate, so a linear function $f(x) = mx + b$ could be used to model the data.

3. The data rises first and then falls over the domain, so a quadratic function $f(x) = ax^2 + bx + c$, $a < 0$ could be used to model the data.

5. The data is increasing at a constant rate, so a linear function $f(x) = mx + b$ could be used to model the data.

7. The data rises and falls several times over the domain. This implies that a polynomial function that is neither quadratic nor linear would be best to model the data.

9. The data is increasing at a constant rate, so a linear function $f(x) = mx + b$ could be used to model the data.

11. a) We will choose the points $(0, 13)$ and $(4, 21)$. First, we find the slope:

$$m = \frac{21 - 13}{4 - 0} = \frac{8}{4} = 2$$

Next, since we chose the y-intercept, we can substitute into the slope-intercept equation.

$y = mx + b$

$y = 2x + 13$

Alternatively, we could have substituted the data points in to the equation $y = mx + b$ to obtain a system of equations.

$13 = m \cdot 0 + b \qquad (1)$

$21 = m \cdot 4 + b \qquad (2)$

Subtracting each side of Equation (1) from each side of Equation (2) we get:

$8 = 4m$

$2 = m$

Now substitute $m = 2$ in for either Equation (1) or (2) and solve for b. We use Equation (1).

$13 = (2)(0) + b$

$13 = b$

Therefore, the linear function that fits the data is $y = 2x + 13$

b) We plot the points in the table in the text and then graph the function found in part (a) on the same set of axes.

$y = 2x + 13$

24

0

10

6

c) In 2010, $x = 2010 - 2000 = 10$

Substituting 10 for x we have:

$y = 2(10) + 13 = 33$

The average co-payment for preferred drugs in 2010 will be approximately $33.

13. a) Consider the general quadratic function $y = ax^2 + bx + c$, where y is the braking distance, in feet, and x is the speed, in miles per hour.

Using the points $(20, 25), (40, 105)$, and $(60, 300)$, we substitute each point into the general quadratic function to get the following system of equations:

$25 = a \cdot 20^2 + b \cdot 20 + c$

$105 = a \cdot 40^2 + b \cdot 40 + c$

$300 = a \cdot 60^2 + b \cdot 60 + c$

or,

$25 = 400a + 20b + c$

$105 = 1600a + 40b + c$

$300 = 3600a + 60b + c$

Solving the system of equations, we get $a = 0.14375$, $b = -4.625$, and $c = 60$.

Therefore, the function is $y = 0.144x^2 - 4.63x + 60$.

b) We substitute 50 for x and compute the value of y.

$y = 0.144(50)^2 - 4.63(50) + 60 = 188.5$

The breaking distance of a car traveling at 50 mph is about 188.5 ft.

c) $\boxed{tw}$

15. a) Answers will vary depending on which points are used to find the function. We will use the points $(30, 1.4)$ and $(70, 53.0)$. We substitute in to the equation $y = mx + b$ to obtain a system of equations.

$1.4 = m \cdot 30 + b$ (1)

$53.0 = m \cdot 70 + b$ (2)

Subtracting each side of Equation (1) from each side of Equation (2) we get:

$51.6 = 40m$

$1.29 = m$

Now substitute $m = 1.29$ in for either Equation (1) or (2) and solve for b. We use Equation (1).

$1.4 = (1.29)(30) + b$

$1.4 = 38.7 + b$

$-37.3 = b$

Therefore, the linear function that fits the data is $y = 1.29x - 37.3$.

b) We plot the points in the table in the text and then graph the function found in part (a) on the same set of axes.

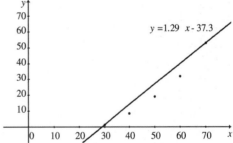

c) Substituting in 55 for x we have:

$y = 1.29(55) - 37.3 = 33.65$.

Approximately 33.65% of 55-yr-old women have high blood pressure.

17. $\boxed{tw}$

19. $\boxed{tw}$

21. a) First, we enter the data into the statistic editor on the calculator, letting L_1 be the values for x and L_2 be the values for y.

L1	L2	L3	2
7	6.25		
8	6.25		
9	6.5		
10	6.75		
11	7		
12	?		

L2(13) =

Using the linear regression feature we get:

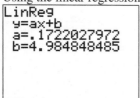

The linear function that fits the data is
$y = 0.172x + 4.98$

b) Substituting 6 in for x we get:

$y = 0.172(6) + 4.98$

$= 6.012$

The prime rate in June 2005 will be
approximately 6.012%

c) The actual prime rate in June of 2005 was
6.0%. So the linear regression on the
calculator gave us a slightly closer
approximation to the actual prime rate for
that month. However, they are both very
close.

d) Using the cubic regression feature on the
calculator results in:

CubicReg
 y=ax³+bx²+cx+d
 a=⁻4.856255ᴇ⁻4
 b=.0112803863
 c=.0974303474
 d=5.106060606

$y = -0.000486x^3 + 0.011x^2 + 0.097x + 5.106$

Substituting in 6 for x we get:

$y = -0.000486(6)^3 + .011(6)^2$

$+ 0.097(6) + 5.106$

$= 5.98$

≈ 5.98

The cubic function estimated the prime rate
to be 5.98% for June of 2005.

e) [tw]

23. a) Letting x be the number of years after 1990,
we enter the data into the statistics editor of
the graphing calculator. We enter the years
since 1990 into L_1 and the trade deficit into
L_2.

L1	L2	L3	2
9	73.4		
10	81.6		
11	69		
12	70		
13	66		
14	75.2		

L2(16) =			

Next, we use the cubic regression feature on
the calculator to fit the data to the model.

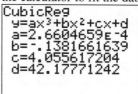

The cubic function that fits the data is:
$y = 0.00027x^3 - 0.1382x^2 + 4.056x + 42.178$

b) To predict the trade deficit in 2006, we
substitute 16 in for x and compute y.

$y = 0.00027(16)^3 - 0.1382(16)^2$

$+ 4.056(16) + 42.178$

≈ 72.80072

≈ 72.8

The trade deficit in 2006 was approximately
72.8 billion dollars.

To predict the trade deficit in 2010, we
substitute 20 in for x and compute y.

$y = 0.00027(20)^3 - 0.1382(20)^2$

$+ 4.056(20) + 42.178$

≈ 70.178

≈ 70.2

The trade deficit in 2010 will be
approximately 70.2 billion dollars.

c) [tw]

Chapter 1

Differentiation

Exercise Set 1.1

---

1. We select x-values closer and closer to 3, and observe the corresponding output.

x	2.99	2.999	3	3.001	3.01
$2x+5$	10.98	10.998	$\to 11 \leftarrow$	11.002	11.02

As x approaches 3, the value of $2x+5$ approaches 11.

3. We solve the equation:
$$-3x = 6$$
$$x = -2$$
Therefore, As x approaches -2, the value of $-3x$ approaches 6.

5. The notation $\lim\limits_{x \to 4} f(x)$ is read "the limit, as x approaches 4, of $f(x)$."

7. The notation $\lim\limits_{x \to 5^-} F(x)$ is read "the limit, as x approaches 5 from the left, of $F(x)$."

9. The notation $\lim\limits_{x \to 2^+}$ is read "the limit, as x approaches 2 from the right."

11. As inputs x approach 3 from the right, outputs $f(x)$ approach 2. Thus the limit from the right is 2. That is,
$$\lim\limits_{x \to 3^+} f(x) = 2.$$

13. As inputs x approach -1 from the left, outputs $f(x)$ approach -3. Thus the limit from the left is -3. That is,
$$\lim\limits_{x \to -1^-} f(x) = -3.$$

15. From Exercise (11) and (12) we k now that $\lim\limits_{x \to 3^-} f(x) = 1$ and $\lim\limits_{x \to 3^+} f(x) = 2$. Since the limit from the left, 1, is not the same as the limit from the right, 2, $\lim\limits_{x \to 3} f(x)$ *does not exist.*

17. As inputs x approach 4 from the left, outputs $f(x)$ approach 3. Thus the limit from the left is 3. That is,
$$\lim\limits_{x \to 4^-} f(x) = 3.$$
As inputs x approach 4 from the right, outputs $f(x)$ approach 3. Thus the limit from the right is 4. That is,
$$\lim\limits_{x \to 4^+} f(x) = 3.$$
Since the limit from the left, 3, is the same as the limit from the right, 3, we have
$$\lim\limits_{x \to 4} f(x) = 3.$$

19. As inputs x approach -2 from the left, outputs $g(x)$ approach 4. Thus the limit from the left is 4. That is,
$$\lim\limits_{x \to -2^-} g(x) = 4.$$

21. As inputs x approach 4 from the right, outputs $g(x)$ approach -1. Thus the limit from the right is -1. That is,
$$\lim\limits_{x \to 4^+} g(x) = -1.$$

23. Since the limit from the left, -1, is the same as the limit from the right, -1, we have
$$\lim\limits_{x \to 4} g(x) = -1.$$

25. As inputs x approach 2 from the left, outputs $g(x)$ approach 0. Thus the limit from the left is 0. That is,
$$\lim\limits_{x \to 2^-} g(x) = 0.$$
As inputs x approach 2 from the right, outputs $g(x)$ approach 0. Thus the limit from the right is 0. That is,
$$\lim\limits_{x \to 2^+} g(x) = 0.$$
Since the limit from the left, 0, is the same as the limit from the right, 0, we have
$$\lim\limits_{x \to 2} g(x) = 0.$$

27. As inputs x approach -3 from the left, outputs $F(x)$ approach 5. Thus the limit from the left is 5. That is,
$$\lim_{x \to -3^-} F(x) = 5.$$
As inputs x approach -3 from the right, outputs $F(x)$ approach 5. Thus the limit from the right is 5. That is,
$$\lim_{x \to -3^+} F(x) = 5.$$

Since the limit from the left, 5, is the same as the limit from the right, 5, we have
$$\lim_{x \to -3} F(x) = 5.$$

29. As inputs x approach -2 from the left, outputs $F(x)$ approach 4. Thus the limit from the left is 4. That is,
$$\lim_{x \to -2^-} F(x) = 4.$$
As inputs x approach -2 from the right, outputs $F(x)$ approach 2. Thus the limit from the right is 2. That is,
$$\lim_{x \to -2^+} F(x) = 2.$$
Since the limit from the left, 4, is not the same as the limit from the right, 2, we have
$$\lim_{x \to -2} F(x) \; does \; not \; exist.$$

31. As inputs x approach 4 from the left, outputs $F(x)$ approach 2. Thus the limit from the left is 2. That is,
$$\lim_{x \to 4^-} F(x) = 2.$$
As inputs x approach 4 from the right, outputs $F(x)$ approach 2. Thus the limit from the right is 2. That is,
$$\lim_{x \to 4^+} F(x) = 2.$$
Since the limit from the left, 2, is the same as the limit from the right, 2, we have
$$\lim_{x \to 4} F(x) = 2.$$

33. As inputs x approach -2 from the right, outputs $F(x)$ approach 2. Thus the limit from the right is 2. That is,
$$\lim_{x \to -2^+} F(x) = 2.$$

35. As inputs x approach -1 from the left, outputs $f(x)$ approach 1. Thus the limit from the left is 1. That is,
$$\lim_{x \to -1^-} f(x) = 1.$$
As inputs x approach -1 from the right, outputs $f(x)$ approach 1. Thus the limit from the right is 1. That is,
$$\lim_{x \to -1^+} f(x) = 1.$$
Since the limit from the left, 1, is the same as the limit from the right, 1, we have
$$\lim_{x \to -1} f(x) = 1.$$

37. As inputs x approach -3 from the left, outputs $f(x)$ increase without bound. We say that the limit from the left is infinity. That is
$$\lim_{x \to -3^-} f(x) = \infty.$$
As inputs x approach -3 from the right, outputs $f(x)$ decrease without bound. We say that limit from the right is negative infinity. That is,
$$\lim_{x \to -3^+} f(x) = -\infty.$$
Since the function values as $x \to 3$ from the left increase without bound, and the function values as $x \to 3$ from the right decrease without bound, the limit does not exist. We have,
$$\lim_{x \to -3} f(x) \; does \; not \; exist.$$

39. As inputs x approach 3 from the left, outputs $f(x)$ approach 0. Thus the limit from the left is 0. That is,
$$\lim_{x \to 3^-} f(x) = 0.$$
As inputs x approach 3 from the right, outputs $f(x)$ approach 0.
Thus the limit from the right is 0. That is,
$$\lim_{x \to 3^+} f(x) = 0.$$
Since the limit from the left, 0, is the same as the limit from the right, 0, we have
$$\lim_{x \to 3} f(x) = 0.$$

41. As inputs x approach -4 from the left, outputs $f(x)$ approach 3. Thus the limit from the left is 3. That is,
$$\lim_{x \to -4^-} f(x) = 3.$$

As inputs x approach -4 from the right, outputs $f(x)$ approach 3. Thus the limit from the right is 3. That is,

$$\lim_{x \to -4^+} f(x) = 3.$$

Since the limit from the left, 3, is the same as the limit from the right, 3, we have

$$\lim_{x \to -4} f(x) = 3.$$

43. As inputs x get larger and larger, outputs $f(x)$ get closer and closer to 1. We have

$$\lim_{x \to \infty} f(x) = 1.$$

45. Defining $f(x) = |x|$ as a piecewise defined function we have:

$$f(x) = \begin{cases} -x, & x < 0 \\ x, & x \geq 0 \end{cases}.$$

We graph the function by creating an input-output table.

x	-2	-1	0	1	2
$f(x)$	2	1	0	1	2

Next, we plot the points from the table and draw the graph at the top of the next page.

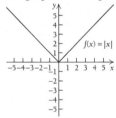

Find $\lim_{x \to 0} f(x)$.

As inputs x approach 0 from the left, outputs $f(x)$ approach 0. We have,

$$\lim_{x \to 0^-} f(x) = 0.$$

As inputs x approach 0 from the right, outputs $f(x)$ approach 0. We have,

$$\lim_{x \to 0^+} f(x) = 0$$

Since the limit from the left is the same as the limit from the right, we have

$$\lim_{x \to 0} f(x) = 0.$$

Find $\lim_{x \to -2} f(x)$.

As inputs x approach -2 from the left, outputs $f(x)$ approach 2. We have,

$$\lim_{x \to -2^-} f(x) = 2.$$

As inputs x approach -2 from the right, outputs $f(x)$ approach 2. We have,

$$\lim_{x \to -2^+} f(x) = 2$$

Since the limit from the left is the same as the limit from the right, we have

$$\lim_{x \to -2} f(x) = 2.$$

47. $g(x) = x^2 - 5$

We graph the function by creating an input-output table.

x	-2	-1	0	1	2
$g(x)$	-1	-4	-5	-4	-1

Next, we plot the points from the table and draw the graph.

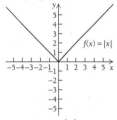

Find $\lim_{x \to 0} g(x)$.

As inputs x approach 0 from the left, outputs $g(x)$ approach -5. We have,

$$\lim_{x \to 0^-} g(x) = -5.$$

As inputs x approach 0 from the right, outputs $g(x)$ approach -5. We have,

$$\lim_{x \to 0^+} g(x) = -5$$

Since the limit from the left is the same as the limit from the right, we have

$$\lim_{x \to 0} g(x) = -5.$$

Find $\lim_{x \to -1} g(x)$.

As inputs x approach -1 from the left, outputs $g(x)$ approach -4. We have,

$$\lim_{x \to -1^-} g(x) = -4.$$

As inputs x approach -1 from the right, outputs $g(x)$ approach -4. We have,

$$\lim_{x \to -1^+} g(x) = -4$$

Since the limit from the left is the same as the limit from the right, we have

$$\lim_{x \to -1} g(x) = -4.$$

49. $F(x) = \dfrac{1}{x-3}$

Since $x = 3$ makes the denominator zero, we exclude the value 3 from the domain. Creating an input-output table we have

x	1	2	2.5	2.9	3.1	3.5	4	5
$F(x)$	$-\frac{1}{2}$	-1	-2	-10	10	2	1	$\frac{1}{2}$

Next we plot the points and draw the graph.

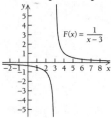

Find $\lim\limits_{x \to 3} F(x)$.

As inputs x approach 3 from the left, outputs $F(x)$ decrease without bound . We have,

$$\lim\limits_{x \to 3^-} F(x) = -\infty.$$

As inputs x approach 3 from the right, outputs $F(x)$ increase without bound. We have,

$$\lim\limits_{x \to 3^+} F(x) = \infty$$

Since the function values as $x \to 3$ from the left decrease without bound, and the function values as $x \to 3$ from the right increase without bound, the limit does not exist. We have,

$$\lim\limits_{x \to 3} F(x) \; does \; not \; exist.$$

Find $\lim\limits_{x \to 4} F(x)$.

As inputs x approach 4 from the left, outputs $F(x)$ approach 1. We have,

$$\lim\limits_{x \to 4^-} F(x) = 1.$$

As inputs x approach 4 from the right, outputs $F(x)$ approach 1. We have,

$$\lim\limits_{x \to 4^+} F(x) = 1$$

Since the limit from the left is the same as the limit from the right, we have

$$\lim\limits_{x \to 4} F(x) = 1.$$

51. $f(x) = \dfrac{1}{x} - 2$

Since $x = 0$ makes the denominator zero, we exclude the value 0 from the domain. Creating an input-output table we have

x	-1	-0.5	-0.1	0.1	0.5	1
$f(x)$	-3	-4	-12	8	0	-1

Next we plot the points and draw the graph.

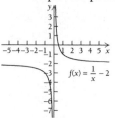

Find $\lim\limits_{x \to \infty} f(x)$.

As inputs x get larger and larger, outputs $f(x)$ get closer and closer to -2. We have,

$$\lim\limits_{x \to \infty} f(x) = -2.$$

Find $\lim\limits_{x \to 0} f(x)$.

As inputs x approach 0 from the left, outputs $f(x)$ decrease without bound. We have,

$$\lim\limits_{x \to 0^-} f(x) = -\infty.$$

As inputs x approach 0 from the right, outputs $f(x)$ increase without bound. We have,

$$\lim\limits_{x \to 0^+} f(x) = \infty$$

Since the function values as $x \to 0$ from the left decrease without bound, and the function values as $x \to 0$ from the right increase without bound, the limit does not exist. We have,

$$\lim\limits_{x \to 0} f(x) \; does \; not \; exist.$$

53. $g(x) = \dfrac{1}{x+2} + 4$

Since $x = -2$ makes the denominator zero, we exclude the value -2 from the domain. Creating an input-output table we have

x	-3	-2.5	-2.1	-1.9	-1.5	-1	0
$g(x)$	3	2	-6	14	6	3	$\frac{9}{2}$

Next we plot the points and draw the graph.

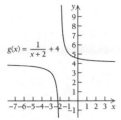

$g(x) = \dfrac{1}{x+2} + 4$

Find $\lim\limits_{x \to \infty} g(x)$.

As inputs x get larger and larger, outputs $g(x)$ get closer and closer to 4. We have

$$\lim\limits_{x \to \infty} g(x) = 4.$$

Find $\lim\limits_{x \to -2} g(x)$.

As inputs x approach -2 from the left, outputs $g(x)$ decrease without bound. We have,

$$\lim\limits_{x \to -2^-} g(x) = -\infty.$$

As inputs x approach -2 from the right, outputs $g(x)$ increase without bound. We have,

$$\lim\limits_{x \to -2^+} g(x) = \infty.$$

Since the function values as $x \to 2$ from the left decrease without bound, and the function values as $x \to 2$ from the right increase without bound, the limit does not exist. We have,

$\lim\limits_{x \to -2} g(x)$ *does not exist.*

55. $F(x) = \begin{cases} 2x+1, & \text{for } x < 1 \\ x, & \text{for } x \geq 1. \end{cases}$

We create an input-output table for each piece of the function.
For $x < 1$

x	-1	0	0.9
$F(x)$	-1	1	2.8

We plot the points and draw the graph. Notice we draw an open circle at the point $(1,3)$ to indicate that the point is not part of the graph.
For $x \geq 1$

x	1	2	3
$F(x)$	1	2	3

We plot the points and draw the graph. Notice we draw a solid circle at the point $(1,1)$ to indicate that the point is part of the graph.

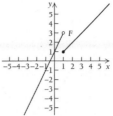

Find $\lim\limits_{x \to 1^-} F(x)$.

As inputs x approach 1 from the left, outputs $F(x)$ approach 3. That is,

$$\lim\limits_{x \to 1^-} F(x) = 3.$$

Find $\lim\limits_{x \to 1^+} F(x)$.

As inputs x approach 1 from the right, outputs $F(x)$ approach 1. That is,

$$\lim\limits_{x \to 1^+} F(x) = 1.$$

Find $\lim\limits_{x \to 1} F(x)$

Since the limit from the left, 3, is not the same as the limit from the right, 1, we have

$\lim\limits_{x \to 1} F(x)$ *does not exist.*

57. $g(x) = \begin{cases} -x+4, & \text{for } x < 3 \\ x-3, & \text{for } x > 3. \end{cases}$

We create an input-output table for each piece of the function.
For $x < 3$

x	0	1	2	2.9
$g(x)$	4	3	2	1.1

We plot the points from the table at the top of the previous page and draw the graph. Notice we draw an open circle at the point $(3,1)$ to indicate that the point is not part of the graph.
For $x > 3$

x	3.1	4	5	6
$g(x)$	0.1	1	2	3

We plot the points and draw the graph at the top of the next page. Notice we draw an open circle at the point $(3,0)$ to indicate that the point is not part of the graph.

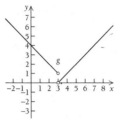

Find $\lim\limits_{x\to 3^-} g(x)$.

As inputs x approach 3 from the left, outputs $g(x)$ approach 1. That is,

$$\lim\limits_{x\to 3^-} g(x) = 1.$$

Find $\lim\limits_{x\to 3^+} g(x)$.

As inputs x approach 3 from the right, outputs $g(x)$ approach 0. That is,

$$\lim\limits_{x\to 3^+} g(x) = 0.$$

Find $\lim\limits_{x\to 3} g(x)$

Since the limit from the left, 1, is not the same as the limit from the right, 0, we have

$$\lim\limits_{x\to 3} g(x) \text{ does not exist.}$$

59. $G(x) = \begin{cases} x^2, & \text{for } x < -1 \\ x+2, & \text{for } x > -1. \end{cases}$

We create an input-output table for each piece of the function.

For $x < -1$

x	-3	-2	-1.1
$G(x)$	9	4	-1.21

We plot the points and draw the graph. Notice we draw an open circle at the point $(-1,1)$ to indicate that the point is not part of the graph.

For $x > -1$

x	-0.9	0	1
$G(x)$	1.1	2	3

We plot the points and draw the graph. Notice we draw an open circle at the point $(-1,1)$ to indicate that the point is not part of the graph.

Find $\lim\limits_{x\to -1} G(x)$.

As inputs x approach -1 from the left, outputs $G(x)$ approach 1. We have,

$$\lim\limits_{x\to -1^-} G(x) = 1.$$

As inputs x approach -1 from the right, outputs $G(x)$ approach 1. We have,

$$\lim\limits_{x\to -1^+} G(x) = 1$$

Since the limit from the left is the same as the limit from the right, we have

$$\lim\limits_{x\to -1} G(x) = 1.$$

61. $\lim\limits_{x\to 0.25^-} C(x) = 3.30$

$\lim\limits_{x\to 0.25^+} C(x) = 3.30$

$\lim\limits_{x\to 0.25} C(x) = 3.30$

63. $\lim\limits_{x\to 0.6^-} C(x) = 3.70$

$\lim\limits_{x\to 0.6^+} C(x) = 4.10$

$\lim\limits_{x\to 0.6} C(x) \text{ does not exist.}$

65. $\lim\limits_{x\to 2^-} p(x) = 0.63$

$\lim\limits_{x\to 2^+} p(x) = 0.87$

$\lim\limits_{x\to 2} p(x) \text{ does not exist.}$

67. $\lim\limits_{x\to 3^-} p(x) = 0.87$

$\lim\limits_{x\to 3^+} p(x) = 1.11$ $(0.87 + 0.24 = 1.11)$

$\lim\limits_{x\to 3} p(x) \text{ does not exist.}$

69. $\lim\limits_{t\to 1.5^-} p(t) = 11.0 \text{ hundred}$

$\lim\limits_{t\to 1.5^+} p(t) = 12.0 \text{ hundred}$

$\lim\limits_{t\to 1.5} p(t) \text{ does not exist.}$

71. $\boxed{tw}$

73. $\lim\limits_{t\to 0.8^-} p(t) = 35$

$\lim\limits_{t\to 0.8^+} p(t) = 34$

$\lim\limits_{t\to 0.8} p(t) \text{ does not exist.}$

75. As inputs x approach 2 from the right, outputs $f(x)$ approach 4. We have,

$$\lim_{x \to 2^+} f(x) = 4 .$$ In order for $\lim_{x \to 2} f(x)$ to exist we need $\lim_{x \to 2^-} f(x) = 4$. We will use the letter c for the unknown in the equation and this gives us

$$\lim_{x \to 2^-} \frac{1}{2}(x) + c = 4$$

Substitute 2 in for x to get the equation:

$$\frac{1}{2}(2) + c = 4$$ and solving for c we get

$$1 + c = 4$$

$$c = 3.$$

Therefore, in order for the limit to exist as x approaches 2, the function must be:

$$f(x) = \begin{cases} \frac{1}{2}x + 3 & \text{for } x < 2 \\ -x + 6 & \text{for } x > 2. \end{cases}$$

77. As inputs x approach 2 from the left, outputs $f(x)$ approach -5 . We have,

$$\lim_{x \to 2^-} f(x) = -5 .$$ In order for $\lim_{x \to 2} f(x)$ to exist we need $\lim_{x \to 2^+} f(x) = -5$. We will use the letter c for the unknown in the equation and this gives us

$$\lim_{x \to 2^+} -x^2 + c = -5$$

Substitute 2 in for x to get the equation:

$$-(2)^2 + c = -5$$ and solving for c we get

$$-(4) + c = -5$$

$$c = -1.$$

Therefore, in order for the limit to exist as x approaches 2, the function must be:

$$f(x) = \begin{cases} x^2 - 9 & \text{for } x < 2 \\ -x^2 + \underline{-1} & \text{for } x > 2. \end{cases}$$

79. Graph $f(x) = \begin{cases} x^2 - 2, & \text{for } x < 0 \\ 2 - x^2, & \text{for } x \ge 0. \end{cases}$

Using the calculator we enter the function into the graphing editor as follows:

```
Plot1  Plot2  Plot3
\Y1 ■X^2-2/(X<0)
\Y2 ■2-X^2/(X≥0)
\Y3 =
\Y4 =
\Y5 =
\Y6 =
\Y7 =
```

When you select the table feature you get:

X	Y1	Y2
-3	7	ERROR
-2	2	ERROR
-1	-1	ERROR
0	ERROR	2
1	ERROR	1
2	ERROR	-2
3	ERROR	-7

X=3

The calculator graphs the function:

Using the trace feature, we find the limits.
Find $\lim_{x \to 0} f(x)$.

As inputs x approach 0 from the left, outputs $f(x)$ approach -2 . We have,

$$\lim_{x \to 0^-} f(x) = -2 .$$

As inputs x approach 0 from the right, outputs $f(x)$ approach 2. We have,

$$\lim_{x \to 0^+} f(x) = 2$$

Since the limit from the left, -2 , is not the same as the limit from the right, 2, we have

$$\lim_{x \to 0} f(x) \text{ does not exist.}$$

Find $\lim_{x \to -2} f(x)$.

As inputs x approach -2 from the left, outputs $f(x)$ approach 2. We have,

$$\lim_{x \to -2^-} f(x) = 2 .$$

As inputs x approach -2 from the right, outputs $f(x)$ approach 2. We have,

$$\lim_{x \to -2^+} f(x) = 2$$

Since the limit from the left is the same as the limit from the right, we have

$$\lim_{x \to -2} f(x) = 2 .$$

81. Graph $f(x) = \dfrac{1}{x^2 - 4x - 5}$

Using the calculator we enter the function into the graphing editor as follows:

```
Plot1  Plot2  Plot3
\Y1 ■1/(X^2-4X-5)

\Y2 =
\Y3 =
\Y4 =
\Y5 =
\Y6 =
```

Using the following window:

The calculator graphs the function:

Using the trace feature on the calculator we find the limits.

Find $\lim_{x \to -1} f(x)$.

As inputs x approach -1 from the left, outputs $f(x)$ increase without bound. We have,

$$\lim_{x \to -1^-} f(x) = \infty.$$

As inputs x approach -1 from the right, outputs $f(x)$ decrease without bound. We have,

$$\lim_{x \to -1^+} f(x) = -\infty$$

Since the function values as $x \to -1$ from the left increase without bound, and the function values as $x \to -1$ from the right decrease without bound, the limit does not exist. We have,

$$\lim_{x \to -1} f(x) \text{ does not exist.}$$

Find $\lim_{x \to 5} f(x)$.

As inputs x approach 5 from the left, outputs $f(x)$ decrease without bound. We have,

$$\lim_{x \to 5^-} f(x) = -\infty.$$

As inputs x approach 5 from the right, outputs $f(x)$ increase without bound. We have,

$$\lim_{x \to 5^+} f(x) = \infty$$

Since the function values as $x \to 5$ from the left decrease without bound, and the function values as $x \to 5$ from the right increase without bound, the limit does not exist. We have,

$$\lim_{x \to 5} f(x) \text{ does not exist.}$$

Exercise Set 1.2

1. By limit principle L1,
$$\lim_{x \to 3} 7 = 7$$
Therefore, The statement is a true.

3. By limit principle L2,
$$\lim_{x \to 1}\left[g(x) \right]^2 = \left[\lim_{x \to 1} g(x) \right]^2 = \left[5 \right]^2 = 25.$$
Therefore, the statement is false.

5. By the definition of continuity, in order for f to be continuous at $x = 2$, $f(2)$ must exist. Therefore, the statement is true.

7. This statement is false. If $\lim_{x \to 4} F(x)$ exists but is not equal to $F(4)$, then F is not continuous.

9. It follows from the Theorem on Limits of Rational Functions that we can find the limit by substitution:
$$\lim_{x \to 1}(3x + 2) = 3(1) + 2$$
$$= 5.$$

11. It follows from the Theorem on Limits of Rational Functions that we can find the limit by substitution:
$$\lim_{x \to -1}(x^2 - 4) = (-1)^2 - 4$$
$$= 1 - 4$$
$$= -3.$$

13. It follows from the Theorem on Limits of Rational Functions that we can find the limit by substitution:
$$\lim_{x \to 3}(x^2 - 4x + 7) = (3)^2 - 4(3) + 7$$
$$= 9 - 12 + 7$$
$$= 4.$$

15. It follows from the Theorem on Limits of Rational Functions that we can find the limit by substitution:
$$\lim_{x \to 2}(2x^4 - 3x^3 + 4x - 1)$$
$$= 2(2)^4 - 3(2)^3 + 4(2) - 1$$
$$= 2(16) - 3(8) + 8 - 1$$
$$= 32 - 24 + 8 - 1$$
$$= 15.$$

17. It follows from the Theorem on Limits of Rational Functions that we can find the limit by substitution:

$$\lim_{x \to 3}\left(\frac{x^2-8}{x-2}\right) = \frac{(3)^2-8}{3-2}$$

$$= \frac{9-8}{3-2}$$

$$= 1.$$

19. We attempt to find the limit by substitution:

$$\lim_{x \to 3}\frac{x^2-9}{x-3} = \frac{(3)^2-9}{(3)-3}$$

$$= \frac{0}{0}$$

This is an indeterminate form. In order to find the limit we will simplify the function by factoring the numerator and dividing out common factors. Then we will apply the Theorem on Limits of Rational Functions to the simplified function.

$$\lim_{x \to 3}\frac{x^2-9}{x-3} = \lim_{x \to 3}\frac{(x-3)(x+3)}{x-3}$$

$$= \lim_{x \to 3}(x+3) \quad \text{simplifying, assuming } x \neq -3$$

$$= 3+3 \qquad \text{substitution}$$

$$= 6.$$

21. $\lim_{x \to 4}\sqrt{x^2-9}$

By limit principle L2,

$$\lim_{x \to 4}\sqrt{x^2-9} = \sqrt{\lim_{x \to 4}(x^2-9)}$$

$$= \sqrt{\lim_{x \to 4}x^2 - \lim_{x \to 4}9} \quad \text{By L3}$$

$$= \sqrt{\left(\lim_{x \to 4}x\right)^2 - 9} \quad \text{By L2 and L1}$$

$$= \sqrt{(4)^2-9}$$

$$= \sqrt{16-9}$$

$$= \sqrt{7}.$$

23. $\lim_{x \to 2}\sqrt{x^2-9}$

By limit principle L2,

$$\lim_{x \to 2}\sqrt{x^2-9} = \sqrt{\lim_{x \to 2}(x^2-9)}$$

$$= \sqrt{\lim_{x \to 2}x^2 - \lim_{x \to 2}9} \quad \text{By L3}$$

$$= \sqrt{\left(\lim_{x \to 2}x\right)^2 - 9} \quad \text{By L2 and L1}$$

$$= \sqrt{(2)^2-9}$$

$$= \sqrt{4-9}$$

$$= \sqrt{-5}.$$

Therefore, $\lim_{x \to 2}\sqrt{x^2-9}$ *does not exist.*

25. $\lim_{x \to 3^+}\sqrt{x^2-9}$

By limit principle L2,

$$\lim_{x \to 3^+}\sqrt{x^2-9} = \sqrt{\lim_{x \to 3^+}(x^2-9)}$$

$$= \sqrt{\lim_{x \to 3^+}x^2 - \lim_{x \to 3^+}9} \quad \text{By L3}$$

$$= \sqrt{\left(\lim_{x \to 3^+}x\right)^2 - 9} \quad \text{By L2 and L1}$$

$$= \sqrt{(3)^2-9}$$

$$= \sqrt{9-9}$$

$$= \sqrt{0}$$

$$= 0.$$

27. The function is not continuous over the interval, because $f(x)$ is not continuous at $x=1$. As x approaches 1 from the left, $f(x)$ approaches 2. However, as x approaches 1 from the right $f(x)$ approaches -1. Therefore, $f(x)$ is not continuous at 1.

29. The function is not continuous over the interval, because $k(x)$ is not continuous at $x=-1$. The function is not defined at $x=-1$, in other words $k(-1)$ does not exist. Therefore, $k(x)$ is not continuous at -1.

31. The function is not continuous over the interval, because it is not continuous at $x=-2$. The limit does not exist as x approaches -2, furthermore, $t(-2)$ does not exist. Therefore the function is not continuous at $x=-2$.

33. a) As inputs x approach 1 from the right, outputs $g(x)$ approach -2. Thus, the limit from the right is -2. $\lim\limits_{x \to 1^+} g(x) = -2$.

As inputs x approach 1 from the left, outputs $g(x)$ approach -2. Thus, the limit from the left is -2. $\lim\limits_{x \to 1^-} g(x) = -2$.

Since the limit from the left, -2, is the same as the limit from the right, -2, we have. $\lim\limits_{x \to 1} g(x) = -2$.

b) When the input is 1, the output $g(1)$ is -2. That is $g(1) = -2$.

c) The function $g(x)$ is continuous at $x = 1$, because

1) $g(1)$ exists, $g(1) = -2$

2) $\lim\limits_{x \to 1} g(x)$ exists, $\lim\limits_{x \to 1} g(x) = -2$, and

3) $\lim\limits_{x \to 1} g(x) = -2 = g(1)$.

d) As inputs x approach -2 from the right, outputs $g(x)$ approach -3. Thus, the limit from the right is -3. $\lim\limits_{x \to -2^+} g(x) = -3$.

As inputs x approach -2 from the left, outputs $g(x)$ approach 4. Thus, the limit from the left is 4. $\lim\limits_{x \to -2^-} g(x) = 4$.

Since the limit from the left, 4, is not the same as the limit from the right, -3, we say $\lim\limits_{x \to -2} g(x)$ does not exist.

e) When the input is -2, the output $g(-2)$ is -3. That is $g(-2) = -3$.

f) Since the limit of $g(x)$ as x approaches -2 does not exist, the function is not continuous at $x = -2$.

35. a) As inputs x approach 1 from the right, outputs $h(x)$ approach 2. Thus, the limit from the right is 2. $\lim\limits_{x \to 1^+} h(x) = 2$.

As inputs x approach 1 from the left, outputs $h(x)$ approach 2. Thus, the limit from the left is 2. $\lim\limits_{x \to 1^-} h(x) = 2$.

Since the limit from the left, 2, is the same as the limit from the right, 2, we have. $\lim\limits_{x \to 1} h(x) = 2$.

b) When the input is 1, the output $h(1)$ is 2. That is $h(1) = 2$.

c) The function $h(x)$ is continuous at $x = 1$, because

1) $h(1)$ exists, $h(1) = 2$

2) $\lim\limits_{x \to 1} h(x)$ exists, $\lim\limits_{x \to 1} h(x) = 2$, and

3) $\lim\limits_{x \to 1} h(x) = 2 = h(1)$.

d) As inputs x approach -2 from the right, outputs $h(x)$ approach 0. Thus, the limit from the right is 0. $\lim\limits_{x \to -2^+} h(x) = 0$.

As inputs x approach -2 from the left, outputs $h(x)$ approach 0. Thus, the limit from the left is 0. $\lim\limits_{x \to -2^-} h(x) = 0$.

Since the limit from the left, 0, is the same as the limit from the right, 0, we say $\lim\limits_{x \to -2} h(x) = 0$.

e) When the input is -2, the output $h(-2)$ is 0. That is $h(-2) = 0$.

f) The function $h(x)$ is continuous at $x = -2$, because

1) $h(-2)$ exists, $h(-2) = 0$

2) $\lim\limits_{x \to -2} h(x)$ exists, $\lim\limits_{x \to -2} h(x) = 0$, and

3) $\lim\limits_{x \to -2} h(x) = 0 = h(-2)$.

37. a) As inputs x approach 3 from the right, outputs $G(x)$ approach 3. Thus, $\lim\limits_{x \to 3^+} G(x) = 3$.

b) As inputs x approach 3 from the left, outputs $G(x)$ approach 1. Thus, $\lim\limits_{x \to 3^-} G(x) = 1$.

c) Since the limit from the left, 1, is not the same as the limit from the right, 3, the limit does not exist. $\lim\limits_{x \to 3} G(x)$ does not exist.

d) $G(3) = 1$

e) The function $G(x)$ is not continuous at $x = 3$ because the limit does not exist as x approaches 3.

f) The function $G(x)$ is continuous at $x = 0$, because

1) $G(0)$ exists,

2) $\lim\limits_{x \to 0} G(x)$ exists, , and

3) $\lim\limits_{x \to 0} G(x) = 0 = G(0)$.

g) The function $G(x)$ is continuous at $x = 2.9$, because

1) $G(2.9)$ exists,

2) $\lim\limits_{x \to 2.9} G(x)$ exists, , and

3) $\lim\limits_{x \to 2.9} G(x) = G(2.9)$.

39. First we find the function value when $x = 5$.

$f(5) = 3(5) - 2 = 13$. Hence, $f(5)$ exists.

Next, we find the limit as x approaches 5. It follows from the Theorem on Limits of Rational Functions that we can find the limit by substitution: $\lim\limits_{x \to 5} f(x) = 3(5) - 2 = 13$

Therefore, $\lim\limits_{x \to 5} f(x) = 13 = f(5)$ and the function is continuous at $x = 5$.

41. The function $G(x) = \dfrac{1}{x}$ is not continuous at

$x = 0$ because $G(0) = \dfrac{1}{0}$ is undefined.

43. First we find the function value when $x = 3$.

$g(3) = \dfrac{1}{3}(3) + 4 = 1 + 4 = 5$. Hence, $g(3)$

exists.

Next, we find the limit as x approaches 3. As the inputs x approach 3 from the right, the outputs $g(x)$ approach 5, that is,

$\lim\limits_{x \to 3^-} g(x) = \dfrac{1}{3}(3) + 4 = 5$.

As the inputs x approach 3 from the left, the outputs $g(x)$ approach 5, that is,

$\lim\limits_{x \to 3^+} g(x) = 2(3) - 1 = 5$.

Since the limit from the left, 5, is the same as the limit from the right, 5. The limit exists. We have:

$\lim\limits_{x \to 3} g(x) = 5$.

Therefore, we have

$\lim\limits_{x \to 3} g(x) = 5 = g(3)$.

Thus the function is continuous at $x = 3$.

45. The function is not continuous at $x = 3$ because the limit does not exist as x approaches 3. To verify this we take the limit as x approaches 3 from the left and the limit as x approaches 3 from the right.

As x approaches 3 from the left we have

$\lim\limits_{x \to 3^-} F(x) = \dfrac{1}{3}(3) + 4 = 5$.

As x approaches 3 from the right we have

$\lim\limits_{x \to 3^+} F(x) = 2(3) - 5 = 1$.

Since the limit from the left, 5, is not the same as the limit from the right, 1, the limit does not exist. $\lim\limits_{x \to 3} F(x)$ does not exist.

47. First we find the function value when $x = 3$.

$f(3) = 2(3) - 1 = 5$. Hence, $f(3)$ exists.

Next, we find the limit as x approaches 3. As the inputs x approach 3 from the left, the outputs $f(x)$ approach 5, that is,

$\lim\limits_{x \to 3^-} f(x) = \dfrac{1}{3}(3) + 4 = 5$.

As the inputs x approach 3 from the left, the outputs $f(x)$ approach 5, that is,

$\lim\limits_{x \to 3^+} f(x) = 2(3) - 1 = 5$.

Since the limit from the left, 5, is the same as the limit from the right, 5. The limit exists. We have:

$\lim\limits_{x \to 3} f(x) = 5$.

Therefore, we have

$\lim\limits_{x \to 3} f(x) = 5 = f(3)$.

Thus the function is continuous at $x = 3$.

49. The function is not continuous at $x = 2$. To verify this, we take the limit as x approaches 2. Using the Theorem on Limits of Rational Functions, we simplify the function near 2 by factoring the numerator and dividing out common factors.

$\lim\limits_{x \to 2} G(x) = \lim\limits_{x \to 2} \dfrac{x^2 - 4}{x - 2}$

$= \lim\limits_{x \to 2} \dfrac{(x-2)(x+2)}{x-2}$

$= \lim\limits_{x \to 2} (x+2)$

$= 2 + 2 = 4$

Therefore,

$\lim\limits_{x \to 2} G(x) = 4$.

However, when $x = 2$, the output $G(2)$ is defined to be 5. That is, $G(2) = 5$. Therefore, $\lim_{x \to 2} G(x) = 4 \neq 5 = G(2)$. Thus the function is not continuous at $x = 2$.

51. First we find the function value when $x = 5$.
$f(5) = (5) + 1 = 6$, $f(5)$ exists.

Next we find the limit as x approaches 5. To find the limit as x approaches 5 from the left, we first simplify the rational function by factoring the numerator and dividing out common factors.

$$\lim_{x \to 5^-} f(x) = \lim_{x \to 5^-} \frac{x^2 - 4x - 5}{x - 5}$$
$$= \lim_{x \to 5^-} \frac{(x - 5)(x + 1)}{x - 5}$$
$$= \lim_{x \to 5^-} (x + 1)$$
$$= 5 + 1$$
$$= 6$$

To find the limit as x approaches 5 from the right, we can use substitution.

$$\lim_{x \to 5^+} f(x) = \lim_{x \to 5^+} x + 1$$
$$= 5 + 1$$
$$= 6$$

Therefore, the limit exists.
$\lim_{x \to 5} f(x) = 6$.

Thus we have,
$\lim_{x \to 5} f(x) = 6 = f(5)$.

Therefore, the function in continuous at $x = 5$.

53. The function is not continuous at $x = 5$ because $g(5)$ does not exist.

$$g(5) = \frac{1}{(5)^2 - 7(5) + 10}$$
$$= \frac{1}{25 - 35 + 10}$$
$$= \frac{1}{0}$$

55. First we find the function value when $x = 4$.

$$F(4) = \frac{1}{(4)^2 - 7(4) + 10}$$
$$= \frac{1}{16 - 28 + 10}$$
$$= \frac{1}{-2}$$
$$= -\frac{1}{2}$$

Hence, $F(4)$ exists.

Next we find the limit. Applying the Theorem on Limits of Rational Functions we have:

$$\lim_{x \to 4} F(x) = \frac{1}{(4)^2 - 7(4) + 10} = -\frac{1}{2}.$$

We now have,

$$\lim_{x \to 4} F(x) = \frac{1}{(4)^2 - 7(4) + 10} = -\frac{1}{2} = F(4).$$

Therefore, the function is continuous at $x = 4$.

57. Yes, the function is continuous over the interval $(-4, 4)$. Since the function is defined for every value in the interval, the Theorem on Limits of Rational Functions tells us $\lim_{x \to a} g(x) = g(a)$ for all values a in the interval. Thus $g(x)$ is continuous over the interval.

59. No, the function is not continuous over the interval $(-7, 7)$ because the function does not exist at $x = 0$. $f(0) = \frac{1}{0} + 3$, which is undefined.

61. Yes, the function is continuous on $\mathbb{R}$. The function is defined for all real numbers, so by the Theorem on Limits of Rational Functions, $\lim_{x \to a} g(x) = g(a)$ for all a in $\mathbb{R}$.

63. As the inputs x approach 0 from the left, the outputs approach -1. We see this by looking at a table:

x	−0.1	−0.01	−0.001
$\dfrac{|x|}{x}$	−1	−1	−1

$$\lim_{x \to 0^-} \frac{|x|}{x} = -1.$$

As the inputs x approach 0 from the right, the outputs approach 1. We see this by looking at a table:

x	0.001	0.01	0.1
$\dfrac{\lvert x\rvert}{x}$	1	1	1

$$\lim_{x\to 0^+}\frac{\lvert x\rvert}{x}=1$$

Since the limit from the left, -1, is not the same as the limit from the right, 1, the limit does not exist. $\displaystyle\lim_{x\to 0}\frac{\lvert x\rvert}{x}$ does not exist.

65. 6

67. -0.2887, or $-\dfrac{1}{2\sqrt{3}}$

69. 0.75, or $\dfrac{3}{4}$

71. 0.25, or $\dfrac{1}{4}$

Exercise Set 1.3

---

1. a) $f(x)=4x^2$

so,

$$f(x+h)=4(x+h)^2 \quad \text{substituing } x+h \text{ for } x$$
$$=4\left(x^2+2xh+h^2\right)$$
$$=4x^2+8xh+4h^2$$

Then,

$$\frac{f(x+h)-f(x)}{h} \quad \text{Difference quotient}$$

$$=\frac{\left(4x^2+8xh+4h^2\right)-4x^2}{h} \quad \text{Substituting}$$

$$=\frac{8xh+4h^2}{h}$$

$$=\frac{h(8x+4h)}{h} \quad \text{Factoring the numerator}$$

$$=\frac{h}{h}\cdot(8x+4h) \quad \text{Removing a factor} = 1.$$

$$=8x+4h, \quad \text{Simplified difference quotient}$$

or $4(2x+h)$

b) The difference quotient column in the table can be completed using the simplified difference quotient.

$8x+4h$ Simplified difference quotient

$$=8(5)+4(2)=48$$

substituiting 5 for x and 2 for h;

$$=8(5)+4(1)=44$$

substituiting 5 for x and 1 for h;

$$=8(5)+4(0.1)=40.4$$

substituiting 5 for x and 0.1 for h;

$$=8(5)+4(0.01)=40.04$$

substituiting 5 for x and 0.01 for h.

The completed table is:

x	h	$\dfrac{f(x+h)-f(x)}{h}$
5	2	48
5	1	44
5	0.1	40.4
5	0.01	40.04

3. a) $f(x)=-4x^2$

so,

$$f(x+h)=-4(x+h)^2 \quad \text{substituing } x+h \text{ for } x$$
$$=-4\left(x^2+2xh+h^2\right)$$
$$=-4x^2-8xh-4h^2$$

Then

$$\frac{f(x+h)-f(x)}{h} \quad \text{Difference quotient}$$

$$=\frac{\left(-4x^2-8xh-4h^2\right)-\left(-4x^2\right)}{h} \quad \text{Substituting}$$

$$=\frac{-8xh-4h^2}{h}$$

$$=\frac{h(-8x-4h)}{h} \quad \text{Factoring the numerator}$$

$$=\frac{h}{h}\cdot(-8x-4h) \quad \text{Removing a factor = 1.}$$

$$=-8x-4h, \quad \text{Simplified difference quotient}$$

$$\text{or} \ -4(2x+h)$$

b) The difference quotient column in the table can be completed using the simplified difference quotient.

$$-8x-4h \quad \text{Simplified difference quotient}$$

$$=-8(5)-4(2)=-48$$

     substitiuting 5 for x and 2 for h;

$$=-8(5)-4(1)=-44$$

     substitiuting 5 for x and 1 for h;

$$=-8(5)-4(0.1)=-40.4$$

     substitiuting 5 for x and 0.1 for h;

$$=-8(5)-4(0.01)=-40.04$$

     substitiuting 5 for x and 0.01 for h.

The completed table is:

x	h	$\dfrac{f(x+h)-f(x)}{h}$
5	2	−48
5	1	−44
5	0.1	−40.4
5	0.01	−40.04

5.   a) $f(x)=x^2+x$

We substitute $x+h$ for x

$$f(x+h)=(x+h)^2+(x+h)$$

$$=\left(x^2+2xh+h^2\right)+x+h$$

$$=x^2+2xh+h^2+x+h$$

Then

$$\frac{f(x+h)-f(x)}{h} \quad \text{Difference quotient}$$

$$=\frac{\left(x^2+2xh+h^2+x+h\right)-\left(x^2+x\right)}{h}$$

$$=\frac{2xh+h^2+h}{h}$$

$$=\frac{h(2x+h+1)}{h} \quad \text{Factoring the numerator}$$

$$=\frac{h}{h}\cdot(2x+h+1) \quad \text{Removing a factor = 1.}$$

$$=2x+h+1 \quad \text{Simplified difference quotient}$$

b) The difference quotient column in the table can be completed using the simplified difference quotient.

$$2x+h+1 \quad \text{Simplified difference quotient}$$

$$=2(5)+(2)+1=13$$

     substitiuting 5 for x and 2 for h;

$$=2(5)+(1)+1=12$$

     substitiuting 5 for x and 1 for h;

$$=2(5)+(0.1)+1=11.1$$

     substitiuting 5 for x and 0.1 for h;

$$=2(5)+(0.01)+1=11.01$$

     substitiuting 5 for x and 0.01 for h.

The completed table is:

x	h	$\dfrac{f(x+h)-f(x)}{h}$
5	2	13
5	1	12
5	0.1	11.1
5	0.01	11.01

7.   a) $f(x)=\dfrac{2}{x}$

We substitute $x+h$ for x

$$f(x+h)=\frac{2}{x+h}$$

Then

$$\frac{f(x+h)-f(x)}{h} \quad \text{Difference quotient}$$

$$= \frac{\left(\dfrac{2}{x+h}\right)-\left(\dfrac{2}{x}\right)}{h}$$

$$= \frac{\left(\dfrac{2}{x+h}\cdot\dfrac{x}{x}\right)-\left(\dfrac{2}{x}\cdot\dfrac{(x+h)}{(x+h)}\right)}{h} \quad \text{multiplying by 1}$$

$$= \frac{\left(\dfrac{2x}{x(x+h)}\right)-\left(\dfrac{2(x+h)}{x(x+h)}\right)}{h}$$

$$= \frac{\dfrac{-2h}{x(x+h)}}{h} \quad \text{adding fractions}$$

$$= \frac{\dfrac{-2h}{x(x+h)}}{\dfrac{h}{1}} \quad h=\frac{h}{1}$$

$$= \frac{-2h}{x(x+h)}\cdot\frac{1}{h} \quad \text{multiplying by the reciprocal}$$

$$= \frac{h}{h}\cdot\left(\frac{-2}{x(x+h)}\right) \quad \text{Removing a factor = 1.}$$

$$= \frac{-2}{x(x+h)} \quad \text{Simplified difference quotient}$$

b) The difference quotient column in the table can be completed using the simplified difference quotient.

$$-\frac{2}{x(x+h)} \quad \text{Simplified difference quotient}$$

$$= -\frac{2}{5(5+2)} = -\frac{2}{35}$$

       substitiuting 5 for x and 2 for h;

$$= -\frac{2}{5(5+1)} = -\frac{2}{30} = -\frac{1}{15}$$

       substitiuting 5 for x and 1 for h;

$$= -\frac{2}{5(5+0.1)} = -\frac{2}{25.5} = -\frac{4}{51}$$

       substitiuting 5 for x and 0.1 for h;

$$= -\frac{2}{5(5+0.01)} = -\frac{2}{25.05} = -\frac{40}{501}$$

       substitiuting 5 for x and 0.01 for h.

The completed table is:

x	h	$\dfrac{f(x+h)-f(x)}{h}$
5	2	$-\dfrac{2}{35}$
5	1	$-\dfrac{1}{15}$
5	0.1	$-\dfrac{4}{51}$
5	0.01	$-\dfrac{40}{501}$

9.   a) $f(x)=-2x+5$

We substitute $x+h$ for x

$$f(x+h)=-2(x+h)+5$$

$$-2x-2h+5$$

Then

$$\frac{f(x+h)-f(x)}{h} \quad \text{Difference quotient}$$

$$= \frac{(-2x-2h+5)-(-2x+5)}{h}$$

$$= \frac{-2h}{h}$$

$$= -2 \quad \text{Simplified difference quotient}$$

b) The difference quotient is -2 for all values of x and h. The completed table is:

x	h	$\dfrac{f(x+h)-f(x)}{h}$
5	2	-2
5	1	-2
5	0.1	-2
5	0.01	-2

11. a) $f(x)=1-x^3$

so,

$$f(x+h)=1-(x+h)^3 \quad \text{substituing } x+h \text{ for } x$$

$$=1-\left(x^3+3x^2h+3xh^2+h^3\right)$$

$$=1-x^3-3x^2h-3xh^2-h^3$$

Then

$$\frac{f(x+h)-f(x)}{h} \quad \text{Difference quotient}$$

$$=\frac{\left(1-x^3-3x^2h-3xh^2-h^3\right)-\left(1-x^3\right)}{h}$$

$$=\frac{-3x^2h-3xh^2-h^3}{h}$$

$$=\frac{h\left(-3x^2-3xh-h^2\right)}{h} \quad \text{Factoring the numerator}$$

$$=\frac{h}{h}\cdot\left(-3x^2-3xh-h^2\right) \quad \text{Removing a factor} = 1.$$

$$=-3x^2-3xh-h^2$$

Simplified difference quotient

b) The difference quotient column in the table can be completed using the simplified difference quotient.

$$-3x^2-3xh-h^2$$

$$=-3(5)^2-3(5)(2)-(2)^2 = -109$$

substitiuting 5 for x and 2 for h;

$$=-3(5)^2-3(5)(1)-(1)^2 = -91$$

substitiuting 5 for x and 1 for h;

$$=-3(5)^2-3(5)(0.1)-(0.1)^2 = -76.51$$

substitiuting 5 for x and 0.1 for h;

$$=-3(5)^2-3(5)(0.01)-(0.01)^2 = -75.1501$$

substitiuting 5 for x and 0.01 for h.

The completed table is:

x	h	$\dfrac{f(x+h)-f(x)}{h}$
5	2	−109
5	1	−91
5	0.1	−76.51
5	0.01	−75.1501

13. a) $f(x)=x^2-3x$

We substitute $x+h$ for x

$$f(x+h)=(x+h)^2-3(x+h)$$

$$=\left(x^2+2xh+h^2\right)-3x-3h$$

$$=x^2+2xh+h^2-3x-3h$$

Then

$$\frac{f(x+h)-f(x)}{h} \quad \text{Difference quotient}$$

$$=\frac{\left(x^2+2xh+h^2-3x-3h\right)-\left(x^2-3x\right)}{h}$$

$$=\frac{2xh+h^2-3h}{h}$$

$$=\frac{h(2x+h-3)}{h} \quad \text{Factoring the numerator}$$

$$=\frac{h}{h}\cdot(2x+h-3) \quad \text{Removing a factor} = 1.$$

$$=2x+h-3 \quad \text{Simplified difference quotient}$$

b) The difference quotient column in the table can be completed using the simplified difference quotient.

$$2x+h-3 \quad \text{Simplified difference quotient}$$

$$=2(5)+(2)-3 = 9$$

substitiuting 5 for x and 2 for h;

$$=2(5)+(1)-3 = 8$$

substitiuting 5 for x and 1 for h;

$$=2(5)+(0.1)-3 = 7.1$$

substitiuting 5 for x and 0.1 for h;

$$=2(5)+(0.01)-3 = 7.01$$

substitiuting 5 for x and 0.01 for h.

Using the values from the previous page, the completed table is:

x	h	$\dfrac{f(x+h)-f(x)}{h}$
5	2	9
5	1	8
5	0.1	7.1
5	0.01	7.01

15. a) $f(x)=x^2+4x-3$

We substitute $x+h$ for x

$$f(x+h)=(x+h)^2+4(x+h)-3$$

$$=\left(x^2+2xh+h^2\right)+4x+4h-3$$

$$=x^2+2xh+h^2+4x+4h-3$$

Then

$$\frac{f(x+h)-f(x)}{h} \quad \text{Difference quotient}$$

$$=\frac{\left(x^2+2xh+h^2+4x+4h-3\right)-\left(x^2+4x-3\right)}{h}$$

$$=\frac{2xh+h^2+4h}{h}$$

$$=\frac{h(2x+h+4)}{h} \quad \text{Factoring the numerator}$$

$$=\frac{h}{h}\cdot(2x+h+4) \quad \text{Removing a factor} = 1$$

$$=2x+h+4 \quad \text{Simplified difference quotient}$$

b) The difference quotient column in the table can be completed using the simplified difference quotient.

$$2x+h+4 \quad \text{Simplified difference quotient}$$

$$=2(5)+(2)+4=16$$

substitiuting 5 for x and 2 for h;

$$=2(5)+(1)+4=15$$

substitiuting 5 for x and 1 for h;

$$=2(5)+(0.1)+4=14.1$$

substitiuting 5 for x and 0.1 for h;

$$=2(5)+(0.01)+4=14.01$$

substitiuting 5 for x and 0.01 for h.

The completed table is:

x	h	$\dfrac{f(x+h)-f(x)}{h}$
5	2	16
5	1	15
5	0.1	14.1
5	0.01	14.01

17. a) To find the average rate of change from 1994 to 2000, we locate the corresponding points $(1994,0)$ and $(2000,20)$. Using these two points we calculate the average rate of change

$$\frac{20-0}{2000-1994}=\frac{20}{6}=\frac{10}{3}=3\tfrac{1}{3}$$

The average rate of change of total employment from 1994 to 2000 increased approximately $3\tfrac{1}{3}\%$ per year.

b) To find the average rate of change from 2000 to 2005, we locate the corresponding points $(2000,20)$ and $(2005,22)$. Using these two points we calculate the average rate of change

$$\frac{22-20}{2005-2000}=\frac{2}{5}=0.4$$

The average rate of change of total employment from 2000 to 2005 increased approximately 0.4% per year.

c) To find the average rate of change from 1994 to 2005, we locate the corresponding points $(1994,0)$ and $(2005,22)$. Using these two points we calculate the average rate of change

$$\frac{22-0}{2005-1994}=\frac{22}{11}=2$$

The average rate of change of total employment from 1994 to 2005 increased approximately 2% per year.

19. a) To find the average rate of change from 1994 to 2000, we locate the corresponding points $(1994,1)$ and $(2000,40)$. Using these two points we calculate the average rate of change

$$\frac{40-1}{2000-1994}=\frac{39}{6}=\frac{13}{2}=6.5$$

The average rate of change of employment for professional services from 1994 to 2000 increased approximately 6.5% per year.

b) To find the average rate of change from 2000 to 2005, we locate the corresponding points $(2000,40)$ and $(2005,50)$.

Using these two points we calculate the average rate of change

$$\frac{50-40}{2005-2000}=\frac{10}{5}=2.0$$

The average rate of change of employment for professional services from 2000 to 2005 increased approximately 2.0% per year.

c) To find the average rate of change from 1994 to 2005, we locate the corresponding points $(1994,1)$ and $(2005,50)$. Using these two points we calculate the average rate of change

$$\frac{50-1}{2005-1994}=\frac{49}{11}\approx4.4545$$

The average rate of change of employment for professional services from 1994 to 2005 increased approximately 4.5% per year.

21. a) To find the average rate of change from 1994 to 2000, we locate the corresponding points $(1994,0)$ and $(2000,17)$. Using these two points we calculate the average rate of change
$$\frac{17-0}{2000-1994}=\frac{17}{6}=2\frac{5}{6}$$
The average rate of change of education employment from 1994 to 2000 increased approximately $2\frac{5}{6}\%$ per year.

b) To find the average rate of change from 2000 to 2005, we locate the corresponding points $(2000,17)$ and $(2005,28)$. Using these two points we calculate the average rate of change
$$\frac{28-17}{2005-2000}=\frac{11}{5}=2\frac{1}{5}=2.2$$
The average rate of change of education employment from 2000 to 2005 increased approximately $2\frac{1}{5}$, or about 2.2% per year.

c) To find the average rate of change from 1994 to 2005, we locate the corresponding points $(1994,0)$ and $(2005,28)$. Using these two points we calculate the average rate of change
$$\frac{28-0}{2005-1994}=\frac{28}{11}=2\frac{6}{11}\approx 2.545455$$
The average rate of change of education employment from 1994 to 2005 increased approximately $2\frac{6}{11}\%$, or about 2.5% per year.

23. a) To find the average rate of change from 1994 to 2000, we locate the corresponding points $(1994,0)$ and $(2000,-10)$.

Using these two points we calculate the average rate of change
$$\frac{-10-0}{2000-1994}=\frac{-10}{6}=-1\frac{2}{3}$$
The average rate of change of natural resources employment from 1994 to 2000 increased approximately $-1\frac{2}{3}\%$ per year.

b) To find the average rate of change from 2000 to 2005, we locate the corresponding points $(2000,-10)$ and $(2005,-4)$. Using these two points we calculate the average rate of change
$$\frac{-4-(-10)}{2005-2000}=\frac{6}{5}=1\frac{1}{5}=1.2$$

The average rate of change of natural resources employment from 2000 to 2005 increased approximately $1\frac{1}{5}$, or about 1.2% per year.

c) To find the average rate of change from 1994 to 2005, we locate the corresponding points $(1994,0)$ and $(2005,-4)$. Using these two points we calculate the average rate of change
$$\frac{-4-0}{2005-1994}=\frac{-4}{11}\approx -0.363$$
The average rate of change of natural resources employment from 1994 to 2005 increased approximately $-\frac{4}{11}\%$, or about -0.4% per year.

25. In order to find the average rate of change from 1960 to 1970, we use the data points $(1960,45.09)$ and $(1970,67.84)$. The average rate of change is
$$\frac{67.84-45.09}{1970-1960}=\frac{22.75}{10}=2.275\approx 2.3.$$
Between the years 1960 to 1970 the average rate of change in U.S. energy consumption increased about 2.3 quadrillion BTUs per year.

In order to find the average rate of change from 1980 to 1990, we use the data points $(1980,78.29)$ and $(1990,84.67)$. The average rate of change is
$$\frac{84.67-78.29}{1990-1980}=\frac{6.38}{10}=0.638\approx 0.6.$$
Between the years 1980 to 1990 the average rate of change in U.S. energy consumption increased about 0.6 quadrillion BTUs per year.

In order to find the average rate of change from 1990 to 2005, we use the data points $(1990,84.67)$ and $(2005,101.85)$. The average rate of change is
$$\frac{101.85-84.67}{2005-1990}=\frac{17.18}{15}=1.1453333\approx 1.1.$$
Between the years 1990 to 205 the average rate of change in U.S. energy consumption increased about 1.1 quadrillion BTUs per year.

27. a) From 0 units to 1 unit the average rate of change is
$$\frac{70-0}{1-0} = 70 \text{ pleasure units per unit.}$$
From 1 unit to 2 units the average rate of change is
$$\frac{109-70}{2-1} = \frac{39}{1} = 39 \text{ pleasure units per unit.}$$
From 2 units to 3 units the average rate of change is
$$\frac{138-109}{3-2} = \frac{29}{1} = 29 \text{ pleasure units per unit.}$$
From 3 units to 4 units the average rate of change is
$$\frac{161-138}{4-3} = \frac{23}{1} = 23 \text{ pleasure units per unit.}$$

b) $\boxed{tw}$

29. $p(x) = 0.03x^2 + 0.56x + 8.63$

a) $p(4) = 0.03(4)^2 + 0.56(4) + 8.63 = 11.35$

b) $p(14) = 0.03(14)^2 + 0.56(14) + 8.63 = 22.35$

c) $P(14) - p(4) = 22.35 - 11.35 = 11$

d) $\dfrac{p(14)-p(4)}{14-4} = \dfrac{11}{10} = 1.1$

This result implies that the average price of a ticket between 1995 ($x = 4$) and 2005 ($x = 14$) grew at an average rate of $1.10 per year.

31. $A(t) = 5000(1.14)^t$

$A(3) = 5000(1.14)^3 = 7407.72$

$A(2) = 5000(1.14)^2 = 6498.00$

$\dfrac{A(3)-A(2)}{3-2} = \dfrac{7407.72 - 6498.00}{3-2} = 909.72$

If unpaid, the debt on the credit card will be growing at an average rate of $909.72 per year between the 2nd and 3rd year.

33. $C(x) = -0.05x^2 + 50x$

First substitute 301 for x.

$C(301) = -0.05(301)^2 + 50(301)$
$$= -4530.05 + 15,050$$
$$= 10,519.95$$

The total cost of producing 301 units is $10,519.95.

Next substitute 300 for x.

$C(300) = -0.05(300)^2 + 50(300)$
$$= -4500 + 15,000$$
$$= 10,500.00$$

The total cost of producing 300 units is $10,500.00.

Now we can substitute to find the average rate of change.

$$\frac{C(301)-C(300)}{301-300} = \frac{10,519.95 - 10,500}{301-300}$$
$$= \frac{19.95}{1}$$
$$= 19.95$$

Total cost will increase $19.95 if the company produces the 301st unit.

35. Note: Answers will vary according to the values estimated from the graph.

a) Locate the points $(0,8)$ and $(12,20)$ on the girls growth median weight chart. Using these points we calculate the average growth rate.
$$\frac{20-8}{12-0} = \frac{12}{12} = 1.$$
The average growth rate of a girl during her first 12 months is 1 pound per month.

b) Locate the points $(12,20)$ and $(24,26.5)$ on the girls growth median weight chart. Using these points we calculate the average growth rate.
$$\frac{26.5-20}{24-12} = \frac{6.5}{12} = 0.541\overline{6} \approx 0.54.$$
The average growth rate of a girl during her second 12 months is approximately 0.54 pounds per month.

c) Locate the points $(0,8)$ and $(24,26.5)$ on the girls growth median weight chart. Using these points we calculate the average growth rate.
$$\frac{26.5-8}{24-0} = \frac{18.5}{24} = 0.7708\overline{3} \approx 0.77.$$
The average growth rate of a girl during her first 24 months is approximately 0.77 pounds per month.

d) We estimate the growth rate of a 12 month old girl to be approximately 0.67 pounds per month. This answer will vary depending upon your tangent line.

e) The graph indicates that the growth rate is fastest during the first 5 months.

37. $H(w) = 0.11w^{1.36}$

a) First we substitute 500 and 700 in for w to find the home range at the respective weights.

$$H(500) = 0.11(500)^{1.36}$$
$$= 0.11(4683.809314)$$
$$= 515.2190246$$
$$\approx 515.22$$

$$H(700) = 0.11(700)^{1.36}$$
$$= 0.11(7401.731628)$$
$$= 814.1904791$$
$$\approx 814.19$$

Next we use the function values to find the average rate at which the mammal's home range will increase

$$\frac{H(700) - H(500)}{700 - 500} = \frac{814.19 - 515.22}{700 - 500}$$
$$= \frac{298.97}{200}$$
$$\approx 1.49485$$

The average rate at which a carnivorous mammal's home range increases as the animal's weight grows from 500 g to 700g is approximately 1.49 hectares per gram.

b) First we substitute 200 and 300 in for w to find the home range at the respective weights.

$$H(200) = 0.11(200)^{1.36}$$
$$= 0.11(1347.102971)$$
$$= 148.1813269$$
$$\approx 148.18$$

$$H(300) = 0.11(300)^{1.36}$$
$$= 0.11(2338.217499)$$
$$= 257.2039249$$
$$\approx 257.20$$

Next we use the function values to find the average rate at which the mammal's home range will increase

$$\frac{H(300) - H(200)}{300 - 200} = \frac{257.20 - 148.18}{300 - 200}$$
$$= \frac{109.02}{100}$$
$$\approx 1.0902$$

The average rate at which a carnivorous mammal's home range increases as the animal's weight grows from 200 g to 300g is approximately 1.09 hectares per gram.

39. a) We locate the points $(0,0)$ and $(8,10)$ on the graph and use them to calculate the average rate of change

$$\frac{10 - 0}{8 - 0} = \frac{10}{8} = \frac{5}{4} = 1.25$$

The average rate of change is 1.25 words per minute.

We locate the points $(8,10)$ and $(16,20)$ on the graph and use them to calculate the average rate of change

$$\frac{20 - 10}{16 - 8} = \frac{10}{8} = \frac{5}{4} = 1.25$$

The average rate of change is 1.25 words per minute.

We locate the points $(16,20)$ and $(24,25)$ on the graph and use them to calculate the average rate of change

$$\frac{25 - 20}{24 - 16} = \frac{5}{8} = 0.625$$

The average rate of change is 0.625 words per minute.

We locate the points $(24,25)$ and $(32,25)$ on the graph and use them to calculate the average rate of change

$$\frac{25 - 25}{32 - 24} = \frac{0}{8} = 0$$

The average rate of change is 0 words per minute.

We locate the points $(32,25)$ and $(36,25)$ on the graph and use them to calculate the average rate of change

$$\frac{25 - 25}{36 - 32} = \frac{0}{4} = 0$$

The average rate of change is 0 words per minute.

b) $\boxed{tw}$

41. $s(t) = 16t^2$

a) First, we find the function values by substituting 3 and 5 in for t respectively.

$$s(3) = 16(3)^2 = 16(9) = 144$$
$$s(5) = 16(5)^2 = 16(25) = 400$$

Next we subtract the function values.

$$s(5) - s(3) = 400 - 144 = 256$$

The object will fall 256 feet in the two second time period between $t = 3$ and $t = 5$.

b) The average rate of change is calculated as

$$\frac{s(5)-s(3)}{5-3}=\frac{400-144}{5-3}$$

$$=\frac{256}{2}$$

$$=128$$

The average velocity of the object during the two second time period from $t=3$ to $t=5$ is 128 feet per second.

43. a) For each curve, as t changes from 0 to 4, $P(t)$ changes from 0 to 500. Thus, the average growth rate for each country is

$$\frac{500-0}{4-0}=\frac{500}{4}=125.$$

The average growth rate for each country is approximately 125 million people per year.

b) tw

c) For Country A:

As t changes from 0 to 1, $P(t)$ changes from 0 to 290. Thus the average growth rate is

$$\frac{290-0}{1-0}=290 \text{ million people per year.}$$

As t changes from 1 to 2, $P(t)$ changes from 290 to 250. Thus the average growth rate is

$$\frac{250-290}{2-1}=-40 \text{ million people per year.}$$

As t changes from 2 to 3, $P(t)$ changes from 250 to 200. Thus the average growth rate is

$$\frac{200-250}{3-2}=-50 \text{ million people per year.}$$

As t changes from 3 to 4, $P(t)$ changes from 200 to 500. Thus the average growth rate is

$$\frac{500-200}{4-3}=300 \text{ million people per year.}$$

For Country B:

As t changes from 0 to 1, $P(t)$ changes from 0 to 125. Thus the average growth rate is

$$\frac{125-0}{1-0}=125 \text{ million people per year.}$$

As t changes from 1 to 2, $P(t)$ changes from 125 to 250. Thus the average growth rate is

$$\frac{250-125}{2-1}=125 \text{ million people per year.}$$

As t changes from 2 to 3, $P(t)$ changes from 250 to 375. Thus the average growth rate is

$$\frac{375-250}{3-2}=125 \text{ million people per year.}$$

As t changes from 3 to 4, $P(t)$ changes from 375 to 500. Thus the average growth rate is

$$\frac{500-375}{4-3}=125 \text{ million people per year.}$$

d) tw

45. a) Tracing along the 4-year private school graph, we see that the largest increase in costs occurred in the 1985-86 year.

b) Tracing along the 4-year public school graph, we see that the largest increases in costs occurred during the 1975-76 year, the 2003-04 year and the 2004-05 year.

c) For the 4-year public school, the cost in 2005 dollars is approximately $6000. To find out what the cost was in 1975 dollars assuming a 3% inflation rate over the 30 years we create the following equation using the simple compound interest formula from section R1 $C(1+r)^{t}=A$.

$$C(1+.03)^{30}=6000$$

$$C(1.03)^{30}=6000$$

$$C=\frac{6000}{(1.03)^{30}}$$

$$C=2471.92$$

$$C\approx 2472$$

The cost of attending a 4 year public school in 1975 was $2472 in 1975 dollars. Likewise, for the 4-year private school, the cost in 2005 dollars is approximately $13,000.

To find out what the cost was in 1975 dollars assuming a 3% inflation rate over the 30 years we create the following equation using the simple compound interest formula from section R1 $C(1+r)^{t}=A$.

$$C(1+.03)^{30}=13,000$$

$$C(1.03)^{30}=13,000$$

$$C=\frac{13,000}{(1.03)^{30}}$$

$$C=5355.83$$

$$C\approx 5356$$

The cost of attending a 4 year private school in 1975 was $5356 in 1975 dollars.

47. $f(x) = ax^2 + bx + c$

Substituting $x + h$ for x we have,

$$f(x+h) = a(x+h)^2 + b(x+h) + c$$
$$= a(x^2 + 2xh + h^2) + bx + bh + c$$
$$= ax^2 + 2axh + ah^2 + bx + bh + c$$

Thus,

$\dfrac{f(x+h) - f(x)}{h}$ Difference quotient

$$= \frac{(ax^2 + 2axh + ah^2 + bx + bh + c) - (ax^2 + bx + c)}{h}$$

$$= \frac{2axh + ah^2 + bh}{h}$$

$$= \frac{h(2ax + ah + b)}{h} \qquad \text{Factoring the numerator}$$

$$= 2ax + ah + b \qquad \text{Simplified difference quotient}$$

49. $f(x) = \sqrt{x}$

Substituting $x + h$ for x we have,

$$f(x+h) = \sqrt{x+h}$$

Thus,

$\dfrac{f(x+h) - f(x)}{h}$ Difference quotient

$$= \frac{\sqrt{x+h} - \sqrt{x}}{h}$$

$$= \frac{\sqrt{x+h} - \sqrt{x}}{h} \cdot \frac{\sqrt{x+h} + \sqrt{x}}{\sqrt{x+h} + \sqrt{x}} \qquad \text{multiplying by 1}$$

$$= \frac{(x+h) - x}{h(\sqrt{x+h} + \sqrt{x})}$$

$$= \frac{h}{h(\sqrt{x+h} + \sqrt{x})}$$

$$= \frac{1}{\sqrt{x+h} + \sqrt{x}} \qquad \text{Simplified difference quotient}$$

51. $f(x) = \dfrac{1}{x^2}$

Substituting $x + h$ for x we have,

$$f(x+h) = \frac{1}{(x+h)^2}$$

Thus,

$\dfrac{f(x+h) - f(x)}{h}$ Difference quotient

$$= \frac{\left(\dfrac{1}{(x+h)^2}\right) - \left(\dfrac{1}{x^2}\right)}{h}$$

Next, we find a common denominator in the numerator.

$$= \frac{\left(\dfrac{1}{(x+h)^2} \cdot \dfrac{x^2}{x^2}\right) - \left(\dfrac{1}{x^2} \cdot \dfrac{(x+h)^2}{(x+h)^2}\right)}{h}$$

$$= \frac{\dfrac{x^2}{x^2(x+h)^2} - \dfrac{x^2 + 2xh + h^2}{x^2(x+h)^2}}{h}$$

$$= \frac{\dfrac{-2xh - h^2}{x^2(x+h)^2}}{h}$$

$$= \frac{-2xh - h^2}{x^2(x+h)^2} \cdot \frac{1}{h}$$

$$= \frac{h(-2x - h)}{x^2(x+h)^2} \cdot \frac{1}{h}$$

$$= \frac{-2x - h}{x^2(x+h)^2} \qquad \text{Simplified difference quotient}$$

53. $f(x) = \dfrac{x}{1+x}$

Substituting $x + h$ for x we have,

$$f(x+h) = \frac{x+h}{1+(x+h)}$$

Then,

$$\frac{f(x+h)-f(x)}{h} \quad \text{Difference quotient}$$

$$=\frac{\dfrac{x+h}{1+(x+h)}-\left(\dfrac{x}{1+x}\right)}{h}$$

Next, we find a common denominator for the numerator.

$$=\frac{\dfrac{x+h}{1+x+h}\cdot\dfrac{1+x}{1+x}-\left(\dfrac{x}{1+x}\cdot\dfrac{1+x+h}{1+x+h}\right)}{h}$$

$$=\frac{\dfrac{(x+h)(1+x)}{(1+x+h)(1+x)}-\dfrac{x(1+x+h)}{(1+x+h)(1+x)}}{h}$$

$$=\frac{\dfrac{x^2+x+h+hx}{(1+x+h)(1+x)}-\dfrac{x+x^2+hx}{(1+x+h)(1+x)}}{h}$$

We continue to simplify the difference quotient as follows.

$$\frac{f(x+h)-f(x)}{h}=\frac{\dfrac{h}{(1+x+h)(1+x)}}{h}$$

$$=\frac{h}{(1+x+h)(1+x)}\cdot\frac{1}{h}$$

$$=\frac{1}{(1+x+h)(1+x)}$$

55. $f(x)=\dfrac{1}{\sqrt{x}}$

Substituting $x+h$ for x we have,

$$f(x+h)=\frac{1}{\sqrt{x+h}}$$

Thus,

$$\frac{f(x+h)-f(x)}{h} \quad \text{Difference quotient}$$

$$=\frac{\left(\dfrac{1}{\sqrt{x+h}}\right)-\left(\dfrac{1}{\sqrt{x}}\right)}{h}$$

Next, we find a common denominator in the numerator.

$$=\frac{\left(\dfrac{1}{\sqrt{x+h}}\cdot\dfrac{\sqrt{x}}{\sqrt{x}}\right)-\left(\dfrac{1}{\sqrt{x}}\cdot\dfrac{\sqrt{x+h}}{\sqrt{x+h}}\right)}{h}$$

$$=\frac{\dfrac{\sqrt{x}}{\sqrt{x}\sqrt{x+h}}-\dfrac{\sqrt{x+h}}{\sqrt{x}\sqrt{x+h}}}{h}$$

$$=\frac{\dfrac{\sqrt{x}-\sqrt{x+h}}{\sqrt{x}\sqrt{x+h}}}{h}$$

$$=\frac{\sqrt{x}-\sqrt{x+h}}{\sqrt{x}\sqrt{x+h}}\cdot\frac{1}{h} \quad \begin{array}{l}\text{Simplifying the}\\\text{complex fraction}\end{array}$$

$$=\frac{\sqrt{x}-\sqrt{x+h}}{h\sqrt{x}\sqrt{x+h}}$$

Next, we rationalize the numerator.

$$=\frac{\sqrt{x}-\sqrt{x+h}}{h\sqrt{x}\sqrt{x+h}}\cdot\frac{\sqrt{x}+\sqrt{x+h}}{\sqrt{x}+\sqrt{x+h}}$$

$$=\frac{x-(x+h)}{h\sqrt{x}\sqrt{x+h}\left(\sqrt{x}+\sqrt{x+h}\right)}$$

$$=\frac{-h}{h\sqrt{x}\sqrt{x+h}\left(\sqrt{x}+\sqrt{x+h}\right)}$$

$$=\frac{-1}{\sqrt{x}\sqrt{x+h}\left(\sqrt{x}+\sqrt{x+h}\right)}$$

Exercise Set 1.4

1. $f(x) = \dfrac{3}{2}x^2$

a), b)

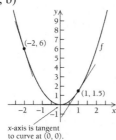

x-axis is tangent
to curve at (0, 0).

c) Find the simplified difference quotient first.

$$\frac{f(x+h) - f(x)}{h}$$

$$= \frac{\dfrac{3}{2}(x+h)^2 - \dfrac{3}{2}x^2}{h}$$

$$= \frac{\dfrac{3}{2}(x^2 + 2xh + h^2) - \dfrac{3}{2}x^2}{h}$$

$$= \frac{\dfrac{3}{2}x^2 + 3xh + \dfrac{3}{2}h^2 - \dfrac{3}{2}x^2}{h}$$

$$= \frac{3xh + \dfrac{3}{2}h^2}{h}$$

$$= \frac{h\left(3x + \dfrac{3}{2}h\right)}{h}$$

$$= 3x + \frac{3}{2}h \qquad \text{Simplified difference quotient}$$

Now we will find the limit of the difference quotient as $h \to 0$ using the simplified difference quotient.

$$\lim_{h\to 0} \frac{f(x+h) - f(x)}{h} = \lim_{h\to 0}\left(3x + \frac{3}{2}h\right)$$
$$= 3x$$

Thus, $f'(x) = 3x$.

d) Find the values of the derivative by making the appropriate substitutions.

$$f'(-2) = 3(-2) = -6 \quad \text{Substituting } -2 \text{ for } x$$

$$f'(0) = 3(0) = 0 \quad \text{Substituting } 0 \text{ for } x$$

$$f'(1) = 3(1) = 3 \quad \text{Substituting } 1 \text{ for } x$$

3. $f(x) = -2x^2$

a), b)

x-axis is tangent
to curve at (0, 0).

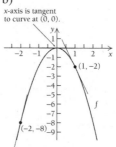

c) Find the simplified difference quotient first.

$$\frac{f(x+h) - f(x)}{h}$$

$$= \frac{-2(x+h)^2 - (-2x^2)}{h}$$

$$= \frac{-2(x^2 + 2xh + h^2) - (-2x^2)}{h}$$

$$= \frac{-2x^2 - 4xh - 2h^2 + 2x^2}{h}$$

$$= \frac{-4xh - 2h^2}{h}$$

$$= \frac{h(-4x - 2h)}{h}$$

$$= -4x - 2h \qquad \text{Simplified difference quotient}$$

Now we will find the limit of the difference quotient as $h \to 0$ using the simplified difference quotient.

$$\lim_{h\to 0} \frac{f(x+h) - f(x)}{h} = \lim_{h\to 0}(-4x - 2h)$$
$$= -4x$$

Thus, $f'(x) = -4x$.

d) Find the values of the derivative by making the appropriate substitutions.

$$f'(-2) = -4(-2) = 8 \quad \text{Substituting } -2 \text{ for } x$$

$$f'(0) = -4(0) = 0 \quad \text{Substituting } 0 \text{ for } x$$

$$f'(1) = -4(1) = -4 \quad \text{Substituting } 1 \text{ for } x$$

5. $f(x) = x^3$

a), b)

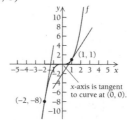

x-axis is tangent
to curve at (0, 0).

c) Find the simplified difference quotient first.

$$\frac{f(x+h)-f(x)}{h}$$

$$=\frac{(x+h)^3-(x^3)}{h}$$

$$=\frac{(x^3+3x^2h+3xh^2+h^3)-(x^3)}{h}$$

$$=\frac{3x^2h+3xh^2+h^3}{h}$$

$$=\frac{h(3x^2+3xh+h^2)}{h}$$

$$=3x^2+3xh+h^2 \quad \text{Simplified difference quotient}$$

Now we will find the limit of the difference quotient as $h \to 0$ using the simplified difference quotient.

$$\lim_{h\to0}\frac{f(x+h)-f(x)}{h}=\lim_{h\to0}(3x^2+3xh+h^2)$$

$$=3x^2$$

Thus, $f'(x)=3x^2$.

d) Find the values of the derivative by making the appropriate substitutions.

$$f'(-2)=3(-2)^2=12 \quad \text{Substituting } -2 \text{ for } x$$

$$f'(0)=3(0)^2=0 \quad \text{Substituting } 0 \text{ for } x$$

$$f'(1)=3(1)^2=3 \quad \text{Substituting } 1 \text{ for } x$$

7. $f(x)=2x+3$

a), b)

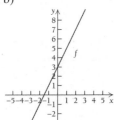

c) Find the simplified difference quotient first.

$$\frac{f(x+h)-f(x)}{h}$$

$$=\frac{2(x+h)+3-(2x+3)}{h}$$

$$=\frac{2x+2h+3-2x-3}{h}$$

$$=\frac{2h}{h}$$

$$=2 \quad \text{Simplified difference quotient}$$

Now we will find the limit of the difference quotient as $h \to 0$ using the simplified difference quotient.

$$\lim_{h\to0}\frac{f(x+h)-f(x)}{h}=\lim_{h\to0}(2)$$

$$=2$$

Thus, $f'(x)=2$.

d) Since the derivative is a constant, the value of the derivative will be 2 regardless of the value of x.

$$f'(-2)=2 \quad \text{Substituting } -2 \text{ for } x$$

$$f'(0)=2 \quad \text{Substituting } 0 \text{ for } x$$

$$f'(1)=2 \quad \text{Substituting } 1 \text{ for } x$$

9. $f(x)=\frac{1}{2}x-3$

a), b)

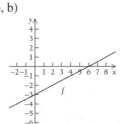

c) Find the simplified difference quotient first.

$$\frac{f(x+h)-f(x)}{h}$$

$$=\frac{\frac{1}{2}(x+h)-3-\left(\frac{1}{2}x-3\right)}{h}$$

$$=\frac{\frac{1}{2}x+\frac{1}{2}h-3-\frac{1}{2}x+3}{h}$$

$$=\frac{\frac{1}{2}h}{h}$$

$$=\frac{1}{2} \quad \text{Simplified difference quotient}$$

Now we will find the limit of the difference quotient as $h \to 0$ using the simplified difference quotient.

$$\lim_{h\to0}\frac{f(x+h)-f(x)}{h}=\lim_{h\to0}\left(\frac{1}{2}\right)$$

$$=\frac{1}{2}$$

Thus, $f'(x)=\frac{1}{2}$.

d) Since the derivative is a constant, the value of the derivative will be $\frac{1}{2}$ regardless of the value of x.

$f'(-2) = \frac{1}{2}$ Substituting -2 for x

$f'(0) = \frac{1}{2}$ Substituting 0 for x

$f'(1) = \frac{1}{2}$ Substituting 1 for x

11. $f(x) = x^2 + x$

a), b)

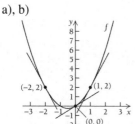

c) Find the simplified difference quotient first.

$$\frac{f(x+h) - f(x)}{h}$$

$$= \frac{(x+h)^2 + (x+h) - (x^2 + x)}{h}$$

$$= \frac{x^2 + 2xh + h^2 + x + h - x^2 - x}{h}$$

$$= \frac{2xh + h^2 + h}{h}$$

$$= \frac{h(2x + h + 1)}{h}$$

$= 2x + h + 1$ Simplified difference quotient

Now we will find the limit of the difference quotient as $h \to 0$ using the simplified difference quotient.

$$\lim_{h \to 0} \frac{f(x+h) - f(x)}{h} = \lim_{h \to 0} (2x + h + 1)$$

$$= 2x + 1$$

Thus, $f'(x) = 2x + 1$.

d) Find the values of the derivative by making the appropriate substitutions.

$f'(-2) = 2(-2) + 1 = -3$

$f'(0) = 2(0) + 1 = 1$

$f'(1) = 2(1) + 1 = 3$

13. $f(x) = 2x^2 + 3x - 2$

a), b)

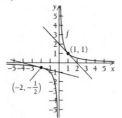

c) Find the simplified difference quotient first.

$$\frac{f(x+h) - f(x)}{h}$$

$$= \frac{\left(2(x+h)^2 + 3(x+h) - 2\right) - \left(2x^2 + 3x - 2\right)}{h}$$

$$= \frac{2\left(x^2 + 2xh + h^2\right) + 3x + 3h - 2 - 2x^2 - 3x + 2}{h}$$

$$= \frac{2x^2 + 4xh + 2h^2 + 3x + 3h - 2 - 2x^2 - 3x + 2}{h}$$

$$= \frac{4xh + 2h^2 + 3h}{h}$$

$$= \frac{h(4x + 2h + 3)}{h}$$

$= 4x + 2h + 3$ Simplified difference quotient

Now we will find the limit of the difference quotient as $h \to 0$ using the simplified difference quotient.

$$\lim_{h \to 0} \frac{f(x+h) - f(x)}{h} = \lim_{h \to 0} (4x + 2h + 3)$$

$$= 4x + 3$$

Thus, $f'(x) = 4x + 3$.

d) Find the values of the derivative by making the appropriate substitutions.

$f'(-2) = 4(-2) + 3 = -5$

$f'(0) = 4(0) + 3 = 3$

$f'(1) = 4(1) + 3 = 7$

15. $f(x) = \frac{1}{x}$

a), b)

There is no tangent line for $x = 0$.

c) Find the simplified difference quotient first.

$$\frac{f(x+h)-f(x)}{h}$$

$$=\frac{\left(\dfrac{1}{x+h}\right)-\left(\dfrac{1}{x}\right)}{h}$$

$$=\frac{\left(\dfrac{1}{x+h}\cdot\dfrac{x}{x}\right)-\left(\dfrac{1}{x}\cdot\dfrac{(x+h)}{(x+h)}\right)}{h}$$

$$=\frac{\left(\dfrac{x}{x(x+h)}\right)-\left(\dfrac{(x+h)}{x(x+h)}\right)}{h}$$

$$=\frac{\dfrac{-h}{x(x+h)}}{h}$$

$$=\frac{-h}{x(x+h)}\cdot\frac{1}{h}$$

$$=\frac{-1}{x(x+h)}\qquad\text{Simplified difference quotient}$$

Now we will find the limit of the difference quotient as $h\to 0$ using the simplified difference quotient.

$$\lim_{h\to 0}\frac{f(x+h)-f(x)}{h}=\lim_{h\to 0}\left(\frac{-1}{x(x+h)}\right)$$

$$=\frac{-1}{x(x+0)}$$

$$=\frac{-1}{x^2}$$

Thus, $f'(x)=\dfrac{-1}{x^2}$.

d) Find the values of the derivative by making the appropriate substitutions.

$$f'(-2)=\frac{-1}{(-2)^2}=-\frac{1}{4}$$

$$f'(0)=\frac{-1}{(0)^2};\text{ Thus, } f'(0)\text{ does not exist.}$$

$$f'(1)=\frac{-1}{(1)^2}=-1$$

17. a) From Example 1 we know that $f'(x)=2x$.

$f'(3)=2(3)=6$, so the slope of the line tangent to the curve at $(3,9)$ is 6. We substitute the point and the slope into the point-slope equation to find the equation of the tangent line.

$$y-y_1=m(x-x_1)$$
$$y-9=6(x-3)$$
$$y-9=6x-18$$
$$y=6x-9$$

b) $f'(-1)=2(-1)=-2$, so the slope of the line tangent to the curve at $(-1,1)$ is -2. We substitute the point and the slope into the point-slope equation to find the equation of the tangent line.

$$y-y_1=m(x-x_1)$$
$$y-1=-2(x-(-1))$$
$$y-1=-2x-2$$
$$y=-2x-1$$

c) $f'(10)=2(10)=20$, so the slope of the line tangent to the curve at $(10,100)$ is 20. We substitute the point and the slope into the point-slope equation to find the equation of the tangent line.

$$y-y_1=m(x-x_1)$$
$$y-100=20(x-10)$$
$$y-100=20x-200$$
$$y=20x-100$$

19. From Exercise 16 we know that $f'(x)=\dfrac{-2}{x^2}$.

a) $f'(1)=\dfrac{-2}{(1)^2}=-2$, so the slope of the line tangent to the curve at $(1,2)$ is -2. We substitute the point and the slope into the point-slope equation to find the equation of the tangent line.

$$y-y_1=m(x-x_1)$$
$$y-2=-2(x-1)$$
$$y-2=-2x+2$$
$$y=-2x+4$$

b) $f'(-1) = \dfrac{-2}{(-1)^2} = -2$, so the slope of the

line tangent to the curve at $(-1, 2)$ is -2.

We substitute the point and the slope into the point-slope equation to find the equation of the tangent line.

$$y - y_1 = m(x - x_1)$$
$$y - (-2) = -2(x - (-1))$$
$$y + 2 = -2x - 2$$
$$y = -2x - 4$$

c) $f'(100) = \dfrac{-2}{(100)^2} = -0.0002$, so the slope of

the line tangent to the curve at $(100, 0.02)$ is

-0.002. We substitute the point and the slope into the point-slope equation to find the equation of the tangent line.

$$y - y_1 = m(x - x_1)$$
$$y - 0.02 = -0.0002(x - 100)$$
$$y - 0.02 = -0.0002x + 0.02$$
$$y = -0.0002x + 0.04$$

21. First, we find $f'(x)$:

$$\frac{f(x+h) - f(x)}{h}$$

$$= \frac{\left(4 - (x+h)^2\right) - \left(4 - x^2\right)}{h}$$

$$= \frac{4 - \left(x^2 + 2xh + h^2\right) - 4 + x^2}{h}$$

$$= \frac{4 - x^2 - 2xh - h^2 - 4 + x^2}{h}$$

$$= \frac{-2xh - h^2}{h}$$

$$= \frac{h(-2x - h)}{h}$$

$$= -2x - h \qquad \text{Simplified difference quotient}$$

$$f'(x) = \lim_{h \to 0} \frac{f(x+h) - f(x)}{h}$$

$$= \lim_{h \to 0}(-2x - h)$$

$$= -2x$$

a) $f'(-1) = -2(-1) = 2$, so the slope of the

line tangent to the curve at $(-1, 3)$ is 2.

We substitute the point and the slope into the point-slope equation to find the equation of the tangent line.

$$y - y_1 = m(x - x_1)$$
$$y - 3 = 2(x - (-1))$$
$$y - 3 = 2x + 2$$
$$y = 2x + 5$$

b) $f'(0) = -2(0) = 0$, so the slope of the line

tangent to the curve at $(0, 4)$ is 0. We

substitute the point and the slope into the point-slope equation to find the equation of the tangent line.

$$y - y_1 = m(x - x_1)$$
$$y - 4 = 0(x - 0)$$
$$y - 4 = 0$$
$$y = 4$$

c) $f'(5) = -2(5) = -10$, so the slope of the

line tangent to the curve at $(5, -21)$ is -10.

We substitute the point and the slope into the point-slope equation to find the equation of the tangent line.

$$y - y_1 = m(x - x_1)$$
$$y - (-21) = -10(x - 5)$$
$$y + 21 = -10x + 50$$
$$y = -10x + 29$$

23. Find the simplified difference quotient for

$f(x) = mx + b$ first.

$$\frac{f(x+h) - f(x)}{h}$$

$$= \frac{m(x+h) + b - (mx + b)}{h}$$

$$= \frac{mx + mh + b - mx - b}{h}$$

$$= \frac{mh}{h}$$

$$= m \qquad \text{Simplified difference quotient}$$

Now we will find the limit of the difference quotient as $h \to 0$ using the simplified difference quotient.

$$\lim_{h \to 0} \frac{f(x+h) - f(x)}{h} = \lim_{h \to 0}(m)$$

$$= m$$

Thus, $f'(x) = m$.

25. If a function has a "sharp point" or a "corner," it will not be differentiable at that point. Thus, the function is not differentiable at x_3, x_4, x_6. The function has a vertical tangent at x_{12}. Vertical lines have undefined slope, hence the function is not differentiable at x_{12}. Also, if a function is discontinuous at some point a, then it is not differentiable at a. The function is discontinuous at the point x_0, thus it is not differentiable at x_0.

Therefore, the graph is not differentiable at the points $x_0, x_3, x_4, x_6, x_{12}$.

27. The following graph is continuous but not differentiable, at $x = 3$.

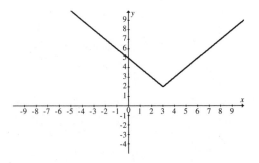

29. The postage function does not have any "sharp points" nor does it have any vertical tangents. However it is discontinuous at all natural numbers. Therefore the postage function is not differentiable for 1, 2, 3, 4, and so on.

31. The ticket price function is continuous everywhere on its domain. It also does not have any "sharp points" or "corners" nor does it have any vertical tangents. Thus, the function is differentiable everywhere on it domain.

33. $\boxed{tw}$

35. $f(x) = \dfrac{1}{x^2}$

Find the difference quotient first.

$$\frac{f(x+h) - f(x)}{h}$$

$$\frac{f(x+h) - f(x)}{h} = \frac{\dfrac{1}{(x+h)^2} - \dfrac{1}{x^2}}{h}$$

$$= \frac{\dfrac{1}{(x+h)^2} \cdot \dfrac{x^2}{x^2} - \dfrac{1}{x^2} \cdot \dfrac{(x+h)^2}{(x+h)^2}}{h}$$

$$= \frac{\dfrac{x^2 - (x+h)^2}{x^2(x+h)^2}}{h}$$

$$= \frac{x^2 - x^2 - 2xh - h^2}{x^2(x+h)^2} \cdot \frac{1}{h}$$

$$= \frac{-2xh - h^2}{x^2(x+h)^2} \cdot \frac{1}{h}$$

$$= \frac{-2x - h}{x^2(x+h)^2}$$

Next, we will find the limit of the difference quotient as $h \to 0$.

$$f'(x) = \lim_{h \to 0} \frac{f(x+h) - f(x)}{h}$$

$$= \lim_{h \to 0} \frac{-2x - h}{x^2(x+h)^2}$$

$$= \frac{-2x}{x^2(x+0)^2}$$

$$= \frac{-2x}{x^4}$$

$$= \frac{-2}{x^3}$$

Thus, $f'(x) = \dfrac{-2}{x^3}$.

37. $f(x) = \dfrac{x}{1+x}$

We found the simplified difference quotient in Exercise 53 of Exercise Set 1.3. We now find the limit of the difference quotient as $h \to 0$.

$$\lim_{h \to 0} \frac{f(x+h) - f(x)}{h} = \lim_{h \to 0} \frac{1}{(x+1)(x+1+h)}$$

$$= \frac{1}{(x+1)(x+1+0)}$$

$$= \frac{1}{(x+1)^2}$$

Thus, $f'(x) = \dfrac{1}{(x+1)^2}$.

39. $f(x) = \sqrt{x}$

We found the simplified difference quotient in Exercise 49 of Exercise Set 1.3. We now find the limit of the difference quotient as $h \to 0$.

$$\lim_{h \to 0} \frac{f(x+h) - f(x)}{h} = \lim_{h \to 0} \frac{1}{\sqrt{x+h} + \sqrt{x}}$$

$$= \frac{1}{\sqrt{x+0} + \sqrt{x}}$$

$$= \frac{1}{2\sqrt{x}}$$

Thus, $f'(x) = \dfrac{1}{2\sqrt{x}}$.

41. a) The domain of the rational function is restricted to those input values that do not result in division by 0. The domain for

$f(x) = \dfrac{x^2 - 9}{x + 3}$ consists of all real numbers

except -3. Since $f(-3)$ does not exist, the function is not continuous at -3. Thus, the function is not differentiable at $x = -3$.

b) $\boxed{tw}$

43-47. Left to the student.

49. There is a vertical tangent at $x = 5$, therefore, $f'(x)$ does not exist at $x = 5$.

Exercise Set 1.5

1. $y = x^7$

$$\frac{dy}{dx} = \frac{d}{dx} x^7$$

$$= 7x^{7-1} \qquad \text{Theorem 1}$$

$$= 7x^6$$

3. $y = -3x$

$$\frac{dy}{dx} = \frac{d}{dx}(-3x)$$

$$= -3 \frac{d}{dx} x \qquad \text{Theorem 3}$$

$$= -3(1x^{1-1}) \qquad \text{Theorem 1}$$

$$= -3(x^0)$$

$$= -3 \qquad\qquad [a^0 = 1]$$

5. $y = 12$ Constant function

$$\frac{dy}{dx} = \frac{d}{dx} 12$$

$$= 0 \qquad\qquad \text{Theorem 2}$$

7. $y = 2x^{15}$

$$\frac{dy}{dx} = \frac{d}{dx}(2x^{15})$$

$$= 2 \frac{d}{dx}(x^{15}) \qquad \text{Theorem 3}$$

$$= 2(15x^{15-1}) \qquad \text{Theorme 1}$$

$$= 30x^{14}$$

9. $y = x^{-6}$

$$\frac{dy}{dx} = \frac{d}{dx} x^{-6}$$

$$= -6x^{-6-1} \qquad \text{Theorem 1}$$

$$= -6x^{-7}$$

11. $y = 4x^{-2}$

$$\frac{dy}{dx} = \frac{d}{dx}(4x^{-2})$$

$$= 4 \frac{d}{dx}(x^{-2}) \qquad \text{Theorem 3}$$

$$= 4(-2x^{-2-1}) \qquad \text{Theorem 1}$$

$$= -8x^{-3}$$

13. $y = x^3 + 3x^2$

$$\frac{dy}{dx} = \frac{d}{dx}(x^3 + 3x^2)$$

$$= \frac{d}{dx} x^3 + \frac{d}{dx} 3x^2 \quad \text{Theorem 4}$$

$$= \frac{d}{dx} x^3 + 3 \frac{d}{dx} x^2 \quad \text{Theorem 3}$$

$$= 3x^{3-1} + 3(2x^{2-1}) \quad \text{Theorem 1}$$

$$= 3x^2 + 6x$$

15. $y = 8\sqrt{x} = 8x^{1/2}$

$\dfrac{dy}{dx} = \dfrac{d}{dx}8x^{1/2}$

$\qquad = 8\dfrac{d}{dx}x^{1/2}$ Theorem 3

$\dfrac{dy}{dx} = 8\left(\dfrac{1}{2}x^{1/2-1}\right)$ Theorem 1

$\qquad = 4x^{-1/2}$

$\qquad = \dfrac{4}{x^{1/2}} = \dfrac{4}{\sqrt{x}}$ Properties of exponents

17. $y = x^{0.9}$

$\dfrac{dy}{dx} = \dfrac{d}{dx}x^{0.9}$

$\qquad = 0.9x^{0.9-1}$ Theorem 1

$\qquad = 0.9x^{-0.1}$

19. $y = \dfrac{1}{2}x^{4/5}$

$\dfrac{dy}{dx} = \dfrac{d}{dx}\left(\dfrac{1}{2}x^{4/5}\right)$

$\qquad = \dfrac{1}{2}\cdot\dfrac{d}{dx}\left(x^{4/5}\right)$ Theorem 3

$\qquad = \dfrac{1}{2}\left(\dfrac{4}{5}x^{4/5-1}\right)$ Theorem 1

$\qquad = \dfrac{2}{5}x^{-1/5}$

21. $y = \dfrac{7}{x^3} = 7x^{-3}$

$\dfrac{dy}{dx} = \dfrac{d}{dx}\left(7x^{-3}\right)$

$\qquad = 7\dfrac{d}{dx}\left(x^{-3}\right)$ Theorem 3

$\qquad = 7\left(-3x^{-3-1}\right)$ Theorem 1

$\qquad = -21x^{-4}$

$\qquad = -\dfrac{21}{x^4}$ Properties of exponents

23. $y = \dfrac{4x}{5} = \dfrac{4}{5}x$

$\dfrac{dy}{dx} = \dfrac{d}{dx}\left(\dfrac{4}{5}x\right)$

$\qquad = \dfrac{4}{5}\cdot\dfrac{d}{dx}(x)$ Theorem 3

$\qquad = \dfrac{4}{5}\cdot\left(1x^{1-1}\right)$ Theorem 1

$\qquad = \dfrac{4}{5}$

25. $\dfrac{d}{dx}\left(\sqrt[4]{x} - \dfrac{3}{x}\right)$

$\qquad = \dfrac{d}{dx}\sqrt[4]{x} - \dfrac{d}{dx}\dfrac{3}{x}$ Theorem 4

$\qquad = \dfrac{d}{dx}x^{1/4} - \dfrac{d}{dx}3x^{-1}$ Properties of exponents

$\qquad = \dfrac{d}{dx}x^{1/4} - 3\dfrac{d}{dx}x^{-1}$ Theorem 3

$\qquad = \dfrac{1}{4}x^{1/4-1} - 3\left(-1x^{-1-1}\right)$ Theorem 1

$\qquad = \dfrac{1}{4}x^{-3/4} + 3x^{-2}$

$\qquad = \dfrac{1}{4x^{3/4}} + \dfrac{3}{x^2}$

$\qquad = \dfrac{1}{4\sqrt[4]{x^3}} + \dfrac{3}{x^2}$

27. $\dfrac{d}{dx}\left(\sqrt{x} - \dfrac{2}{\sqrt{x}}\right)$

$\qquad = \dfrac{d}{dx}\sqrt{x} - \dfrac{d}{dx}\dfrac{2}{\sqrt{x}}$ Theorem 4

$\qquad = \dfrac{d}{dx}x^{1/2} - \dfrac{d}{dx}2x^{-1/2}$ Properties of exponents

$\qquad = \dfrac{d}{dx}x^{1/2} - 2\dfrac{d}{dx}x^{-1/2}$ Theorem 3

$\qquad = \dfrac{1}{2}x^{1/2-1} - 2\left(-\dfrac{1}{2}x^{-1/2-1}\right)$ Theorem 1

$\qquad = \dfrac{1}{2}x^{-1/2} + x^{-3/2}$

$\qquad = \dfrac{1}{2x^{1/2}} + \dfrac{1}{x^{3/2}}$

$\qquad = \dfrac{1}{2\sqrt{x}} + \dfrac{1}{\sqrt{x^3}}$

29. $\dfrac{d}{dx}\left(-2\sqrt[3]{x^5}\right)$

$\quad = -2\dfrac{d}{dx}\left(\sqrt[3]{x^5}\right)$ Theorem 3

$\quad = -2\dfrac{d}{dx}\left(x^{5/3}\right)$

$\quad = -2\left(\dfrac{5}{3}x^{5/3-1}\right)$ Theorem 1

$\quad = -\dfrac{10}{3}x^{2/3} = \dfrac{10\sqrt[3]{x^2}}{3}$

31. $\dfrac{d}{dx}\left(5x^2 - 7x + 3\right)$

$\quad = \dfrac{d}{dx}5x^2 - \dfrac{d}{dx}7x + \dfrac{d}{dx}3$ Theorem 4

$\quad = 5\dfrac{d}{dx}x^2 - 7\dfrac{d}{dx}x + \dfrac{d}{dx}3$ Theorem 3

$\quad = 5\left(2x^{2-1}\right) - 7\left(1x^{1-1}\right) + 0$ Theorems 1 and 2

$\quad = 10x - 7$

33. $f(x) = 0.6x^{1.5}$

$\quad f'(x) = \dfrac{d}{dx}0.6x^{1.5}$

$\qquad = 0.6\dfrac{d}{dx}x^{1.5}$ Theorem 3

$\qquad = 0.6\left(1.5x^{1.5-1}\right)$ Theorem 1

$\qquad = 0.9x^{0.5}$

35. $f(x) = \dfrac{2x}{3} = \dfrac{2}{3}x$

$\quad f'(x) = \dfrac{d}{dx}\left(\dfrac{2}{3}x\right)$

$\qquad = \dfrac{2}{3}\dfrac{d}{dx}(x)$

$\qquad = \dfrac{2}{3}\left(1x^{1-1}\right)$

$\qquad = \dfrac{2}{3}$

37. $f(x) = \dfrac{4}{7x^3} = \dfrac{4x^{-3}}{7} = \dfrac{4}{7}x^{-3}$

$\quad f'(x) = \dfrac{d}{dx}\left(\dfrac{4}{7}x^{-3}\right)$

$\qquad = \dfrac{4}{7}\dfrac{d}{dx}\left(x^{-3}\right)$

$\qquad = \dfrac{4}{7}\left(-3x^{-3-1}\right)$

$\qquad = \dfrac{-12}{7}x^{-4}$

$\qquad = -\dfrac{12}{7x^4}$

39. $f(x) = \dfrac{5}{x} - x^{2/3} = 5x^{-1} - x^{2/3}$

$\quad f'(x) = \dfrac{d}{dx}\left(5x^{-1} - x^{2/3}\right)$

$\qquad = \dfrac{d}{dx}\left(5x^{-1}\right) - \dfrac{d}{dx}\left(x^{2/3}\right)$

$\qquad = 5\dfrac{d}{dx}\left(x^{-1}\right) - \dfrac{d}{dx}\left(x^{2/3}\right)$

$\qquad = 5\left(-1x^{-1-1}\right) - \dfrac{2}{3}x^{2/3-1}$

$\qquad = -5x^{-2} - \dfrac{2}{3}x^{-1/3}$

$\qquad = -\dfrac{5}{x^2} - \dfrac{2}{3}x^{-1/3}$

41. $f(x) = 4x - 7$

$\quad f'(x) = \dfrac{d}{dx}(4x - 7)$

$\qquad = \dfrac{d}{dx}(4x) - \dfrac{d}{dx}(7)$

$\quad f'(x) = 4\dfrac{d}{dx}(x) - \dfrac{d}{dx}(7)$

$\qquad = 4\left(1x^{1-1}\right) - 0$

$\qquad = 4$

43. $f(x) = \dfrac{x^{4/3}}{4} = \dfrac{1}{4}x^{4/3}$

$\quad f'(x) = \dfrac{d}{dx}\left(\dfrac{1}{4}x^{4/3}\right)$

$\qquad = \dfrac{1}{4}\dfrac{d}{dx}\left(x^{4/3}\right)$

$\qquad = \dfrac{1}{4}\left(\dfrac{4}{3}x^{4/3-1}\right)$

$\qquad = \dfrac{1}{3}x^{1/3}$, or $\dfrac{\sqrt[3]{x}}{3}$

45. $f(x) = -0.01x^2 - 0.5x + 70$

$$f'(x) = \frac{d}{dx}\left(-0.01x^2 - 0.5x + 70\right)$$

$$= \frac{d}{dx}\left(-0.01x^2\right) - \frac{d}{dx}(0.5x) + \frac{d}{dx}(70)$$

$$= -0.01\frac{d}{dx}\left(x^2\right) - 0.5\frac{d}{dx}(x) + \frac{d}{dx}(70)$$

$$= -0.01\left(2x^{2-1}\right) - 0.5\left(1x^{1-1}\right) + 0$$

$$= -0.02x - 0.5$$

47. $y = 3x^{-\frac{2}{3}} + x^{\frac{3}{4}} + x^{\frac{6}{5}} + \dfrac{8}{x^3}$

$$y = 3x^{-\frac{2}{3}} + x^{\frac{3}{4}} + x^{\frac{6}{5}} + 8x^{-3}$$

$$y' = \frac{d}{dx}\left(3x^{-\frac{2}{3}} + x^{\frac{3}{4}} + x^{\frac{6}{5}} + 8x^{-3}\right)$$

$$= \frac{d}{dx}\left(3x^{-\frac{2}{3}}\right) + \frac{d}{dx}\left(x^{\frac{3}{4}}\right) + \frac{d}{dx}\left(x^{\frac{6}{5}}\right) + \frac{d}{dx}\left(8x^{-3}\right)$$

$$= 3\left(\frac{-2}{3}x^{-\frac{2}{3}-1}\right) + \left(\frac{3}{4}x^{\frac{3}{4}-1}\right) +$$

$$\left(\frac{6}{5}x^{\frac{6}{5}-1}\right) + 8\left(-3x^{-3-1}\right)$$

$$= -2x^{-\frac{5}{3}} + \frac{3}{4}x^{-\frac{1}{4}} + \frac{6}{5}x^{\frac{1}{5}} - 24x^{-4}$$

$$= -2x^{-\frac{5}{3}} + \frac{3}{4}x^{-\frac{1}{4}} + \frac{6}{5}x^{\frac{1}{5}} - \frac{24}{x^4}$$

49. $y = \dfrac{2}{x} - \dfrac{x}{2} = 2x^{-1} - \dfrac{1}{2}x$

$$y' = \frac{d}{dx}\left(2x^{-1} - \frac{1}{2}x\right)$$

$$= \frac{d}{dx}\left(2x^{-1}\right) - \frac{d}{dx}\left(\frac{1}{2}x\right)$$

$$= 2\frac{d}{dx}\left(x^{-1}\right) - \frac{1}{2}\frac{d}{dx}(x)$$

$$= 2\left(-1x^{-1-1}\right) - \frac{1}{2}\left(1x^{1-1}\right)$$

$$= -2x^{-2} - \frac{1}{2}$$

$$= -\frac{2}{x^2} - \frac{1}{2}$$

51. $f(x) = x^2 + 4x - 5$

First, we find $f'(x)$

$$f'(x) = \frac{d}{dx}\left(x^2 + 4x - 5\right)$$

$$= \frac{d}{dx}\left(x^2\right) + 4\frac{d}{dx}(x) - \frac{d}{dx}5$$

$$= \left(2x^{2-1}\right) + 4\left(1x^{1-1}\right) - 0$$

$$= 2x + 4$$

Therefore,

$$f'(10) = 2(10) + 4$$

$$= 24$$

53. $y = \dfrac{4}{x^2} = 4x^{-2}$

Find $\dfrac{dy}{dx}$ first.

$$\frac{dy}{dx} = \frac{d}{dx}\left(4x^{-2}\right)$$

$$= 4\frac{d}{dx}\left(x^{-2}\right)$$

$$= 4\left(-2x^{-2-1}\right)$$

$$= -8x^{-3}$$

$$= -\frac{8}{x^3}$$

Therefore,

$$\left.\frac{dy}{dx}\right|_{x=-2} = -\frac{8}{(-2)^3}$$

$$= -\frac{8}{(-8)}$$

$$= 1$$

55. We will need the derivative to find the slope of the tangent line at each of the indicated points. We find the derivative first.

$$f(x) = x^3 - 2x + 1$$

$$f'(x) = \frac{d}{dx}\left(x^3 - 2x + 1\right)$$

$$= \frac{d}{dx}\left(x^3\right) - \frac{d}{dx}(2x) + \frac{d}{dx}(1)$$

$$= \left(3x^{3-1}\right) - 2\left(1x^{1-1}\right) + 0$$

$$= 3x^2 - 2$$

a) Using the derivative, we find the slope of the line tangent to the curve at point $(2,5)$ by evaluating the derivative at $x = 2$.

$f'(2) = 3(2)^2 - 2 = 10$. Therefore the slope of the tangent line is 10. We use the point-slope equation to find the equation of the tangent line on the next page.

$$y - y_1 = m(x - x_1)$$
$$y - 5 = 10(x - 2)$$
$$y - 5 = 10x - 20$$
$$y = 10x - 15$$

b) Using the derivative, we find the slope of the line tangent to the curve at point $(-1,2)$ by evaluating the derivative at $x = -1$.

$f'(-1) = 3(-1)^2 - 2 = 1$. Therefore the slope of the tangent line is 1. We use the point-slope equation to find the equation of the tangent line.

$$y - y_1 = m(x - x_1)$$
$$y - 2 = 1(x - (-1))$$
$$y - 2 = x + 1$$
$$y = x + 3$$

c) Using the derivative, we find the slope of the line tangent to the curve at point $(0,1)$ by evaluating the derivative at $x = 0$.

$f'(0) = 3(0)^2 - 2 = -2$. Therefore the slope of the tangent line is -2. We use the point-slope equation to find the equation of the tangent line.

$$y - y_1 = m(x - x_1)$$
$$y - 1 = -2(x - 0)$$
$$y - 1 = -2x$$
$$y = -2x + 1$$

57. $y = x^2 - 3$

A horizontal tangent line has slope equal to 0, so we first find the values of x that make

$$\frac{dy}{dx} = 0.$$

First, we find the derivative.

$$\frac{dy}{dx} = \frac{d}{dx}(x^2 - 3)$$
$$\frac{dy}{dx} = \frac{d}{dx}x^2 - \frac{d}{dx}3$$
$$= 2x - 0$$
$$= 2x$$

Next, we set the derivative equal to zero and solve for x.

$$\frac{dy}{dx} = 0$$
$$2x = 0$$
$$x = \frac{0}{2} = 0$$

So the horizontal tangent will occur when $x = 0$. Next we find the point on the graph. For $x = 0$, $y = (0)^2 - 3 = -3$, so there is a horizontal tangent at the point $(0, -3)$.

59. $y = -x^3 + 1$

A horizontal tangent line has slope equal to 0, so we first find the values of x that make

$$\frac{dy}{dx} = 0.$$

First, we find the derivative.

$$\frac{dy}{dx} = \frac{d}{dx}(-x^3 + 1)$$
$$= -\frac{d}{dx}x^3 + \frac{d}{dx}1$$
$$= -3x^2 - 0$$
$$= -3x^2$$

Next, we set the derivative equal to zero and solve for x.

$$\frac{dy}{dx} = 0$$
$$-3x^2 = 0$$
$$x^2 = 0$$
$$x = 0$$

So the horizontal tangent will occur when $x = 0$. Next we find the point on the graph. For $x = 0$, $y = -(0)^3 + 1 = 1$, so there is a horizontal tangent at the point $(0,1)$.

61. $y = 3x^2 - 5x + 4$

A horizontal tangent line has slope equal to 0, so we need to find the values of x that make

$$\frac{dy}{dx} = 0.$$

First, we find the derivative.

$$\frac{dy}{dx} = \frac{d}{dx}\left(3x^2 - 5x + 4\right)$$

$$= \frac{d}{dx}3x^2 - \frac{d}{dx}5x + \frac{d}{dx}4$$

$$= 3\left(2x^{2-1}\right) - 5\left(1x^{1-1}\right) + 0$$

$$= 6x - 5$$

Next, we set the derivative equal to zero and solve for x.

$$\frac{dy}{dx} = 0$$

$$6x - 5 = 0$$

$$6x = 5$$

$$x = \frac{5}{6}$$

So the horizontal tangent will occur when

$x = \frac{5}{6}$. Next we find the point on the graph.

For $x = \frac{5}{6}$,

$$y = 3\left(\frac{5}{6}\right)^2 - 5\left(\frac{5}{6}\right) + 4$$

$$= 3\left(\frac{25}{36}\right) - \frac{25}{6} + 4$$

$$= \frac{25}{12} - \frac{25}{6} + 4$$

$$= \frac{25}{12} - \frac{25}{6} \cdot \frac{2}{2} + \frac{4}{1} \cdot \frac{12}{12}$$

$$= \frac{25}{12} - \frac{50}{12} + \frac{48}{12}$$

$$= \frac{25 - 50 + 48}{12}$$

$$= \frac{23}{12}$$

Therefore, there is a horizontal tangent at the point $\left(\frac{5}{6}, \frac{23}{12}\right)$.

63. $y = -0.01x^2 - 0.5x + 70$

A horizontal tangent line has slope equal to 0, so we need to find the values of x that make

$$\frac{dy}{dx} = 0.$$

First, we find the derivative.

$$\frac{dy}{dx} = \frac{d}{dx}\left(-0.01x^2 - 0.5x + 70\right)$$

$$= -0.02x - 0.5 \qquad \text{See Exercise 45.}$$

Next, we set the derivative equal to zero and solve for x.

$$\frac{dy}{dx} = 0$$

$$-0.02x - 0.5 = 0$$

$$-0.02x = 0.5$$

$$x = \frac{0.5}{-0.02}$$

$$x = -25$$

So the horizontal tangent will occur when $x = -25$. Next we find the point on the graph.

For $x = -25$,

$$y = -0.01(-25)^2 - 0.5(-25) + 70$$

$$= -0.01(625) + 12.5 + 70$$

$$= -6.25 + 12.5 + 70$$

$$= 76.25$$

Therefore, there is a horizontal tangent at the point $(-25, 76.25)$.

65. $y = 2x + 4$ \qquad\qquad Linear function

$$\frac{dy}{dx} = 2 \qquad\qquad \text{Slope is 2}$$

There are no values of x for which $\frac{dy}{dx} = 0$, so there are no points on the graph at which there is a horizontal tangent.

67. $y = 4$ \qquad\qquad Constant Function

$$\frac{dy}{dx} = 0 \qquad\qquad \text{Theorem 2}$$

$\frac{dy}{dx} = 0$ for all values of x, so the tangent line is horizontal for all points on the graph.

69. $y = -x^3 + x^2 + 5x - 1$

A horizontal tangent line has slope equal to 0, so we need to find the values of x that make $\frac{dy}{dx} = 0$.

First, we find the derivative.

$$\frac{dy}{dx} = \frac{d}{dx}\left(-x^3 + x^2 + 5x - 1\right)$$

$$= -\frac{d}{dx}\left(x^3\right) + \frac{d}{dx}\left(x^2\right) + \frac{d}{dx}(5x) - \frac{d}{dx}(1)$$

$$= -3x^2 + 2x + 5$$

Next, we set the derivative equal to zero and solve for x.

$$\frac{dy}{dx} = 0$$

$$-3x^2 + 2x + 5 = 0$$

$$3x^2 - 2x - 5 = 0 \qquad \text{Multiply both sides by -1.}$$

$$(3x - 5)(x + 1) = 0 \qquad \text{Factor the left hand side.}$$

$$3x - 5 = 0 \quad \text{or} \quad x + 1 = 0$$

$$3x = 5 \quad \text{or} \quad x = -1$$

$$x = \frac{5}{3} \quad \text{or} \quad x = -1$$

There are two horizontal tangents. One at $x = \frac{5}{3}$ and one at $x = -1$.

Next we find the points on the graph where the horizontal tangents occur.

For $x = -1$

$$y = -(-1)^3 + (-1)^2 + 5(-1) - 1$$

$$y = -(-1) + (1) - 5 - 1$$

$$y = -4$$

For $x = \frac{5}{3}$

$$y = -\left(\frac{5}{3}\right)^3 + \left(\frac{5}{3}\right)^2 + 5\left(\frac{5}{3}\right) - 1$$

$$y = -\left(\frac{125}{27}\right) + \left(\frac{25}{9}\right) + \frac{25}{3} - 1$$

$$y = -\frac{125}{27} + \frac{75}{27} + \frac{225}{27} - \frac{27}{27}$$

$$y = \frac{148}{27} = 5\tfrac{13}{27}$$

Therefore, there are horizontal tangents at the points $\left(\frac{5}{3}, 5\tfrac{13}{27}\right)$ and $(-1, -4)$.

71. $y = \frac{1}{3}x^3 - 3x + 2$

A horizontal tangent line has slope equal to 0, so we need to find the values of x that make $\frac{dy}{dx} = 0$.

First, we find the derivative.

$$\frac{dy}{dx} = \frac{d}{dx}\left(\frac{1}{3}x^3 - 3x + 2\right)$$

$$= \frac{d}{dx}\left(\frac{1}{3}x^3\right) - \frac{d}{dx}(3x) + \frac{d}{dx}(2)$$

$$= x^2 - 3$$

Next, we set the derivative equal to zero and solve for x.

$$\frac{dy}{dx} = 0$$

$$x^2 - 3 = 0$$

$$x^2 = 3$$

$$x = \pm\sqrt{3}$$

There are two horizontal tangents. One at $x = -\sqrt{3}$ and one at $x = \sqrt{3}$. Next we find the points on the graph where the horizontal tangents occur.

For $x = -\sqrt{3}$

$$y = \frac{1}{3}\left(-\sqrt{3}\right)^3 - 3\left(-\sqrt{3}\right) + 2$$

$$= \frac{1}{3}\left(-3\sqrt{3}\right) - 3\left(-\sqrt{3}\right) + 2$$

$$= -\sqrt{3} + 3\sqrt{3} + 2$$

$$= 2 + 2\sqrt{3}$$

For $x = \sqrt{3}$

$$y = \frac{1}{3}\left(\sqrt{3}\right)^3 - 3\left(\sqrt{3}\right) + 2$$

$$= \frac{1}{3}\left(3\sqrt{3}\right) - 3\left(\sqrt{3}\right) + 2$$

$$= +\sqrt{3} - 3\sqrt{3} + 2$$

$$= 2 - 2\sqrt{3}$$

Therefore, there are horizontal tangents at the points $\left(-\sqrt{3}, 2 + 2\sqrt{3}\right)$ and $\left(\sqrt{3}, 2 - 2\sqrt{3}\right)$.

73. $y = \frac{1}{3}x^3 + \frac{1}{2}x^2 - 2$

A horizontal tangent line has slope equal to 0, so we need to find the values of x that make

$\frac{dy}{dx} = 0$.

First, we find the derivative.

$$\frac{dy}{dx} = \frac{d}{dx}\left(\frac{1}{3}x^3 + \frac{1}{2}x^2 - 2\right)$$

$$= \frac{d}{dx}\left(\frac{1}{3}x^3\right) + \frac{d}{dx}\left(\frac{1}{2}x^2\right) - \frac{d}{dx}(2)$$

$$= x^2 + x$$

Next, we set the derivative equal to zero and solve for x.

$$\frac{dy}{dx} = 0$$

$$x^2 + x = 0$$

$$x(x+1) = 0$$

$$x = 0 \quad \text{or} \quad x+1 = 0$$

$$x = 0 \quad \text{or} \quad x = -1$$

There are two horizontal tangents. One at $x = 0$ and one at $x = -1$.

Next we find the points on the graph where the horizontal tangents occur.

For $x = 0$

$$y = \frac{1}{3}(0)^3 + \frac{1}{2}(0)^2 - 2$$

$$y = -2$$

For $x = -1$

$$y = \frac{1}{3}(-1)^3 + \frac{1}{2}(-1)^2 - 2$$

$$= -\frac{1}{3} + \frac{1}{2} - 2$$

$$= -\frac{2}{6} + \frac{3}{6} - \frac{12}{6}$$

$$= -\frac{11}{6}$$

Therefore, there are horizontal tangents at the points $(0, -2)$ and $\left(-1, -\frac{11}{6}\right)$.

75. $y = 20x - x^2$

To find the tangent line that has slope equal to 1, so we need to find the values of x that make

$\frac{dy}{dx} = 1$.

First, we find the derivative.

$$\frac{dy}{dx} = \frac{d}{dx}\left(20x - x^2\right)$$

$$= \frac{d}{dx}20x - \frac{d}{dx}x^2$$

$$= 20 - 2x$$

Next, we set the derivative equal to 1 and solve for x.

$$\frac{dy}{dx} = 1$$

$$20 - 2x = 1$$

$$-2x = 1 - 20$$

$$-2x = -19$$

$$x = \frac{19}{2}$$

So the tangent will occur when $x = \frac{19}{2}$.

Next we find the point on the graph.

For $x = \frac{19}{2}$,

$$y = 20\left(\frac{19}{2}\right) - \left(\frac{19}{2}\right)^2$$

$$y = 190 - \left(\frac{361}{4}\right)$$

$$= \frac{760}{4} - \frac{361}{4}$$

$$= \frac{399}{4}$$

The tangent line has slope 1 at the point $\left(\frac{19}{2}, \frac{399}{4}\right)$.

77. $y = -0.025x^2 + 4x$

To find the tangent line that has slope equal to 1, we need to find the values of x that make

$\frac{dy}{dx} = 1$.

First, we find the derivative.

$$\frac{dy}{dx} = \frac{d}{dx}\left(-0.025x^2 + 4x\right)$$

$$= \frac{d}{dx} - 0.025x^2 + \frac{d}{dx}4x$$

$$= -0.025(2x) + 4$$

$$= -0.05x + 4$$

Next, we set the derivative equal to 1 and solve for x.

$$\frac{dy}{dx} = 1$$

$$-0.05x + 4 = 1$$

$$-0.05x = -3$$

$$x = \frac{-3}{-0.05}$$

$$x = 60$$

So the tangent will occur when $x = 60$. Next we find the point on the graph.

For $x = 60$,

$$y = -0.025(60)^2 + 4(60)$$

$$= -0.025(3600) + 240$$

$$= -90 + 240$$

$$= 150$$

The tangent line has slope 1 at the point $(60, 150)$.

79. $y = \frac{1}{3}x^3 + 2x^2 + 2x$

To find the tangent line that has slope equal to 1, we need to find the values of x that make $\frac{dy}{dx} = 1$.

First, we find the derivative.

$$\frac{dy}{dx} = \frac{d}{dx}\left(\frac{1}{3}x^3 + 2x^2 + 2x\right)$$

$$= \frac{d}{dx}\left(\frac{1}{3}x^3\right) + \frac{d}{dx}2x^2 + \frac{d}{dx}2x$$

$$= x^2 + 4x + 2$$

Next, we set the derivative equal to 1 and solve for x.

$$\frac{dy}{dx} = 1$$

$$x^2 + 4x + 2 = 1$$

$$x^2 + 4x + 1 = 0$$

This is a quadratic equation, not readily factorable, so we use the quadratic formula where $a = 1, b = 4, \text{and}, c = 1$.

$$x = \frac{-b \pm \sqrt{b^2 - 4ac}}{2a}$$

$$x = \frac{-(4) \pm \sqrt{(4)^2 - 4(1)(1)}}{2(1)} \quad \text{Substituting}$$

$$= \frac{-4 \pm \sqrt{12}}{2}$$

$$= \frac{-4 \pm 2\sqrt{3}}{2} \qquad \left[\sqrt{12} = \sqrt{4 \cdot 3} = 2\sqrt{3}\right]$$

$$= \frac{2\left(-2 \pm \sqrt{3}\right)}{2}$$

$$= -2 \pm \sqrt{3}$$

There are two tangent lines that have slope equal to 1. The first one occurs at $x = -2 + \sqrt{3}$ and the second one occurs at $x = -2 - \sqrt{3}$. We use the original equation to find the point on the graph.

For $x = -2 + \sqrt{3}$,

$$y = \frac{1}{3}\left(-2 + \sqrt{3}\right)^3 + 2\left(-2 + \sqrt{3}\right)^2 + 2\left(-2 + \sqrt{3}\right)$$

$$= \frac{1}{3}\left(-26 + 15\sqrt{3}\right) + 2\left(7 - 4\sqrt{3}\right) - 4 + 2\sqrt{3}$$

$$= -\frac{26}{3} + 5\sqrt{3} + 14 - 8\sqrt{3} - 4 + 2\sqrt{3}$$

$$= \frac{4}{3} - \sqrt{3}$$

For $x = -2 - \sqrt{3}$,

$$y = \frac{1}{3}\left(-2 - \sqrt{3}\right)^3 + 2\left(-2 - \sqrt{3}\right)^2 + 2\left(-2 - \sqrt{3}\right)$$

$$= \frac{1}{3}\left(-26 - 15\sqrt{3}\right) + 2\left(7 + 4\sqrt{3}\right) - 4 - 2\sqrt{3}$$

$$= -\frac{26}{3} - 5\sqrt{3} + 14 + 8\sqrt{3} - 4 - 2\sqrt{3}$$

$$= \frac{4}{3} + \sqrt{3}$$

The tangent lines have slope 1 at the points $\left(-2 + \sqrt{3}, \frac{4}{3} - \sqrt{3}\right)$ and $\left(-2 - \sqrt{3}, \frac{4}{3} + \sqrt{3}\right)$.

81. a) In order to find the rate of change of the area with respect to the radius, we must find the derivative of the function with respect to r.

$$A'(r) = \frac{d}{dr}\left(3.14r^2\right)$$

$$= 3.14\left(2r^{2-1}\right)$$

$$= 6.28r$$

b) $\boxed{tw}$

83. $w(t) = 8.15 + 1.82t - 0.0596t^2 + 0.000758t^3$

a) In order to find the rate of change of weight with respect to time, we take the derivative of the function with respect to t.

$w'(t)$

$= \dfrac{d}{dt}\left(8.15 + 1.82t - 0.0596t^2 + 0.000758t^3\right)$

$= 0 + 1.82 - 0.0596(2t) + 0.000758(3t^2)$

$= 1.82 - 0.1192t + 0.002274t^2$

Therefore, the rate of change of weight with respect to time is given by:

$w'(t) = 1.82 - 0.1192t + 0.002274t^2$

b) The weight of the baby at age 10 months can be found by evaluating the function when $t = 10$.

$w(10) = 8.15 + 1.82(10) - 0.0596(10)^2$

$\qquad\qquad + 0.000758(10)^3$

$\approx 21.148 \qquad$ Using a calculator

Therefore, a 10 month old boy weighs approximately 21.148 pounds.

c) The rate of change of the baby's weight with respect to time at age of 10 months can be found by evaluating the derivative when $t = 10$.

$w'(10)$

$= 1.82 - 0.1192(10) + 0.002274(10)^2$

≈ 0.8554

A 10 month old boys weight will be increasing at a rate of 0.8554 pounds per month.

85. $R(v) = \dfrac{6000}{v} = 6000v^{-1}$

a) Using the power rule, we take the derivative of R with respect to v.

$R'(v) = 6000\left(-1v^{-1-1}\right)$

$\qquad = -6000v^{-2}$

$\qquad = -\dfrac{6000}{v^2}$

The rate of change of heart rate with respect to the output per beat is

$R'(v) = -\dfrac{6000}{v^2}$.

b) To find the heart rate at $v = 80$ ml per beat, we evaluate the function $R(v)$ when $v = 80$.

$R(80) = \dfrac{6000}{80} = 75$.

The heart rate is 75 beats per minute when the output per beat is 80 ml per beat.

c) To find the rate of change of the heart beat at $v = 80$ ml per beat, we evaluate the derivative $R'(v)$ at $v = 80$.

$R'(80) = -\dfrac{6000}{80^2}$

$R'(80) = -\dfrac{15}{16}$

$\qquad\quad \approx -0.9375$

The heart rate is decreasing at a rate of 0.9375 beats per minute when the output per beat is 80 ml per beat.

87. a) Using the power rule, we find the growth rate $\dfrac{dP}{dt}$.

$\dfrac{dP}{dt} = \dfrac{d}{dt}\left(100,000 + 2000t^2\right)$

$\qquad = 0 + 2000(2t)$

$\qquad = 4000t$

b) Evaluate the function P when $t = 10$.

$P(10) = 100,000 + 2000(10)^2$

$\qquad = 100,000 + 2000(100)$

$\qquad = 300,000$

The population of the city will be 300,000 people after 10 years.

c) Evaluate the derivative $P'(t)$ when $t = 10$.

$\dfrac{dP}{dt}\bigg|_{t=10} = P'(10) = 4000(10) = 40,000$

The population's growth rate after 10 years is 40,000 people per year.

d) $\boxed{tw}$

89. $V = 1.22\sqrt{h} = 1.22h^{1/2}$

a) Using the power rule,

$$\frac{dV}{dt} = \frac{d}{dt}\left(1.22h^{1/2}\right)$$

$$= 1.22\left(\frac{1}{2}h^{1/2-1}\right)$$

$$= 0.61h^{-1/2}$$

$$= \frac{0.61}{h^{1/2}} = \frac{0.61}{\sqrt{h}}$$

b) Evaluate the function V when $h = 40,000$.

$$V = 1.22\sqrt{40,000}$$

$$= 244$$

A person would be able to see 244 miles to the horizon from a height of 40,000 feet.

c) Evaluate the derivative $\dfrac{dV}{dh}$ when

$h = 40,000$.

$$\left.\frac{dV}{dh}\right|_{h=40,000} = \frac{0.61}{\sqrt{40,000}}$$

$$= \frac{0.61}{200}$$

$$\approx 0.00305$$

The rate of change at $h = 40,000$ is 0.00305 miles per foot.

d) $\boxed{tw}$

91. $f(x) = x^2 - 4x + 1$

The derivative is positive when $f'(x) > 0$.

Find $f'(x)$.

$$f'(x) = \frac{d}{dx}\left(x^2 - 4x + 1\right)$$

$$= 2x - 4$$

Next, we solve the inequality

$$f'(x) > 0$$

$$2x - 4 > 0$$

$$2x > 4$$

$$x > 2$$

Therefore, the interval for which $f'(x)$ is

positive is $(2, \infty)$.

93. $f(x) = \dfrac{1}{3}x^3 - x^2 - 3x + 5$

The derivative is positive when $f'(x) > 0$.

Find $f'(x)$.

$$f'(x) = \frac{d}{dx}\left(\frac{1}{3}x^3 - x^2 - 3x + 5\right)$$

$$= x^2 - 2x - 3$$

Next, we solve the inequality

$$f'(x) > 0$$

$$x^2 - 2x - 3 > 0$$

First we find where the quadratic is equal to zero, in order to determine the intervals that we will need to test.

$$x^2 - 2x - 3 = 0$$

$$(x-3)(x+1) = 0$$

$$x + 1 = 0 \quad \text{or} \quad x - 3 = 0$$

$$x = -1 \quad \text{or} \quad x = 3$$

Now we will test a value to the left of -1, between -1 and 3 and to the right of 3 to determine where the quadratic is positive or negative. We choose the values $x = -2$, $x = 0$, and, $x = 4$ to test.

When $x = -2$, the derivative $f'(x)$ is

$$f'(-2) = (-2)^2 - 2(-2) - 3 = 5.$$

When $x = 0$, the derivative $f'(x)$ is

$$f'(0) = (0)^2 - 2(0) - 3 = -3.$$

When $x = 4$, the derivative $f'(x)$ is

$$f'(4) = (4)^2 - 2(4) - 3 = 5.$$

We organize the results in the table below.

Test point x	Test $x = -2$	-1	Test $x = 0$	3	Test $x = 4$
$f'(x)$	$f'(-2) = 5$	0	$f'(0) = -3$	0	$f'(4) = 5$

From the table, we can see that $f'(x)$ is

positive on the intervals. $(-\infty, -1) \cup (3, \infty)$.

95. $y = 2x^6 - x^4 - 2$

A horizontal tangent line has slope equal to 0, so we need to find the values of x that make

$\dfrac{dy}{dx} = 0$.

First, we find the derivative.

$\dfrac{dy}{dx} = \dfrac{d}{dx}\left(2x^6 - x^4 - 2\right)$

$= \dfrac{d}{dx}2x^6 - \dfrac{d}{dx}x^4 - \dfrac{d}{dx}2$

$= 2\left(6x^{6-1}\right) - \left(4x^{4-1}\right) + 0$

$= 12x^5 - 4x^3$

Next, we set the derivative equal to zero and solve for x.

$\dfrac{dy}{dx} = 0$

$12x^5 - 4x^3 = 0$

$4x^3\left(3x^2 - 1\right) = 0$

$4x^3 = 0 \quad \text{or} \quad 3x^2 - 1 = 0$

$x = 0 \quad \text{or} \quad 3x^2 = 1$

$x = 0 \quad \text{or} \quad x^2 = \dfrac{1}{3}$

$x = 0 \quad \text{or} \quad x = \pm\sqrt{\dfrac{1}{3}} = \pm\dfrac{1}{\sqrt{3}}$

So the horizontal tangent will occur when

$x = 0, \quad x = \dfrac{1}{\sqrt{3}}, \quad \text{and } x = -\dfrac{1}{\sqrt{3}}$. Next we find

the points on the graph.

For $x = 0$,

$y = 2(0)^6 - (0) - 2$

$= -2$

For $x = \dfrac{1}{\sqrt{3}}$,

$y = 2\left(\dfrac{1}{\sqrt{3}}\right)^6 - \left(\dfrac{1}{\sqrt{3}}\right)^4 - 2$

$= 2\left(\dfrac{1}{27}\right) - \dfrac{1}{9} - 2$

$= \dfrac{2}{27} - \dfrac{3}{27} - \dfrac{54}{27}$

$= -\dfrac{55}{27}$

For $x = -\dfrac{1}{\sqrt{3}}$,

$y = 2\left(-\dfrac{1}{\sqrt{3}}\right)^6 - \left(-\dfrac{1}{\sqrt{3}}\right)^4 - 2$

$= 2\left(\dfrac{1}{27}\right) - \dfrac{1}{9} - 2$

$= \dfrac{2}{27} - \dfrac{3}{27} - \dfrac{54}{27}$

$= -\dfrac{55}{27}$

Therefore, there are horizontal tangents at the

points $(0, -2)$, $\left(\dfrac{1}{\sqrt{3}}, -\dfrac{55}{27}\right)$, and, $\left(-\dfrac{1}{\sqrt{3}}, -\dfrac{55}{27}\right)$.

97. $y = (x - 1)(x + 1) = x^2 - 1$

First, we multiply the two binomials Therefore,

$\dfrac{dy}{dx} = \dfrac{d}{dx}\left(x^2 - 1\right)$

$= \dfrac{d}{dx}\left(x^2\right) - \dfrac{d}{dx}(1)$.

$= 2x$

99. $y = \dfrac{5x^2 - 8x + 3}{8}$

First we separate the fraction.

$y = \dfrac{5x^2}{8} - \dfrac{8x}{8} + \dfrac{3}{8}$

$= \dfrac{5}{8}x^2 - x + \dfrac{3}{8}$

Therefore,

$\dfrac{dy}{dx} = \dfrac{d}{dx}\left(\dfrac{5}{8}x^2 - x + \dfrac{3}{8}\right)$

$= \dfrac{d}{dx}\left(\dfrac{5}{8}x^2\right) - \dfrac{d}{dx}(x) + \dfrac{d}{dx}\left(\dfrac{3}{8}\right)$

$= \dfrac{5}{8}\left(2x^{2-1}\right) - 1 + 0$

$= \dfrac{5}{4}x - 1$

101. $y = \dfrac{x^5 - 3x^4 + 2x + 4}{x^2}$

First, we separate the fraction.

$y = \dfrac{x^5}{x^2} - \dfrac{3x^4}{x^2} + \dfrac{2x}{x^2} + \dfrac{4}{x^2}$

$= x^{5-2} - 3x^{4-2} + 2x^{1-2} + 4x^{-2}$

$= x^3 - 3x^2 + 2x^{-1} + 4x^{-2}$

Therefore,

$\dfrac{dy}{dx} = \dfrac{d}{dx}\left(x^3 - 3x^2 + 2x^{-1} + 4x^{-2}\right)$

$= \dfrac{d}{dx}\left(x^3\right) - \dfrac{d}{dx}\left(3x^2\right) + \dfrac{d}{dx}\left(2x^{-1}\right) + \dfrac{d}{dx}\left(4x^{-2}\right)$

$= 3x^2 - 3\left(2x^1\right) + 2\left(-1x^{-1-1}\right) + 4\left(-2x^{-2-1}\right)$

$= 3x^2 - 6x - 2x^{-2} - 8x^{-3}$

$= 3x^2 - 6x - \dfrac{2}{x^2} - \dfrac{8}{x^3}$

103. $y = \sqrt{7x}$

First, we simplify the radical.

$y = \sqrt{7 \cdot x}$

$= \sqrt{7}\sqrt{x} \qquad \left[\sqrt{m \cdot n} = \sqrt{m}\sqrt{n}\right]$

$= \sqrt{7}\left(x^{1/2}\right) \qquad \left[\sqrt[m]{a} = a^{1/m}; m = 2\right]$

Therefore,

$\dfrac{dy}{dx} = \dfrac{d}{dx}\left(\sqrt{7}\left(x\right)^{1/2}\right)$

$= \sqrt{7}\dfrac{d}{dx}\left(x^{1/2}\right)$

$= \sqrt{7}\left(\dfrac{1}{2}x^{1/2-1}\right)$

$= \dfrac{\sqrt{7}}{2}x^{-1/2}$

$= \dfrac{\sqrt{7}}{2x^{1/2}}$

$= \dfrac{\sqrt{7}}{2\sqrt{x}}$

105. $y = (x-3)^2$

First square the binomial.

$y = (x-3)^2$

$= (x-3)(x-3)$

$= x^2 - 6x + 9$

Therefore,

$\dfrac{dy}{dx} = \dfrac{d}{dx}\left(x^2 - 6x + 9\right)$

$= \dfrac{d}{dx}\left(x^2\right) - \dfrac{d}{dx}\left(6x\right) + \dfrac{d}{dx}\left(9\right)$

$= 2x - 6$

107. $y = \left(\sqrt{x} + \sqrt[3]{x}\right)^2$

First rewrite the radicals as expressions with rational exponents, and then square the binomial.

$y = \left(x^{1/2} + x^{1/3}\right)^2$

$= \left(x^{1/2} + x^{1/3}\right)\left(x^{1/2} + x^{1/3}\right)$

$= x^{1/2+1/2} + 2x^{1/2+1/3} + x^{1/3+1/3}$

$= x + 2x^{5/6} + x^{2/3}$

Therefore,

$\dfrac{dy}{dx} = \dfrac{d}{dx}\left(x + 2x^{5/6} + x^{2/3}\right)$

$= \dfrac{d}{dx}\left(x\right) + \dfrac{d}{dx}\left(2x^{5/6}\right) + \dfrac{d}{dx}\left(x^{2/3}\right)$

$= 1 + 2\left(\dfrac{5}{6}x^{5/6-1}\right) + \dfrac{2}{3}x^{2/3-1}$

$= 1 + \dfrac{5}{3}x^{-1/6} + \dfrac{2}{3}x^{-1/3}$

109. First, we rewrite 1 using properties of exponents. We know $x^0 = 1$, therefore,

$\dfrac{d}{dx}(1) = \dfrac{d}{dx}\left(x^0\right)$. Applying theorem 1, we have

$\dfrac{d}{dx}(1) = \dfrac{d}{dx}\left(x^0\right)$

$= 0x^{0-1}$

$= 0x^{-1}$

$= 0$

111. $\boxed{tw}$

113. $f(x) = 1.6x^3 - 2.3x - 3.7$

First we enter the equation into the graphing editor on the calculator.

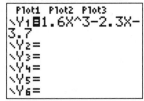

Using the window:

We get the graph:

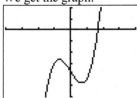

The horizontal tangents occur at the turning points of this function. Using the trace feature, or the minimum/maximum feature on the calculator, we find the turning points. We estimate the x-values at which the tangent lines are horizontal are
$x = -0.692$ and $x = 0.692$.

115. $f(x) = \dfrac{5x^2 + 8x - 3}{3x^2 + 2}$

First we enter the equation into the graphing editor on the calculator.

Using the window:

We get the graph:

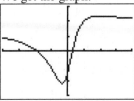

The horizontal tangents occur at the turning points of this function. Using the trace feature, or the minimum/maximum feature on the calculator, we find the turning points. We estimate the x-values at which the tangent lines are horizontal are
$x = -0.346$ and $x = 1.929$.

117. $f(x) = x^4 - 3x^2 + 1$

Using the calculator, we graph the function and the derivative in the same window. We can use the nDeriv feature to graph the derivative without actually calculating the derivative. The syntax is shown in the screen shot at the top of the next column.

Using the window:

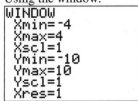

We get the graph:

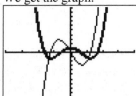

Note, the function $f(x)$ is the thicker graph.

Using the calculator, we can find the derivative of the function when $x = 1$.

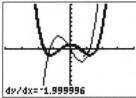

We have $f'(1) = -2$.

119. $f(x) = x^4 - x^3$

Using the calculator, we graph the function and the derivative in the same window. We can use the nDeriv feature to graph the derivative without actually calculating the derivative.

Using the window:

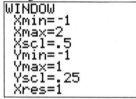

We get the graph:

Note, the function $f(x)$ is the thicker graph.

Using the calculator, we can find the derivative of the function when $x = 1$.

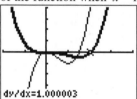

We have $f'(1) = 1$.

121. $f(x) = 20x^3 - 3x^5$

Using the calculator, we graph the function and the derivative in the same window. We can use the nDeriv feature to graph the derivative without actually calculating the derivative.

Using the window:

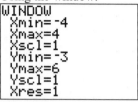

We get the graph:

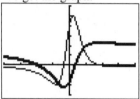

Note, the function $f(x)$ is the thicker graph.

Using the calculator, we can find the derivative of the function when $x = 1$.

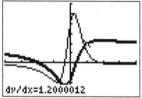

We have $f'(1) = 1.2$.

Exercise Set 1.6

1. Differentiate $y = x^5 \cdot x^6$ using the Product Rule (Theorem 5).

$$\frac{dy}{dx} = \frac{d}{dx}\left(x^5 \cdot x^6\right)$$

$$= x^5 \cdot \frac{d}{dx}\left(x^6\right) + x^6 \frac{d}{dx}\left(x^5\right)$$

$$= x^5 \cdot 6x^5 + x^6 \cdot 5x^4$$

$$= 6x^{10} + 5x^{10}$$

$$= 11x^{10}$$

Differentiate $y = x^5 \cdot x^6 = x^{11}$ using the Power Rule (Theorem 1).

$$\frac{dy}{dx} = \frac{d}{dx}x^{11}$$

$$= 11x^{11-1}$$

$$= 11x^{10}$$

The two results are equivalent.

3. Differentiate $f(x) = (2x+5)(3x-4)$ using the Product Rule (Theorem 5).

$$f'(x) = \frac{d}{dx}\Big[(2x+5)(3x-4)\Big]$$

$$= (2x+5)\cdot\frac{d}{dx}(3x-4)+$$

$$(3x-4)\cdot\frac{d}{dx}(2x+5)$$

$$= (2x+5)\cdot 3 + (3x-4)\cdot 2$$

$$= 6x+15+6x-8$$

$$= 12x+7$$

Differentiate $f(x) = (2x+5)(3x-4)$ using the Power Rule (Theorem 1). First, we multiply the binomial terms in the function.

$$f(x) = (2x+5)(3x-4)$$

$$= 6x^2 + 7x - 20$$

Therefore, by Theorem 1 and Theorem 4 we have:

$$f'(x) = \frac{d}{dx}(6x^2 + 7x - 20)$$

$$= \frac{d}{dx}(6x^2) + \frac{d}{dx}(7x) - \frac{d}{dx}(20) \quad \text{Theorem 4}$$

$$= 12x+7 \qquad\qquad\qquad \text{Theorem 1}$$

The two results are equivalent.

5. Differentiate $G(x) = 4x^2(x^3+5x)$ using the Product Rule.

$$G'(x) = \frac{d}{dx}\Big[4x^2(x^3+5x)\Big]$$

$$= 4x^2\cdot\frac{d}{dx}(x^3+5x)+(x^3+5x)\cdot\frac{d}{dx}(4x^2)$$

$$= 4x^2\cdot(3x^2+5)+(x^3+5x)\cdot(8x)$$

$$= 12x^4 + 20x^2 + 8x^4 + 40x^2$$

$$= 20x^4 + 60x^2$$

Differentiate $G(x) = 4x^2(x^3+5x)$ using the Power Rule. First, we multiply the function.

$$G(x) = 4x^2(x^3+5x)$$

$$= 4x^5 + 20x^3$$

Therefore, we have:

$$G'(x) = \frac{d}{dx}(4x^5 + 20x^3)$$

$$= \frac{d}{dx}(4x^5) + \frac{d}{dx}(20x^3) \quad \text{Theorem 4}$$

$$= 4(5x^4) + 20(3x^2) \qquad \begin{array}{l}\text{Theorem 1}\\\text{Theorem 3}\end{array}$$

$$= 20x^4 + 60x^2$$

The two results are equivalent.

7. Differentiate $y = \left(3\sqrt{x}+2\right)x^2$ using the Product Rule.

$$\frac{dy}{dx} = \frac{d}{dx}\Big[\left(3\sqrt{x}+2\right)x^2\Big]$$

$$= \left(3\sqrt{x}+2\right)\cdot\frac{d}{dx}(x^2)+x^2\cdot\frac{d}{dx}\left(3\sqrt{x}+2\right)$$

$$= \left(3x^{1/2}+2\right)\cdot\frac{d}{dx}(x^2)+x^2\cdot\frac{d}{dx}\left(3x^{1/2}+2\right)$$

$$= \left(3x^{1/2}+2\right)\cdot 2x + x^2\cdot\frac{3}{2}x^{-1/2} \quad \text{Theorem 1}$$

$$= 6x^{3/2} + 4x + \frac{3}{2}x^{3/2}$$

$$= \frac{15}{2}x^{3/2} + 4x$$

Differentiate $y = \left(3\sqrt{x}+2\right)x^2$ using the Power Rule. First, we multiply the function.

$$y = \left(3\sqrt{x}+2\right)x^2$$

$$= 3x^{1/2+2} + 2x^2$$

$$= 3x^{5/2} + 2x^2$$

Therefore, we have:

$$\frac{dy}{dx} = \frac{d}{dx}\left(3x^{5/2} + 2x^2\right)$$

$$= \frac{d}{dx}\left(3x^{5/2}\right) + \frac{d}{dx}(2x^2) \quad \text{Theorem 4}$$

$$= 3\left(\frac{5}{2}x^{5/2-1}\right) + 2(2x^1) \qquad \begin{array}{l}\text{Theorem 1}\\\text{Theorem 3}\end{array}$$

$$= \frac{15}{2}x^{3/2} + 4x$$

The two results are equivalent.

9. Differentiate $g(x) = (4x-3)(2x^2+3x+5)$ using the Product Rule.

$$g'(x) = \frac{d}{dx}\Big[(4x-3)(2x^2+3x+5)\Big]$$

$$= (4x-3)\cdot\frac{d}{dx}(2x^2+3x+5)+$$

$$(2x^2+3x+5)\cdot\frac{d}{dx}(4x-3)$$

$$= (4x-3)\cdot(4x+3)+(2x^2+3x+5)\cdot 4$$

$$= 16x^2 - 9 + 8x^2 + 12x + 20$$

$$= 24x^2 + 12x + 11$$

Differentiate $g(x) = (4x-3)(2x^2 + 3x + 5)$ using the Power Rule. First, we multiply the terms in the function.

$g(x) = (4x-3)(2x^2 + 3x + 5)$

$\quad = 8x^3 + 6x^2 + 11x - 15$

Therefore, we have:

$g'(x) = \dfrac{d}{dx}(8x^3 + 6x^2 + 11x - 15)$

$\quad = \dfrac{d}{dx}(8x^3) + \dfrac{d}{dx}(6x^2) +$

$\qquad\qquad \dfrac{d}{dx}(11x) - \dfrac{d}{dx}(15)$

$\quad = 24x^2 + 12x + 11$

The two results are equivalent.

11. Differentiate $F(t) = (\sqrt{t} + 2)(3t - 4\sqrt{t} + 7)$ using the Product Rule.

$F'(t) = \dfrac{d}{dt}\left[(\sqrt{t} + 2)(3t - 4\sqrt{t} + 7)\right]$

$\quad = (t^{1/2} + 2) \cdot \dfrac{d}{dt}(3t - 4t^{1/2} + 7) +$

$\qquad (3t - 4t^{1/2} + 7) \cdot \dfrac{d}{dt}(t^{1/2} + 2) \quad \left[\sqrt{t} = t^{1/2}\right]$

$\quad = (t^{1/2} + 2) \cdot \left(3 - 4\left(\dfrac{1}{2}t^{-1/2}\right)\right) +$

$\qquad (3t - 4t^{1/2} + 7) \cdot \left(\dfrac{1}{2}t^{-1/2}\right)$

$\quad = (t^{1/2} + 2) \cdot \left(3 - 2t^{-1/2}\right) +$

$\qquad (3t - 4t^{1/2} + 7) \cdot \left(\dfrac{1}{2}t^{-1/2}\right)$

$\quad = 3t^{1/2} - 2 + 6 - 4t^{-1/2} + \dfrac{3}{2}t^{1/2} - 2 + \dfrac{7}{2}t^{-1/2}$

$\quad = \dfrac{9}{2}t^{1/2} - \dfrac{1}{2}t^{-1/2} + 2$

$\quad = \dfrac{9\sqrt{t}}{2} - \dfrac{1}{2\sqrt{t}} + 2$

Differentiate $F(t) = (\sqrt{t} + 2)(3t - 4\sqrt{t} + 7)$ using the Power Rule

$F(t) = (\sqrt{t} + 2)(3t - 4\sqrt{t} + 7)$

$\quad = 3t^{3/2} - 4t + 7t^{1/2} + 6t - 8t^{1/2} + 14$

$\quad = 3t^{3/2} - t^{1/2} + 2t + 14$

Therefore, we have:

$F'(t) = \dfrac{d}{dt}\left(3t^{3/2} - t^{1/2} + 2t + 14\right)$

$\quad = \dfrac{d}{dt}\left(3t^{3/2}\right) - \dfrac{d}{dt}\left(t^{1/2}\right) + \dfrac{d}{dt}(2t) + \dfrac{d}{dt}(14)$

$\quad = \dfrac{9t^{1/2}}{2} - \dfrac{1}{2}t^{-1/2} + 2$

$\quad = \dfrac{9\sqrt{t}}{2} - \dfrac{1}{2\sqrt{t}} + 2$

The two results are equivalent.

13. Differentiate $y = \dfrac{x^7}{x^3}$ using the Quotient Rule (Theorem 6).

$\dfrac{dy}{dx} = \dfrac{d}{dx}\left(\dfrac{x^7}{x^3}\right)$

$\quad = \dfrac{x^3 \dfrac{d}{dx}(x^7) - x^7 \dfrac{d}{dx}(x^3)}{(x^3)^2}$

$\quad = \dfrac{x^3(7x^6) - x^7(3x^2)}{x^6}$

$\dfrac{dy}{dx} = \dfrac{7x^9 - 3x^9}{x^6}$

$\quad = \dfrac{4x^9}{x^6}$

$\quad = 4x^3, \quad$ for $x \neq 0$

Differentiate $y = \dfrac{x^7}{x^3} = x^4$ using the Power Rule.

$\dfrac{dy}{dx} = \dfrac{d}{dx}x^4$

$\quad = 4x^{4-1}$

$\quad = 4x^3,$ for $x \neq 0$

The two results are equivalent.

15. Differentiate $f(x) = \dfrac{2x^5 + x^2}{x}$ using the Quotient Rule.

$$f'(x) = \frac{d}{dx}\left(\frac{2x^5 + x^2}{x}\right)$$

$$= \frac{x\dfrac{d}{dx}(2x^5 + x^2) - (2x^5 + x^2)\dfrac{d}{dx}(x)}{(x)^2}$$

$$f'(x) = \frac{x(10x^4 + 2x) - (2x^5 + x^2)(1)}{x^2}$$

$$= \frac{10x^5 + 2x^2 - 2x^5 - x^2}{x^2}$$

$$= \frac{8x^5 + x^2}{x^2}$$

$$= \frac{x^2(8x^3 + 1)}{x^2}$$

$$= 8x^3 + 1, \qquad \text{for } x \neq 0$$

Differentiate $f(x) = \dfrac{2x^5 + x^2}{x}$ using the Power Rule.

First, factor the numerator and divide the common factors.

$$f(x) = \frac{2x^5 + x^2}{x}$$

$$= \frac{x(2x^4 + x)}{x}$$

$$= 2x^4 + x$$

$$f'(x) = \frac{d}{dx}(2x^4 + x)$$

$$= 8x^3 + 1, \qquad \text{for } x \neq 0$$

The two results are equivalent.

17. Differentiate $G(x) = \dfrac{8x^3 - 1}{2x - 1}$ using the Quotient Rule.

$$G'(x) = \frac{d}{dx}\left(\frac{8x^3 - 1}{2x - 1}\right)$$

$$= \frac{(2x-1)\dfrac{d}{dx}(8x^3 - 1) - (8x^3 - 1)\dfrac{d}{dx}(2x - 1)}{(2x - 1)^2}$$

$$= \frac{(2x-1)(24x^2) - (8x^3 - 1)(2)}{(2x - 1)^2}$$

$$= \frac{(2x-1)(24x^2) - (2x-1)(4x^2 + 2x + 1)(2)}{(2x - 1)^2}$$

$$= \frac{(2x-1)\left[(24x^2) - (4x^2 + 2x + 1)(2)\right]}{(2x - 1)^2}$$

$$= \frac{(2x-1)\left[(24x^2) - (8x^2 + 4x + 2)\right]}{(2x - 1)^2}$$

$$= \frac{\left[16x^2 - 4x - 2\right]}{(2x - 1)}$$

$$= \frac{(2x-1)(8x+2)}{(2x - 1)}$$

$$= 8x + 2; \qquad x \neq \tfrac{1}{2}$$

Differentiate $G(x) = \dfrac{8x^3 - 1}{2x - 1}$ using the Power Rule.

First, factor the numerator and divide the common factors.

$$G(x) = \frac{8x^3 - 1}{2x - 1}$$

$$= \frac{(2x-1)(4x^2 + 2x + 1)}{2x - 1} \qquad \text{Difference of cubes}$$

$$= 4x^2 + 2x + 1$$

$$G'(x) = \frac{d}{dx}(4x^2 + 2x + 1)$$

$$= 8x + 2; \qquad x \neq \tfrac{1}{2}$$

The two results are equivalent.

19. Differentiate $y = \dfrac{t^2 - 16}{t+4}$ using the Quotient Rule.

$$\frac{dy}{dt} = \frac{d}{dt}\left(\frac{t^2-16}{t+4}\right)$$

$$= \frac{(t+4)\frac{d}{dt}(t^2-16) - (t^2-16)\frac{d}{dt}(t+4)}{(t+4)^2}$$

$$= \frac{(t+4)(2t) - (t^2-16)(1)}{(t+4)^2}$$

$$= \frac{2t^2 + 8t - t^2 + 16}{(t+4)^2}$$

$$= \frac{t^2 + 8t + 16}{(t+4)^2}$$

$$\frac{dy}{dt} = \frac{(t+4)^2}{(t+4)^2}$$

$$= 1; \qquad t \neq -4$$

Differentiate $y = \dfrac{t^2-16}{t+4}$ using the Power Rule. First, factor the numerator and divide the common factors.

$$y = \frac{t^2-16}{t+4}$$

$$= \frac{(t+4)(t-4)}{t+4} \qquad \text{Difference of squares}$$

$$= t-4$$

$$\frac{dy}{dt} = \frac{d}{dt}(x-4)$$

$$= 1, \qquad \text{for } t \neq -4$$

The two results are equivalent.

21. $f(x) = (3x^2 - 2x + 5)(4x^2 + 3x - 1)$

Using the Product Rule, we have:

$$f'(x) = \frac{d}{dx}\left[(3x^2 - 2x + 5)(4x^2 + 3x - 1)\right]$$

$$= (3x^2 - 2x + 5)\cdot\frac{d}{dx}(4x^2 + 3x - 1) +$$

$$\quad (4x^2 + 3x - 1)\cdot\frac{d}{dx}(3x^2 - 2x + 5)$$

$$= (3x^2 - 2x + 5)\cdot(8x + 3) +$$

$$\quad (4x^2 + 3x - 1)\cdot(6x - 2)$$

Simplifying, we get

$$= (24x^3 - 7x^2 + 34x + 15) +$$

$$\quad (24x^3 + 10x^2 - 12x + 2)$$

$$= 48x^3 + 3x^2 + 22x + 17$$

23. $y = \dfrac{5x^2 - 1}{2x^3 + 3}$

Using the Quotient Rule.

$$\frac{dy}{dx} = \frac{d}{dx}\left(\frac{5x^2-1}{2x^3+3}\right)$$

$$= \frac{(2x^3+3)\frac{d}{dx}(5x^2-1) - (5x^2-1)\frac{d}{dx}(2x^3+3)}{(2x^3+3)^2}$$

$$= \frac{(2x^3+3)(10x) - (5x^2-1)(6x^2)}{(2x^3+3)^2}$$

$$= \frac{20x^4 + 30x - 30x^4 + 6x^2}{(2x^3+3)^2}$$

$$= \frac{-10x^4 + 6x^2 + 30x}{(2x^3+3)^2}$$

$$= \frac{-2x(5x^3 - 3x - 15)}{(2x^3+3)^2}$$

25. $G(x) = (8x + \sqrt{x})(5x^2 + 3)$

$$G(x) = (8x + x^{1/2})(5x^2 + 3) \qquad \left[\sqrt{x} = x^{1/2}\right]$$

Using the Product Rule, we have:

$$G'(x) = \frac{d}{dx}\left[(8x + x^{1/2})(5x^2 + 3)\right]$$

$$= (8x + x^{1/2})\cdot\frac{d}{dx}(5x^2 + 3) +$$

$$\quad (5x^2 + 3)\cdot\frac{d}{dx}(8x + x^{1/2})$$

$$= (8x + x^{1/2})\cdot(10x) +$$

$$\quad (5x^2 + 3)\cdot\left(8 + \frac{1}{2}x^{-1/2}\right)$$

$$= \left(80x^2 + 10x^{3/2}\right) +$$

$$\quad \left(40x^2 + \frac{5}{2}x^{3/2} + \frac{3}{2}x^{-1/2} + 24\right)$$

$$= 120x^2 + \frac{25x^{3/2}}{2} + \frac{3}{2x^{1/2}} + 24$$

27. $g(t) = \dfrac{t}{3-t} + 5t^3$

Differentiating we have:

$g'(t) = \dfrac{d}{dt}\left(\dfrac{t}{3-t} + 5t^3\right)$

$\quad = \dfrac{d}{dt}\left(\dfrac{t}{3-t}\right) + \dfrac{d}{dt}\left(5t^3\right)$

We will apply the Quotient Rule to the first term, and the Power Rule to the second term.

$g'(t) = \underbrace{\dfrac{(3-t)\cdot\dfrac{d}{dt}(t) - t\cdot\dfrac{d}{dt}(3-t)}{(3-t)^2}}_{\text{Quotient Rule}} + 15t^2$

$\quad = \dfrac{(3-t)(1) - t(-1)}{(3-t)^2} + 15t^2$

$\quad = \dfrac{3}{(3-t)^2} + 15t^2$

29. $F(x) = (x+3)^2 = (x+3)(x+3)$

Using the Product Rule, we have

$F'(x) = \dfrac{d}{dx}\left[(x+3)(x+3)\right]$

$\quad = (x+3)\cdot\dfrac{d}{dx}(x+3) + (x+3)\cdot\dfrac{d}{dx}(x+3)$

$\quad = (x+3)\cdot(1) + (x+3)\cdot(1)$

$\quad = 2x+6$

$\quad = 2(x+3)$

31. $y = (x^3 - 4x)^2 = (x^3 - 4x)(x^3 - 4x)$

Using the Product Rule, we have

$\dfrac{dy}{dx} = \dfrac{d}{dx}\left[(x^3 - 4x)(x^3 - 4x)\right]$

$\quad = (x^3 - 4x)\cdot\dfrac{d}{dx}(x^3 - 4x) +$

$\qquad (x^3 - 4x)\cdot\dfrac{d}{dx}(x^3 - 4x)$

$\quad = (x^3 - 4x)\cdot(3x^2 - 4) +$

$\qquad (x^3 - 4x)\cdot(3x^2 - 4)$

$\quad = 2(x^3 - 4x)(3x^2 - 4)$

33. $g(x) = 5x^{-3}\left(x^4 - 5x^3 + 10x - 2\right)$

Using the Product Rule:

$g'(x) = 5x^{-3}\dfrac{d}{dx}\left(x^4 - 5x^3 + 10x - 2\right) +$

$\qquad \left(x^4 - 5x^3 + 10x - 2\right)\dfrac{d}{dx}\left(5x^{-3}\right)$

$\quad = \left(5x^{-3}\right)\left(4x^3 - 15x^2 + 10\right) +$

$\qquad \left(x^4 - 5x^3 + 10x - 2\right)\left(-15x^{-4}\right)$

$\quad = 20 - 75x^{-1} + 50x^{-3} - 15 + 75x^{-1} -$

$\qquad 150x^{-3} + 30x^{-4}$

$\quad = 5 - 100x^{-3} + 30x^{-4}$

35. $F(t) = \left(t + \dfrac{2}{t}\right)\left(t^2 - 3\right) = \left(t + 2t^{-1}\right)\left(t^2 - 3\right)$

Using the Product Rule, we have:

$f'(t) = \dfrac{d}{dt}\left[\left(t + 2t^{-1}\right)\left(t^2 - 3\right)\right]$

$\quad = \left(t + 2t^{-1}\right)\cdot\dfrac{d}{dt}\left(t^2 - 3\right) +$

$\qquad \left(t^2 - 3\right)\cdot\dfrac{d}{dt}\left(t + 2t^{-1}\right)$

$f'(t) = \left(t + 2t^{-1}\right)\cdot(2t) + \left(t^2 - 3\right)\cdot\left(1 - 2t^{-2}\right)$

$\quad = 2t^2 + 4 + \left(t^2 - 2 - 3 + 6t^{-2}\right)$

$\quad = 3t^2 - 1 + 6t^{-2}$

$\quad = 3t^2 - 1 + \dfrac{6}{t^2}$

37. $y = \dfrac{x^2 + 1}{x^3 - 1} - 5x^2$

Differentiating we have:

$\dfrac{dy}{dx} = \dfrac{d}{dx}\left(\dfrac{x^2 + 1}{x^3 - 1} - 5x^2\right)$

$\quad = \dfrac{d}{dx}\left(\dfrac{x^2 + 1}{x^3 - 1}\right) - \dfrac{d}{dx}\left(5x^2\right)$

We will apply the Quotient Rule to the first term, and the Power Rule to the second term.

$\dfrac{dy}{dx} = \underbrace{\dfrac{\left(x^3 - 1\right)\cdot(2x) - \left(x^2 + 1\right)\cdot\left(3x^2\right)}{\left(x^3 - 1\right)^2}}_{\text{Quotient Rule}} - 10x$

Simplifying, we get

$\quad = \dfrac{2x^4 - 2x - \left(3x^4 + 3x^2\right)}{\left(x^3 - 1\right)^2} - 10x$

$\quad = \dfrac{-x^4 - 3x^2 - 2x}{\left(x^3 - 1\right)^2} - 10x$

39. $y = \dfrac{\sqrt[3]{x} - 7}{\sqrt{x} + 3} = \dfrac{x^{1/3} - 7}{x^{1/2} + 3}$

Using the Quotient Rule.

$\dfrac{dy}{dx} = \dfrac{d}{dx}\left(\dfrac{x^{1/3} - 7}{x^{1/2} + 3} \right)$

$= \dfrac{\left(x^{1/2} + 3\right)\dfrac{d}{dx}\left(x^{1/3} - 7\right) - \left(x^{1/3} - 7\right)\dfrac{d}{dx}\left(x^{1/2} + 3\right)}{\left(x^{1/2} + 3\right)^2}$

$= \dfrac{\left(x^{1/2} + 3\right)\left(\dfrac{1}{3}x^{-2/3}\right) - \left(x^{1/3} - 7\right)\left(\dfrac{1}{2}x^{-1/2}\right)}{\left(x^{1/2} + 3\right)^2}$

Note, the previous derivative can be simplified as follows

$\dfrac{dy}{dx} = \dfrac{\left(x^{1/2} + 3\right)\left(\dfrac{1}{3}x^{-2/3}\right) - \left(x^{1/3} - 7\right)\left(\dfrac{1}{2}x^{-1/2}\right)}{\left(x^{1/2} + 3\right)^2}$

$= \dfrac{\dfrac{1}{3}x^{-1/6} + x^{-2/3} - \dfrac{1}{2}x^{-1/6} + \dfrac{7}{2}x^{-1/2}}{\left(x^{1/2} + 3\right)^2}$

$= \dfrac{\dfrac{1}{3}x^{-1/6} + x^{-2/3} - \dfrac{1}{2}x^{-1/6} + \dfrac{7}{2}x^{-1/2}}{\left(x^{1/2} + 3\right)^2}$

$= \dfrac{x^{-2/3} - \dfrac{1}{6}x^{-1/6} + \dfrac{7}{2}x^{-1/2}}{\left(x^{1/2} + 3\right)^2} \cdot \dfrac{6x^{2/3}}{6x^{2/3}}$

$= \dfrac{6 - \sqrt{x} + 21x^{1/6}}{6x^{2/3}\left(\sqrt{x} + 3\right)^2}$

41. $f(x) = \dfrac{x}{x^{-1} + 1}$

Using the Quotient Rule, we have

$f'(x) = \dfrac{d}{dx}\left(\dfrac{x}{x^{-1} + 1} \right)$

$= \dfrac{\left(x^{-1} + 1\right)\dfrac{d}{dx}(x) - (x)\dfrac{d}{dx}\left(x^{-1} + 1\right)}{\left(x^{-1} + 1\right)^2}$

$= \dfrac{\left(x^{-1} + 1\right)(1) - (x)\left(-1x^{-2}\right)}{\left(x^{-1} + 1\right)^2}$

$= \dfrac{x^{-1} + 1 + x^{-1}}{\left(x^{-1} + 1\right)^2}$

$= \dfrac{2x^{-1} + 1}{\left(x^{-1} + 1\right)^2}, \qquad \text{for } x \neq 0 \text{ and } x \neq -1$

Note, the previous derivative could be simplified as follows:

$f'(x) = \dfrac{2x^{-1} + 1}{\left(x^{-1} + 1\right)^2}$

$= \dfrac{\dfrac{2}{x} + 1}{\left(\dfrac{1}{x} + 1\right)^2}$

$= \dfrac{\dfrac{2 + x}{x}}{\left(\dfrac{1 + x}{x}\right)^2}$

$f'(x) = \dfrac{2 + x}{x} \cdot \dfrac{x^2}{(1 + x)^2}$

$= \dfrac{x(x + 2)}{(1 + x)^2}, \qquad \text{for } x \neq 0 \text{ and } x \neq -1$

43. $F(t) = \dfrac{1}{t - 4}$

Using the Quotient Rule, we have

$F'(t) = \dfrac{d}{dt}\left(\dfrac{1}{t - 4} \right)$

$= \dfrac{(t - 4)\dfrac{d}{dt}(1) - (1)\dfrac{d}{dt}(t - 4)}{(t - 4)^2}$

$= \dfrac{(t - 4)(0) - (1)(1)}{(t - 4)^2}$

$= \dfrac{-1}{(t - 4)^2}$

45. $f(x) = \dfrac{3x^2 + 2x}{x^2 + 1}$

Using the Quotient Rule, we have

$f'(x) = \dfrac{d}{dx}\left(\dfrac{3x^2 + 2x}{x^2 + 1}\right)$

$= \dfrac{\left(x^2 + 1\right)\dfrac{d}{dx}\left(3x^2 + 2x\right) - \left(3x^2 + 2x\right)\dfrac{d}{dx}\left(x^2 + 1\right)}{\left(x^2 + 1\right)^2}$

$= \dfrac{\left(x^2 + 1\right)(6x + 2) - \left(3x^2 + 2x\right)(2x)}{\left(x^2 + 1\right)^2}$

$= \dfrac{6x^3 + 2x^2 + 6x + 2 - \left(6x^3 + 4x^2\right)}{\left(x^2 + 1\right)^2}$

$= \dfrac{-2x^2 + 6x + 2}{\left(x^2 + 1\right)^2}$

$= \dfrac{-2\left(x^2 - 3x - 1\right)}{\left(x^2 + 1\right)^2}$

47. $g(t) = \dfrac{-t^2 + 3t + 5}{t^2 - 2t + 4}$

the Quotient Rule, we have

$g'(t) = \dfrac{d}{dt}\left(\dfrac{-t^2 + 3t + 5}{t^2 - 2t + 4}\right)$

$= \dfrac{\left(t^2 - 2t + 4\right)\dfrac{d}{dt}\left(-t^2 + 3t + 5\right)}{\left(t^2 - 2t + 4\right)^2} -$

$\dfrac{\left(-t^2 + 3t + 5\right)\dfrac{d}{dt}\left(t^2 - 2t + 4\right)}{\left(t^2 - 2t + 4\right)^2}$

$= \dfrac{\left(t^2 - 2t + 4\right)(-2t + 3) - \left(-t^2 + 3t + 5\right)(2t - 2)}{\left(t^2 - 2t + 4\right)^2}$

Note, the previous derivative could be simplified as follows:

$g'(t) = \dfrac{-2t^3 + 7t^2 - 14t + 12 - \left(-2t^3 + 8t^2 + 4t - 10\right)}{\left(t^2 - 2t + 4\right)^2}$

$= \dfrac{-t^2 - 18t + 22}{\left(t^2 - 2t + 4\right)^2}$

49. – 96. Left to the student.

97. $y = \dfrac{8}{x^2 + 4}$

$\dfrac{dy}{dx} = \dfrac{\left(x^2 + 4\right)(0) - 8(2x)}{\left(x^2 + 4\right)^2}$

$\dfrac{dy}{dx} = \dfrac{-16x}{\left(x^2 + 4\right)^2}$

When $x = 0$, $\dfrac{dy}{dx} = \dfrac{-16(0)}{\left(0^2 + 4\right)^2} = 0$, so the slope of

the tangent line at $(0,2)$ is 0. The equation of

the horizontal line passing through $(0,2)$ is

$y = 2$.

When $x = -2$, $\dfrac{dy}{dx} = \dfrac{-16(-2)}{\left((-2)^2 + 4\right)^2} = \dfrac{32}{64} = \dfrac{1}{2}$, so

the slope of the tangent line at $(-2,1)$ is $\dfrac{1}{2}$.

Using the point-slope equation, we have:

$y - y_1 = m\left(x - x_1\right)$

$y - 1 = \dfrac{1}{2}\left(x - (-2)\right)$

$y - 1 = \dfrac{1}{2}x + 1$

$y = \dfrac{1}{2}x + 2$

99. $y = x^2 + \dfrac{3}{x - 1}$

$\dfrac{dy}{dx} = 2x + \dfrac{(x - 1)(0) - 3(1)}{(x - 1)^2}$

$= 2x - \dfrac{3}{(x - 1)^2}$

When $x = 2$, $y = (2)^2 + \dfrac{3}{2 - 1} = 4 + 3 = 7$, and

$\dfrac{dy}{dx} = 2(2) - \dfrac{3}{(2 - 1)^2} = 4 - 3 = 1$.

Therefore, the slope of the tangent line at $(2,7)$ is 1.

Using the point-slope equation, we have:

$y - y_1 = m\left(x - x_1\right)$

$y - 7 = 1(x - 2)$

$y - 7 = x - 2$

$y = x + 5$

When $x = 3$, $y = (3)^2 + \dfrac{3}{3-1} = 9 + \dfrac{3}{2} = \dfrac{21}{2}$, and

$$\dfrac{dy}{dx} = 2(3) - \dfrac{3}{(3-1)^2} = 6 - \dfrac{3}{4} = \dfrac{21}{4}.$$

Therefore, the slope of the tangent line at

$\left(3, \dfrac{21}{2}\right)$ is $\dfrac{21}{4}$.

Using the point-slope equation, we have:

$$y - y_1 = m(x - x_1)$$

$$y - \dfrac{21}{2} = \dfrac{21}{4}(x - 3)$$

$$y - \dfrac{21}{2} = \dfrac{21}{4}x - \dfrac{63}{4}$$

$$y = \dfrac{21}{4}x - \dfrac{21}{4}$$

101. The average cost of producing x items

is $A_C(x) = \dfrac{C(x)}{x}$. Therefore,

$$A_C(x) = \dfrac{950 + 15\sqrt{x}}{x}.$$

Next, we take the derivative using the Quotient Rule to find the rate at which average cost is changing.

$$A_C{}'(x) = \dfrac{d}{dx}\left(\dfrac{950 + 15x^{\frac{1}{2}}}{x}\right)$$

$$A_C{}'(x) = \dfrac{x\left(\dfrac{15}{2}x^{-\frac{1}{2}}\right) - \left(950 + 15x^{\frac{1}{2}}\right)(1)}{(x)^2}$$

$$= \dfrac{\dfrac{15}{2}x^{\frac{1}{2}} - 950 - 15x^{\frac{1}{2}}}{x^2}$$

$$= \dfrac{-\dfrac{15}{2}x^{\frac{1}{2}} - 950}{x^2}$$

$$= \dfrac{-\dfrac{15}{2}\sqrt{x} - 950}{x^2}$$

Substituting 400 for x, we have

$$A_C{}'(400) = \dfrac{-\dfrac{15}{2}\sqrt{400} - 950}{(400)^2}$$

$$= \dfrac{-150 - 950}{160,000}$$

$$\approx -0.006875$$

Therefore, when 400 jackets have been produced, average cost is changing at a rate of -0.0069 dollars per jacket.

103. The average revenue of producing x items

is $A_R(x) = \dfrac{R(x)}{x}$. Therefore,

$$A_R(x) = \dfrac{85\sqrt{x}}{x} = \dfrac{85}{x^{\frac{1}{2}}}.$$

Next, we take the derivative using the Quotient Rule to find the rate at which average revenue is changing.

$$A_R{}'(x) = \dfrac{d}{dx}\left(\dfrac{85}{\sqrt{x}}\right)$$

$$= \dfrac{x^{\frac{1}{2}}(0) - (85)\left(\dfrac{1}{2}x^{-\frac{1}{2}}\right)}{\left(x^{\frac{1}{2}}\right)^2}$$

$$= \dfrac{-\dfrac{85}{2}x^{-\frac{1}{2}}}{x}$$

$$= -\dfrac{85}{2x^{\frac{3}{2}}}$$

Substituting 400 for x, we have

$$A_R{}'(400) = -\dfrac{85}{2(400)^{\frac{3}{2}}}$$

$$= -\dfrac{85}{16000}$$

$$= -0.00531250$$

Therefore, when 400 jackets have been produced, average revenue is changing at a rate of -0.0053 dollars per jacket.

105. $A_p(x) = \dfrac{P(x)}{x} = \dfrac{R(x) - C(x)}{x}$

From Exercises 101 and 103, we know that

$$A_p(x) = \frac{85x^{\frac{1}{2}} - \left(950 + 15x^{\frac{1}{2}}\right)}{x} = \frac{70x^{\frac{1}{2}} - 950}{x}$$

Using the Quotient Rule to take the derivative, we have:

$$A_p'(x) = \frac{x\left(\dfrac{70}{2}x^{-\frac{1}{2}}\right) - \left(70x^{\frac{1}{2}} - 950\right)(1)}{(x)^2}$$

$$= \frac{35x^{\frac{1}{2}} - 70x^{\frac{1}{2}} + 950}{x^2}$$

$$= \frac{-35x^{\frac{1}{2}} + 950}{x^2}$$

Substituting 400 for x, we have:

$$A_p'(400) = \frac{-35(400)^{\frac{1}{2}} + 950}{(400)^2}$$

$$= \frac{-700 + 950}{16{,}000}$$

$$= \frac{250}{16{,}000}$$

$$\approx 0.015625$$

When 400 jackets have been produced and sold, the average profit is changing at a rate of 0.0156 dollars per jacket.

Alternatively, we could have used the information in Exercises 101 and 103 to find the rate of change of average profit when 400 jackets are produced and sold. Notice that

$$A_p'(x) = A_R'(x) - A_C'(x)$$

$$= -0.0053125 - (-0.006875)$$

$$= 0.00156250$$

107. The average profit of producing x items is

$A_p(x) = \dfrac{R(x) - C(x)}{x}$. Therefore,

$$A_p(x) = \frac{65x^{0.9} - \left(4300 + 2.1x^{0.6}\right)}{x}$$

$$= \frac{65x^{0.9} - 2.1x^{0.6} - 4300}{x}.$$

Using the Quotient Rule to take the derivative, we have

$A_p'(x)$

$$= \frac{x\left(65\left(0.9x^{-0.1}\right) - 2.1\left(0.6x^{-0.4}\right)\right) - \left(65x^{0.9} - 2.1x^{0.6} - 4300\right)(1)}{(x)^2}$$

$$= \frac{58.5x^{0.9} - 1.26x^{0.6} - \left(65x^{0.9} - 2.1x^{0.6} - 4300\right)}{x^2}$$

$$= \frac{-6.5x^{0.9} + 0.84x^{0.6} + 4300}{x^2}$$

Substituting 50 for x, we have

$$A_p'(50) = \frac{-6.5(50)^{0.9} + 0.84(50)^{0.6} + 4300}{(50)^2}$$

$$= \frac{4089.00428745}{2500}$$

$$= 1.63560171$$

$$\approx 1.64$$

Therefore, when 50 vases have been produced and sold, the average profit is changing at rate of 1.64 dollars per vase.

109. $P(t) = 567 + t\left(36t^{0.6} - 104\right)$

a) Using the Product Rule and remembering that the derivative of a constant is 0, we have

$$P'(t) = 0 + t\left(36\left(0.6t^{-0.4}\right)\right) + \left(36t^{0.6} - 104\right)(1)$$

$$= 21.6t^{0.6} + 36t^{0.6} - 104$$

$$= 57.6t^{0.6} - 104$$

b) Substituting 45 for t, we have:

$$P'(45) = 57.6(45)^{0.6} - 104$$

$$= 565.39243291 - 104$$

$$= 461.39243291$$

c) $\boxed{tw}$

111. $T(t) = \dfrac{4t}{t^2 + 1} + 98.6$

a) $T'(t) = \dfrac{\left(t^2 + 1\right)(4) - (4t)(2t)}{\left(t^2 + 1\right)^2} + 0$

$$= \frac{4t^2 + 4 - 8t^2}{\left(t^2 + 1\right)^2}$$

$$= \frac{-4t^2 + 4}{\left(t^2 + 1\right)^2}$$

b) $T(2) = \dfrac{4(2)}{(2)^2 + 1} + 98.6 = \dfrac{8}{5} + 98.6 = 100.2$

After 2 hours, the temperature of the ill person is approximately 100.2 degrees Fahrenheit.

c) $T'(2) = \dfrac{-4(2)^2 + 4}{\left((2)^2 + 1\right)^2} = \dfrac{-12}{25} = -0.48$

After 2 hours, the person's temperature is changing at rate of -0.48 degrees per hour.

113. $y(t) = 5t(t-1)(2t+3)$

First, group the factors of $y(t)$ in order to apply the product rule.

$$y(t) = \left[5t(t-1)\right] \cdot (2t+3)$$

Now we can take the derivative. Notice that when we take the derivative of the first term, $\left[5t(t-1)\right]$ we will have to apply the Product Rule again.

$$y'(t) = \left[5t(t-1)\right](2) + (2t+3)\underbrace{\left[(5t)(1) + (t-1)(5)\right]}_{\text{Product Rule for } \left[5t(t-1)\right]}$$

$$= 10t(t-1) + (2t+3)\left[5t + 5t - 5\right]$$

$$= 10t(t-1) + (2t+3)(10t - 5)$$

The previous derivative can be simplified as follows:

$$y'(t) = 10t^2 - 10t + 20t^2 + 30t - 10t - 15$$

$$= 30t^2 + 10t - 15$$

115. $g(x) = (x^3 - 8) \cdot \dfrac{x^2 + 1}{x^2 - 1}$

We will begin by applying the Product Rule.

$$g'(x) = (x^3 - 8)\dfrac{d}{dx}\left(\dfrac{x^2+1}{x^2-1}\right) + \dfrac{x^2+1}{x^2-1} \cdot \dfrac{d}{dx}(x^3 - 8)$$

Notice, that we will have to apply the Quotient Rule to take the derivative of $\dfrac{x^2+1}{x^2-1}$.

$$g'(x) = (x^3 - 8)\dfrac{(x^2-1)(2x) - (x^2+1)(2x)}{(x^2-1)^2} +$$

$$\left(\dfrac{x^2+1}{x^2-1}\right) \cdot (3x^2)$$

$$= (x^3 - 8)\dfrac{-4x}{(x^2-1)^2} + \left(\dfrac{x^2+1}{x^2-1}\right) \cdot (3x^2)$$

The derivative can be simplified as follows.

$$g'(x) = \dfrac{-4x(x^3 - 8)}{(x^2-1)^2} + \dfrac{3x^2(x^2+1)}{x^2-1}$$

$$= \dfrac{-4x^4 + 32x}{(x^2-1)^2} + \dfrac{3x^4 + 3x^2}{x^2-1} \cdot \dfrac{x^2-1}{x^2-1}$$

$$= \dfrac{-4x^4 + 32x + 3x^6 - 3x^2}{(x^2-1)^2}$$

$$= \dfrac{3x^6 - 4x^4 - 3x^2 + 32x}{(x^2-1)^2}$$

117. $f(x) = \dfrac{(x-1)(x^2+x+1)}{x^4 - 3x^3 - 5}$

First we will group the numerator, to apply the Quotient Rule. Remember that we will have to apply the Product Rule when taking the derivative of the numerator.

$$f(x) = \dfrac{\left[(x-1)(x^2+x+1)\right]}{x^4 - 3x^3 - 5}$$

$f'(x)$

$$= \dfrac{(x^4 - 3x^3 - 5)\left[(x-1)(2x+1) + (x^2+x+1)(1)\right]}{(x^4 - 3x^3 - 5)^2} -$$

$$\dfrac{\left[(x-1)(x^2+x+1)\right](4x^3 - 9x^2)}{(x^4 - 3x^3 - 5)^2}$$

$$= \dfrac{(x^4 - 3x^3 - 5)\left[2x^2 - x - 1 + x^2 + x + 1\right]}{(x^4 - 3x^3 - 5)^2} -$$

$$\dfrac{\left[x^3 - 1\right](4x^3 - 9x^2)}{(x^4 - 3x^3 - 5)^2}$$

$$= \dfrac{(x^4 - 3x^3 - 5)\left[3x^2\right] - \left[x^3 - 1\right](4x^3 - 9x^2)}{(x^4 - 3x^3 - 5)^2}$$

$$= \dfrac{3x^6 - 9x^5 - 15x^2 - 4x^6 + 9x^5 + 4x^3 - 9x^2}{(x^4 - 3x^3 - 5)^2}$$

$$= \dfrac{-x^6 + 4x^3 - 24x^2}{(x^4 - 3x^3 - 5)^2}$$

119. $f(x) = \dfrac{x^2}{x^2-1}$ and $g(x) = \dfrac{1}{x^2-1}$

a) $f'(x) = \dfrac{(x^2-1)(2x) - x^2(2x)}{(x^2-1)^2} = \dfrac{-2x}{(x^2-1)^2}$

b) $g'(x) = \dfrac{(x^2-1)(0) - 1(2x)}{(x^2-1)^2} = \dfrac{-2x}{(x^2-1)^2}$

c) $\boxed{tw}$

121. $\boxed{tw}$

123. a) Definition of the derivative.
   b) Adding and subtracting the same quantity is the same as adding 0.
   c) The limit of a sum is the sum of the limits.
   d) Factoring common factors.
   e) The limit of a product is the product of the limits and $\lim\limits_{h \to 0} f(x+h) = f(x)$.
   f) Definition of the derivative.
   g) Using Leibniz's notation.

125. The break even point occurs when $P(x) = 0$.

$$P(x) = R(x) - C(x)$$
$$= 85x^{1/2} - \left(950 + 15x^{1/2}\right)$$
$$= 70x^{1/2} - 950$$

Using the window:

```
WINDOW
 Xmin=0
 Xmax=500
 Xscl=100
 Ymin=-500
 Ymax=500
 Yscl=100
 Xres=1
```

We graph the profit function on the calculator.

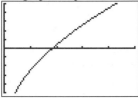

The break even point will be the zero of the function. Using the calculator we have:

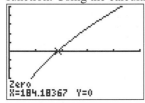

We see that the break-even point occurs at $x = 184$ jackets.
The profit is changing at rate of

$$P'(x) = 70\left(\frac{1}{2}x^{-1/2}\right) = \frac{70}{2x^{1/2}} = \frac{70}{2\sqrt{x}}$$

Substituting 184 for x we have:

$$P'(184) = \frac{70}{2\sqrt{184}}$$
$$= 2.580234233$$
$$\approx 2.58$$

Therefore, at the break-even point, profit is increasing at a rate of 2.58 dollars per jacket.

From Exercise 105 we know that:

$$A_P'(x) = \frac{-35x^{1/2} + 950}{x^2}$$

Substituting 184 for x we get:

$$A_P'(184) = \frac{-35(184)^{1/2} + 950}{(184)^2}$$
$$\approx 0.014037$$
$$\approx 0.014$$

At the break-even point, average profit is changing at a rate of 0.014 dollars per jacket.

127. $f(x) = \left(x + \dfrac{2}{x}\right)\left(x^2 - 3\right)$

Using the calculator, we graph the function and the derivative in the same window. We can use the nDeriv feature to graph the derivative without actually calculating the derivative.

```
Plot1 Plot2 Plot3
\Y1◻(X+2/X)(X^2-
3)
\Y2◻nDeriv(Y1,X,
X)
\Y3=
\Y4=
\Y5=
```

Using the window:

```
WINDOW
 Xmin=-4
 Xmax=4
 Xscl=1
 Ymin=-20
 Ymax=20
 Yscl=5
 Xres=1
```

The graph is shown at the top of the next page.

We get the graph:

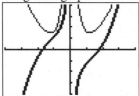

Note, the function $f(x)$ is the thicker graph.

The horizontal tangents occur at the turning points of this function, or at the x-intercepts of the derivative. We can see that the derivative never intersects the x-axis, therefore, there are no points at which the tangent line is horizontal.

129. $f(x) = \dfrac{0.3x}{0.04 + x^2}$

Using the calculator, we graph the function and the derivative in the same window. We can use the nDeriv feature to graph the derivative without actually calculating the derivative.

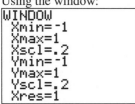

Using the window:

```
WINDOW
 Xmin=-1
 Xmax=1
 Xscl=.2
 Ymin=-1
 Ymax=1
 Yscl=.2
 Xres=1
```

We get the graph:

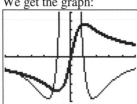

Note, the function $f(x)$ is the thicker graph.

The horizontal tangents occur at the turning points of this function, or at the x-intercepts of the derivative. Using the trace feature, the minimum/maximum feature on the function, or the zero feature on the derivative on the calculator, we find the points of horizontal tangency.

We estimate the points at which the tangent lines are horizontal

are $(-0.2, -0.75)$ and $(0.2, 0.75)$.

131. $f(x) = \dfrac{4x}{x^2 + 1}$

Using the calculator, we graph the function and the derivative in the same window. We can use the nDeriv feature to graph the derivative without actually calculating the derivative.

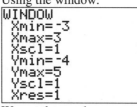

Using the window:

```
WINDOW
 Xmin=-3
 Xmax=3
 Xscl=1
 Ymin=-4
 Ymax=5
 Yscl=1
 Xres=1
```

We get the graph:

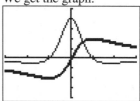

Note, the function $f(x)$ is the thicker graph.

The horizontal tangents occur at the turning points of this function, or at the x-intercepts of the derivative. Using the trace feature, the minimum/maximum feature on the function, or the zero feature on the derivative on the calculator, we find the points of horizontal tangency.

We estimate the points at which the tangent lines are horizontal are $(-1, -2)$ and $(1, 2)$.

Exercise Set 1.7

1. $y = (2x+1)^2$

Using the Extended Power Rule:

$$\frac{dy}{dx} = \frac{d}{dx}\left[(2x+1)^2\right]$$

$$= 2(2x+1)^{2-1} \cdot \frac{d}{dx}(2x+1)$$

$$= 2(2x+1)(2)$$

$$= 8x+4$$

Simplifying the function first, we have:

$$y = (2x+1)^2$$

$$= (2x+1)(2x+1)$$

$$= 4x^2 + 4x + 1$$

Now we take the derivative using the Power Rule.

$$\frac{dy}{dx} = \frac{d}{dx}(4x^2 + 4x + 1)$$

$$= \frac{d}{dx}(4x^2) + \frac{d}{dx}(4x) + \frac{d}{dx}(1)$$

$$= 8x+4$$

The results are the same.

3. $y = (7-x)^{55}$

Using the Extended Power Rule:

$$\frac{dy}{dx} = \frac{d}{dx}\left[(7-x)^{55}\right]$$

$$= 55(7-x)^{55-1} \cdot \frac{d}{dx}(7-x)$$

$$\frac{dy}{dx} = 55(7-x)^{54}(-1)$$

$$= -55(7-x)^{54}$$

5. $y = \sqrt{1+8x} = (1+8x)^{\frac{1}{2}}$

Using the Extended Power Rule

$$\frac{dy}{dx} = \frac{d}{dx}\left[(1+8x)^{\frac{1}{2}}\right]$$

$$= \frac{1}{2}(1+8x)^{-\frac{1}{2}}\frac{d}{dx}(1+8x)$$

$$= \frac{1}{2(1+8x)^{\frac{1}{2}}} \cdot (8)$$

$$= \frac{4}{\sqrt{1+8x}}$$

7. $y = \sqrt{3x^2-4} = (3x^2-4)^{\frac{1}{2}}$

Using the Extended Power Rule

$$\frac{dy}{dx} = \frac{d}{dx}\left[(3x^2-4)^{\frac{1}{2}}\right]$$

$$= \frac{1}{2}(3x^2-4)^{-\frac{1}{2}}\frac{d}{dx}(3x^2-4)$$

$$= \frac{1}{2(3x^2-4)^{\frac{1}{2}}} \cdot (6x)$$

$$= \frac{3x}{\sqrt{3x^2-4}}$$

9. $y = (8x^2-6)^{-40}$

Using the Extended Power Rule

$$\frac{dy}{dx} = \frac{d}{dx}\left[(8x^2-6)^{-40}\right]$$

$$= -40(8x^2-6)^{-40-1}\frac{d}{dx}(8x^2-6)$$

$$= -40(8x^2-6)^{-41} \cdot (16x)$$

$$= -640x(8x^2-6)^{-41}$$

$$= \frac{-640x}{(8x^2-6)^{41}}$$

11. $y = (x-4)^8(2x+3)^6$

Using the Product Rule, we have

$$\frac{dy}{dx} = \frac{d}{dx}\left[(x-4)^8(2x+3)^6\right]$$

$$= (x-4)^8\frac{d}{dx}(2x+3)^6 + (2x+3)^6\frac{d}{dx}(x-4)^8$$

Next, we will apply the Extended Power Rule.

$$\frac{dy}{dx} = (x-4)^8\left[6(2x+3)^{6-1}\frac{d}{dx}(2x+3)\right] +$$

$$(2x+3)^6\left[8(x-4)^{8-1}\frac{d}{dx}(x-4)\right]$$

$$= (x-4)^8\left[6(2x+3)^5(2)\right] +$$

$$(2x+3)^6\left[8(x-4)^7(1)\right]$$

$$= 12(x-4)^8(2x+3)^5 + 8(2x+3)^6(x-4)^7$$

Factoring out common factors, we have:

$$\frac{dy}{dx} = 4(x-4)^7(2x+3)^5\left[3(x-4)+2(2x+3)\right]$$

$$= 4(x-4)^7(2x+3)^5\left[3x-12+4x+6\right]$$

$$= 4(x-4)^7(2x+3)^5(7x-6)$$

13. $y = \dfrac{1}{(3x+8)^2} = (3x+8)^{-2}$

Using the Extended Power Rule

$\dfrac{dy}{dx} = \dfrac{d}{dx}\left[(3x+8)^{-2}\right]$

$\qquad = -2(3x+8)^{-2-1}\dfrac{d}{dx}(3x+8)$

$\qquad = -2(3x+8)^{-3}\cdot(3)$

$\qquad = -6(3x+8)^{-3}$

$\qquad = \dfrac{-6}{(3x+8)^3}$

15. $y = \dfrac{4x^2}{(7-5x)^3}$

First, we use the Quotient Rule.

$\dfrac{dy}{dx} = \dfrac{d}{dx}\left[\dfrac{4x^2}{(7-5x)^3}\right]$

$\qquad = \dfrac{(7-5x)^3\,\dfrac{d}{dx}(4x^2) - 4x^2\,\dfrac{d}{dx}(7-5x)^3}{\left((7-5x)^3\right)^2}$

Next, using the Extended Power Rule, we have:

$\dfrac{dy}{dx} = \dfrac{(7-5x)^3(8x) - 4x^2\left[3(7-5x)^2(-5)\right]}{(7-5x)^6}$

$\qquad = \dfrac{8x(7-5x)^3 + 60x^2(7-5x)^2}{\left((7-5x)^3\right)^2}$

$\qquad = \dfrac{(7-5x)^2\left[8x(7-5x)+60x^2\right]}{(7-5x)^6}$ Factoring

$\qquad = \dfrac{56x - 40x^2 + 60x^2}{(7-5x)^4}$ Dividing common factors

$\qquad = \dfrac{20x^2 + 56x}{(7-5x)^4}$

$\qquad = \dfrac{4x(5x+14)}{(7-5x)^4}$

17. $f(x) = \left(1+x^3\right)^3 - \left(2+x^8\right)^4$

Using the Difference Rule and then the Extended Power Rule we have:

$f'(x) = \dfrac{d}{dx}\left[\left(1+x^3\right)^3 - \left(2+x^8\right)^4\right]$

$\qquad = \dfrac{d}{dx}\left(1+x^3\right)^3 - \dfrac{d}{dx}\left(2+x^8\right)^4$

$\qquad = 3\left(1+x^3\right)^{3-1}\left(\dfrac{d}{dx}\left(1+x^3\right)\right) -$

$\qquad\qquad 4\left(2+x^8\right)^{4-1}\left(\dfrac{d}{dx}\left(2+x^8\right)\right)$

$\qquad = 3\left(1+x^3\right)^2\left(3x^2\right) - 4\left(2+x^8\right)^3\left(8x^7\right)$

$\qquad = 9x^2\left(1+x^3\right)^2 - 32x^7\left(2+x^8\right)^3$

19. $f(x) = x^2 + (200-x)^2$

Using the Sum Rule and the Extended Power Rule, we have:

$f'(x) = \dfrac{d}{dx}\left[x^2 + (200-x)^2\right]$

$\qquad = \dfrac{d}{dx}\left(x^2\right) + \dfrac{d}{dx}(200-x)^2$

$\qquad = 2x + 2(200-x)^{2-1}\left[\dfrac{d}{dx}(200-x)\right]$

$\qquad = 2x + 2(200-x)(-1)$

$\qquad = 2x + 2x - 400$

$\qquad = 4x - 400$

21. $g(x) = \sqrt{x} + (x-3)^3 = x^{1/2} + (x-3)^3$

Using the Sum Rule and the Extended Power Rule, we have:

$g'(x) = \dfrac{d}{dx}\left[x^{1/2} + (x-3)^3\right]$

$\qquad = \dfrac{d}{dx}\left(x^{1/2}\right) + \dfrac{d}{dx}(x-3)^3$

$\qquad = \dfrac{1}{2}x^{1/2-1} + 3(x-3)^{3-1}\left[\dfrac{d}{dx}(x-3)\right]$

$\qquad = \dfrac{1}{2}x^{-1/2} + 3(x-3)^2(1)$

$\qquad = \dfrac{1}{2x^{1/2}} + 3(x-3)^2$

$\qquad = \dfrac{1}{2\sqrt{x}} + 3(x-3)^2$

23. $f(x) = -5x(2x-3)^4$

Using the Product Rule, we have

$$f'(x) = \frac{d}{dx}\left[-5x(2x-3)^4\right]$$

$$= -5x\frac{d}{dx}\left[(2x-3)^4\right] + (2x-3)^4\frac{d}{dx}(-5x)$$

Using the Extended Power Rule, we have

$$f'(x) = -5x\left[4(2x-3)^3\left(\frac{d}{dx}(2x-3)\right)\right] +$$

$$(2x-3)^4(-5)$$

$$= -5x\left[4(2x-3)^3(2)\right] + (2x-3)^4(-5)$$

$$= -40x(2x-3)^3 - 5(2x-3)^4$$

$$= -5(2x-3)^3\left[8x + (2x-3)\right] \quad \text{Factoring}$$

$$= -5(2x-3)^3(10x-3)$$

25. $g(x) = (3x-1)^7(2x+1)^5$

Using the Product Rule and the Extended Power Rule, we have

$$g'(x) = \frac{d}{dx}\left[(3x-1)^7(2x+1)^5\right]$$

$$= (3x-1)^7\frac{d}{dx}(2x+1)^5 + (2x+1)^5\frac{d}{dx}(3x-1)^7$$

$$= (3x-1)^7\left[5(2x+1)^4(2)\right] +$$

$$(2x+1)^5\left[7(3x-1)^6(3)\right]$$

$$= 10(3x-1)^7(2x+1)^4 + 21(2x+1)^5(3x-1)^6$$

$$= (3x-1)^6(2x+1)^4\left[10(3x-1) + 21(2x+1)\right]$$

$$= (3x-1)^6(2x+1)^4\left[30x-10+42x+21\right]$$

$$= (3x-1)^6(2x+1)^4(72x+11)$$

27. $f(x) = x^2\sqrt{4x-1} = x^2(4x-1)^{1/2}$

Using the Product Rule and the Extended Power Rule, we have

$$f'(x) = \frac{d}{dx}\left[x^2(4x-1)^{1/2}\right]$$

$$= x^2\left[\frac{1}{2}(4x-1)^{-1/2}(4)\right] + (4x-1)^{1/2}(2x)$$

$$f'(x) = \frac{2x^2}{(4x-1)^{1/2}} + 2x(4x-1)^{1/2}$$

$$= \frac{2x^2}{\sqrt{(4x-1)}} + 2x\sqrt{(4x-1)}$$

The derivative can be simplified as follows:

$$f'(x) = \frac{2x^2}{\sqrt{4x-1}} + \frac{2x\sqrt{4x-1}}{1}\cdot\frac{\sqrt{4x-1}}{\sqrt{4x-1}}$$

$$= \frac{2x^2}{\sqrt{4x-1}} + \frac{2x(4x-1)}{\sqrt{4x-1}}$$

$$= \frac{2x^2}{\sqrt{4x-1}} + \frac{8x^2-2x}{\sqrt{4x-1}}$$

$$= \frac{10x^2-2x}{\sqrt{4x-1}}$$

$$= \frac{2x(5x-1)}{\sqrt{4x-1}}$$

29. $G(x) = \sqrt[3]{x^5+6x} = (x^5+6x)^{1/3}$

Using the Extended Power Rule, we have

$$G'(x) = \frac{d}{dx}\left[(x^5+6x)^{1/3}\right]$$

$$= \frac{1}{3}(x^5+6x)^{1/3-1}\frac{d}{dx}(x^5+6x)$$

$$= \frac{1}{3}(x^5+6x)^{-2/3}(5x^4+6)$$

$$= \frac{5x^4+6}{3(x^5+6x)^{2/3}}$$

$$= \frac{5x^4+6}{3\cdot\sqrt[3]{(x^5+6x)^2}}$$

31. $f(x) = \left(\dfrac{3x-1}{5x+2}\right)^4$

Using the Extended Power Rule, we have

$$f'(x) = 4\left(\frac{3x-1}{5x+2}\right)^3\frac{d}{dx}\left[\frac{3x-1}{5x+2}\right]$$

Using the Quotient Rule, we have

$$f'(x) = 4\left(\frac{3x-1}{5x+2}\right)^3\left[\frac{(5x+2)(3)-(3x-1)(5)}{(5x+2)^2}\right]$$

$$= 4\left(\frac{3x-1}{5x+2}\right)^3\left[\frac{15x+6-15x+5}{(5x+2)^2}\right]$$

$$= 4\left(\frac{3x-1}{5x+2}\right)^3\left[\frac{11}{(5x+2)^2}\right]$$

$$= \frac{44(3x-1)^3}{(5x+2)^5}$$

33. $g(x) = \sqrt{\dfrac{4-x}{3+x}} = \left(\dfrac{4-x}{3+x}\right)^{\frac{1}{2}}$

Using the Extended Power Rule, we have

$g'(x) = \dfrac{1}{2}\left(\dfrac{4-x}{3+x}\right)^{\frac{1}{2}-1} \dfrac{d}{dx}\left[\dfrac{4-x}{3+x}\right]$

Using the Quotient Rule, we have

$g'(x) = \dfrac{1}{2}\left(\dfrac{4-x}{3+x}\right)^{-\frac{1}{2}}\left[\dfrac{(3+x)(-1)-(4-x)(1)}{(3+x)^2}\right]$

$= \dfrac{1}{2}\left(\dfrac{4-x}{3+x}\right)^{-\frac{1}{2}}\left[\dfrac{-3-x-4+x}{(3+x)^2}\right]$

$= \dfrac{1}{2}\left(\dfrac{3+x}{4-x}\right)^{\frac{1}{2}}\left[\dfrac{-7}{(3+x)^2}\right]$

$= \dfrac{-7}{2(3+x)^{\frac{3}{2}}(4-x)^{\frac{1}{2}}}$

$= \dfrac{-7}{2\sqrt{(3+x)^3}\cdot\sqrt{4-x}}$

35. $f(x) = \left(2x^3 - 3x^2 + 4x + 1\right)^{100}$

Using the Extended Power Rule, we have

$f'(x) = 100\left(2x^3 - 3x^2 + 4x + 1\right)^{99}\left(6x^2 - 6x + 4\right)$

$= 200\left(3x^2 - 3x + 2\right)\left(2x^3 - 3x^2 + 4x + 1\right)^{99}$

37. $g(x) = \left(\dfrac{2x+3}{5x-1}\right)^{-4} = \left(\dfrac{5x-1}{2x+3}\right)^{4}$

Using the Extended Power Rule, we have

$g'(x) = \dfrac{d}{dx}\left[\left(\dfrac{5x-1}{2x+3}\right)^4\right]$

$= 4\left(\dfrac{5x-1}{2x+3}\right)^{4-1}\left[\dfrac{d}{dx}\left(\dfrac{5x-1}{2x+3}\right)\right]$

Next, using the Quotient Rule, we have

$g'(x) = 4\left(\dfrac{5x-1}{2x+3}\right)^3\left[\dfrac{(2x+3)(5)-(5x-1)(2)}{(2x+3)^2}\right]$

$= 4\left(\dfrac{5x-1}{2x+3}\right)^3\left[\dfrac{10x+15-10x+2}{(2x+3)^2}\right]$

$= 4\left(\dfrac{5x-1}{2x+3}\right)^3\left[\dfrac{17}{(2x+3)^2}\right]$

$= \dfrac{68(5x-1)^3}{(2x+3)^5}$

39. $f(x) = \sqrt{\dfrac{x^2+x}{x^2-x}} = \left(\dfrac{x^2+x}{x^2-x}\right)^{\frac{1}{2}}$

Using the Extended Power Rule, we have

$f'(x) = \dfrac{1}{2}\left(\dfrac{x^2+x}{x^2-x}\right)^{\frac{1}{2}-1}\dfrac{d}{dx}\left[\dfrac{x^2+x}{x^2-x}\right]$

Using the Quotient Rule, we have

$f'(x)$

$= \dfrac{1}{2}\left(\dfrac{x^2+x}{x^2-x}\right)^{-\frac{1}{2}}\left[\dfrac{(x^2-x)(2x+1)-(x^2+x)(2x-1)}{(x^2-x)^2}\right]$

$= \dfrac{1}{2}\left(\dfrac{x^2+x}{x^2-x}\right)^{-\frac{1}{2}}\left[\dfrac{2x^3-x^2-x-2x^3-x^2+x}{(x^2-x)^2}\right]$

$= \dfrac{1}{2}\left(\dfrac{x^2-x}{x^2+x}\right)^{\frac{1}{2}}\left(\dfrac{-2x^2}{(x^2-x)^2}\right)$

$= \dfrac{-x^2}{(x^2-x)^{\frac{3}{2}}(x^2+x)^{\frac{1}{2}}}$

The previous derivative can be simplified as follows:

$f'(x) = \dfrac{-x^2}{(x^2-x)^{\frac{3}{2}}(x^2+x)^{\frac{1}{2}}}$

$= \dfrac{-x^2}{x^{\frac{3}{2}}(x-1)^{\frac{3}{2}}x^{\frac{1}{2}}(x+1)^{\frac{1}{2}}}$　　Factoring

$= \dfrac{-x^2}{x^2(x-1)^{\frac{3}{2}}(x+1)^{\frac{1}{2}}}$

$= \dfrac{-1}{(x-1)^{\frac{3}{2}}(x+1)^{\frac{1}{2}}}$

41. $f(x) = \dfrac{(2x+3)^4}{(3x-2)^5}$

Using the Quotient Rule and the Extended Power Rule, we have:

$f'(x) = \dfrac{d}{dx}\left[\dfrac{(2x+3)^4}{(3x-2)^5}\right]$

$= \dfrac{(3x-2)^5\left[4(2x+3)^3(2)\right] - (2x+3)^4\left[5(3x-2)^4(3)\right]}{\left((3x-2)^5\right)^2}$

$= \dfrac{8(2x+3)^3(3x-2)^5 - 15(2x+3)^4(3x-2)^4}{(3x-2)^{10}}$

$= \dfrac{(2x+3)^3(3x-2)^4\left[8(3x-2) - 15(2x+3)\right]}{(3x-2)^{10}}$

$= \dfrac{(2x+3)^3\left[24x - 16 - 30x - 45\right]}{(3x-2)^6}$

$= \dfrac{(2x+3)^3\left[-6x - 61\right]}{(3x-2)^6}$

$= \dfrac{-(2x+3)^3\left[6x + 61\right]}{(3x-2)^6}$

$f'(x) = \dfrac{-(2x+3)^3\left[6x + 61\right]}{(3x-2)^6}$

43. $f(x) = 12(2x+1)^{\frac{2}{3}}(3x-4)^{\frac{5}{4}}$

Using the Product Rule and the Extended Power Rule, we have:

$f'(x) = 12\dfrac{d}{dx}\left[(2x+1)^{\frac{2}{3}}(3x-4)^{\frac{5}{4}}\right]$

$= 12\left[(2x+1)^{\frac{2}{3}}\left[\dfrac{5}{4}(3x-4)^{\frac{1}{4}}(3)\right]\right] +$

$\qquad 12\left[(3x-4)^{\frac{5}{4}}\left[\dfrac{2}{3}(2x+1)^{-\frac{1}{3}}(2)\right]\right]$

$= 12\left[\dfrac{15}{4}(2x+1)^{\frac{2}{3}}(3x-4)^{\frac{1}{4}}\right] +$

$\qquad 12\left[\dfrac{4}{3}(3x-4)^{\frac{5}{4}}(2x+1)^{-\frac{1}{3}}\right]$

$= 45(2x+1)^{\frac{2}{3}}(3x-4)^{\frac{1}{4}} + \dfrac{16(3x-4)^{\frac{5}{4}}}{(2x+1)^{\frac{1}{3}}}$

Simplifying, we have

$f'(x)$

$= \dfrac{45(2x+1)^{\frac{2}{3}}(3x-4)^{\frac{1}{4}}}{1} \cdot \dfrac{(2x+1)^{\frac{1}{3}}}{(2x+1)^{\frac{1}{3}}} + \dfrac{16(3x-4)^{\frac{5}{4}}}{(2x+1)^{\frac{1}{3}}}$

$= \dfrac{45(2x+1)(3x-4)^{\frac{1}{4}}}{(2x+1)^{\frac{1}{3}}} + \dfrac{16(3x-4)^{\frac{5}{4}}}{(2x+1)^{\frac{1}{3}}}$

$= \dfrac{45(2x+1)(3x-4)^{\frac{1}{4}} + 16(3x-4)^{\frac{5}{4}}}{(2x+1)^{\frac{1}{3}}}$

$= \dfrac{(3x-4)^{\frac{1}{4}}\left[45(2x+1) + 16(3x-4)^{\frac{4}{4}}\right]}{(2x+1)^{\frac{1}{3}}}$

$= \dfrac{(3x-4)^{\frac{1}{4}}\left[45(2x+1) + 16(3x-4)\right]}{(2x+1)^{\frac{1}{3}}}$

$= \dfrac{(3x-4)^{\frac{1}{4}}\left[90x + 45 + 48x - 64\right]}{(2x+1)^{\frac{1}{3}}}$

$= \dfrac{(3x-4)^{\frac{1}{4}}\left[138x - 19\right]}{(2x+1)^{\frac{1}{3}}}$

$f'(x) = \dfrac{(3x-4)^{\frac{1}{4}}\left[138x - 19\right]}{(2x+1)^{\frac{1}{3}}}$

45. $y = \sqrt{u} = u^{\frac{1}{2}}$ and $u = x^2 - 1$

$\dfrac{dy}{du} = \dfrac{1}{2}u^{\frac{1}{2}-1} = \dfrac{1}{2}u^{-\frac{1}{2}} = \dfrac{1}{2\sqrt{u}}$

$\dfrac{du}{dx} = 2x^{2-1} = 2x$

Applying the Chain Rule, we have:

$\dfrac{dy}{dx} = \dfrac{dy}{du} \cdot \dfrac{du}{dx}$

$\qquad = \dfrac{1}{2\sqrt{u}} \cdot 2x$

$\qquad = \dfrac{2x}{2\sqrt{x^2-1}}$ Substituting $x^2 - 1$ for u.

$\qquad = \dfrac{x}{\sqrt{x^2-1}}$ Simplifying

47. $y = u^{50}$ and $u = 4x^3 - 2x^2$

$\dfrac{dy}{du} = 50u^{50-1} = 50u^{49}$

$\dfrac{du}{dx} = 4\left(3x^{3-1}\right) - 2\left(2x^{2-1}\right) = 12x^2 - 4x$

$\dfrac{dy}{dx} = \dfrac{dy}{du} \cdot \dfrac{du}{dx}$

$\qquad = 50u^{49} \cdot \left(12x^2 - 4x\right)$

$\qquad$ Substituting $4x^3 - 2x^2$ for u.

$\qquad = 50\left(4x^3 - 2x^2\right)^{49} \cdot \left(12x^2 - 4x\right)$

$\qquad = 200x\left(3x-1\right)\left(4x^3 - 2x^2\right)^{49}$ Simplifying

49. $y = u(u+1)$ and $u = x^3 - 2x$

$\dfrac{dy}{du} = u(1) + (u+1)(1)$ Product Rule

$\qquad = 2u + 1$

$\dfrac{du}{dx} = 3x^2 - 2$

$\dfrac{dy}{dx} = \dfrac{dy}{du} \cdot \dfrac{du}{dx}$

$\qquad = (2u+1) \cdot \left(3x^2 - 2\right)$

$\qquad$ Substituting $x^3 - 2x$ for u.

$\qquad = \left(2\left(x^3 - 2x\right)+1\right) \cdot \left(3x^2 - 2\right)$

$\qquad = \left(2x^3 - 4x + 1\right) \cdot \left(3x^2 - 2\right)$ Simplifying

51. $y = 5u^2 + 3u$ and $u = x^3 + 1$

$\dfrac{dy}{du} = 10u + 3$

$\dfrac{du}{dx} = 3x^2$

Applying the Chain Rule, we have:

$\dfrac{dy}{dx} = \dfrac{dy}{du} \cdot \dfrac{du}{dx}$

$\qquad = (10u+3) \cdot \left(3x^2\right)$

$\qquad = \left(10\left(x^3+1\right)+3\right) \cdot \left(3x^2\right)$ Substituting for u.

$\qquad = 3x^2\left(\left(10x^3 + 10\right)+3\right)$

$\qquad = 3x^2\left(10x^3 + 13\right)$

53. $y = \sqrt[3]{2u+5} = (2u+5)^{1/3}$ and $u = x^2 - x$

$\dfrac{dy}{du} = \dfrac{1}{3}(2u+5)^{-2/3}(2)$ Extended Power Rule

$\qquad = \dfrac{2}{3(2u+5)^{2/3}}$

$\dfrac{du}{dx} = 2x - 1$

$\dfrac{dy}{dx} = \dfrac{dy}{du} \cdot \dfrac{du}{dx}$

$\qquad = \left(\dfrac{2}{3(2u+5)^{2/3}}\right) \cdot (2x-1)$

$\qquad = \left(\dfrac{2}{3\left(2\left(x^2-x\right)+5\right)^{2/3}}\right) \cdot (2x-1)$ Substituting

$\qquad = \dfrac{2(2x-1)}{3\left(2x^2 - 2x + 5\right)^{2/3}}$

55. $y = \dfrac{1}{u^2 + u}$ and $u = 5 + 3t$

$\dfrac{dy}{du} = \dfrac{\left(u^2+u\right)(0) - (1)(2u+1)}{\left(u^2+u\right)^2}$ Quotient Rule

$\qquad = \dfrac{-(2u+1)}{\left(u^2+u\right)^2}$

$\dfrac{du}{dt} = 3$

$\dfrac{dy}{dt} = \dfrac{dy}{du} \cdot \dfrac{du}{dt}$

$\qquad = \left(\dfrac{-(2u+1)}{\left(u^2+u\right)^2}\right) \cdot (3)$

$\qquad = \left(\dfrac{-(2(5+3t)+1)}{\left((5+3t)^2 + (5+3t)\right)^2}\right) \cdot (3)$ Substituting

$\qquad = \dfrac{-3(10+6t+1)}{\left((5+3t)^2 + (5+3t)\right)^2}$

$\qquad = \dfrac{-3(6t+11)}{(5+3t)^2\left((5+3t)+1\right)^2}$ Factoring

$\qquad = \dfrac{-3(6t+11)}{(5+3t)^2(6+3t)^2}$

57. $y = \sqrt{x^2 + 3x} = (x^2 + 3x)^{\frac{1}{2}}$

First, we find the derivative using the Extended Power Rule.

$\dfrac{dy}{dx} = \dfrac{1}{2}(x^2 + 3x)^{\frac{1}{2}-1}(2x + 3)$

$= \dfrac{2x + 3}{2\sqrt{x^2 + 3x}}$

When $x = 1$,

$\dfrac{dy}{dx} = \dfrac{2(1) + 3}{2\sqrt{(1)^2 + 3(1)}} = \dfrac{5}{2\sqrt{4}} = \dfrac{5}{2 \cdot 2} = \dfrac{5}{4}$

Thus, the slope of the tangent line at $(1, 2)$ is $\dfrac{5}{4}$.

Using the point-slope equation, we find the equation of the tangent line.

$y - y_1 = m(x - x_1)$

$y - 2 = \dfrac{5}{4}(x - 1)$

$y - 2 = \dfrac{5}{4}x - \dfrac{5}{4}$

$y = \dfrac{5}{4}x + \dfrac{3}{4}$

59. $y = x\sqrt{2x + 3} = x(2x + 3)^{\frac{1}{2}}$

First, we find the derivative using the Product Rule and the Extended Power Rule.

$\dfrac{dy}{dx} = x\left(\dfrac{1}{2}(2x + 3)^{\frac{1}{2}-1}(2)\right) + (2x + 3)^{\frac{1}{2}}(1)$

$= \dfrac{x}{\sqrt{2x + 3}} + \sqrt{2x + 3}$

When $x = 3$,

$\dfrac{dy}{dx} = \dfrac{(3)}{\sqrt{2(3) + 3}} + \sqrt{2(3) + 3}$

$= \dfrac{3}{\sqrt{9}} + \sqrt{9}$

$= \dfrac{3}{3} + 3$

$= 1 + 3$

$= 4$

Thus, the slope of the tangent line at $(3, 9)$ is 4.

Using the point-slope equation, we find the equation of the tangent line.

$y - y_1 = m(x - x_1)$

$y - 9 = 4(x - 3)$

$y - 9 = 4x - 12$

$y = 4x - 3$

61. $f(x) = \dfrac{x^2}{(1 + x)^5}$

a) Using the Quotient Rule and the Extended Power Rule, we have:

$f'(x) = \dfrac{(1 + x)^5 (2x) - x^2\left[5(1 + x)^4 (1)\right]}{\left((1 + x)^5\right)^2}$

$= \dfrac{2x(1 + x)^5 - 5x^2 (1 + x)^4}{(1 + x)^{10}}$

$= \dfrac{(1 + x)^4 \left(2x(1 + x) - 5x^2\right)}{(1 + x)^{10}}$ Factoring

$= \dfrac{2x + 2x^2 - 5x^2}{(1 + x)^6}$

$= \dfrac{2x - 3x^2}{(1 + x)^6}$

$= \dfrac{x(2 - 3x)}{(1 + x)^6}$

b) Using the Product Rule and the Extended Power Rule on $f(x) = x^2 (1 + x)^{-5}$, we have

$f'(x) = x^2 \left(-5(1 + x)^{-6} (1)\right) + (1 + x)^{-5} (2x)$

$= \dfrac{-5x^2}{(1 + x)^6} + \dfrac{2x}{(1 + x)^5}$

$= \dfrac{-5x^2}{(1 + x)^6} + \dfrac{2x}{(1 + x)^5} \cdot \dfrac{(1 + x)}{(1 + x)}$

$= \dfrac{-5x^2 + 2x + 2x^2}{(1 + x)^6}$

$= \dfrac{2x - 3x^2}{(1 + x)^6}$

$= \dfrac{x(2 - 3x)}{(1 + x)^6}$

c) The results are the same.

63. $h(x) = \left(3x^2 - 7\right)^5$

Let $f(x) = x^5$ and $g(x) = 3x^2 - 7$.

$(f \circ g)(x) = f(g(x)) = f(3x^2 - 7) = \left(3x^2 - 7\right)^5$

Thus, $h(x) = (f \circ g)(x)$.

Answers may vary.

65. $h(x) = \dfrac{x^3 + 1}{x^3 - 1}$

Let $f(x) = \dfrac{x+1}{x-1}$ and $g(x) = x^3$.

$(f \circ g)(x) = f(g(x)) = f(x^3) = \dfrac{x^3 + 1}{x^3 - 1}$

Thus, $h(x) = (f \circ g)(x)$.

Answers may vary.

67. Using the Chain Rule:

$f(u) = u^3, \; g(x) = u = 2x^4 + 1$

First find $f'(u)$ and $g'(x)$.

$f'(u) = 3u^2$

$f'(g(x)) = 3\left(2x^4 + 1\right)^2$ Substituting $g(x)$ for u.

$g'(x) = 8x^3$

The Chain Rule states

$(f \circ g)'(x) = f'(g(x)) \cdot g'(x)$

Substituting, we have:

$(f \circ g)'(x) = 3\left(2x^4 + 1\right)^2 \cdot (8x)$

$\qquad = 24x\left(2x^4 + 1\right)^2$

Therefore,

$(f \circ g)'(-1) = 24(-1)\left(2(-1)^4 + 1\right)^2$

$\qquad = -24\left(2 + 1\right)^2$

$\qquad = -24 \cdot 9$

$\qquad = -216$

Finding $f(g(x))$ first, we have:

$f(g(x)) = f(2x^4 + 1) = \left(2x^4 + 1\right)^3$

By the Extended Power Rule:

$f'(g(x)) = 3\left(2x^4 + 1\right)^2 (8x)$

$\qquad = 24x\left(2x^4 + 1\right)^2$

Therefore, $f'(g(-1)) = -216$ as above.

69. Using the Chain Rule:

$f(u) = \sqrt[3]{u} = u^{1/3}, \; g(x) = u = 1 + 3x^2$

First find $f'(u)$ and $g'(x)$.

$f'(u) = \dfrac{1}{3} u^{-2/3} = \dfrac{1}{3 \cdot \sqrt[3]{u^2}}$

$f'(g(x)) = \dfrac{1}{3 \cdot \sqrt[3]{\left(1 + 3x^2\right)^2}}$ Substituting $g(x)$ for u.

$g'(x) = 6x$

The Chain Rule states

$(f \circ g)'(x) = f'(g(x)) \cdot g'(x)$

Substituting, we have:

$(f \circ g)'(x) = \dfrac{1}{3 \cdot \sqrt[3]{\left(1 + 3x^2\right)^2}} \cdot (6x)$

$\qquad = \dfrac{2x}{\sqrt[3]{\left(1 + 3x^2\right)^2}}$

Therefore,

$(f \circ g)'(2) = \dfrac{2(2)}{\sqrt[3]{\left(1 + 3(2)^2\right)^2}}$

$\qquad = \dfrac{4}{\sqrt[3]{(13)^2}}$

$\qquad \approx 0.72348760$

Finding $f(g(x))$ first, we have:

$f(g(x)) = f(1 + 3x^2) = \left(1 + 3x^2\right)^{1/3}$

By the Extended Power Rule:

$f'(g(x)) = \dfrac{1}{3}\left(1 + 3x^2\right)^{-2/3} (6x)$

$\qquad = \dfrac{2x}{\left(1 + 3x^2\right)^{2/3}}$

Therefore $(f \circ g)'(2) = \dfrac{2}{\sqrt[3]{(13)^2}} \approx 0.72348760$

as above.

71. $R(x) = 1000\sqrt{x^2 - 0.1x} = 1000(x^2 - 0.1x)^{\frac{1}{2}}$

Using the Extended Power Rule, we have

$R'(x) = 1000\left[\frac{1}{2}(x^2 - 0.1x)^{-\frac{1}{2}}(2x - 0.1)\right]$

$= 500\left[\frac{2x - 0.1}{(x^2 - 0.1x)^{\frac{1}{2}}}\right]$

$= \frac{500(2x - 0.1)}{\sqrt{x^2 - 0.1x}}$

Substituting 20 for x, we have

$R'(20) = \frac{500(2(20) - 0.1)}{\sqrt{(20)^2 - 0.1(20)}}$

$= 1000.00314070$

≈ 1000

When 20 items have been sold, revenue is changing at a rate of 1,000 thousand of dollars per item, or 1,000,000 dollars per item.

73. $P(x) = R(x) - C(x)$ and

$P'(x) = R'(x) - C'(x)$

Since we are trying to find the rate at which total profit is changing as a function of x, we can use the derivatives found in Exercise 71 and 72 to find the derivative of the profit function. There is no need to find the Profit function first and then take the derivative.

$P'(x) = R'(x) - C'(x)$

$= \frac{500(2x - 0.1)}{\sqrt{x^2 - 0.1x}} - \frac{4000x}{3(x^2 + 2)^{\frac{2}{3}}}$

75. Let x be the number of years since 1995.

$C(x) = 0.21x^4 - 5.92x^3 + 50.53x^2 -$

$\qquad 18.92x + 1114.93$

a) $\dfrac{dC}{dx} = 0.21(4x^3) - 5.92(3x^2) +$

$\qquad 50.53(2x) - 18.92$

$= 0.84x^3 - 17.76x^2 + 101.06x - 18.92$

b) $\boxed{tw}$

c) Substitute 15 for x. $[2010 - 1995 = 15]$

$\dfrac{dC}{dx} = 0.84(15)^3 - 17.76(15)^2 +$

$\qquad 101.06(15) - 18.92$

$= 2835 - 3996 + 1515.9 - 18.92$

$= 335.98$

Outstanding consumer credit will be rising at a rate of 335.98 billion of dollars per year in 2010.

77. $A = 1000(1 + i)^3$

a) Using the Extended Power Rule, we have

$\dfrac{dA}{di} = 1000(3)(1 + i)^2(1)$

$= 3000(1 + i)^2$

b) $\boxed{tw}$

79. $D(p) = \dfrac{80,000}{p}$ and $p = 1.6t + 9$

a) Substitute $1.6t + 9$ in for p in the demand function.

$D(t) = \dfrac{80,000}{1.6t + 9}$

b) Using the Quotient Rule, we have

$D'(t) = \dfrac{(1.6t + 9)(0) - (80,000)(1.6)}{(1.6t + 9)^2}$

$= \dfrac{-128,000}{(1.6t + 9)^2}$

Therefore,

$D'(100) = \dfrac{-128,000}{(1.6(100) + 9)^2}$

$= \dfrac{-128,000}{28,561}$

≈ -4.48163580

After 100 days, quantity demanded is changing -4.482 units per day.

81. $D = 0.85A(c + 25)$ and $c = (140 - y)\dfrac{w}{72x}$

a) Substituting 5 for A we have:

$D(c) = 0.85(5)(c + 25)$

$= 4.25(c + 25)$

$= 4.25c + 106.25$

Substituting 0.6 for x, 45 for y, we have:

$$c(w) = (140 - 45)\frac{w}{72(0.6)}$$

$$= 95\frac{w}{43.2}$$

$$= \frac{95w}{43.2} \approx 2.199w$$

b) $\dfrac{dD}{dc} = \dfrac{d}{dc}(4.25c + 106.25) = 4.25$

The dosage changes at a rate of 4.25 mg per unit of creatine clearance.

c) $\dfrac{dc}{dw} = \dfrac{d}{dw}\left(\dfrac{95w}{43.2}\right) = \dfrac{95}{43.2} \approx 2.199$

The creatine clearance changes at a rate of 2.199 unit of creatine clearance per kilogram.

d) By the Chain Rule:

$$\frac{dD}{dw} = \frac{dD}{dc} \cdot \frac{dc}{dw} = (4.25)\left(\frac{95}{43.2}\right) \approx 9.346$$

The dosage changes at a rate of 9.35 milligrams per kilogram.

e) $\boxed{tw}$

83. $y = \sqrt[3]{x^3 + 6x + 1} \cdot x^5 = (x^3 + 6x + 1)^{\frac{1}{3}} \cdot x^5$

Using the Product Rule and the Extended Power Rule, we have

$$\frac{dy}{dx} = (x^3 + 6x + 1)^{\frac{1}{3}} \cdot (5x^4) +$$

$$x^5 \left[\frac{1}{3}(x^3 + 6x + 1)^{-\frac{2}{3}}(3x^2 + 6)\right]$$

$$= 5x^4(x^3 + 6x + 1)^{\frac{1}{3}} + \frac{3x^5(x^2 + 2)}{3(x^3 + 6x + 1)^{\frac{2}{3}}}$$

The derivative can be further simplified by finding a common denominator and combining the fractions.

$$\frac{dy}{dx} = \frac{(x^3 + 6x + 1)^{\frac{2}{3}}}{(x^3 + 6x + 1)^{\frac{2}{3}}} \cdot \frac{5x^4(x^3 + 6x + 1)^{\frac{1}{3}}}{1} +$$

$$\frac{x^5(x^2 + 2)}{(x^3 + 6x + 1)^{\frac{2}{3}}}$$

$$\frac{dy}{dx} = \frac{5x^4(x^3 + 6x + 1)}{(x^3 + 6x + 1)^{\frac{2}{3}}} + \frac{x^5(x^2 + 2)}{(x^3 + 6x + 1)^{\frac{2}{3}}}$$

$$= \frac{5x^7 + 30x^5 + 5x^4 + x^7 + 2x^5}{(x^3 + 6x + 1)^{\frac{2}{3}}}$$

$$= \frac{6x^7 + 32x^5 + 5x^4}{(x^3 + 6x + 1)^{\frac{2}{3}}}$$

85. $y = \left(\dfrac{x}{\sqrt{x-1}}\right)^3$

Using the Extended Power Rule and the Quotient Rule, we have:

$$\frac{dy}{dx} = 3\left(\frac{x}{(x-1)^{\frac{1}{2}}}\right)^{3-1} \frac{d}{dx}\left(\frac{x}{(x-1)^{\frac{1}{2}}}\right)$$

$$= 3\left(\frac{x}{\sqrt{x-1}}\right)^2 \cdot \frac{(x-1)^{\frac{1}{2}}(1) - x\left(\frac{1}{2}(x-1)^{-\frac{1}{2}} \cdot 1\right)}{\left(\sqrt{x-1}\right)^2}$$

$$= \frac{3x^2}{x-1}\left(\frac{(x-1)^{\frac{1}{2}} - \dfrac{x}{2(x-1)^{\frac{1}{2}}}}{x-1}\right)$$

$$= \frac{3x^2}{x-1}\left(\frac{\dfrac{2x-2}{2(x-1)^{\frac{1}{2}}} - \dfrac{x}{2(x-1)^{\frac{1}{2}}}}{x-1}\right)$$

$$= \frac{3x^2}{(x-1)^2}\left(\frac{x-2}{(x-1)^{\frac{1}{2}}}\right)$$

$$= \frac{3x^2(x-1)}{(x-1)^{\frac{5}{2}}}$$

87. $y = \dfrac{\sqrt{1-x^2}}{1-x} = \dfrac{\left(1-x^2\right)^{\frac{1}{2}}}{1-x}$

Using the Quotient Rule and the Extended Power Rule, we have:

$$\frac{dy}{dx} = \frac{\left(1-x\right)\left(\frac{1}{2}\left(1-x^2\right)^{-\frac{1}{2}}\left(-2x\right)\right)-\left(1-x^2\right)^{\frac{1}{2}}\left(-1\right)}{\left(1-x\right)^2}$$

$$\frac{dy}{dx} = \frac{\dfrac{x\left(x-1\right)}{\left(1-x^2\right)^{\frac{1}{2}}}+\left(1-x^2\right)^{\frac{1}{2}}}{\left(1-x\right)^2}$$

$$= \frac{\dfrac{x^2-x}{\left(1-x^2\right)^{\frac{1}{2}}}+\dfrac{\left(1-x^2\right)^{\frac{1}{2}}}{1}\cdot\dfrac{\left(1-x^2\right)^{\frac{1}{2}}}{\left(1-x^2\right)^{\frac{1}{2}}}}{\left(1-x\right)^2}$$

$$= \frac{\dfrac{x^2-x}{\left(1-x^2\right)^{\frac{1}{2}}}+\dfrac{\left(1-x^2\right)}{\left(1-x^2\right)^{\frac{1}{2}}}}{\left(1-x\right)^2}$$

$$= \frac{1-x}{\left(1-x^2\right)^{\frac{1}{2}}\left(1-x\right)^2}$$

$$= \frac{1}{\left(1-x^2\right)^{\frac{1}{2}}\left(1-x\right)}$$

89. $y = \left(\dfrac{x^2-x-1}{x^2+1}\right)^3$

Using the Extended Power Rule and the Quotient Rule, we have

$$\frac{dy}{dx}$$

$$= 3\left(\frac{x^2-x-1}{x^2+1}\right)^2\left[\frac{\left(x^2+1\right)\left(2x-1\right)-\left(x^2-x-1\right)\left(2x\right)}{\left(x^2+1\right)^2}\right]$$

$$= 3\left(\frac{x^2-x-1}{x^2+1}\right)^2\left[\frac{x^2+4x-1}{\left(x^2+1\right)^2}\right]$$

$$= \frac{3\left(x^2-x-1\right)^2\left(x^2+4x-1\right)}{\left(x^2+1\right)^4}$$

91. $f(t) = \sqrt{3t+\sqrt{t}} = \left(3t+t^{\frac{1}{2}}\right)^{\frac{1}{2}}$

Using the Extended Power Rule, we have

$$f'(t) = \frac{1}{2}\left(3t+t^{\frac{1}{2}}\right)^{-\frac{1}{2}}\left(3+\frac{1}{2}t^{-\frac{1}{2}}\right)$$

$$= \frac{3+\dfrac{1}{2\sqrt{t}}}{2\sqrt{3t+\sqrt{t}}}$$

$$= \frac{\dfrac{3}{1}\cdot\dfrac{2\sqrt{t}}{2\sqrt{t}}+\dfrac{1}{2\sqrt{t}}}{2\sqrt{3t+\sqrt{t}}}$$

Simplifying, we have:

$$f'(t) = \frac{\dfrac{6\sqrt{t}+1}{2\sqrt{t}}}{2\sqrt{3t+\sqrt{t}}}$$

$$= \frac{6\sqrt{t}+1}{4\sqrt{t}\sqrt{3t+\sqrt{t}}}$$

93. $\boxed{tw}$ Let $Q(x) = \dfrac{N(x)}{D(x)}$.

Then. $Q(x) = N(x)\left[D(x)\right]^{-1}$

Therefore,

$$Q'(x) = N(x)\frac{d}{dx}\left[D(x)\right]^{-1}+\left[D(x)\right]^{-1}\frac{d}{dx}N(x)$$

Product Rule

$$= N(x)(-1)\left[D(x)\right]^{-2}\cdot D'(x)+\left[D(x)\right]^{-1}N'(x)$$

Extended Power Rule

$$= \frac{-N(x)\cdot D'(x)}{\left[D(x)\right]^2}+\frac{N'(x)}{D(x)}$$

$$= \frac{-N(x)\cdot D'(x)}{\left[D(x)\right]^2}+\frac{N'(x)\cdot D(x)}{\left[D(x)\right]^2}$$

$$= \frac{D(x)\cdot N'(x)-N(x)\cdot D'(x)}{\left[D(x)\right]^2}$$

95. $g(x) = \sqrt{6x^3 - 3x^2 - 48x + 45}$, $[-5, 5]$

Using the calculator, we graph the function and the derivative in the same window. We can use the nDeriv feature to graph the derivative without actually calculating the derivative.

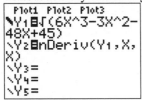

Using the window:

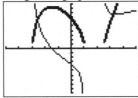

We get the graph:

Note, the function $f(x)$ is the thicker graph.

The horizontal tangents occur at the turning points of this function, or at the x-intercepts of the derivative. Using the trace feature, the minimum/maximum feature on the function, or the zero feature on the derivative on the calculator, we find the points of horizontal tangency.

We estimate the point at which the tangent line is horizontal to be $(-1.47481, 9.4878)$.

97. $g(x) = \dfrac{4x}{\sqrt{x-10}} = \dfrac{4x}{(x-10)^{1/2}}$

Using the Quotient Rule and the Extended Power Rule, we have

$$g'(x) = \frac{(x-10)^{1/2}(4) - 4x\left(\dfrac{1}{2}(x-10)^{-1/2}(1)\right)}{\left((x-10)^{1/2}\right)^2}$$

$$= \frac{4(x-10)^{1/2} - \dfrac{2x}{(x-10)^{1/2}}}{(x-10)}$$

$$= \frac{\dfrac{4(x-10)}{(x-10)^{1/2}} - \dfrac{2x}{(x-10)^{1/2}}}{(x-10)}$$

$$= \frac{2x-40}{(x-10)^{3/2}}$$

Using the window:

The graph of the function and the derivative are shown below.

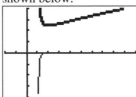

Note: The graph of the function is the thicker graph. The graph of the derivative is difficult to see in this window. For values greater than $x = 20$ the derivative is positive and increasing, but very close to 0.

Exercise Set 1.8

1. $y = x^5 + 9$

$\dfrac{dy}{dx} = 5x^{5-1} = 5x^4$ First Derivative

$\dfrac{d^2y}{dx^2} = 5(4x^{4-1}) = 20x^3$ Second Derivative

3. $\quad y = 2x^4 - 5x$

$\dfrac{dy}{dx} = 2(4x^3) - 5 =$

$\quad = 8x^3 - 5 \qquad$ First Derivative

$\dfrac{d^2y}{dx^2} = 8(3x^2)$

$\quad = 24x^2 \qquad$ Second Derivative

5. $\quad y = 4x^2 + 3x - 1$

$\dfrac{dy}{dx} = 4(2x^{2-1}) + 3 - 0$

$\quad = 8x + 3 \qquad$ First Derivative

$\dfrac{d^2y}{dx^2} = 8 \qquad$ Second Derivative

7. $\quad y = 7x + 2$

$\dfrac{dy}{dx} = 7 \qquad$ First Derivative

$\dfrac{d^2y}{dx^2} = 0 \qquad$ Second Derivative

9. $\quad y = \dfrac{1}{x^2} = x^{-2}$

$\dfrac{dy}{dx} = -2x^{-2-1}$

$\quad = -2x^{-3}$

$\quad = \dfrac{-2}{x^3} \qquad$ First Derivative

$\dfrac{d^2y}{dx^2} = -2(-3x^{-3-1})$

$\quad = 6x^{-4}$

$\quad = \dfrac{6}{x^4} \qquad$ Second Derivative

11. $\quad y = \sqrt{x} = x^{\frac{1}{2}}$

$\dfrac{dy}{dx} = \dfrac{1}{2}x^{\frac{1}{2}-1}$

$\quad = \dfrac{1}{2}x^{-\frac{1}{2}}$

$\quad = \dfrac{1}{2x^{\frac{1}{2}}} = \dfrac{1}{2\sqrt{x}} \qquad$ First Derivative

$\dfrac{d^2y}{dx^2} = \dfrac{1}{2} \cdot \left(-\dfrac{1}{2}x^{-\frac{1}{2}-1}\right)$

$\quad = -\dfrac{1}{4}x^{-\frac{3}{2}}$

$\quad = -\dfrac{1}{4x^{\frac{3}{2}}} = -\dfrac{1}{4\sqrt{x^3}} \qquad$ Second Derivative

13. $\quad f(x) = x^4 + \dfrac{3}{x}$

$f'(x) = 4x^{4-1} + 3(-1x^{-1-1})$

$\quad = 4x^3 - 3x^{-2}$

$\quad = 4x^3 - \dfrac{3}{x^2} \qquad$ First Derivative

$f''(x) = 4(3x^{3-1}) - 3(-2x^{-2-1})$

$\quad = 12x^2 + 6x^{-3}$

$\quad = 12x^2 + \dfrac{6}{x^3} \qquad$ Second Derivative

15. $\quad f(x) = x^{\frac{1}{5}}$

$f'(x) = \dfrac{1}{5}x^{\frac{1}{5}-1}$

$\quad = \dfrac{1}{5}x^{-\frac{4}{5}}$

$\quad = \dfrac{1}{5x^{\frac{4}{5}}} \qquad$ First Derivative

$f''(x) = \dfrac{1}{5}\left(-\dfrac{4}{5}\right)x^{-\frac{4}{5}-1}$

$\quad = -\dfrac{4}{25}x^{-\frac{9}{5}}$

$\quad = -\dfrac{4}{25x^{\frac{9}{5}}} \qquad$ Second Derivative

17. $\quad f(x) = 4x^{-3}$

$f'(x) = 4(-3x^{-3-1})$

$\quad = -12x^{-4}$

$\quad = -\dfrac{12}{x^4} \qquad$ First Derivative

$f''(x) = -12(-4x^{-4-1})$

$\quad = 48x^{-5}$

$\quad = \dfrac{48}{x^5} \qquad$ Second Derivative

19. $f(x) = \left(x^2 + 3x\right)^7$

$f'(x) = 7\left(x^2 + 3x\right)^{7-1}(2x+3)$ Theorem 7

$\quad = 7(2x+3)\left(x^2+3x\right)^6$ First Derivative

$f''(x) = 7(2x+3)\left(6\left(x^2+3x\right)^{6-1}(2x+3)\right) +$

$\quad\quad 7\left(x^2+6x\right)^6(2)$ Theorem 5

$\quad = 42(2x+3)^2\left(x^2+3x\right)^5 + 14\left(x^2+3x\right)^6$

We can simplify the second derivative by factoring out common factors.

$f''(x)$

$= 14\left(x^2+3x\right)^5\left[3(2x+3)^2 + \left(x^2+3x\right)\right]$

$= 14\left(x^2+3x\right)^5\left[3\left(4x^2+12x+9\right)+\left(x^2+3x\right)\right]$

$= 14\left(x^2+3x\right)^5\left[12x^2+36x+27+x^2+3x\right]$

$= 14\left(x^2+3x\right)^5\left(13x^2+39x+27\right)$ $\underset{\text{Derivative}}{\text{Second}}$

21. $f(x) = \left(2x^2-3x+1\right)^{10}$

$f'(x) = 10\left(2x^2-3x+1\right)^{10-1}(4x-3)$ Theorem 7

$\quad = 10\left(2x^2-3x+1\right)^9(4x-3)$ $\underset{\text{Derivative}}{\text{First}}$

$f''(x) = 10\left(2x^2-3x+1\right)^9(4) +$ Theorem 5

$\quad 10(4x-3)\cdot 9\left(2x^2-3x+1\right)^8(4x-3)$

$\quad = 40\left(2x^2-3x+1\right)^9 +$

$\quad 90(4x-3)^2\left(2x^2-3x+1\right)^8$

$\quad = 10\left(2x^2-3x+1\right)^8\left(4\left(2x^2-3x+1\right)+\right.$

$\quad\quad\quad\quad \left. 9\left(16x^2-24x+9\right)\right)$

$\quad = 10\left(2x^2-3x+1\right)^8\left(8x^2-12x+4+\right.$

$\quad\quad\quad\quad \left. 144x^2-216x+81\right)$

$\quad = 10\left(2x^2-3x+1\right)^8\left(152x^2-228x+85\right)$

$\quad\quad\quad\quad$ Second Derivative

23. $f(x) = \sqrt[4]{\left(x^2+1\right)^3} = \left(x^2+1\right)^{3/4}$

$f'(x) = \dfrac{3}{4}\left(x^2+1\right)^{-1/4}(2x)$ Theorem 7

$\quad = \dfrac{3}{2}x\left(x^2+1\right)^{-1/4}$ First Derivative

$f''(x) = \dfrac{3}{2}x\cdot\dfrac{-1}{4}\left(x^2+1\right)^{-5/4}(2x) +$

$\quad\quad \dfrac{3}{2}\left(x^2+1\right)^{-1/4}(1)$ Theorem 5

$\quad = -\dfrac{3}{4}x^2\left(x^2+1\right)^{-5/4} + \dfrac{3}{2}\left(x^2+1\right)^{-1/4}$

$\quad = \dfrac{-3x^2}{4\left(x^2+1\right)^{5/4}} + \dfrac{3}{2\left(x^2+1\right)^{1/4}}$

We can simplify the second derivative by finding a common denominator and combining the fractions.

$f''(x) = \dfrac{-3x^2}{4\left(x^2+1\right)^{5/4}} + \dfrac{3}{2\left(x^2+1\right)^{1/4}}\cdot\dfrac{2\left(x^2+1\right)}{2\left(x^2+1\right)}$

$\quad = \dfrac{-3x^2}{4\left(x^2+1\right)^{5/4}} + \dfrac{6\left(x^2+1\right)}{4\left(x^2+1\right)^{5/4}}$

$\quad = \dfrac{3x^2+6}{4\left(x^2+1\right)^{5/4}}$

$\quad = \dfrac{3\left(x^2+2\right)}{4\left(x^2+1\right)^{5/4}}$

25. $y = x^{2/3} + 4x$

$y' = \dfrac{2}{3}x^{2/3-1} + 4$

$\quad = \dfrac{2}{3}x^{-1/3} + 4$ First Derivative

$y'' = \dfrac{2}{3}\cdot\dfrac{-1}{3}x^{-1/3-1}$

$\quad = -\dfrac{2}{9}x^{-4/3}$

$\quad = -\dfrac{2}{9x^{4/3}}$ Second Derivative

27. $y = \left(x^3-x\right)^{3/4}$

$y' = \dfrac{3}{4}\left(x^3-x\right)^{3/4-1}\left(3x^2-1\right)$ Theorem 7

$\quad = \dfrac{3}{4}\left(x^3-x\right)^{-1/4}\left(3x^2-1\right)$ First Derivative

$$y'' = \frac{3}{4}(x^3 - x)^{-\frac{1}{4}}(6x) + \qquad \text{Theorem 5}$$

$$\frac{3}{4}(3x^2 - 1) \cdot \frac{-1}{4}(x^3 - x)^{-\frac{1}{4}-1}(3x^2 - 1)$$

$$= \frac{9}{2}x(x^3 - x)^{-\frac{1}{4}} +$$

$$\frac{-3}{16}(3x^2 - 1)^2(x^3 - x)^{-\frac{5}{4}}$$

$$= \frac{9x}{2(x^3 - x)^{\frac{1}{4}}} - \frac{3(3x^2 - 1)^2}{16(x^3 - x)^{\frac{5}{4}}}$$

The second derivative can be simplified by finding a common denominator and combining the fractions.

$$y'' = \frac{9x}{2(x^3 - x)^{\frac{1}{4}}} \cdot \frac{8(x^3 - x)}{8(x^3 - x)} - \frac{3(3x^2 - 1)^2}{16(x^3 - x)^{\frac{5}{4}}}$$

$$= \frac{72x^4 - 72x^2}{16(x^3 - x)^{\frac{5}{4}}} - \frac{3(9x^4 - 6x^2 + 1)}{16(x^3 - x)^{\frac{5}{4}}}$$

$$= \frac{72x^4 - 72x^2 - 27x^4 + 18x^2 - 3}{16(x^3 - x)^{\frac{5}{4}}}$$

$$= \frac{45x^4 - 54x^2 - 3}{16(x^3 - x)^{\frac{5}{4}}} \qquad \text{Second Derivative}$$

29. $\qquad y = 2x^{\frac{5}{4}} + x^{\frac{1}{2}}$

$$y' = 2 \cdot \frac{5}{4}x^{\frac{5}{4}-1} + \frac{1}{2}x^{\frac{1}{2}-1}$$

$$= \frac{5}{2}x^{\frac{1}{4}} + \frac{1}{2}x^{-\frac{1}{2}} \qquad \text{First Derivative}$$

$$y'' = \frac{5}{2} \cdot \frac{1}{4}x^{\frac{1}{4}-1} + \frac{1}{2} \cdot \frac{-1}{2}x^{-\frac{1}{2}-1}$$

$$= \frac{5}{8}x^{-\frac{3}{4}} - \frac{1}{4}x^{-\frac{3}{2}}$$

$$= \frac{5}{8x^{\frac{3}{4}}} - \frac{1}{4x^{\frac{3}{2}}} \qquad \text{Second Derivative}$$

31. $\qquad y = \frac{2}{x^3} + \frac{1}{x^2} = 2x^{-3} + x^{-2}$

$$y' = 2(-3x^{-3-1}) + (-2x^{-2-1})$$

$$= -6x^{-4} - 2x^{-3} \qquad \text{First Derivative}$$

$$y'' = -6(-4x^{-4-1}) - 2(-3x^{-3-1})$$

$$= 24x^{-5} + 6x^{-4}$$

$$= \frac{24}{x^5} + \frac{6}{x^4} \qquad \text{Second Derivative}$$

33. $\qquad y = (x^2 + 3)(4x - 1)$

$$y' = (x^2 + 3)(4) + (4x - 1)(2x) \quad \text{Theorem 5}$$

$$= 4x^2 + 12 + 8x^2 - 2x$$

$$= 12x^2 - 2x + 12 \qquad \text{First Derivative}$$

$$y'' = 12(2x^{2-1}) - 2$$

$$= 24x - 2 \qquad \text{Second Derivative}$$

35. $\qquad y = \frac{3x + 1}{2x - 3}$

$$y' = \frac{(2x - 3)(3) - (3x + 1)(2)}{(2x - 3)^2} \qquad \text{Theorem 6}$$

$$= \frac{6x - 9 - 6x - 2}{(2x - 3)^2}$$

$$= \frac{-11}{(2x - 3)^2} \qquad \text{First Derivative}$$

$$y'' = \frac{(2x - 3)^2(0) - (-11)(2(2x - 3)^{2-1}(2))}{((2x - 3)^2)^2}$$

$$\qquad \text{Theorem 6 and Theorem 7}$$

$$= \frac{44(2x - 3)}{(2x - 3)^4}$$

$$= \frac{44}{(2x - 3)^3} \qquad \text{Second Derivative}$$

37. $\qquad y = x^4$

$$\frac{dy}{dx} = 4x^{4-1} = 4x^3 \qquad \text{First Derivative}$$

$$\frac{d^2y}{dx^2} = 4(3x^{3-1}) = 12x^2 \qquad \text{Second Derivative}$$

$$\frac{d^3y}{dx^3} = 12(2x^{2-1}) = 24x \qquad \text{Third Derivative}$$

$$\frac{d^4y}{dx^4} = 24 \qquad \text{Fourth Derivative}$$

39. $\quad y = x^6 - x^3 + 2x$

$\dfrac{dy}{dx} = 6x^{6-1} - 3x^{3-1} + 2$

$\qquad = 6x^5 - 3x^2 + 2 \qquad$ First Derivative

$\dfrac{d^2 y}{dx^2} = 6\left(5x^{5-1}\right) - 3\left(2x^{2-1}\right)$

$\qquad = 30x^4 - 6x \qquad$ Second Derivative

$\dfrac{d^3 y}{dx^3} = 30\left(4x^{4-1}\right) - 6$

$\qquad = 120x^3 - 6 \qquad$ Third Derivative

$\dfrac{d^4 y}{dx^4} = 120\left(3x^{3-1}\right)$

$\qquad = 360x^2 \qquad$ Fourth Derivative

$\dfrac{d^5 y}{dx^5} = 360\left(2x^{2-1}\right)$

$\qquad = 720x \qquad$ Fifth Derivative

41. $\quad f(x) = x^{-2} - x^{1/2}$

$f'(x) = -2x^{-2-1} - \dfrac{1}{2}x^{1/2 - 1}$

$\qquad = -2x^{-3} - \dfrac{1}{2}x^{-1/2} \qquad \begin{matrix}\text{First}\\\text{Derivative}\end{matrix}$

$f''(x) = -2\left(-3x^{-3-1}\right) - \dfrac{1}{2} \cdot \dfrac{-1}{2}x^{-1/2 - 1}$

$\qquad = 6x^{-4} + \dfrac{1}{4}x^{-3/2} \qquad \begin{matrix}\text{Second}\\\text{Derivative}\end{matrix}$

$f'''(x) = 6\left(-4x^{-4-1}\right) + \dfrac{1}{4} \cdot \dfrac{-3}{2}x^{-3/2 - 1}$

$\qquad = -24x^{-5} - \dfrac{3}{8}x^{-5/2} \qquad \begin{matrix}\text{Third}\\\text{Derivative}\end{matrix}$

$f^{(4)}(x) = -24\left(-5x^{-5-1}\right) - \dfrac{3}{8} \cdot \dfrac{-5}{2}x^{-5/2 - 1}$

$\qquad = 120x^{-6} + \dfrac{15}{16}x^{-7/2} \qquad \begin{matrix}\text{Fourth}\\\text{Derivative}\end{matrix}$

43. $\quad g(x) = x^4 - 3x^3 - 7x^2 - 6x + 9$

$g'(x) = 4x^{4-1} - 3\left(3x^{3-1}\right) - 7\left(2x^{2-1}\right) - 6$

$\qquad = 4x^3 - 9x^2 - 14x - 6 \qquad \begin{matrix}\text{First}\\\text{Derivative}\end{matrix}$

$g''(x) = 4\left(3x^{3-1}\right) - 9\left(2x^{2-1}\right) - 14$

$\qquad = 12x^2 - 18x - 14 \qquad \begin{matrix}\text{Second}\\\text{Derivative}\end{matrix}$

$g'''(x) = 12\left(2x^{2-1}\right) - 18$

$\qquad = 24x - 18 \qquad \begin{matrix}\text{Third}\\\text{Derivative}\end{matrix}$

$g^{(4)}(x) = 24 \qquad \begin{matrix}\text{Fourth}\\\text{Derivative}\end{matrix}$

$g^{(5)}(x) = 0 \qquad \begin{matrix}\text{Fifth}\\\text{Derivative}\end{matrix}$

$g^{(6)}(x) = 0 \qquad \begin{matrix}\text{Sixth}\\\text{Derivative}\end{matrix}$

45. $\quad s(t) = t^3 + t$

a) $\quad v(t) = s'(t) = 3t^2 + 1$

b) $\quad a(t) = v'(t) = s''(t) = 6t$

c) When $t = 4$

$\quad v(4) = 3(4)^2 + 1 = 49$

$\quad a(4) = 6(4) = 24$

After 4 seconds, the velocity is 49 feet per second, and the acceleration is 24 feet per second squared.

47. $\quad s(t) = 3t + 10$

a) $\quad v(t) = s'(t) = 3$

b) $\quad a(t) = v'(t) = s''(t) = 0$

c) When $t = 2$

$\quad v(2) = 3$

$\quad a(2) = 0$

After 2 hours, the velocity is 3 miles per hour, and the acceleration is 0 miles per hour squared.

d) $\boxed{tw}$

49. $\quad s(t) = 16t^2$

a) When $t = 3$, $s(3) = 16(3)^2 = 144$.

The hammer falls 144 feet in 3 seconds.

b) $\quad v(t) = s'(t) = 32t$

When $t = 3$, $v(3) = 32(3) = 96$

the hammer is falling at 96 feet per second after 3 seconds.

c) $a(t) = v'(t) = s''(t) = 32$

When $t = 3$, $a(3) = 32$

the hammer is accelerating at 32 feet per second squared after 3 seconds.

51. $s(t) = 4.905t^2$

The velocity and acceleration are given by:

$v(t) = s'(t) = 9.81t$

$a(t) = v'(t) = s''(t) = 9.81$

After 2 seconds, we have

$v(2) = 9.81(2) = 19.62$

The stone is falling at 19.62 meters per second.

$a(2) = 9.81$

The stone is accelerating at 9.81 meters per second squared.

53. a) The plane's velocity is greater at $t = 20$ seconds. We know this because the slope of the tangent line is greater at $t = 20$ then it is at $t = 6$.

b) The plane's acceleration is positive, since the velocity (slope of the tangent lines) is increasing over time.

55. $S(t) = 2t^3 - 40t^2 + 220t + 160$

a) $S'(t) = 6t^2 - 80t + 220$

When $t = 1$,

$S'(1) = 6(1)^2 - 80(1) + 220 = 146$

After 1 month, sales are increasing at 146 thousand (146,000)dollars per month.

When $t = 2$,

$S'(2) = 6(2)^2 - 80(2) + 220 = 84$

After 2 month, sales are increasing at 84 thousand (84,000) dollars per month.

When $t = 4$,

$S'(4) = 6(4)^2 - 80(4) + 220 = -4$

After 4 months, sales are changing at a rate of -4 thousand (-4000) dollars per month.

b) $S''(t) = 12t - 80$

When $t = 1$, $S''(1) = 12(1) - 80 = -68$

After 1 month, the rate of change of sales are changing at a rate of -68 thousand $(-68,000)$ dollars per month squared.

When $t = 2$,

$S''(2) = 12(2) - 80 = -56$

After 2 months, the rate of change of sales are changing at a rate of -56 thousand $(-56,000)$ dollars per month squared.

When $t = 4$, $S''(4) = 12(4) - 80 = -32$

After 4 months, the rate of change of sales are changing at a rate of -32 thousand $(-32,000)$ dollars per month squared.

c) ☐ *tw*

57. $y = \dfrac{1}{(1-x)} = (1-x)^{-1}$

$y' = -1(1-x)^{-1-1}(-1) = 1(1-x)^{-2}$

$y'' = -2(1-x)^{-2-1}(-1) = 2(1-x)^{-3}$

$y''' = 2(-3)(1-x)^{-3-1}(-1) = 6(1-x)^{-4}$

Therefore,

$y''' = \dfrac{6}{(1-x)^4}$

59. $y = \dfrac{1}{\sqrt{2x+1}} = (2x+1)^{-\frac{1}{2}}$

$y' = \dfrac{-1}{2}(2x+1)^{-\frac{1}{2}-1}(2)$

$\quad = -(2x+1)^{-\frac{3}{2}}$

$y'' = -\dfrac{-3}{2}(2x+1)^{-\frac{3}{2}-1}(2)$

$\quad = 3(2x+1)^{-\frac{5}{2}}$

$y''' = 3\left(\dfrac{-5}{2}\right)(2x+1)^{-\frac{5}{2}-1}(2)$

$\quad = -15(2x+1)^{-\frac{7}{2}}$

$\quad = -\dfrac{15}{(2x+1)^{\frac{7}{2}}}$

61. $y = \dfrac{\sqrt{x}+1}{\sqrt{x}-1} = \dfrac{x^{\frac{1}{2}}+1}{x^{\frac{1}{2}}-1}$

$$y' = \frac{\left(x^{\frac{1}{2}}-1\right)\left(\frac{1}{2}x^{-\frac{1}{2}}\right)-\left(x^{\frac{1}{2}}+1\right)\left(\frac{1}{2}x^{-\frac{1}{2}}\right)}{\left(x^{\frac{1}{2}}-1\right)^2}$$

$$= \frac{\frac{1}{2}-\frac{1}{2}x^{-\frac{1}{2}}-\frac{1}{2}-\frac{1}{2}x^{-\frac{1}{2}}}{\left(x^{\frac{1}{2}}-1\right)^2}$$

$$= \frac{-x^{-\frac{1}{2}}}{\left(x^{\frac{1}{2}}-1\right)^2}$$

$$y'' = \frac{\left(x^{\frac{1}{2}}-1\right)^2\left(\frac{1}{2}x^{-\frac{3}{2}}\right)-\left(-x^{-\frac{1}{2}}\right)\left[2\left(x^{\frac{1}{2}}-1\right)\frac{1}{2}x^{-\frac{1}{2}}\right]}{\left(\left(x^{\frac{1}{2}}-1\right)^2\right)^2}$$

$$= \frac{\frac{1}{2}x^{-\frac{3}{2}}\left(x^{\frac{1}{2}}-1\right)^2+x^{-1}\left(x^{\frac{1}{2}}-1\right)}{\left(x^{\frac{1}{2}}-1\right)^4}$$

$$= \frac{\left(x^{\frac{1}{2}}-1\right)\left(\frac{1}{2}x^{-\frac{3}{2}}\left(x^{\frac{1}{2}}-1\right)+x^{-1}\right)}{\left(x^{\frac{1}{2}}-1\right)^4}$$

$$= \frac{\frac{1}{2}x^{-1}-\frac{1}{2}x^{-\frac{3}{2}}+x^{-1}}{\left(x^{\frac{1}{2}}-1\right)^3}$$

$$= \frac{\left(\frac{3}{2}x^{-1}-\frac{1}{2}x^{-\frac{3}{2}}\right)}{\left(x^{\frac{1}{2}}-1\right)^3}$$

$$= \frac{\frac{3}{2x}-\frac{1}{2x^{\frac{3}{2}}}}{\left(x^{\frac{1}{2}}-1\right)^3}$$

$$= \frac{\frac{3x^{\frac{1}{2}}}{2x^{\frac{3}{2}}}-\frac{1}{2x^{\frac{3}{2}}}}{\left(x^{\frac{1}{2}}-1\right)^3}$$

$$= \frac{3x^{\frac{1}{2}}-1}{2x^{\frac{3}{2}}\left(x^{\frac{1}{2}}-1\right)^3}$$

63. $y = x^k$

$$\frac{dy}{dx} = kx^{k-1}$$

$$\frac{d^2y}{dx^2} = k(k-1)x^{k-2}$$

$$\frac{d^3y}{dx^3} = k(k-1)(k-2)x^{k-3}$$

$$\frac{d^4y}{dx^4} = k(k-1)(k-2)(k-3)x^{k-4}$$

$$\frac{d^5y}{dx^5} = k(k-1)(k-2)(k-3)(k-4)x^{k-5}$$

65. $f(x) = \dfrac{x-1}{x+2}$

$$f'(x) = \frac{(x+2)(1)-(x-1)(1)}{(x+2)^2}$$

$$= \frac{3}{(x+2)^2}$$

Notice, $f'(x) = \dfrac{3}{(x+2)^2} = 3(x+2)^{-2}$

$$f''(x) = 3(-2)(x+2)^{-2-1}$$

$$= -6(x+2)^{-3}$$

$$= -\frac{6}{(x+2)^3}$$

Notice, $f''(x) = -\dfrac{6}{(x+2)^3} = -6(x+2)^{-3}$

$$f'''(x) = -6(-3)(x+2)^{-3-1}(1)$$

$$= 18(x+2)^{-4}$$

$$= \frac{18}{(x+2)^4}$$

Notice, $f'''(x) = \dfrac{18}{(x+2)^4} = 18(x+2)^{-4}$

$$f^{(4)}(x) = 18(-4)(x+2)^{-4-1}(1)$$

$$= -72(x+2)^{-5}$$

$$= -\frac{72}{(x+2)^5}$$

67. $s(t) = 16t^2$

Find the velocity function, $v(t) = s'(t) = 32t$.

This velocity function is in feet per second, we need to convert 50 miles per hour to feet per second. There are 5280 feet in one mile and 3600 seconds in one hour. Therefore,

$$\left(50\frac{\text{mi}}{\text{hr}}\right)\left(\frac{5280\text{ft}}{1\text{ mi}}\right)\left(\frac{1\text{ hr}}{3600\text{ sec}}\right) = \frac{220}{3}\frac{\text{ft}}{\text{sec}}$$

Now we solve the equation:

$$v(t) = \frac{220}{3}$$

$$32t = \frac{220}{3}$$

$$t = \frac{220}{3} \cdot \frac{1}{32}$$

$$t = \frac{55}{24} \approx 2.29$$

The ball will have to fall approximately 2.29 seconds to reach a speed of 50 mi/hr.

69. $s(t) = -t^3 + 3t;$ $\quad [-3,3]$

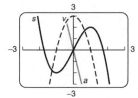

From the graph we see that $v(t)$ switches at $t = 0$.

71. $s(t) = t^3 - 3t^2 + 2;$ $\quad [-2,4]$

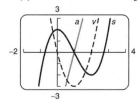

From the graph we see that $v(t)$ switches at $t = 1$.

Chapter 2

Applications of Differentiation

1. $f(x) = x^2 + 4x + 5$

First, find the critical points.
$f'(x) = 2x + 4$

$f'(x)$ exists for all real numbers. We solve
$f'(x) = 0$
$2x + 4 = 0$
$\quad 2x = -4$
$\quad\ x = -2$

The only critical value is -2. We use -2 to divide the real number line into two intervals,
A: $(-\infty, -2)$ and B: $(-2, \infty)$:

We use a test value in each interval to determine the sign of the derivative in each interval.

A: Test $-3, f'(-3) = 2(-3) + 4 = -2 < 0$

B: Test $0, \quad f'(0) = 2(0) + 4 = 4 > 0$

We see that $f(x)$ is increasing on $(-\infty, -2)$ and decreasing on $(-2, \infty)$, and the change from decreasing to increasing indicates that a relative minimum occurs at $x = -2$. We substitute into the original equation to find $f(-2)$:

$$f(-2) = (-2)^2 + 4(-2) + 5 = 1$$

Thus, there is a relative minimum at $(-2, 1)$. We use the information obtained to sketch the graph. Other function values are listed below.

x	$f(x)$
-5	10
-4	5
-3	2
-2	1
-1	2
0	5
1	10

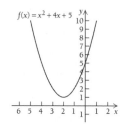

3. $f(x) = 5 - x - x^2$

First, find the critical points.
$f'(x) = -1 - 2x$

$f'(x)$ exists for all real numbers. We solve
$f'(x) = 0$
$-1 - 2x = 0$
$\quad -2x = 1$
$\qquad x = -\dfrac{1}{2}$

The only critical value is $-\dfrac{1}{2}$. We use $-\dfrac{1}{2}$ to divide the real number line into two intervals,

A: $\left(-\infty, -\dfrac{1}{2}\right)$ and B: $\left(-\dfrac{1}{2}, \infty\right)$:

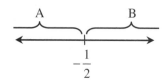

We use a test value in each interval to determine the sign of the derivative in each interval.

A: Test $-1, f'(-1) = -1 - 2(-1) = 1 > 0$

B: Test $0, \quad f'(0) = -1 - 2(0) = -1 < 0$

We see that $f(x)$ is increasing on $\left(-\infty, -\dfrac{1}{2}\right)$

and decreasing on $\left(-\dfrac{1}{2}, \infty\right)$, and the change from increasing to decreasing indicates that a

relative maximum occurs at $x = -\dfrac{1}{2}$. We

substitute into the original equation to find

$f\left(-\dfrac{1}{2}\right)$:

$$f\left(-\dfrac{1}{2}\right) = 5 - \left(-\dfrac{1}{2}\right) - \left(-\dfrac{1}{2}\right)^2 = \dfrac{21}{4}$$

Thus, there is a relative maximum at $\left(-\dfrac{1}{2}, \dfrac{21}{4}\right)$.

We use the information obtained to sketch the graph.

Other function values are listed below.

x	$f(x)$
-3	-1
-2	3
-1	5
$-\frac{1}{2}$	$\frac{21}{4}$
0	5
1	3
2	-1

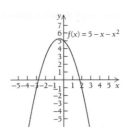

Other function values are listed below.

x	$g(x)$
-4	25
-3	10
-2	1
-1	-2
0	1
1	10
2	25

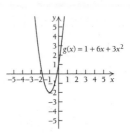

5. $g(x) = 1 + 6x + 3x^2$

First, find the critical points.

$g'(x) = 6 + 6x$

$g'(x)$ exists for all real numbers. We solve:

$g'(x) = 0$

$6 + 6x = 0$

$6x = -6$

$x = -1$

The only critical value is -1. We use -1 to divide the real number line into two intervals,

A: $(-\infty, -1)$ and B: $(-1, \infty)$:

We use a test value in each interval to determine the sign of the derivative in each interval.

A: Test -2, $g'(-2) = 6 + 6(-2) = -6 < 0$

B: Test 0, $g'(0) = 6 + 6(0) = 6 > 0$

We see that $g'(x)$ is decreasing on $(-\infty, -1)$

and increasing on $(-1, \infty)$, and the change from decreasing to increasing indicates that a relative minimum occurs at $x = -1$. We substitute into the original equation to find $g(-1)$:

$g(-1) = 1 + 6(-1) + 3(-1)^2 = -2$

Thus, there is a relative minimum at $(-1, -2)$.

We use the information obtained to sketch the graph.

7. $G(x) = x^3 - x^2 - x + 2$

First, find the critical points.

$G'(x) = 3x^2 - 2x - 1$

$G'(x)$ exists for all real numbers. We solve

$$G'(x) = 0$$

$$3x^2 - 2x - 1 = 0$$

$$(3x + 1)(x - 1) = 0$$

$3x + 1 = 0$ or $x - 1 = 0$

$3x = -1$ or $x = 1$

$x = -\frac{1}{3}$ or $x = 1$

The critical values are $-\frac{1}{3}$ and 1. We use them to divide the real number line into three intervals,

A: $\left(-\infty, -\frac{1}{3}\right)$, B: $\left(-\frac{1}{3}, 1\right)$, and C: $(1, \infty)$.

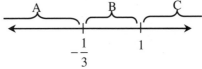

We use a test value in each interval to determine the sign of the derivative in each interval.

A: Test -1,

$$G'(-1) = 3(-1)^2 - 2(-1) - 1 = 4 > 0$$

B: Test 0,

$$G'(0) = 3(0)^2 - 2(0) - 1 = -1 < 0$$

C: Test 2,

$$G'(2) = 3(2)^2 - 2(2) - 1 = 7 > 0$$

We see that $G(x)$ is increasing on $\left(-\infty, -\dfrac{1}{3}\right)$, decreasing on $\left(-\dfrac{1}{3}, 1\right)$, and increasing on $(1, \infty)$. So there is a relative maximum at $x = -\dfrac{1}{3}$ and a relative minimum at $x = 1$.

We find $G\left(-\dfrac{1}{3}\right)$:

$$G\left(-\frac{1}{3}\right) = \left(-\frac{1}{3}\right)^3 - \left(-\frac{1}{3}\right)^2 - \left(-\frac{1}{3}\right) + 2$$

$$= -\frac{1}{27} - \frac{1}{9} + \frac{1}{3} + 2$$

$$= \frac{59}{27}$$

Then we find $G(1)$:

$$G(1) = (1)^3 - (1)^2 - (1) + 2$$

$$= 1 - 1 - 1 + 2$$

$$= 1$$

There is a relative maximum at $\left(-\dfrac{1}{3}, \dfrac{59}{27}\right)$, and there is a relative minimum at $(1, 1)$. We use the information obtained to sketch the graph. Other function values are listed below.

x	$G(x)$
-2	-8
-1	1
0	2
2	4
3	17

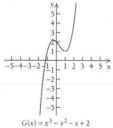

$G(x) = x^3 - x^2 - x + 2$

9. $f(x) = x^3 - 3x + 6$

First, find the critical points.

$f'(x) = 3x^2 - 3$

$f'(x)$ exists for all real numbers. We solve

$$f'(x) = 0$$

$$3x^2 - 3 = 0$$

$$3x^2 = 3$$

$$x^2 = 1$$

$$x = \pm 1$$

The critical values are -1 and 1. We use them to divide the real number line into three intervals,

A: $(-\infty, -1)$, B: $(-1, 1)$, and C: $(1, \infty)$.

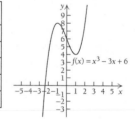

We use a test value in each interval to determine the sign of the derivative in each interval.

A: Test -3, $f'(-3) = 3(-3)^2 - 3 = 24 > 0$

B: Test 0, $f'(0) = 3(0)^2 - 3 = -3 < 0$

C: Test 2, $f'(2) = 3(2)^2 - 3 = 9 > 0$

We see that $f(x)$ is increasing on $(-\infty, -1)$, decreasing on $(-1, 1)$, and increasing on $(1, \infty)$. So there is a relative maximum at $x = -1$ and a relative minimum at $x = 1$.

We find $f(-1)$:

$$f(-1) = (-1)^3 - 3(-1) + 6 = -1 + 3 + 6 = 8$$

Then we find $f(1)$:

$$f(1) = (1)^3 - 3(1) + 6 = 1 - 3 + 6 = 4$$

There is a relative maximum at $(-1, 8)$, and there is a relative minimum at $(1, 4)$. We use the information obtained to sketch the graph. Other function values are listed below.

x	$f(x)$
-3	-12
-2	4
0	6
2	8
3	24

$f(x) = x^3 - 3x + 6$

11. $f(x) = 3x^2 + 2x^3$

First, find the critical points.

$f'(x) = 6x + 6x^2$

$f'(x)$ exists for all real numbers. We solve

$$f'(x) = 0$$

$$6x + 6x^2 = 0$$

$$6x(1 + x) = 0$$

$$6x = 0 \qquad \text{or} \qquad x + 1 = 0$$

$$x = 0 \qquad \text{or} \qquad x = -1$$

The critical values are -1 and 0.

We use the critical values to divide the real number line into three intervals,

A: $(-\infty,-1)$, B: $(-1,0)$, and C:$(0,\infty)$.

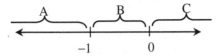

We use a test value in each interval to determine the sign of the derivative in each interval.

A: Test -2,

$$f'(-2)=6(-2)+6(-2)^2=12>0$$

B: Test $-\dfrac{1}{2}$,

$$f'\left(-\frac{1}{2}\right)=6\left(-\frac{1}{2}\right)+6\left(-\frac{1}{2}\right)^2=-\frac{3}{2}<0$$

C: Test 1,

$$f'(1)=6(1)+6(1)^2=12>0$$

We see that $f(x)$ is increasing on $(-\infty,-1)$, decreasing on $(-1,0)$, and increasing on $(0,\infty)$. So there is a relative maximum at $x=-1$ and a relative minimum at $x=0$.

We find $f(-1)$:

$$f(-1)=3(-1)^2+2(-1)^3=1$$

Then we find $f(0)$:

$$f(0)=3(0)^2+2(0)^3=0$$

There is a relative maximum at $(-1,1)$, and there is a relative minimum at $(0,0)$. We use the information obtained to sketch the graph. Other function values are listed below.

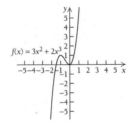

x	$f(x)$
-3	-27
-2	-4
$\frac{1}{2}$	1
2	28

13. $g(x)=2x^3-16$

First, find the critical points.

$$g'(x)=6x^2$$

$g'(x)$ exists for all real numbers. We solve

$$g'(x)=0$$

$$6x^2=0$$

$$x=0$$

The only critical value is 0. We use 0 to divide the real number line into two intervals,

A: $(-\infty,0)$, and B: $(0,\infty)$.

We use a test value in each interval to determine the sign of the derivative in each interval.

A: Test $-1, g'(-1)=6(-1)^2=6>0$

B: Test 1, $\quad g'(1)=6(1)^2=6>0$

We see that $g(x)$ is increasing on $(-\infty,0)$ and increasing on $(0,\infty)$, so the function has no relative extema. We use the information obtained to sketch the graph. Other function values are listed below.

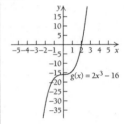

x	$g(x)$
-2	-32
-1	-18
0	-16
1	-14
2	0
3	38

15. $G(x)=x^3-6x^2+10$

First, find the critical points.

$$G'(x)=3x^2-12x$$

$G'(x)$ exists for all real numbers. We solve

$$G'(x)=0$$

$$x^2-4x=0 \qquad \text{Dividing by 3}$$

$$x(x-4)=0$$

$$x=0 \qquad \text{or} \qquad x-4=0$$

$$x=0 \qquad \text{or} \qquad x=4$$

The critical values are 0 and 4. We use them to divide the real number line into three intervals,

A: $(-\infty,0)$, B: $(0,4)$, and C:$(4,\infty)$.

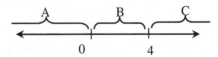

We use a test value in each interval to determine the sign of the derivative in each interval.

A: Test $-1, G'(-1)=3(-1)^2-12(-1)=15>0$

B: Test 1, $\quad G'(1)=3(1)^2-12(1)=-9<0$

C: Test 5, $\quad G'(5)=3(5)^2-12(5)=15>0$

We see that $G(x)$ is increasing on $(-\infty, 0)$, decreasing on $(0, 4)$, and increasing on $(4, \infty)$. So there is a relative maximum at $x = 0$ and a relative minimum at $x = 4$.

We find $G(0)$:

$$G(0) = (0)^3 - 6(0)^2 + 10$$
$$= 10$$

Then we find $G(4)$:

$$G(4) = (4)^3 - 6(4)^2 + 10$$
$$= 64 - 96 + 10$$
$$= -22$$

There is a relative maximum at $(0, 10)$, and there is a relative minimum at $(4, -22)$. We use the information obtained to sketch the graph. Other function values are listed below.

x	$G(x)$
-2	-22
-1	3
1	5
2	-6
3	-17

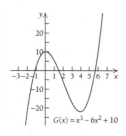

17. $g(x) = x^3 - x^4$

First, find the critical points.

$$g'(x) = 3x^2 - 4x^3$$

$g'(x)$ exists for all real numbers. We solve

$$g'(x) = 0$$
$$3x^2 - 4x^3 = 0$$
$$x^2(3 - 4x) = 0$$
$$x^2 = 0 \quad \text{or} \quad 3 - 4x = 0$$
$$x = 0 \quad \text{or} \quad -4x = -3$$
$$x = 0 \quad \text{or} \quad x = \frac{3}{4}$$

The critical values are 0 and $\frac{3}{4}$. We use them to divide the real number line into three intervals,

A: $(-\infty, 0)$, B: $\left(0, \frac{3}{4}\right)$, and C: $\left(\frac{3}{4}, \infty\right)$.

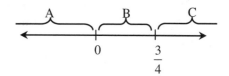

We use a test value in each interval to determine the sign of the derivative in each interval.

A: Test -1, $g'(-1) = 3(-1)^2 - 4(-1)^3 = 7 > 0$

B: Test $\frac{1}{2}$, $g'\left(\frac{1}{2}\right) = 3\left(\frac{1}{2}\right)^2 - 4\left(\frac{1}{2}\right)^3$

$$= 3\left(\frac{1}{4}\right) - 4\left(\frac{1}{8}\right) = \frac{1}{4} > 0$$

C: Test 1, $g'(1) = 3(1)^2 - 4(1)^3 = -1 < 0$

We see that $g(x)$ is increasing on $(-\infty, 0)$ and $\left(0, \frac{3}{4}\right)$, and is decreasing on $\left(\frac{3}{4}, \infty\right)$. So there is no relative extrema at $x = 0$ but there is a relative maximum at $x = \frac{3}{4}$.

We find $g\left(\frac{3}{4}\right)$:

$$g\left(\frac{3}{4}\right) = \left(\frac{3}{4}\right)^3 - \left(\frac{3}{4}\right)^4 = \frac{27}{64} - \frac{81}{256} = \frac{27}{256}$$

There is a relative maximum at $\left(\frac{3}{4}, \frac{27}{256}\right)$. We use the information obtained to sketch the graph. Other function values are listed below.

x	$g(x)$
-2	-24
-1	-2
0	0
$\frac{1}{2}$	$\frac{1}{16}$
1	0
2	-8

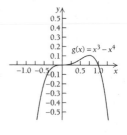

19. $f(x) = \dfrac{1}{3}x^3 - 2x^2 + 4x - 1$

First, find the critical points.

$f'(x) = x^2 - 4x + 4$

$f'(x)$ exists for all real numbers. We solve

$$f'(x) = 0$$
$$x^2 - 4x + 4 = 0$$
$$(x-2)^2 = 0$$
$$x - 2 = 0$$
$$x = 2$$

The only critical value is 2.

We divide the real number line into two intervals,

A: $(-\infty, 2)$ and B: $(2, \infty)$.

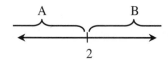

We use a test value in each interval to determine the sign of the derivative in each interval.

A: Test 0, $f'(0) = (0)^2 - 4(0) + 4 = 4 > 0$

B: Test 3, $f'(3) = (3)^2 - 4(3) + 4 = 1 > 0$

We see that $f(x)$ is increasing on both $(-\infty, 2)$ and $(2, \infty)$. Therefore, there are no relative extrema.

We use the information obtained to sketch the graph. Other function values are listed below.

x	$f(x)$
-3	-40
-2	$-\frac{59}{3}$
-1	$-\frac{22}{3}$
0	-1
1	$\frac{4}{3}$
2	$\frac{5}{3}$
3	2

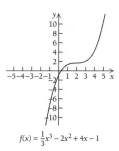

$f(x) = \frac{1}{3}x^3 - 2x^2 + 4x - 1$

21. $g(x) = 2x^4 - 20x^2 + 18$

First, find the critical points.

$g'(x) = 8x^3 - 40x$

$g'(x)$ exists for all real numbers. We solve

$$g'(x) = 0$$
$$8x^3 - 40x = 0$$
$$8x(x^2 - 5) = 0$$

$8x = 0$ or $x^2 - 5 = 0$

$x = 0$ or $x^2 = 5$

$x = 0$ or $x = \pm\sqrt{5}$

The critical values are 0, $\sqrt{5}$ and $-\sqrt{5}$. We use them to divide the real number line into four intervals,

A: $\left(-\infty, -\sqrt{5}\right)$, B: $\left(-\sqrt{5}, 0\right)$,

C: $\left(0, \sqrt{5}\right)$, and D: $\left(\sqrt{5}, \infty\right)$.

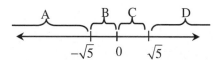

We use a test value in each interval to determine the sign of the derivative in each interval.

A: Test -3,

$$g'(-3) = 8(-3)^3 - 40(-3) = -96 < 0$$

B: Test -1,

$$g'(-1) = 8(-1)^3 - 40(-1) = 32 > 0$$

C: Test 1,

$$g'(1) = 8(1)^3 - 40(1) = -32 < 0$$

D: Test 3,

$$g'(3) = 8(3)^3 - 40(3) = 96 > 0$$

We see that $g(x)$ is decreasing on $\left(-\infty, -\sqrt{5}\right)$, increasing on $\left(-\sqrt{5}, 0\right)$, decreasing again on $\left(0, \sqrt{5}\right)$, and increasing again on $\left(\sqrt{5}, \infty\right)$.

Thus, there is a relative minimum at $x = -\sqrt{5}$, a relative maximum at $x = 0$, and another relative minimum at $x = \sqrt{5}$.

We find $g\left(-\sqrt{5}\right)$:

$$g\left(-\sqrt{5}\right) = 2\left(-\sqrt{5}\right)^4 - 20\left(-\sqrt{5}\right)^2 + 18 = -32$$

Then we find $g(0)$:

$$g(0) = 2(0)^4 - 20(0)^2 + 18 = 18$$

Then we find $g(\sqrt{5})$

$$g(\sqrt{5}) = 2(\sqrt{5})^4 - 20(\sqrt{5})^2 + 18 = -32$$

There are relative minima at $(-\sqrt{5}, -32)$ and $(\sqrt{5}, -32)$. There is a relative maximum at $(0, 18)$ We use the information obtained to sketch the graph. Other function values are listed below.

x	$g(x)$
−4	210
−3	0
−1	0
1	0
3	0
4	210

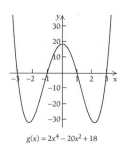

$g(x) = 2x^4 - 20x^2 + 18$

23. $F(x) = \sqrt[3]{x-1} = (x-1)^{\frac{1}{3}}$

First, find the critical points.

$$F'(x) = \frac{1}{3}(x-1)^{-\frac{2}{3}}(1)$$

$$= \frac{1}{3(x-1)^{\frac{2}{3}}}$$

$F'(x)$ does not exist when

$3(x-1)^{\frac{2}{3}} = 0$, which means that $F'(x)$ does not exist when $x = 1$. The equation $F'(x) = 0$ has no solution, therefore, the only critical value is $x = 1$.

We use 1 to divide the real number line into two intervals,

   A: $(-\infty, 1)$ and B: $(1, \infty)$:

We use a test value in each interval to determine the sign of the derivative in each interval.

A: Test $0, F'(0) = \dfrac{1}{3(0-1)^{\frac{2}{3}}} = \dfrac{1}{3} > 0$

B: Test $2, F'(2) = \dfrac{1}{3(2-1)^{\frac{2}{3}}} = \dfrac{1}{3} > 0$

We see that $F(x)$ is increasing on both $(-\infty, 1)$ and $(1, \infty)$. Thus, there are no relative extrema for $F(x)$. We use the information obtained to sketch the graph. Other function values are listed.

x	$F(x)$
−7	−2
0	−1
1	0
2	1
9	2

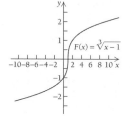

25. $f(x) = 1 - x^{\frac{2}{3}}$

First, find the critical points.

$$f'(x) = \frac{-2}{3}x^{-\frac{1}{3}}$$

$$= \frac{-2}{3\sqrt[3]{x}}$$

$f'(x)$ does not exist when

$3\sqrt[3]{x} = 0$, which means that $f'(x)$ does not exist when $x = 0$. The equation $f'(x) = 0$ has no solution, therefore, the only critical value is $x = 0$.

We use 0 to divide the real number line into two intervals,

   A: $(-\infty, 0)$ and B: $(0, \infty)$:

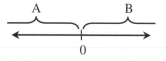

We use a test value in each interval to determine the sign of the derivative in each interval.

A: Test $-1, f'(-1) = -\dfrac{2}{3\sqrt[3]{-1}} = \dfrac{2}{3} > 0$

B: Test $1, f'(1) = -\dfrac{2}{3\sqrt[3]{1}} = -\dfrac{2}{3} < 0$

We see that $f(x)$ is increasing on $(-\infty, 0)$ and decreasing on $(0, \infty)$. Thus, there is a relative maximum at $x = 0$.

We find $f(0)$:

$$f(0) = 1 - (0)^{\frac{2}{3}} = 1.$$

Therefore, there is a relative maximum at $(0, 1)$.

We use the information obtained to sketch the graph. Other function values are listed below.

x	$f(x)$
-8	-3
-1	0
1	0
8	-3

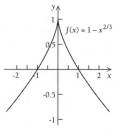

27. $G(x) = \dfrac{-8}{x^2+1} = -8(x^2+1)^{-1}$

First, find the critical points.

$G'(x) = -8(-1)(x^2+1)^{-2}(2x)$

$\qquad = \dfrac{16x}{(x^2+1)^2}$

$G'(x)$ exists for all real numbers. We solve

$G'(x) = 0$

$\dfrac{16x}{(x^2+1)^2} = 0$

$16x = 0$

$x = 0$

The only critical value is 0.

We use 0 to divide the real number line into two intervals,

A: $(-\infty, 0)$ and B: $(0, \infty)$:

A   B

0

We use a test value in each interval to determine the sign of the derivative in each interval.

A: Test -1, $G'(-1) = \dfrac{16(-1)}{((-1)^2+1)^2} = \dfrac{-16}{4} = -4 < 0$

B: Test 1, $G'(1) = \dfrac{16(1)}{((1)^2+1)^2} = \dfrac{16}{4} = 4 > 0$

We see that $G(x)$ is decreasing on $(-\infty, 0)$ and increasing on $(0, \infty)$. Thus, a relative minimum occurs at $x = 0$.

We find $G(0)$:

$G(0) = \dfrac{-8}{(0)^2+1} = -8$

Thus, there is a relative minimum at $(0, -8)$.

We use the information obtained to sketch the graph. Other function values are listed below.

x	$G(x)$
-3	$-\frac{4}{5}$
-2	$-\frac{8}{5}$
-1	-4
1	-4
2	$-\frac{8}{5}$
3	$-\frac{4}{5}$

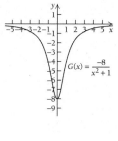

29. $g(x) = \dfrac{4x}{x^2+1}$

First, find the critical points.

$g'(x) = \dfrac{(x^2+1)(4) - 4x(2x)}{(x^2+1)^2}$ Quotient Rule

$\qquad = \dfrac{4x^2 + 4 - 8x^2}{(x^2+1)^2}$

$\qquad = \dfrac{4 - 4x^2}{(x^2+1)^2}$

$g'(x)$ exists for all real numbers. We solve

$g'(x) = 0$

$\dfrac{4-4x^2}{(x^2+1)^2} = 0$

$4 - 4x^2 = 0$ Multiplying by $(x^2+1)^2$

$x^2 - 1 = 0$ Dividing by -4

$x^2 = 1$

$x = \pm\sqrt{1}$

$x = \pm 1$

The critical values are -1 and 1. We use them to divide the real number line into three intervals,

A: $(-\infty, -1)$, B: $(-1, 1)$, and C: $(1, \infty)$.

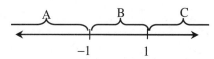

We use a test value in each interval to determine the sign of the derivative in each interval.

A: Test $-2, g'(-2) = \dfrac{4-4(-2)^2}{\left((-2)^2+1\right)^2} = -\dfrac{12}{25} < 0$

B: Test $0, \quad g'(0) = \dfrac{4-4(0)^2}{\left((0)^2+1\right)^2} = 4 > 0$

C: Test $2, \quad g'(2) = \dfrac{4-4(2)^2}{\left((2)^2+1\right)^2} = -\dfrac{12}{25} < 0$

We see that $g(x)$ is decreasing on $(-\infty,-1)$, increasing on $(-1,1)$, and decreasing again on $(1,\infty)$. So there is a relative minimum at $x=-1$ and a relative maximum at $x=1$. We find $g(-1)$:

$$g(-1) = \frac{4(-1)}{(-1)^2+1} = \frac{-4}{2} = -2$$

Then we find $g(1)$:

$$g(1) = \frac{4(1)}{(1)^2+1} = \frac{4}{2} = 2$$

There is a relative minimum at $(-1,-2)$, and there is a relative maximum at $(1,2)$. We use the information obtained to sketch the graph. Other function values are listed below.

x	$g(x)$
-3	$-\frac{6}{5}$
-2	$-\frac{8}{5}$
0	0
2	$\frac{8}{5}$
3	$\frac{6}{5}$

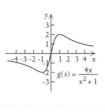

31. $f(x) = \sqrt[3]{x} = (x)^{1/3}$

First, find the critical points.

$$f'(x) = \frac{1}{3}(x)^{-2/3}$$
$$= \frac{1}{3(x)^{2/3}} = \frac{1}{3 \cdot \sqrt[3]{x^2}}$$

$f'(x)$ does not exist when $x=0$. The equation $f'(x)=0$ has no solution, therefore, the only critical value is $x=0$.

We use 0 to divide the real number line into two intervals,
A: $(-\infty,0)$ and B: $(0,\infty)$:

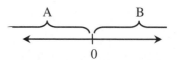

We use a test value in each interval to determine the sign of the derivative in each interval.

A: Test $-1, f'(-1) = \dfrac{1}{3\sqrt[3]{(-1)^2}} = \dfrac{1}{3} > 0$

B: Test $1, f'(1) = \dfrac{1}{3\left(\sqrt[3]{(1)^2}\right)} = \dfrac{1}{3} > 0$

We see that $f(x)$ is increasing on both $(-\infty,0)$ and $(0,\infty)$. Thus, there are no relative extrema for $f(x)$. We use the information obtained to sketch the graph. Other function values are listed below.

x	$f(x)$
-8	-2
-1	-1
0	0
1	1
8	2

33. $g(x) = \sqrt{x^2+2x+5} = \left(x^2+2x+5\right)^{1/2}$

First, find the critical points.

$$g'(x) = \frac{1}{2}\left(x^2+2x+5\right)^{-1/2}(2x+2)$$
$$= \frac{2(x+1)}{2\left(x^2+2x+5\right)^{-1/2}}$$
$$= \frac{x+1}{\sqrt{x^2+2x+5}}$$

The equation $x^2+2x+5=0$ has no solution, so $g'(x)$ exists for all real numbers. Next we find out where the derivative is zero. We solve

$$g'(x) = 0$$
$$\frac{x+1}{\sqrt{x^2+2x+5}} = 0$$
$$x+1 = 0$$
$$x = -1$$

The only critical value is -1.

We use -1 to divide the real number line into two intervals,

A: $(-\infty, -1)$ and B: $(-1, \infty)$:

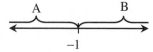

We use a test value in each interval to determine the sign of the derivative in each interval.

A: Test -2,

$$g'(-2) = \frac{(-2)+1}{\sqrt{(-2)^2 + 2(-2) + 5}} = \frac{-1}{\sqrt{5}} < 0$$

B: Test 0,

$$g'(0) = \frac{(0)+1}{\sqrt{(0)^2 + 2(0) + 5}} = \frac{1}{\sqrt{5}} > 0$$

We see that $g(x)$ is decreasing on $(-\infty, -1)$ and increasing on $(-1, \infty)$, and the change from decreasing to increasing indicates that a relative minimum occurs at $x = -1$. We substitute into the original equation to find $g(-1)$:

$$g(-1) = \sqrt{(-1)^2 + 2(-1) + 5} = \sqrt{4} = 2$$

Thus, there is a relative minimum at $(-1, 2)$.

We use the information obtained to sketch the graph. Other function values are listed below.

x	$g(x)$
-4	3.61
-2	2.24
0	2.24
1	2.83
3	4.47

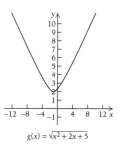

$$g(x) = \sqrt{x^2 + 2x + 5}$$

35. – 67.   Left to the student.

69.   Answers may vary, one such graph is:

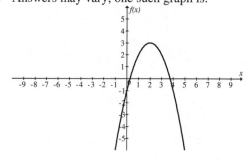

71.   Answers may vary, one such graph is:

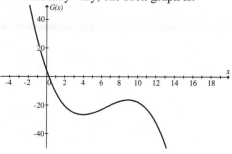

73.   Answers may vary, one such graph is:

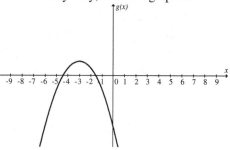

75.   Answers may vary, one such graph is:

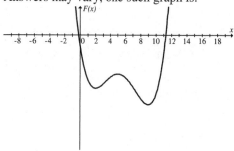

77.   Answers may vary, one such graph is:

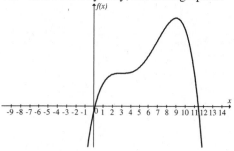

79. $T(t) = -0.1t^2 + 1.2t + 98.6, \quad 0 \le t \le 12$

First, we find the critical points.

$T'(t) = -0.2t + 1.2$

$T'(t)$ exists for all real numbers. Solve

$$T'(t) = 0$$

$$-0.2t + 1.2 = 0$$

$$-0.2t = -1.2$$

$$t = 6$$

The only critical value is 6. We use it to divide the interval $[0,12]$ into two intervals:

A: $[0,6)$ and B: $(6,12]$

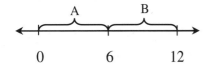

Next, we test a point in each interval to determine the sign of the derivative.

A: Test $0, T'(0) = -0.2(0) + 1.2 = 1.2 > 0$

B: Test $7, T'(7) = -0.2(7) + 1.2 = -0.2 < 0$

Since, $T(t)$ is increasing on $[0,6)$ and decreasing on $(6,12]$, there is a relative maximum at $t = 6$.

$T(6) = -0.1(6)^2 + 1.2(6) + 98.6 = 102.2$

There is a relative maximum at $(6, 102.2)$. We sketch the graph.

t	$T(t)$
0	98.6
3	101.3
5	102.1
7	102.1
8	101.8
12	98.6

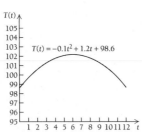

81. $A(t) = 0.0265t^3 - 0.453t^2 + 1.796t + 7.47,$

$0 \le t \le 10$

First, find the critical points.

$A'(t) = 0.0795t^2 - 0.906t + 1.796$

$A'(t)$ exists for all real numbers. We solve

$$A'(t) = 0$$

$$0.0795t^2 - 0.906t + 1.796 = 0$$

This is a quadratic equation with
$a = 0.0795, b = -0.906,$ and $c = 1.796$

Applying the Quadratic formula we have:

$$t = \frac{-b \pm \sqrt{b^2 - 4ac}}{2a}$$

$$= \frac{-(-0.906) \pm \sqrt{(-0.906)^2 - 4(0.0795)(1.796)}}{2(0.0795)}$$

$$= \frac{0.906 \pm \sqrt{0.249708}}{1.59}$$

$t = 2.5553$ or $t = 8.8409$.

The critical values are 2.5553 and 8.8409. We use them to divide the real number line into three intervals, A: $(-\infty, 2.5553)$,

B: $(2.5553, 8.8409)$, and C: $(8.8409, \infty)$.

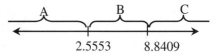

We use a test value in each interval to determine the sign of the derivative in each interval.

A: Test 0,

$$A'(0) = 0.0795(0)^2 - 0.906(0) + 1.796$$

$$= 1.796 > 0$$

B: Test 3,

$$A'(3) = 0.0795(3)^2 - 0.906(3) + 1.796$$

$$= -0.2065 < 0$$

C: Test 2,

$$A'(9) = 0.0795(9)^2 - 0.906(9) + 1.796$$

$$= 0.0815 > 0$$

We see that $A(t)$ is increasing on $(-\infty, 2.5553)$, decreasing on $(2.5553, 8.8409)$, and increasing again on $(8.8409, \infty)$. So there is a relative maximum at $t = 2.5553$ and a relative maximum at $t = 8.8409$.

We find $A(2.5553)$:

$$A(2.5553) = 0.0265(2.5553)^3 -$$

$$0.453(2.5553)^2 + 1.796(2.5553) +$$

$$7.47$$

$$\approx 9.5436$$

Then we find $A(8.8409)$:

$$A(8.8409) = 0.0265(8.8409)^3 -$$
$$0.453(8.8409)^2 + 1.796(8.8409) +$$
$$7.47$$
$$\approx 6.2531$$

There is a relative maximum at $(2.5553, 9.5436)$ and there is a relative maximum at $(8.8409, 6.2531)$. We use the information obtained to sketch the graph. Other function values are listed below.

x	$A(t)$
0	7.47
4	9.102
6	7.662
10	6.63

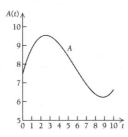

83. [tw]

85. The derivative is negative over the interval $(-\infty, -1)$ and positive over the interval $(-1, \infty)$. Furthermore it is equal to zero when $x = -1$. This means that the function is decreasing over the interval $(-\infty, -1)$, increasing over the interval $(-1, \infty)$ and has a horizontal tangent at $x = -1$. A possible graph is shown below.

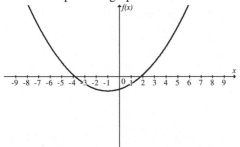

87. The derivative is positive over the interval $(-\infty, 1)$ and negative over the interval $(1, \infty)$. Furthermore it is equal to zero when $x = 1$. This means that the function is increasing over the interval $(-\infty, 1)$, decreasing over the interval $(1, \infty)$ and has a horizontal tangent at $x = 1$. A possible graph is shown at the top of the next column.

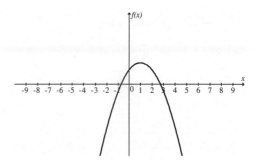

89. The derivative is positive over the interval $(-4, 2)$ and negative over the intervals $(-\infty, -4)$ and $(2, \infty)$. Furthermore it is equal to zero when $x = -4$ and $x = 2$. This means that the function is decreasing over the interval $(-\infty, -4)$, then increasing over the interval $(-4, 2)$, and then decreasing again over the interval $(2, \infty)$. The function has horizontal tangents at $x = -4$ and $x = 2$. A possible graph is shown on the next page.

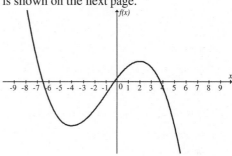

91. $f(x) = -x^6 - 4x^5 + 54x^4 + 160x^3 - 641x^2$
$$-828x + 1200$$

Using the calculator we enter the function into the graphing editor as follows:

```
Plot1 Plot2 Plot3
\Y1◻-X^6-4X^5+54
X^4+160X^3-641X^
2-828X+1200
\Y2=
\Y3=
\Y4=
\Y5=
```

Using the following window:

```
WINDOW
 Xmin=-8
 Xmax=8
 Xscl=1
 Ymin=-3000
 Ymax=7000
 Yscl=1000
 Xres=1
```

The graph of the function is:

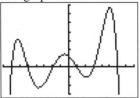

We find the relative extrema using the minimum/maximum feature on the calculator. There are relative minima at $\left(-3.683,-2288.03\right)$ and $\left(2.116,-1083.08\right)$.

There are relative maxima at $\left(-6.262,3213.8\right)$, $\left(-0.559,1440.06\right)$, and $\left(5.054,6674.12\right)$.

93. $f\left(x\right)=\sqrt[3]{\left|4-x^2\right|}+1$

Using the calculator we enter the function into the graphing editor as follows:

Using the following window:

```
WINDOW
 Xmin=-10
 Xmax=10
 Xscl=1
 Ymin=0
 Ymax=6
 Yscl=.5
 Xres=1
```

The graph of the function is:

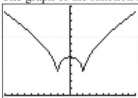

We find the relative extrema using the minimum/maximum feature on the calculator. There are relative minima at $\left(-2,1\right)$ and $\left(2,1\right)$.

There is a relative maximum at $\left(0,2.587\right)$.

95. $\boxed{tw}$

97. $\boxed{tw}$

Exercise Set 2.2

1. $f\left(x\right)=5-x^2$

First, find $f'\left(x\right)$ and $f''\left(x\right)$.

$f'\left(x\right)=-2x$

$f''\left(x\right)=-2$

Next, find the critical points of $f\left(x\right)$. Since $f'\left(x\right)$ exists for all real numbers x, the only critical points occur when $f'\left(x\right)=0$.

$f'\left(x\right)=0$

$-2x=0$

$x=0$

We find the function value at $x=0$.

$f\left(0\right)=5-\left(0\right)^2=5$.

The critical point is $\left(0,5\right)$.

Next, we apply the Second Derivative test.

$f''\left(x\right)=-2$

$f''\left(0\right)=-2<0$

Therefore, $f\left(0\right)=5$ is a relative maximum.

3. $f\left(x\right)=x^2-x$

First, find $f'\left(x\right)$ and $f''\left(x\right)$.

$f'\left(x\right)=2x-1$

$f''\left(x\right)=2$

Next, find the critical points of $f\left(x\right)$. Since $f'\left(x\right)$ exists for all real numbers x, the only critical points occur when $f'\left(x\right)=0$.

$f'\left(x\right)=0$

$2x-1=0$

$2x=1$

$x=\dfrac{1}{2}$

We find the function value at $x=\dfrac{1}{2}$.

$f\left(\dfrac{1}{2}\right)=\left(\dfrac{1}{2}\right)^2-\left(\dfrac{1}{2}\right)=\dfrac{1}{4}-\dfrac{1}{2}=-\dfrac{1}{4}$.

The critical point is $\left(\dfrac{1}{2},-\dfrac{1}{4}\right)$.

Next, we apply the Second Derivative test.

$$f''(x) = 2$$

$$f''\left(\frac{1}{2}\right) = 2 > 0$$

Therefore, $f\left(\frac{1}{2}\right) = -\frac{1}{4}$ is a relative minimum.

5. $f(x) = -5x^2 + 8x - 7$

First, find $f'(x)$ and $f''(x)$.

$$f'(x) = -10x + 8$$

$$f''(x) = -10$$

Next, find the critical points of $f(x)$. Since $f'(x)$ exists for all real numbers x, the only critical points occur when $f'(x) = 0$.

$$f'(x) = 0$$

$$-10x + 8 = 0$$

$$-10x = -8$$

$$x = \frac{-8}{-10}$$

$$x = \frac{4}{5}$$

We find the function value at $x = \frac{4}{5}$.

$$f\left(\frac{4}{5}\right) = -5\left(\frac{4}{5}\right)^2 + 8\left(\frac{4}{5}\right) - 7$$

$$= -5\left(\frac{16}{25}\right) + \frac{32}{5} - \frac{7}{1}$$

$$= -\frac{16}{5} + \frac{32}{5} - \frac{35}{5}$$

$$= -\frac{19}{5}$$

The critical point is $\left(\frac{4}{5}, -\frac{19}{5}\right)$.

Next, we apply the Second Derivative test.

$$f''(x) = -10$$

$$f''\left(\frac{4}{5}\right) = -10 < 0$$

Therefore, $f\left(\frac{4}{5}\right) = -\frac{19}{5}$ is a relative maximum.

7. $f(x) = 8x^3 - 6x + 1$

First, find $f'(x)$ and $f''(x)$.

$$f'(x) = 24x^2 - 6$$

$$f''(x) = 48x$$

Next, find the critical points of $f(x)$. Since $f'(x)$ exists for all real numbers x, the only critical points occur when $f'(x) = 0$.

$$f'(x) = 0$$

$$24x^2 - 6 = 0$$

$$4x^2 - 1 = 0 \qquad \text{Dividing by 6}$$

$$x^2 = \frac{1}{4}$$

$$x = \pm\sqrt{\frac{1}{4}}$$

$$x = \pm\frac{1}{2}$$

There are two critical values $x = -\frac{1}{2}$ and $x = \frac{1}{2}$.

We find the function value at $x = -\frac{1}{2}$

$$f\left(-\frac{1}{2}\right) = 8\left(-\frac{1}{2}\right)^3 - 6\left(-\frac{1}{2}\right) + 1$$

$$= 8\left(-\frac{1}{8}\right) + 3 + 1$$

$$= -1 + 3 + 1$$

$$= 3$$

The critical point is $\left(-\frac{1}{2}, 3\right)$.

Next, we apply the Second Derivative test.

$$f''(x) = 48x$$

$$f''\left(-\frac{1}{2}\right) = 48\left(-\frac{1}{2}\right) = -24 < 0$$

Therefore, $f\left(-\frac{1}{2}\right) = 3$ is a relative maximum.

Now, we find the function value at $x = \frac{1}{2}$.

$$f\left(\frac{1}{2}\right) = 8\left(\frac{1}{2}\right)^3 - 6\left(\frac{1}{2}\right) + 1$$

$$= 8\left(\frac{1}{8}\right) - 3 + 1$$

$$= 1 - 3 + 1$$

$$= -1$$

The critical point is $\left(\frac{1}{2}, -1\right)$.

Next, we apply the Second Derivative test.

$$f''(x) = 48x$$

$$f''\left(\frac{1}{2}\right) = 48\left(\frac{1}{2}\right) = 24 > 0$$

Therefore, $f\left(\frac{1}{2}\right) = -1$ is a relative minimum.

9. $f(x) = x^3 - 12x$

a) Find $f'(x)$ and $f''(x)$.

$$f'(x) = 3x^2 - 12$$

$$f''(x) = 6x$$

The domain of f is $\mathbb{R}$.

b) Find the critical points of $f(x)$. Since $f'(x)$ exists for all real numbers x, the only critical points occur when $f'(x) = 0$.

$$f'(x) = 0$$

$$3x^2 - 12 = 0$$

$$3x^2 = 12$$

$$x^2 = 4$$

$$x = \pm 2$$

There are two critical values $x = -2$ and $x = 2$.

We find the function value at $x = -2$

$$f(-2) = (-2)^3 - 12(-2) = 16.$$

The critical point on the graph is $(-2, 16)$.

Next, we find the function value at $x = 2$.

$$f(2) = (2)^3 - 12(2) = -16.$$

The critical point on the graph is $(2, -16)$.

c) Apply the Second Derivative test to the critical points.

For $x = -2$

$$f''(x) = 6x$$

$$f''(-2) = 6(-2) = -12 < 0$$

The critical point $(-2, 16)$ is a relative maximum.

For $x = 2$

$$f''(x) = 6x$$

$$f''(2) = 6(2) = 12 > 0$$

The critical point $(2, -16)$ is a relative minimum.

If we use the critical values $x = -2$ and $x = 2$ to divide the real line into three intervals, $(-\infty, -2), (-2, 2)$, and $(2, \infty)$, we know from the extrema above, that $f(x)$ is increasing over the interval $(-\infty, -2)$, decreasing over the interval $(-2, 2)$ and then increasing again over the interval $(2, \infty)$.

d) Find the points of inflection. $f''(x)$ exists for all real numbers, so we solve the equation

$$f''(x) = 0$$

$$6x = 0$$

$$x = 0$$

Therefore, a possible inflection point occurs at $x = 0$.

$$f(0) = (0)^3 - 12(0) = 0.$$

This gives the point $(0, 0)$ on the graph.

e) To determine concavity, we use the possible inflection point to divide the real number line into two intervals $A : (-\infty, 0)$ and $B : (0, \infty)$. We test a point in each interval

A: Test -1: $f''(-1) = 6(-1) = -6 < 0$

B: Test 1: $f''(1) = 6(1) = 6 > 0$

Then, $f(x)$ is concave down on the interval $(-\infty, 0)$ and concave up on the interval $(0, \infty)$, so $(0, 0)$ is an inflection point.

f) Finally, we use the preceding information to sketch the graph of the function. Additional function values can also be calculated as needed.

x	$f(x)$
-3	9
-1	11
1	-11
3	-9

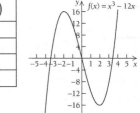

11. $f(x) = 3x^3 - 36x - 3$

a) First, find $f'(x)$ and $f''(x)$.

$$f'(x) = 9x^2 - 36$$

$$f''(x) = 18x$$

The domain of f is $\mathbb{R}$.

b) Find the critical points of $f(x)$. Since $f'(x)$ exists for all real numbers x, the only critical points occur when $f'(x) = 0$.

$$f'(x) = 0$$

$$9x^2 - 36 = 0$$

$$9x^2 = 36$$

$$x^2 = 4$$

$$x = \pm 2$$

There are two critical values $x = -2$ and $x = 2$.

We find the function value at $x = -2$

$$f(-2) = 3(-2)^3 - 36(-2) - 3 = 45.$$

The critical point on the graph is $(-2, 45)$.

Next, we find the function value at $x = 2$.

$$f(2) = 3(2)^3 - 36(2) - 3 = -51.$$

The critical point on the graph is $(2, -51)$.

c) Apply the Second Derivative test to the critical points.

For $x = -2$

$$f''(x) = 18x$$

$$f''(-2) = 18(-2) = -36 < 0$$

The critical point $(-2, 45)$ is a relative maximum.

For $x = 2$

$$f''(x) = 18x$$

$$f''(2) = 18(2) = 36 > 0$$

The critical point $(2, -51)$ is a relative minimum.

c) We use the critical values $x = -2$ and $x = 2$ to divide the real line into three intervals, A:$(-\infty, -2)$, B: $(-2, 2)$, and C: $(2, \infty)$, we know from the extrema above, that $f(x)$ is increasing over the interval $(-\infty, -2)$, decreasing over the interval $(-2, 2)$ and then increasing again over the interval $(2, \infty)$.

d) Find the points of inflection. $f''(x)$ exists for all real numbers, so we solve the equation

$$f''(x) = 0$$

$$18x = 0$$

$$x = 0$$

Therefore, a possible inflection point occurs at $x = 0$.

$$f(0) = 3(0)^3 - 36(0) - 3 = -3.$$

This gives the point $(0, -3)$ on the graph.

e) To determine concavity, we use the possible inflection point to divide the real number line into two intervals

$A:(-\infty, 0)$ and B:$(0, \infty)$. We test a point in each interval

A: Test -1: $f''(-1) = 18(-1) = -18 < 0$

B: Test 1: $\quad f''(1) = 18(1) = 18 > 0$

Then, $f(x)$ is concave down on the interval $(-\infty, 0)$ and concave up on the interval $(0, \infty)$, so $(0, -3)$ is an inflection point.

f) We use the preceding information to sketch the graph of the function. Additional function values can also be calculated as needed.

x	$f(x)$
-3	24
-1	30
1	-36
3	-30

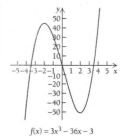

$f(x) = 3x^3 - 36x - 3$

13. $f(x) = \dfrac{8}{3}x^3 - 2x + \dfrac{1}{3}$

a) Find $f'(x)$ and $f''(x)$.

$$f'(x) = 8x^2 - 2$$

$$f''(x) = 16x$$

The domain of f is $\mathbb{R}$.

b) Next, find the critical points of $f(x)$. Since $f'(x)$ exists for all real numbers x, the only critical points occur when $f'(x) = 0$.

$$f'(x) = 0$$
$$8x^2 - 2 = 0$$
$$8x^2 = 2$$
$$x^2 = \frac{1}{4}$$
$$x = \pm\frac{1}{2}$$

There are two critical values
$x = -\dfrac{1}{2}$ and $x = \dfrac{1}{2}$.

We find the function value at $x = -\dfrac{1}{2}$

$$f\left(-\frac{1}{2}\right) = \frac{8}{3}\left(-\frac{1}{2}\right)^3 - 2\left(-\frac{1}{2}\right) + \frac{1}{3}$$

$$= \frac{8}{3}\left(-\frac{1}{8}\right) + 1 + \frac{1}{3}$$

$$= -\frac{1}{3} + 1 + \frac{1}{3}$$

$$= 1$$

The critical point on the graph is $\left(-\dfrac{1}{2}, 1\right)$.

Next, we find the function value at $x = \dfrac{1}{2}$.

$$f\left(\frac{1}{2}\right) = \frac{8}{3}\left(\frac{1}{2}\right)^3 - 2\left(\frac{1}{2}\right) + \frac{1}{3}$$

$$= \frac{8}{3}\left(\frac{1}{8}\right) - 1 + \frac{1}{3}$$

$$= \frac{1}{3} - 1 + \frac{1}{3}$$

$$= -\frac{1}{3}$$

The critical point on the graph is $\left(\dfrac{1}{2}, -\dfrac{1}{3}\right)$.

c) Apply the Second Derivative test to the critical points.

For $x = -\dfrac{1}{2}$

$$f''(x) = 16x$$

$$f''\left(-\frac{1}{2}\right) = 16\left(-\frac{1}{2}\right) = -8 < 0$$

The critical point $\left(-\dfrac{1}{2}, 1\right)$ is a relative

maximum.

For $x = \dfrac{1}{2}$

$$f''(x) = 16x$$

$$f''\left(\frac{1}{2}\right) = 16\left(\frac{1}{2}\right) = 8 > 0$$

The critical point $\left(\dfrac{1}{2}, -\dfrac{1}{3}\right)$ is a relative

minimum.

We use the critical values

$x = -\dfrac{1}{2}$, and $x = \dfrac{1}{2}$ to divide the real line

into three intervals,

$A: \left(-\infty, -\dfrac{1}{2}\right)$, $B: \left(-\dfrac{1}{2}, \dfrac{1}{2}\right)$, and $C: \left(\dfrac{1}{2}, \infty\right)$,

we know from the extrema above, that $f(x)$ is increasing over the interval

$\left(-\infty, -\dfrac{1}{2}\right)$, decreasing over the interval

$\left(-\dfrac{1}{2}, \dfrac{1}{2}\right)$ and then increasing again over the

interval $\left(\dfrac{1}{2}, \infty\right)$.

d) We find the points of inflection.
$f''(x)$ exists for all real numbers, so we
solve the equation
$$f''(x) = 0$$
$$16x = 0$$
$$x = 0$$
Therefore, a possible inflection point occurs
at $x = 0$.

$$f(0) = \frac{8}{3}(0)^3 - 2(0) + \frac{1}{3} = \frac{1}{3}.$$

This gives the point $\left(0, \dfrac{1}{3}\right)$ on the graph.

e) To determine concavity, we use the possible
inflection point to divide the real number
line into two intervals $A: (-\infty, 0)$ and

$B: (0, \infty)$. We test a point in each interval

A: Test -1: $f''(-1) = 16(-1) = -16 < 0$

B: Test 1: $f''(1) = 16(1) = 16 > 0$

Then, $f(x)$ is concave down on the interval

$(-\infty, 0)$ and concave up on the interval

$(0, \infty)$, so $\left(0, \dfrac{1}{3}\right)$ is an inflection point.

f) We use the preceding information to sketch the graph of the function. Additional function values can also be calculated as needed.

x	$f(x)$
-2	-17
-1	$-\frac{1}{3}$
1	1
2	$\frac{53}{3}$

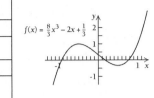

$f(x) = \frac{8}{3}x^3 - 2x + \frac{1}{3}$

15. $f(x) = -x^3 + 3x^2 - 4$

a) First, find $f'(x)$ and $f''(x)$.

$$f'(x) = -3x^2 + 6x$$

$$f''(x) = -6x + 6$$

The domain of f is $\mathbb{R}$.

b) Find the critical points of $f(x)$. Since $f'(x)$ exists for all real numbers x, the only critical points occur when $f'(x) = 0$.

$$f'(x) = 0$$

$$-3x^2 + 6x = 0$$

$$-3x(x-2) = 0$$

$$-3x = 0 \quad \text{or} \quad x - 2 = 0$$

$$x = 0 \quad \text{or} \quad x = 2$$

There are two critical values $x = 0$ and $x = 2$.

We find the function value at $x = 0$

$$f(0) = -(0)^3 + 3(0)^2 - 4 = -4$$

The critical point on the graph is $(0, -4)$.

Next, we find the function value at $x = 6$.

$$f(2) = -(2)^3 + 3(2)^2 - 4$$

$$= -8 + 12 - 4$$

$$= 0$$

The critical point on the graph is $(2, 0)$.

c) Apply the Second Derivative test to the critical points.

For $x = 0$

$$f''(x) = -6x + 6$$

$$f''(0) = -6(0) + 6 = 6 > 0$$

The critical point $(0, -4)$ is a relative minimum.

For $x = 2$

$$f''(x) = -6x + 6$$

$$f''(2) = -6(2) + 6 = -6 < 0$$

The critical point $(2, 0)$ is a relative maximum.

We use the critical values $x = 0$ and $x = 2$ to divide the real line into three intervals, A: $(-\infty, 0)$, B: $(0, 2)$, and C: $(2, \infty)$, we know from the extrema above, that $f(x)$ is decreasing over the interval $(-\infty, 0)$, increasing over the interval $(0, 2)$ and then decreasing again over the interval $(2, \infty)$.

d) Find the points of inflection. $f''(x)$ exists for all real numbers, so we solve the equation

$$f''(x) = 0$$

$$-6x + 6 = 0$$

$$-6x = -6$$

$$x = 1$$

Therefore, a possible inflection point occurs at $x = 1$.

$$f(1) = -(1)^3 + 3(1)^2 - 4$$

$$= -1 + 3 - 4$$

$$= -2$$

This gives the point $(1, -2)$ on the graph.

e) To determine concavity, we use the possible inflection point to divide the real number line into two intervals A: $(-\infty, 1)$ and B: $(1, \infty)$. We test a point in each interval

A: Test 0: $f''(0) = -6(0) + 6 = 6 > 0$

B: Test 2: $f''(2) = -6(2) + 6 = -2 < 0$

Then, $f(x)$ is concave up on the interval $(-\infty, 1)$ and concave down on the interval $(1, \infty)$, so $(1, -2)$ is an inflection point.

f) We use the preceding information to sketch the graph of the function. Additional function values can also be calculated as needed.

x	$f(x)$
-2	16
-1	0
3	-4
4	-20

17. $f(x) = 3x^4 - 16x^3 + 18x^2$

a) First, find $f'(x)$ and $f''(x)$.

$$f'(x) = 12x^3 - 48x^2 + 36x$$
$$f''(x) = 36x^2 - 96x + 36$$

The domain of f is $\mathbb{R}$.

b) Next, find the critical points of $f(x)$. Since $f'(x)$ exists for all real numbers x, the only critical points occur when $f'(x) = 0$.

$$f'(x) = 0$$
$$12x^3 - 48x^2 + 36x = 0$$
$$12x(x^2 - 4x + 3) = 0$$
$$12x(x-1)(x-3) = 0$$
$$12x = 0 \quad \text{or} \quad x - 1 = 0 \quad \text{or} \quad x - 3 = 0$$
$$x = 0 \quad \text{or} \quad x = 1 \quad \text{or} \quad x = 3$$

There are three critical values $x = 0$, $x = 1$, and $x = 3$.

Then

$$f(0) = 3(0)^4 - 16(0)^3 + 18(0)^2 = 0$$
$$f(1) = 3(1)^4 - 16(1)^3 + 18(1)^2 = 5$$
$$f(3) = 3(3)^4 - 16(3)^3 + 18(3)^2 = -27$$

Thus, the critical points $(0,0)$, $(1,5)$, and $(3,-27)$ are on the graph.

c) Apply the Second Derivative test to the critical points.

$$f''(0) = 36(0)^2 - 96(0) + 36 = 36 > 0$$

The critical point $(0,0)$ is a relative minimum.

$$f''(1) = 36(1)^2 - 96(1) + 36 = -24 < 0$$

The critical point $(1,5)$ is a relative maximum.

$$f''(3) = 36(3)^2 - 96(3) + 36 = 72 > 0$$

The critical point $(3,-27)$ is a relative minimum.

We use the critical values $0, 1,$ and 3 to divide the real line into four intervals, $A: (-\infty, 0)$, $B: (0,1)$, $C: (1,3)$ and $D: (3, \infty)$, we know from the extrema above, that $f(x)$ is decreasing over the intervals $(-\infty, 0)$ and $(1,3)$ and $f(x)$ increasing over the intervals $(0,1)$ and $(3, \infty)$.

d) Find the points of inflection. $f''(x)$ exists for all real numbers, so we solve the equation $f''(x) = 0$.

$$f''(x) = 0$$
$$36x^2 - 96x + 36 = 0$$
$$12(3x^2 - 8x + 3) = 0$$
$$3x^2 - 8x + 3 = 0$$

Using the quadratic formula, we find that $x = \dfrac{4 \pm \sqrt{7}}{3}$, so $x \approx 0.451$ or $x \approx 2.215$ are possible inflection points.

$$f(0.451) \approx 2.321$$
$$f(2.215) \approx -13.358$$

So, $(0.451, 2.321)$ and $(2.215, -13.358)$ are two more points on the graph.

e) To determine concavity, we use the possible inflection point to divide the real number line into three intervals $A: (-\infty, 0.451)$, $B: (0.451, 2.215)$, and $C: (2.215, \infty)$. We test a point in each interval

A: Test 0:
$$f''(0) = 36(0)^2 - 96(0) + 36 = 36 > 0$$

B: Test 1:
$$f''(1) = 36(1)^2 - 96(1) + 36 = -24 < 0$$

C: Test 3:
$$f''(3) = 36(3)^2 - 96(3) + 36 = 72 > 0$$

Then, $f(x)$ is concave up on the interval $(-\infty, 0.451)$ and concave down on the interval $(0.451, 2.215)$ and concave up on the interval $(2.215, \infty)$, so $(0.451, 2.321)$ and $(2.215, -13.358)$ are inflection points.

f) We use the preceding information to sketch the graph of the function. Additional function values can also be calculated as needed.

x	$f(x)$
-1	37
2	-8
4	32

$f(x) = 3x^4 - 16x^3 + 18x^2$

19. $f(x) = x^4 - 6x^2$

a) First, find $f'(x)$ and $f''(x)$.

$$f'(x) = 4x^3 - 12x$$

$$f''(x) = 12x^2 - 12$$

The domain of f is $\mathbb{R}$.

b) Find the critical points of $f(x)$. Since $f'(x)$ exists for all real numbers x, the only critical points occur when $f'(x) = 0$.

$$f'(x) = 0$$

$$4x^3 - 12x = 0$$

$$4x(x^2 - 3) = 0$$

$$4x = 0 \quad \text{or} \quad x^2 - 3 = 0$$

$$x = 0 \quad \text{or} \quad x = \pm\sqrt{3}$$

There are three critical values $-\sqrt{3}, 0,$ and $\sqrt{3}$.

Then

$$f(-\sqrt{3}) = (-\sqrt{3})^4 - 6(-\sqrt{3})^2$$

$$= 9 - 6(3)$$

$$= -9$$

$$f(0) = (0)^4 - 6(0)^2 = 0$$

$$f(\sqrt{3}) = (\sqrt{3})^4 - 6(\sqrt{3})^2$$

$$= 9 - 6(3)$$

$$= -9$$

Thus, the critical points $(-\sqrt{3}, -9), (0, 0),$ $(\sqrt{3}, -9)$ and are on the graph.

c) Apply the Second Derivative test to the critical points.

$$f''(-\sqrt{3}) = 12(-\sqrt{3})^2 - 12$$

$$= 12(3) - 12 = 24 > 0$$

The critical point $(-\sqrt{3}, -9)$ is a relative minimum.

$$f''(0) = 12(0)^2 - 12 = -12 < 0$$

The critical point $(0, 0)$ is a relative maximum.

$$f''(\sqrt{3}) = 12(\sqrt{3})^2 - 12$$

$$= 12(3) - 12 = 24 > 0$$

The critical point $(\sqrt{3}, -9)$ is a relative minimum.

If we use the critical values $-\sqrt{3}, 0,$ and $\sqrt{3}$ to divide the real line into four intervals, A : $(-\infty, -\sqrt{3})$, B : $(-\sqrt{3}, 0)$, C : $(0, \sqrt{3})$, and D : $(\sqrt{3}, \infty)$

Then $f(x)$ is decreasing over the intervals $(-\infty, -\sqrt{3})$ and $(0, \sqrt{3})$, and $f(x)$ increasing over the intervals $(-\sqrt{3}, 0)$ and $(\sqrt{3}, \infty)$.

d) Find the points of inflection. $f''(x)$ exists for all real numbers, so we solve the equation

$$f''(x) = 0$$

$$12x^2 - 12 = 0$$

$$x^2 - 1 = 0$$

$$x^2 = 1$$

$$x = \pm 1$$

So $x = -1$ or $x = 1$ are possible inflection points.

$$f(-1) = (-1)^4 - 6(-1)^2 = 1 - 6 = -5$$

$$f(1) = (1)^4 - 6(1)^2 = 1 - 6 = -5$$

So, $(-1, -5)$ and $(1, -5)$ are two more points on the graph.

e) To determine concavity, we use the possible inflection point to divide the real number line into three intervals $A : (-\infty, -1)$, B: $(-1, 1)$, and C: $(1, \infty)$. We test a point in each interval

A: Test -2:

$$f''(-2) = 12(-2)^2 - 12 = 36 > 0$$

B: Test 0:

$$f''(0) = 12(0)^2 - 12 = -12 < 0$$

C: Test 2:

$$f''(2) = 12(2)^2 - 12 = 36 > 0$$

Then, $f(x)$ is concave up on the intervals $(-\infty, -1)$ and $(1, \infty)$ and concave down on the interval $(-1, 1)$, so $(-1, -5)$ and $(1, -5)$ are inflection points.

f) We use the preceding information to sketch the graph of the function. Additional function values can also be calculated as needed.

x	$f(x)$
-3	27
-2	-8
2	-8
3	27

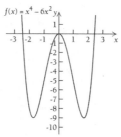

21. $f(x) = x^3 - 2x^2 - 4x + 3$

a) $f'(x) = 3x^2 - 4x - 4$

$f''(x) = 6x - 4$

The domain of f is $\mathbb{R}$.

b) $f'(x)$ exists for all values of x, so the only critical points of f are where $f'(x) = 0$.

$$3x^2 - 4x - 4 = 0$$
$$(3x + 2)(x - 2) = 0$$
$$3x + 2 = 0 \quad \text{or} \quad x - 2 = 0$$
$$x = -\frac{2}{3} \quad \text{or} \quad x = 2$$

The critical values are $-\frac{2}{3}$ and 2.

We find the function values for each of the critical values.

$$f\left(-\frac{2}{3}\right) = \left(-\frac{2}{3}\right)^3 - 2\left(-\frac{2}{3}\right)^2 - 4\left(-\frac{2}{3}\right) + 3$$

$$= -\frac{8}{27} - \frac{8}{9} + \frac{8}{3} + 3$$

$$= -\frac{8}{27} - \frac{24}{27} + \frac{72}{27} + \frac{81}{27}$$

$$= \frac{121}{27}$$

$$f(2) = (2)^3 - 2(2)^2 - 4(2) + 3$$
$$= 8 - 8 - 8 + 3$$
$$= -5$$

The critical points $\left(-\frac{2}{3}, \frac{121}{7}\right)$ and $(2, -5)$ are on the graph.

c) Applying the Second Derivative Test, we have:

$$f''\left(-\frac{2}{3}\right) = 6\left(-\frac{2}{3}\right) - 4 = -4 - 4 = -8 < 0$$

So $\left(-\frac{2}{3}, \frac{121}{7}\right)$ is a relative maximum.

$$f''(2) = 6(2) - 4 = 12 - 4 = 8 > 0$$

So $(2, -5)$ is a relative minimum.

Then, if we use the points $-\frac{2}{3}$ and 2 to divide the real number line into three intervals, $\left(-\infty, -\frac{2}{3}\right)$, $\left(-\frac{2}{3}, 2\right)$, and $(2, \infty)$, we know that f is increasing on $\left(-\infty, -\frac{2}{3}\right)$ and on $(2, \infty)$ and f is decreasing on $\left(-\frac{2}{3}, 2\right)$.

d) Find the points of inflection. $f''(x)$ exists for all values of x, so the only possible inflection points occur when $f''(x) = 0$.

$$6x - 4 = 0$$
$$6x = 4$$
$$x = \frac{4}{6} = \frac{2}{3}$$

There is a possible inflection point at $\frac{2}{3}$.

$$f\left(\frac{2}{3}\right) = \left(\frac{2}{3}\right)^3 - 2\left(\frac{2}{3}\right)^2 - 4\left(\frac{2}{3}\right) + 3$$

$$= \frac{8}{27} - \frac{8}{9} - \frac{8}{3} + 3$$

$$= \frac{8}{27} - \frac{24}{27} - \frac{72}{27} + \frac{81}{27}$$

$$= -\frac{7}{27}$$

Another point on the graph is $\left(\frac{2}{3}, -\frac{7}{27}\right)$.

e) To determine concavity we use $\frac{2}{3}$ to divide the real number line into two intervals,

A : $\left(-\infty, \frac{2}{3}\right)$ and B: $\left(\frac{2}{3}, \infty\right)$, Then test a point in each interval.

A: Test $0, f''(0) = 6(0) - 4 = -4 < 0$

B: Test $1, f''(1) = 6(1) - 4 = 2 > 0$

We see that f is concave down on

$\left(-\infty, \frac{2}{3}\right)$ and concave up on $\left(\frac{2}{3}, \infty\right)$, so

$\left(\frac{2}{3}, -\frac{7}{27}\right)$ is an inflection point.

f) We sketch the graph using the preceding information. Additional function values may also be calculated as necessary.

x	$f(x)$
-2	-5
-1	4
0	3
1	-2
3	0
4	19

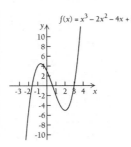

23. $f(x) = 3x^4 + 4x^3$

a) $f'(x) = 12x^3 + 12x^2$

   $f''(x) = 36x^2 + 24x$

   The domain of f is $\mathbb{R}$.

b) $f'(x)$ exists for all values of x, so the only critical points of f are where $f'(x) = 0$.

   $12x^3 + 12x^2 = 0$

   $12x^2(x+1) = 0$

   $12x^2 = 0$ or $x + 1 = 0$

   $x = 0$ or $x = -1$

The critical values are -1 and 0.

$f(-1) = 3(-1)^4 + 4(-1)^3 = -1$

$f(0) = 3(0)^4 + 4(0)^3 = 0$

The critical points $(-1, -1)$ and $(0, 0)$ are on the graph.

c) Applying the Second Derivative Test, we have:

$f''(-1) = 36(-1)^2 + 24(-1) = 36 - 24$

   $= 12 > 0$

So $(-1, -1)$ is a relative minimum.

$f''(0) = 36(0)^2 + 24(0) = 0$

The test fails. We will use the First Derivative Test. We use 0 to divide the interval $(-1, \infty)$ into two intervals,

A: $(-1, 0)$ and B: $(0, \infty)$, and test a point in each interval.

A: Test $-\frac{1}{2}$,

$$f'\left(-\frac{1}{2}\right) = 12\left(-\frac{1}{2}\right)^3 + 12\left(-\frac{1}{2}\right)^2$$

$$= \frac{3}{2} > 0$$

B: Test 2,

$$f'(2) = 12(2)^3 + 12(2)^2 = 144 > 0$$

f is increasing on both intervals $(-1, 0)$ and $(0, \infty)$. Therefore, $(0, 0)$ is not a relative extremum. Since $(-1, -1)$ is a relative minimum, we know that f is decreasing on $(-\infty, -1)$.

d) Find the points of inflection. $f''(x)$ exists for all values of x, so the only possible inflection points occur when $f''(x) = 0$.

$$f''(x) = 0$$

$$36x^2 + 24x = 0$$

$$12x(3x + 2) = 0$$

$$12x = 0 \quad \text{or} \quad 3x + 2 = 0$$

$$x = 0 \quad \text{or} \quad x = -\frac{2}{3}$$

There are a possible inflection points at

$-\frac{2}{3}$ and 0.

$$f\left(-\frac{2}{3}\right) = 3\left(-\frac{2}{3}\right)^4 + 4\left(-\frac{2}{3}\right)^3$$

$$= 3\left(\frac{16}{81}\right) + 4\left(-\frac{8}{27}\right)$$

$$= \frac{16}{27} - \frac{32}{27}$$

$$= -\frac{16}{27}$$

$$f(0) = 3(0)^4 + 4(0)^3 = 0$$

This gives one additional point $\left(-\frac{2}{3}, -\frac{16}{27}\right)$ on the graph.

e) To determine concavity we use $-\frac{2}{3}$ and 0 to divide the real number line into three intervals, A: $\left(-\infty, -\frac{2}{3}\right)$, B: $\left(-\frac{2}{3}, 0\right)$, and C: $(0, \infty)$. Then test a point in each interval.

A: Test -1,

$$f''(-1) = 36(-1)^2 + 24(-1) = 12 > 0$$

B: Test $-\frac{1}{2}$,

$$f''\left(-\frac{1}{2}\right) = 36\left(-\frac{1}{2}\right)^2 + 24\left(-\frac{1}{2}\right)$$

$$= -3 < 0$$

C: Test 1,

$$f''(1) = 36(1)^2 + 24(1) = 60 > 0$$

We see that f is concave up on the intervals $\left(-\infty, -\frac{2}{3}\right)$ and $(0, \infty)$, and concave down on the interval $\left(-\frac{2}{3}, 0\right)$, so both $\left(-\frac{2}{3}, -\frac{16}{27}\right)$ and $(0, 0)$ are inflection points.

f) We sketch the graph using the preceding information. Additional function values may also be calculated as necessary.

x	$f(x)$
-2	16
1	7
2	80

$f(x) = 3x^4 + 4x^3$

25. $f(x) = x^3 - 6x^2 - 135x$

a) $f'(x) = 3x^2 - 12x - 135$

$f''(x) = 6x - 12$

The domain of f is $\mathbb{R}$.

b) $f'(x)$ exists for all values of x, so the only critical points of f are where $f'(x) = 0$.

$$3x^2 - 12x - 135 = 0$$

$$x^2 - 4x - 45 = 0$$

$$(x - 9)(x + 5) = 0$$

$$x - 9 = 0 \quad \text{or} \quad x + 5 = 0$$

$$x = 9 \quad \text{or} \quad x = -5$$

The critical values are -5 and 9.

$$f(-5) = (-5)^3 - 6(-5)^2 - 135(-5)$$

$$= -125 - 150 + 675$$

$$= 400$$

$$f(9) = (9)^3 - 6(9)^2 - 135(9)$$

$$= 729 - 486 - 1215$$

$$= -972$$

The critical points $(-5, 400)$ and $(9, -972)$ are on the graph.

c) Applying the Second Derivative Test, we have:

$$f''(-5) = 6(-5) - 12 = -30 - 12$$

$$= -42 < 0$$

The critical point $(-5, 400)$ is a relative maximum.

$$f''(9) = 6(9) - 12 = 54 - 12$$

$$= 42 > 0$$

The critical point $(9, -972)$ is a relative minimum.

If we use the points -5 and 9 to divide the real number line into three intervals $(-\infty, -5)$, $(-5, 9)$, and $(9, \infty)$ we see that $f(x)$ is increasing on the intervals $(-\infty, -5)$ and $(9, \infty)$ and $f(x)$ is decreasing on the interval $(-5, 9)$.

d) Find the points of inflection. $f''(x)$ exists for all values of x, so the only possible inflection points occur when $f''(x) = 0$.

$$6x - 12 = 0$$

$$6x = 12$$

$$x = 2$$

The only possible inflection point is 2.

$$f(2) = (2)^3 - (2)^2 - 135(2)$$
$$= 8 - 24 - 270$$
$$= -286$$

The point $(2, -286)$ is a possible inflection point on the graph.

e) To determine concavity we use 2 to divide the real number line into two intervals,

A : $(-\infty, 2)$ and B: $(2, \infty)$, Then test a point in each interval.

A: Test 0, $f''(0) = 6(0) - 12 = -12 < 0$

B: Test 3, $f''(3) = 6(3) - 12 = 6 > 0$

We see that f is concave down on the interval $(-\infty, 2)$ and concave up on the interval $(2, \infty)$, Therefore $(2, -286)$ is an inflection point.

f) We sketch the graph using the preceding information. Additional function values may also be calculated as necessary.

x	$f(x)$
-11	-572
-10	-250
-3	324
0	0
2	-286
5	-700
15	0
16	400

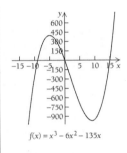

$f(x) = x^3 - 6x^2 - 135x$

27. $f(x) = x^4 - 4x^3 + 10$

a) $f'(x) = 4x^3 - 12x^2$

   $f''(x) = 12x^2 - 24x$

   The domain of f is $\mathbb{R}$.

b) $f'(x)$ exists for all values of x, so the only critical points of f are where $f'(x) = 0$.

   $4x^3 - 12x^2 = 0$

   $4x^2(x - 3) = 0$

   $4x^2 = 0$ or $x - 3 = 0$

   $x = 0$ or $x = 3$

   The critical values are 0 and 3.

   $f(0) = (0)^4 - 4(0)^3 + 10 = 10$

   $f(3) = (3)^4 - 4(3)^3 + 10 = -17$

   The critical points $(0, 10)$ and $(3, -17)$ are on the graph.

c) Applying the Second Derivative Test, we have:

   $f''(0) = 12(0)^2 - 24(0) = 0$

   The test fails, we will use the First Derivative Test.

   Divide $(-\infty, 3)$ into two intervals,

   A: $(-\infty, 0)$ and B: $(0, 3)$, and test a point in each interval.

   A: Test -1,

   $f'(-1) = 4(-1)^3 - 12(-1)^2 = -16 < 0$

   B: Test 1, $f'(1) = 4(1)^3 - 12(1)^2 = -8 < 0$

   Since, f is decreasing on both intervals, $(0, 10)$ is not a relative extremum.

   We use the Second Derivative Test for $x = 3$.

   $f''(3) = 12(3)^2 - 24(3) = 36 > 0$

   The critical point $(3, -17)$ is a relative minimum.

   When we applied the First Derivative Test, we saw that $f(x)$ was decreasing on the intervals $(-\infty, 0)$ and $(0, 3)$. Since $(3, -17)$ is a relative minimum, we know that $f(x)$ is increasing on $(3, \infty)$.

d) Find the points of inflection. $f''(x)$ exists for all values of x, so the only possible inflection points occur when $f''(x) = 0$.

   $12x^2 - 24x = 0$

   $12x(x - 2) = 0$

   $12x = 0$ or $x - 2 = 0$

   $x = 0$ or $x = 2$

   $f(0) = (0)^4 - 4(0)^3 + 10 = 10$

   $f(2) = (2)^4 - 4(2)^3 + 10 = -6$

   The points $(0, 10)$ and $(2, -6)$ are possible inflection points on the graph.

e) To determine concavity we use 0 and 2 to divide the real number line into three intervals,

   A : $(-\infty, 0)$, B: $(0, 2)$, and C: $(2, \infty)$, Then test a point in each interval.

A: Test -1,

$$f''(-1) = 12(-1)^2 - 24(-1) = 36 > 0$$

B: Test 1,

$$f''(1) = 12(1)^2 - 24(1) = -12 < 0$$

C: Test 3,

$$f''(3) = 12(3)^2 - 24(3) = 36 > 0$$

We see that f is concave up on the intervals $(-\infty, 0)$ and $(2, \infty)$ and concave down on the interval $(0,2)$. Therefore both $(0,10)$ and $(2,-6)$ are inflection points.

f) We sketch the graph using the preceding information. Additional function values may also be calculated as necessary.

x	$f(x)$
-2	58
-1	15
1	7
4	10
5	135

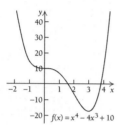

29. $f(x) = x^3 - 6x^2 + 12x - 6$

a) $f'(x) = 3x^2 - 12x + 12$

$f''(x) = 6x - 12$

The domain of f is $\mathbb{R}$.

b) $f'(x)$ exists for all values of x, so the only critical points of f are where $f'(x) = 0$.

$$3x^2 - 12x + 12 = 0$$
$$x^2 - 4x + 4 = 0 \qquad \text{Dividing by 3}$$
$$(x-2)^2 = 0$$
$$x - 2 = 0$$
$$x = 2$$

The critical value is 2.

$$f(2) = (2)^3 - 6(2)^2 + 12(2) - 6 = 2$$

The critical point $(2,2)$ is on the graph.

c) Applying the Second Derivative Test, we have:

$$f''(2) = 6(2) - 12 = 0$$

The test fails, we will use the First Derivative Test.

Divide the real line into two intervals,

A: $(-\infty, 2)$ and B: $(2, \infty)$, and test a point in each interval.

A: Test 0,

$$f'(0) = 3(0)^2 - 12(0) + 12 = 12 > 0$$

B: Test 3,

$$f'(3) = 3(3)^2 - 12(3) + 12 = 3 > 0$$

Since, f is increasing on both intervals, $(2,2)$ is not a relative extremum.

When we applied the First Derivative Test, we saw that $f(x)$ was increasing on the intervals $(-\infty, 2)$ and $(2, \infty)$.

d) Find the points of inflection. $f''(x)$ exists for all values of x, so the only possible inflection points occur when $f''(x) = 0$.

$$6x - 12 = 0$$
$$6x = 12$$
$$x = 2$$

We have already seen that $f(2) = 2$, so the point $(2,2)$ is a possible inflection point on the graph.

e) To determine concavity we use 2 to divide the real number line into two intervals, A : $(-\infty, 2)$ and B: $(2, \infty)$, Then test a point in each interval.

A: Test 0, $f''(0) = 6(0) - 12 = -12 < 0$

B: Test 3, $f''(3) = 6(3) - 12 = 6 > 0$

We see that $f(x)$ is concave down on the interval $(-\infty, 2)$ and concave up on the interval $(2, \infty)$. Therefore, the point $(2,2)$ is an inflection point.

f) We sketch the graph using the preceding information. Additional function values may also be calculated as necessary.

x	$f(x)$
-1	-25
0	-6
1	1
3	3
4	10

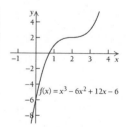

31. $f(x) = 5x^3 - 3x^5$

a) $f'(x) = 15x^2 - 15x^4$

$f''(x) = 30x - 60x^3$

The domain of f is $\mathbb{R}$.

b) $f'(x)$ exists for all values of x, so the only critical points of f are where $f'(x) = 0$.

$15x^2 - 15x^4 = 0$

$15x^2(1 - x^2) = 0$

$15x^2 = 0 \quad$ or $\quad 1 - x^2 = 0$

$x = 0 \quad$ or $\quad x = \pm 1$

The critical values are -1, 0, and 1.

$f(-1) = 5(-1)^3 - 3(-1)^5 = -2$

$f(0) = 5(0)^3 - 3(0)^5 = 0$

$f(1) = 5(1)^3 - 3(1)^5 = 2$

The critical points $(-1, -2)$, $(0, 0)$ and $(1, 2)$ are on the graph.

c) Applying the Second Derivative Test, we have:

$f''(-1) = 30(-1) - 60(-1)^3 = 30 > 0$

So, the critical point $(-1, -2)$ is a relative minimum.

$f''(0) = 30(0) - 60(0)^3 = 0$

The test fails, we will use the First Derivative Test.

Divide $(-1, 1)$ into two intervals, A: $(-1, 0)$ and B: $(0, 1)$, and test a point in each interval.

A: Test $-\dfrac{1}{2}$,

$f'\left(-\dfrac{1}{2}\right) = 15\left(-\dfrac{1}{2}\right)^2 - 15\left(-\dfrac{1}{2}\right)^4$

$= \dfrac{45}{16} > 0$

B: Test $\dfrac{1}{2}$,

$f'\left(\dfrac{1}{2}\right) = 15\left(\dfrac{1}{2}\right)^2 - 15\left(\dfrac{1}{2}\right)^4$

$= \dfrac{45}{16} > 0$

Since, f is increasing on both intervals, $(0, 0)$ is not a relative extremum.

We use the Second Derivative Test for $x = 1$.

$f''(1) = 30(1) - 60(1)^3 = -30 < 0$

The critical point $(1, 2)$ is a relative maximum.

When we applied the First Derivative Test, we saw that $f(x)$ was increasing on the intervals $(-1, 0)$ and $(0, 1)$. Since $(-1, -2)$ is a relative minimum, we know that $f(x)$ is decreasing on $(-\infty, -1)$. Since $(1, 2)$ is a relative maximum, we know that $f(x)$ is decreasing on $(1, \infty)$.

d) Find the points of inflection. $f''(x)$ exists for all values of x, so the only possible inflection points occur when $f''(x) = 0$.

$30x - 60x^3 = 0$

$30x(1 - 2x^2) = 0$

$30x = 0 \quad$ or $\quad 1 - 2x^2 = 0$

$x = 0 \quad$ or $\quad x^2 = \dfrac{1}{2}$

$x = 0 \quad$ or $\quad x = \pm\sqrt{\dfrac{1}{2}} = \pm\dfrac{1}{\sqrt{2}}$

$f\left(-\dfrac{1}{\sqrt{2}}\right) = 5\left(-\dfrac{1}{\sqrt{2}}\right)^3 - 3\left(-\dfrac{1}{\sqrt{2}}\right)^5$

$= -1.237$

$f(0) = 5(0)^3 - 3(0)^5 = 0$

$f\left(\dfrac{1}{\sqrt{2}}\right) = 5\left(\dfrac{1}{\sqrt{2}}\right)^3 - 3\left(\dfrac{1}{\sqrt{2}}\right)^5$

$= 1.237$

The points $\left(-\dfrac{1}{\sqrt{2}}, -1.237\right)$, $(0, 0)$ and $\left(\dfrac{1}{\sqrt{2}}, 1.237\right)$ are possible inflection points on the graph.

e) To determine concavity we use $-\dfrac{1}{\sqrt{2}}$, 0, and $\dfrac{1}{\sqrt{2}}$ to divide the real number line into four intervals, A: $\left(-\infty, -\dfrac{1}{\sqrt{2}}\right)$,

B: $\left(-\dfrac{1}{\sqrt{2}}, 0\right)$, C: $\left(0, \dfrac{1}{\sqrt{2}}\right)$, and

D: $\left(\dfrac{1}{\sqrt{2}}, \infty\right)$.

Then test a point in each interval.

A: Test -1, $\quad f''(-1) = 30(-1) - 60(-1)^3$

$= 30 > 0$

B: Test $-\dfrac{1}{2}$,

$$f''\left(-\frac{1}{2}\right) = 30\left(-\frac{1}{2}\right) - 60\left(-\frac{1}{2}\right)^3$$

$$= -\frac{15}{2} < 0$$

C: Test $\dfrac{1}{2}$,

$$f''\left(\frac{1}{2}\right) = 30\left(\frac{1}{2}\right) - 60\left(\frac{1}{2}\right)^3$$

$$= \frac{15}{2} > 0$$

D: Test 1 ,

$$f''(1) = 30(1) - 60(1)^3$$

$$= -30 < 0$$

We see that f is concave up on the intervals

$\left(-\infty, -\dfrac{1}{\sqrt{2}}\right)$ and $\left(0, \dfrac{1}{\sqrt{2}}\right)$ and concave

down on the intervals

$\left(-\dfrac{1}{\sqrt{2}}, 0\right)$ and $\left(\dfrac{1}{\sqrt{2}}, \infty\right)$. Therefore, the

points $\left(-\dfrac{1}{\sqrt{2}}, -1.237\right)$, $(0,0)$ and

$\left(\dfrac{1}{\sqrt{2}}, 1.237\right)$ are inflection points.

e)   We sketch the graph using the preceding
     information. Additional function values
     may also be calculated as necessary.

x	$f(x)$
-2	56
$-\frac{1}{2}$	$-\frac{17}{32}$
$\frac{1}{2}$	$\frac{17}{32}$
2	-56

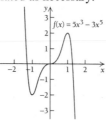

33. $f(x) = x^2 (3-x)^2$

$$= x^2 (9 - 6x + x^2)$$

$$= 9x^2 - 6x^3 + x^4$$

a) $f'(x) = 18x - 18x^2 + 4x^3$

$f''(x) = 18 - 36x + 12x^2$

The domain of f is $\mathbb{R}$.

b) $f'(x)$ exists for all values of x, so the only
     critical points of f are where $f'(x) = 0$.

$$18x - 18x^2 + 4x^3 = 0$$

$$2x(9 - 9x + 2x^2) = 0$$

$$2x(3 - 2x)(3 - x) = 0$$

$2x = 0$ or $3 - 2x = 0$ or $3 - x = 0$

$x = 0$ or $x = \dfrac{3}{2}$ or $x = 3$

The critical values are 0, $\dfrac{3}{2}$, and 3 .

$$f(0) = (0)^2 (3 - (0))^2 = 0$$

$$f\left(\frac{3}{2}\right) = \left(\frac{3}{2}\right)^2 \left(3 - \left(\frac{3}{2}\right)\right)^2 = \frac{81}{16}$$

$$f(3) = (3)^2 (3 - (3))^2 = 0$$

The critical points $(0,0)$, $\left(\dfrac{3}{2}, \dfrac{81}{16}\right)$, and

$(3,0)$ are on the graph.

c)   Applying the Second Derivative Test, we
     have:

$$f''(0) = 18 - 36(0) + 12(0)^2 = 18 > 0$$

So, the critical point $(0,0)$ is a relative
minimum.

$$f''\left(\frac{3}{2}\right) = 18 - 36\left(\frac{3}{2}\right) + 12\left(\frac{3}{2}\right)^2 = -9 < 0$$

So, the critical point $\left(\dfrac{3}{2}, \dfrac{81}{16}\right)$ is a relative

maximum.

$$f''(3) = 18 - 36(3) + 12(3)^2 = 18 > 0$$

So, the critical point $(3,0)$ is a relative
minimum.

We use the points 0, $\dfrac{3}{2}$, and 3 to divide the

real number line into four intervals,

$(-\infty, 0)$, $\left(0, \dfrac{3}{2}\right)$, $\left(\dfrac{3}{2}, 3\right)$, and $(3, \infty)$, we

know that $f(x)$ is decreasing on the

intervals $(-\infty, 0)$ and $\left(\dfrac{3}{2}, 3\right)$, and $f(x)$ is

increasing on the intervals

$\left(0, \dfrac{3}{2}\right)$ and $(3, \infty)$.

d) Find the points of inflection. $f''(x)$ exists for all values of x, so the only possible inflection points occur when $f''(x) = 0$.

$$18 - 36x + 12x^2 = 0$$

$$3 - 6x + 2x^2 = 0 \qquad \text{Dividing by 6}$$

Using the quadratic formula we have:

$$x = \frac{3 \pm \sqrt{3}}{2}$$

$$x \approx 0.634 \text{ or } x \approx 2.366$$

$$f(0.634) \approx 2.250$$

$$f(2.366) \approx 2.250$$

The points, $(0.634, 2.250)$ and $(2.366, 2.250)$ are possible inflection points on the graph.

e) To determine concavity we use 0.634 and 2.366 to divide the real number line into three intervals,

A: $(-\infty, 0.634)$, B: $(0.634, 2.366)$,

and C: $(2.366, \infty)$

Then test a point in each interval.
A: Test 0,

$$f''(0) = 18 - 36(0) + 12(0)^2$$
$$= 18 > 0$$

B: Test 1,

$$f''(1) = 18 - 36(1) + 12(1)^2$$
$$= -6 < 0$$

C: Test 3,

$$f''(3) = 18 - 36(3) + 12(3)^2$$
$$= 18 > 0$$

We see that f is concave up on the intervals $(-\infty, 0.634)$ and $(2.366, \infty)$ and concave down on the interval $(0.634, 2.366)$.

Therefore, the points $(0.634, 2.250)$ and $(2.366, 2.250)$ are inflection points.

f) We sketch the graph using the preceding information. Additional function values may also be calculated as necessary.

x	$f(x)$
-2	100
-1	16
1	4
2	4
4	16

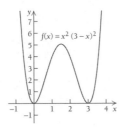

35. $f(x) = (x+1)^{\frac{2}{3}}$

a) $f'(x) = \frac{2}{3}(x+1)^{-\frac{1}{3}} = \frac{2}{3(x+1)^{\frac{1}{3}}}$

$f''(x) = -\frac{2}{9}(x+1)^{-\frac{4}{3}} = -\frac{2}{9(x+1)^{\frac{4}{3}}}$

The domain of f is $\mathbb{R}$.

b) $f'(x)$ does not exist for $x = -1$. The equation $f'(x) = 0$ has no solution, therefore, $x = -1$ is the only critical point.
$f(-1) = (-1+1)^{\frac{2}{3}} = 0$.
So, the critical point, $(-1, 0)$ is on the graph.

c) We apply the First Derivative Test. We use -1 to divide the real number line into two intervals A: $(-\infty, -1)$ and B: $(-1, \infty)$ and then we test a point in each interval.
A: Test -2,

$$f'(-2) = \frac{2}{3((-2)+1)^{\frac{1}{3}}} = -\frac{2}{3} < 0$$

B: Test 0,

$$f'(0) = \frac{2}{3((0)+1)^{\frac{1}{3}}} = \frac{2}{3} > 0$$

Thus, $(-1, 0)$ is a relative minimum. We also know that $f(x)$ is decreasing on the interval $(-\infty, -1)$ and increasing on the interval $(-1, \infty)$.

d) Find the points of inflection. $f''(x)$ does not exist when $x = -1$. The equation $f''(x) = 0$ has no solution, so $x = -1$ is the only possible inflection point. We know that $f(-1) = 0$.

e) To determine concavity, we divide the real number line into two intervals, A: $(-\infty, -1)$ and B: $(-1, \infty)$ and then we test a point in each interval.
A: Test -2,

$$f''(-2) = -\frac{2}{9((-2)+1)^{\frac{4}{3}}} = -\frac{2}{9} < 0$$

B: Test 0,

$$f''(0) = -\frac{2}{9((0)+1)^{\frac{4}{3}}} = -\frac{2}{9} < 0$$

Thus, $f(x)$ is concave down on the interval $(-\infty, -1)$ and on the interval $(-1, \infty)$. Therefore, the point $(-1, 0)$ is not an inflection point.

f) We sketch the graph using the preceding information. Additional function values may also be calculated as necessary.

x	$f(x)$
-9	4
-2	1
0	1
7	4

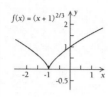

$f(x) = (x+1)^{2/3}$

37. $f(x) = (x-3)^{\frac{1}{3}} - 1$

a) $f'(x) = \frac{1}{3}(x-3)^{-\frac{2}{3}} = \frac{1}{3(x-3)^{\frac{2}{3}}}$

$f''(x) = -\frac{2}{9}(x-3)^{-\frac{5}{3}} = -\frac{2}{9(x-3)^{\frac{5}{3}}}$

The domain of f is $\mathbb{R}$.

b) $f'(x)$ does not exist for $x = 3$. The equation $f'(x) = 0$ has no solution, therefore, $x = 3$ is the only critical point.

$f(3) = ((3)-3)^{\frac{1}{3}} - 1 = -1$.

So, the critical point, $(3, -1)$ is on the graph.

c) We apply the First Derivative Test. We use 3 to divide the real number line into two intervals $A: (-\infty, 3)$ and $B: (3, \infty)$ and then we test a point in each interval.

A: Test 2, $f'(2) = \frac{1}{3((2)-3)^{\frac{2}{3}}} = \frac{1}{3} > 0$

B: Test 4, $f'(4) = \frac{1}{3((4)-3)^{\frac{2}{3}}} = \frac{1}{3} > 0$

$f(x)$ is increasing on both intervals $(-\infty, 3)$ and $(3, \infty)$, therefore $(3, -1)$ is not a relative extremum.

d) Find the points of inflection. $f''(x)$ does not exist when $x = 3$. The equation $f''(x) = 0$ has no solution, so $x = 3$ is the only possible inflection point. We know that $f(3) = -1$.

e) To determine concavity, we divide the real number line into two intervals, $A: (-\infty, 3)$ and B: $(3, \infty)$ and then we test a point in each interval.

A: Test 2, $f''(2) = -\frac{2}{9((2)-3)^{\frac{5}{3}}} = \frac{2}{9} > 0$

B: Test 4, $f''(4) = -\frac{2}{9((4)-3)^{\frac{5}{3}}} = -\frac{2}{9} < 0$

Thus, $f(x)$ is concave up on the interval $(-\infty, 3)$ and $f(x)$ is concave down on the interval $(3, \infty)$. Therefore, the point $(3, -1)$ is an inflection point.

f) We sketch the graph using the preceding information. Additional function values may also be calculated as necessary.

x	$f(x)$
-5	-3
2	-2
4	0
11	1

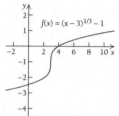

$f(x) = (x-3)^{1/3} - 1$

39. $f(x) = -2(x-4)^{\frac{2}{3}} + 5$

a) $f'(x) = -\frac{4}{3}(x-4)^{-\frac{1}{3}} = -\frac{4}{3(x-4)^{\frac{1}{3}}}$

$f''(x) = \frac{4}{9}(x-4)^{-\frac{4}{3}} = \frac{4}{9(x-4)^{\frac{4}{3}}}$

The domain of f is $\mathbb{R}$.

b) $f'(x)$ does not exist for $x = 4$. The equation $f'(x) = 0$ has no solution, therefore, $x = 4$ is the only critical point.

$f(4) = -2((4)-4)^{\frac{2}{3}} + 5 = 5$.

So, the critical point $(4, 5)$, is on the graph.

c) We apply the First Derivative Test. We use 4 to divide the real number line into two intervals $A: (-\infty, 4)$ and B: $(4, \infty)$ and then we test a point in each interval.

A: Test 3, $f'(3) = -\frac{4}{3((3)-4)^{\frac{1}{3}}} = \frac{4}{3} > 0$

B: Test 5, $f'(5) = -\frac{4}{3((5)-4)^{\frac{1}{3}}} = -\frac{4}{3} < 0$

Thus, $(4,5)$ is a relative maximum. We also know that $f(x)$ is increasing on the interval $(-\infty, 4)$ and decreasing on the interval $(4, \infty)$.

d) Find the points of inflection. $f''(x)$ does not exist when $x = 4$. The equation $f''(x) = 0$ has no solution, so $x = 4$ is the only possible inflection point. We know that $f(4) = 5$.

e) To determine concavity, we divide the real number line into two intervals, $A : (-\infty, 4)$ and $B : (4, \infty)$ and then we test a point in each interval.

A: Test 3, $f''(3) = \dfrac{4}{9\left((3)-4\right)^{4/3}} = \dfrac{4}{9} > 0$

B: Test 5, $f''(5) = -\dfrac{4}{9\left((5)-4\right)^{4/3}} = \dfrac{4}{9} > 0$

Thus, $f(x)$ is concave up on both intervals $(-\infty, 4)$ and $(4, \infty)$ Therefore, the point $(4,5)$ is not an inflection point.

f) We sketch the graph using the preceding information. Additional function values may also be calculated as necessary.

x	$f(x)$
-4	-3
3	3
5	3
12	-3

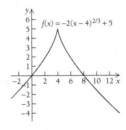

$f(x) = -2(x-4)^{2/3} + 5$

41. $f(x) = x\sqrt{4-x^2} = x\left(4-x^2\right)^{1/2}$

a) $f'(x) = x \cdot \dfrac{1}{2}\left(4-x^2\right)^{-1/2}(-2x)$

$\qquad + \left(4-x^2\right)^{1/2} \cdot (1)$

Next, we simplify the derivative.

$f'(x) = \dfrac{-x^2}{\left(4-x^2\right)^{1/2}} + \left(4-x^2\right)^{1/2}$

$\qquad = \dfrac{-x^2 + 4 - x^2}{\left(4-x^2\right)^{1/2}}$

$\qquad = \dfrac{4 - 2x^2}{\left(4-x^2\right)^{1/2}}$

$\qquad = \left(4 - 2x^2\right)\left(4-x^2\right)^{-1/2}$

$f''(x)$

$= \left(4 - 2x^2\right)\left(-\dfrac{1}{2}\right)\left(4-x^2\right)^{-3/2}(-2x) +$

$\qquad \left(4-x^2\right)^{-1/2}(-4x)$

$= \dfrac{x\left(4 - 2x^2\right)}{\left(4-x^2\right)^{3/2}} - \dfrac{4x}{\left(4-x^2\right)^{1/2}}$

$= \dfrac{4x - 2x^3 - 4x\left(4-x^2\right)}{\left(4-x^2\right)^{3/2}}$

$= \dfrac{4x - 2x^3 - 16x + 4x^3}{\left(4-x^2\right)^{3/2}}$

$= \dfrac{2x^3 - 12x}{\left(4-x^2\right)^{3/2}}$

The domain of $f(x)$ is $[-2, 2]$.

b) First, we find the critical points. $f'(x)$ does not exist when $4 - x^2 = 0$.

Solve:

$4 - x^2 = 0$

$\qquad x^2 = 4$

$\qquad x = \pm\sqrt{4}$

$\qquad x = \pm 2$

Since the domain of $f(x)$ is $[-2, 2]$, relative extrema can not occur at $x = -2$ or $x = 2$ because there is not an open interval containing -2 or 2 on which the function is defined. For this reason, we do not consider -2 or 2 in our discussion of relative extrema.

The other critical points occur where $f'(x) = 0$.

We set the derivative equal to zero and solve for x.

$$f'(x) = 0$$

$$\frac{4-2x^2}{\sqrt{4-x^2}} = 0$$

$$4 - 2x^2 = 0$$

$$2x^2 = 4$$

$$x^2 = 2$$

$$x = \pm\sqrt{2}$$

The critical values are $-\sqrt{2}$ and $\sqrt{2}$.

$$f\left(-\sqrt{2}\right) = -\sqrt{2}\sqrt{4-\left(-\sqrt{2}\right)^2} = -\sqrt{2}\sqrt{2} = -2$$

$$f\left(\sqrt{2}\right) = \sqrt{2}\sqrt{4-\left(\sqrt{2}\right)^2} = \sqrt{2}\sqrt{2} = 2$$

Therefore, $\left(-\sqrt{2},-2\right)$ and $\left(\sqrt{2},2\right)$ are critical points on the graph.

c) We use the Second Derivative Test.

$$f''\left(-\sqrt{2}\right) = \frac{2\left(-\sqrt{2}\right)^3 - 12\left(-\sqrt{2}\right)}{\left[4-\left(-\sqrt{2}\right)^2\right]^{\frac{3}{2}}}$$

$$= \frac{-4\sqrt{2}+12\sqrt{2}}{2^{\frac{3}{2}}} = \frac{8\sqrt{2}}{2\sqrt{2}} = 4 > 0$$

The critical point $\left(-\sqrt{2},-2\right)$ is a relative minimum.

$$f''\left(\sqrt{2}\right) = \frac{2\left(\sqrt{2}\right)^3 - 12\left(\sqrt{2}\right)}{\left[4-\left(\sqrt{2}\right)^2\right]^{\frac{3}{2}}}$$

$$= \frac{4\sqrt{2}-12\sqrt{2}}{2^{\frac{3}{2}}} = \frac{-8\sqrt{2}}{2\sqrt{2}} = -4 < 0$$

The critical point $\left(\sqrt{2},2\right)$ is a relative maximum.

If we use the points $-\sqrt{2}$ and $\sqrt{2}$ to divide the interval $\left[-2,2\right]$ into three intervals $\left[-2,-\sqrt{2}\right)$, $\left(-\sqrt{2},\sqrt{2}\right)$, and $\left(\sqrt{2},2\right]$, we see that $f(x)$ is decreasing on the intervals $\left[-2,-\sqrt{2}\right]$ and $\left[\sqrt{2},2\right]$ and $f(x)$ is increasing on the interval $\left[-\sqrt{2},\sqrt{2}\right]$.

d) Find the points of inflection. $f''(x)$ does not exist where $4 - x^2 = 0$. We know that this occurs at $x = -2$ and $x = 2$. However, just as relative extrema cannot occur at $\left(-2,0\right)$ and $\left(2,0\right)$, they can not be inflection points either. Inflection points could occur where $f''(x) = 0$.

$$\frac{2x^3-12x}{\left(4-x^2\right)^{\frac{3}{2}}} = 0$$

$$2x^3 - 12x = 0$$

$$2x\left(x^2-6\right) = 0$$

$$2x = 0 \quad \text{or} \quad x^2 - 6 = 0$$

$$x = 0 \quad \text{or} \quad x^2 = 6$$

$$x = 0 \quad \text{or} \quad x = \pm\sqrt{6}$$

Note that $f(x)$ is not defined for $x = \pm\sqrt{6}$. Therefore, the only possible inflection point is $x = 0$.

$$f(0) = (0)\sqrt{4-(0)^2} = 0.$$

Therefore, $(0,0)$ is a possible inflection point on the graph.

e) To determine concavity, we use 0 to divide the interval $(-2,2)$ into two intervals, A: $(-2,0)$ and B: $(0,2)$ and then we test a point in each interval.

A: Test -1,

$$f''(-1) = \frac{2(-1)^3 - 12(-1)}{\left[4-(-1)^2\right]^{\frac{3}{2}}} = \frac{10}{3^{\frac{3}{2}}} > 0$$

B: Test 1,

$$f''(1) = \frac{2(1)^3 - 12(1)}{\left[4-(1)^2\right]^{\frac{3}{2}}} = \frac{-10}{3^{\frac{3}{2}}} < 0$$

Thus, $f(x)$ is concave up on the interval $(-2,0)$ and $f(x)$ is concave down on the interval $(0,2)$. Therefore, the point $(0,0)$ is an inflection point.

f) We sketch the graph using the preceding information. Additional function values may also be calculated as necessary.

x	$f(x)$
-1	$-\sqrt{3}$
1	$\sqrt{3}$

43. $f(x) = \dfrac{x}{x^2+1}$

a) $f'(x) = \dfrac{(x^2+1)(1) - x(2x)}{(x^2+1)^2}$ Quotient Rule

$= \dfrac{x^2+1-2x^2}{(x^2+1)^2}$

$= \dfrac{1-x^2}{(x^2+1)^2}$

$f''(x)$

$= \dfrac{(x^2+1)^2(-2x) - (1-x^2)\left[2(x^2+1)^1(2x)\right]}{\left((x^2+1)^2\right)^2}$

$= \dfrac{(x^2+1)\left[-2x(x^2+1) - 4x(1-x^2)\right]}{(x^2+1)^4}$

$= \dfrac{-2x^3 - 2x - 4x + 4x^3}{(x^2+1)^3}$

$= \dfrac{2x^3 - 6x}{(x^2+1)^3}$

The domain of f is $\mathbb{R}$.

b) Since $f'(x)$ exists for all real numbers, the only critical values are where $f'(x) = 0$.

$\dfrac{1-x^2}{(x^2+1)^2} = 0$

$1 - x^2 = 0$ Multiplying by $(x^2+1)^2$

$x^2 = 1$

$x = \pm\sqrt{1} = \pm 1$

The two critical values are $x = -1$ and $x = 1$.

$f(-1) = \dfrac{-1}{(-1)^2+1} = -\dfrac{1}{2}$

$f(1) = \dfrac{1}{(1)^2+1} = \dfrac{1}{2}$

The critical points $\left(-1, -\dfrac{1}{2}\right)$ and $\left(1, \dfrac{1}{2}\right)$ are on the graph.

c) We use the Second Derivative Test.

$f''(-1) = \dfrac{2(-1)^3 - 6(-1)}{\left[(-1)^2+1\right]^3} = \dfrac{4}{8} = \dfrac{1}{2} > 0$

So the point $\left(-1, -\dfrac{1}{2}\right)$ is a relative minimum.

$f''(1) = \dfrac{2(1)^3 - 6(1)}{\left[(1)^2+1\right]^3} = \dfrac{-4}{8} = -\dfrac{1}{2} < 0$

So the point $\left(1, \dfrac{1}{2}\right)$ is a relative maximum.

We use -1 and 1 to divide the real number line into three intervals $(-\infty, -1)$, $(-1, 1)$, and $(1, \infty)$. $f(x)$ is decreasing on the intervals $(-\infty, 1]$ and $[1, \infty)$, and $f(x)$ is increasing on the interval $[-1, 1]$.

d) Find the points of inflection. $f''(x)$ exists for all real numbers, so the only possible points of inflection occur when $f''(x) = 0$.

$\dfrac{2x^3 - 6x}{(x^2+1)^3} = 0$

$2x^3 - 6x = 0$

$2x(x^2 - 3) = 0$

$2x = 0$ or $x^2 - 3 = 0$

$x = 0$ or $x^2 = 3$

$x = 0$ or $x = \pm\sqrt{3}$

There are three possible inflection points $-\sqrt{3}, 0$, and $\sqrt{3}$.

$f(-\sqrt{3}) = \dfrac{-\sqrt{3}}{(-\sqrt{3})^2+1} = -\dfrac{\sqrt{3}}{4}$

$$f(0) = \frac{\sqrt{0}}{\left(\sqrt{0}\right)^2 + 1} = \frac{0}{1} = 0$$

$$f\left(\sqrt{3}\right) = \frac{\sqrt{3}}{\left(\sqrt{3}\right)^2 + 1} = \frac{\sqrt{3}}{4}$$

The points $\left(-\sqrt{3}, -\dfrac{\sqrt{3}}{4}\right)$, $(0,0)$, and

$\left(\sqrt{3}, \dfrac{\sqrt{3}}{4}\right)$ are three possible inflection

points on the graph.

e) To determine concavity we use
$-\sqrt{3}, 0,$ and $\sqrt{3}$ to divide the real number
line into four intervals,

A: $\left(-\infty, -\sqrt{3}\right)$, B: $\left(-\sqrt{3}, 0\right)$, C: $\left(0, \sqrt{3}\right)$,

and D: $\left(\sqrt{3}, \infty\right)$

Then test a point in each interval.

A: Test -2, $f''(-2) = -\dfrac{4}{125} < 0$

B: Test -1, $f''(-1) = \dfrac{1}{2} > 0$

C: Test 1, $f''(1) = -\dfrac{1}{2} < 0$

D: Test 2, $f''(2) = \dfrac{4}{125} > 0$

We see that f is concave down on the
intervals $\left(-\infty, -\sqrt{3}\right)$ and $\left(0, \sqrt{3}\right)$ and
concave up on the
intervals $\left(-\sqrt{3}, 0\right)$ and $\left(\sqrt{3}, \infty\right)$. Therefore

the points $\left(-\sqrt{3}, -\dfrac{\sqrt{3}}{4}\right)$, $(0,0)$, and

$\left(\sqrt{3}, \dfrac{\sqrt{3}}{4}\right)$ are inflection points.

f) We sketch the graph using the preceding
information. Additional function values
may also be calculated as necessary.

x	$f(x)$
-3	$-\dfrac{3}{10}$
-2	$-\dfrac{2}{5}$
2	$\dfrac{2}{5}$
3	$\dfrac{3}{10}$

45. $f(x) = \dfrac{3}{x^2 + 1} = 3\left(x^2 + 1\right)^{-1}$

a) $f'(x) = 3(-1)\left(x^2 + 1\right)^{-2}(2x)$

$= -6x\left(x^2 + 1\right)^{-2}$

$= \dfrac{-6x}{\left(x^2 + 1\right)^2}$

$f''(x)$

$= \dfrac{\left(x^2 + 1\right)^2(-6) - (-6x)\left(2\left(x^2 + 1\right)(2x)\right)}{\left(\left(x^2 + 1\right)^2\right)^2}$

$= \dfrac{\left(x^2 + 1\right)\left[\left(x^2 + 1\right)(-6) - (-6x)(2)(2x)\right]}{\left(x^2 + 1\right)^4}$

$= \dfrac{-6x^2 - 6 + 24x^2}{\left(x^2 + 1\right)^3}$

$= \dfrac{18x^2 - 6}{\left(x^2 + 1\right)^3}$

The domain of f is $\mathbb{R}$.

b) Since $f'(x)$ exists for all real numbers, the
only critical values are where $f'(x) = 0$.

$\dfrac{-6x}{\left(x^2 + 1\right)^2} = 0$

$-6x = 0$ Multiplying
by $\left(x^2 + 1\right)^2$

$x = 0$

The critical value is $x = 0$.

$f(0) = \dfrac{3}{(0)^2 + 1} = 3$

The critical point $(0, 3)$ is on the graph.

c) We use the Second Derivative Test.

$f''(0) = \dfrac{18(0) - 6}{\left((0)^2 + 1\right)^2} = \dfrac{-6}{1} = -6 < 0$

So the point $(0, 3)$ is a relative maximum.
We use 0 to divide the real number line into
two intervals $(-\infty, 0)$ and $(0, \infty)$. $f(x)$ is
increasing on the interval $(-\infty, 0]$, and
$f(x)$ is decreasing on the interval $[0, \infty)$.

d) Find the points of inflection. $f''(x)$ exists
for all real numbers, so the only possible
points of inflection occur when $f''(x) = 0$.

$$\frac{18x^2-6}{\left(x^2+1\right)^3}=0$$

$$18x^2-6=0$$

$$18x^2=6$$

$$x^2=\frac{1}{3}$$

$$x=\pm\frac{1}{\sqrt{3}}$$

There are two possible inflection points

$-\dfrac{1}{\sqrt{3}}$ and $\dfrac{1}{\sqrt{3}}$.

$$f\left(-\frac{1}{\sqrt{3}}\right)=\frac{3}{\left(-\frac{1}{\sqrt{3}}\right)^2+1}$$

$$=\frac{3}{\frac{1}{3}+1}=\frac{3}{\frac{4}{3}}=\frac{9}{4}$$

$$f\left(\frac{1}{\sqrt{3}}\right)=\frac{3}{\left(\frac{1}{\sqrt{3}}\right)^2+1}$$

$$=\frac{3}{\frac{1}{3}+1}=\frac{3}{\frac{4}{3}}=\frac{9}{4}$$

The points $\left(-\dfrac{1}{\sqrt{3}},\dfrac{9}{4}\right)$ and $\left(\dfrac{1}{\sqrt{3}},\dfrac{9}{4}\right)$ are

possible inflection points on the graph.

e) To determine concavity we use

$-\dfrac{1}{\sqrt{3}}$ and $\dfrac{1}{\sqrt{3}}$ to divide the real number

line into three intervals,

$$A:\left(-\infty,-\frac{1}{\sqrt{3}}\right),$$

$$B:\left(-\frac{1}{\sqrt{3}},\frac{1}{\sqrt{3}}\right),\text{ and C: }\left(\frac{1}{\sqrt{3}},\infty\right)$$

Then test a point in each interval.

A: Test $-1,\ f''(-1)=\dfrac{18(-1)^2-6}{\left((-1)^2+1\right)^3}=\dfrac{3}{2}>0$

B: Test $0,\ f''(0)=\dfrac{18(0)^2-6}{\left((0)^2+1\right)^3}=-6<0$

C: Test $1,\ f''(1)=\dfrac{18(1)^2-6}{\left((1)^2+1\right)^3}=\dfrac{3}{2}>0$

We see that f is concave up on the intervals

$\left(-\infty,-\dfrac{1}{\sqrt{3}}\right)$ and $\left(\dfrac{1}{\sqrt{3}},\infty\right)$ and concave

down on the interval $\left(-\dfrac{1}{\sqrt{3}},\dfrac{1}{\sqrt{3}}\right)$. Therefore

the points $\left(-\dfrac{1}{\sqrt{3}},\dfrac{9}{4}\right)$ and $\left(\dfrac{1}{\sqrt{3}},\dfrac{9}{4}\right)$ are

inflection points.

f) We sketch the graph using the preceding
information. Additional function values
may also be calculated as necessary.

x	$f(x)$
-3	$\frac{3}{10}$
-1	$\frac{3}{2}$
1	$\frac{3}{2}$
3	$\frac{3}{10}$

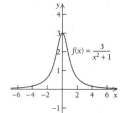

47. Answers may vary, one possible graph is:

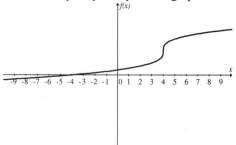

49. Answers may vary, one possible graph is:

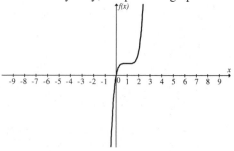

51. Answers may vary, one possible graph is:

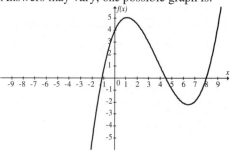

53. Answers may vary, one possible graph is:

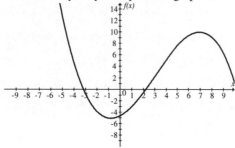

55. – 99.   Left to the Student.

101. $R(x) = 50x - 0.5x^2$

$C(x) = 4x + 10$

$P(x) = R(x) - C(x)$

$\quad = (50x - 0.5x^2) - (4x + 10)$

$\quad = -0.5x^2 + 46x - 10$

We will restrict the domains of all three functions to $x \geq 0$ since a negative number of units cannot be produced and sold.

First graph $R(x) = 50x - 0.5x^2$

$R'(x) = 50 - x$

$R''(x) = -1$

Since $R'(x)$ exists for all $x \geq 0$, the only critical points are where $R'(x) = 0$.

$50 - x = 0$

$\quad 50 = x$ Critical Value

Find the function value at $x = 50$.

$R(50) = 50(50) - 0.5(50)^2$

$\quad\quad = 2500 - 1250$

$\quad\quad = 1250$

This critical point $(50, 1250)$ is on the graph. We use the Second Derivative Test:

$R''(50) = -1 < 0$

The point $(50, 1250)$ is a relative maximum.

We use 50 to divide the interval $[0, \infty)$ into two intervals, $(0, 50)$ and $(50, \infty)$, we know that R is increasing on $[0, 50]$ and decreasing on $[50, \infty)$.

Next, find the inflection points. Since $R''(x)$ exists for all $x \geq 0$, and $R''(x) = -1$, there are no possible inflection points. Furthermore, since $R''(x) < 0$ for all $x \geq 0$, R is concave down over the interval $(0, \infty)$.

Sketch the graph using the preceding information. The x-intercepts of R are found by solving $R(x) = 0$.

$50x - 0.5x^2 = 0$

$0.5x(100 - x) = 0$

$0.5x = 0$ or $100 - x = 0$

$\quad x = 0$ or $100 = x$

The x-intercepts are $(0, 0)$ and $(100, 0)$.

Next, we graph $C(x) = 4x + 10$. This is a linear function with slope 4 and y-intercept $(0, 10)$.

$C(x)$ is increasing over the entire domain $x \geq 0$ and has no relative extrema or points of inflection.

Finally, we graph $P(x) = -0.5x^2 + 46x - 10$

$P'(x) = -x + 46$

$P''(x) = -1$

Since $P'(x)$ exists for all $x \geq 0$, the only critical points occur when $P'(x) = 0$.

$-x + 46 = 0$

$\quad 46 = x$ Critical Value

Find the function value at $x = 46$.

$P(46) = -0.5(46)^2 + 46(46) - 10$

$\quad\quad = -1058 + 2116 - 10$

$\quad\quad = 1048$

The critical point $(46, 1048)$ is on the graph. We use the Second Derivative Test:

$P''(46) = -1 < 0$

The point $(46, 1048)$ is a relative maximum.

We use 46 to divide the interval $[0, \infty)$ into two intervals, $(0, 46)$ and $(46, \infty)$, we know that P is increasing on $[0, 46]$ and decreasing on $[46, \infty)$.

Next, find the inflection points. Since $P''(x)$ exists for all $x \geq 0$, and $P''(x) = -1$, there are no possible inflection points. Furthermore, since $P''(x) < 0$ for all $x \geq 0$, P is concave down over the interval $(0, \infty)$.

Sketch the graph using the preceding information.

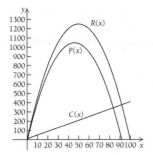

103. $p(x) = \dfrac{13x^3 - 240x^2 - 2460x + 585,000}{75,000}$

$p'(x) = \dfrac{39x^2 - 480x - 2460}{75,000}$

$p''(x) = \dfrac{78x - 480}{75,000}$

Since $p'(x)$ exists for all real numbers, the only critical points are where $p'(x) = 0$.

$\dfrac{39x^2 - 480x - 2460}{75,000} = 0$

$39x^2 - 480x - 2460 = 0$

We use the quadratic formula.

$x = \dfrac{-b \pm \sqrt{b^2 - 4ac}}{2a}$

$= \dfrac{-(-480) \pm \sqrt{(-480)^2 - 4(39)(-2460)}}{2(39)}$

$= \dfrac{480 \pm \sqrt{614,160}}{78}$

$x \approx -3.89$ or $x \approx 16.20$ Critical values

Since the domain of the function is $0 \le x \le 40$, we consider only $x \approx 16.20$

$p(16.20)$

$= \dfrac{13(16.20)^3 - 240(16.20)^2 - 2460(16.20) - 585,000}{75,000}$

≈ 7.17

The critical point $(16.20, 7.17)$ is on the graph.

We use the Second Derivative Test:

$p''(x) = \dfrac{78(16.20) - 480}{75,000} \approx 0.01 > 0$

The point $(16.20, 7.17)$ is a relative minimum.

If we use the point 16.20 to divide the domain into two intervals, $[0, 16.20)$ and $(16.20, 40]$, we know that p is decreasing on $[0, 16.20]$ and increasing on $[16.20, 40]$.

Next, we find the inflection points. $p''(x)$ exists for all real numbers, so the only possible inflection points are where $p''(x) = 0$

$\dfrac{78x - 480}{75,000} = 0$

$78x - 480 = 0$

$78x = 480$

$x \approx 6.15$

$p(6.15)$

$= \dfrac{13(6.15)^3 - 240(6.15)^2 - 2460(6.15) - 585,000}{75,000}$

≈ 7.52

The point $(6.15, 7.52)$ is a possible inflection point.

To determine concavity, we use 6.15 to divide the domain into two intervals

A: $[0, 6.15)$ and B: $(6.15, 40]$ and test a point in each interval.

A: Test 1, $p''(1) = \dfrac{78(1) - 480}{75,000} = -0.005 < 0$

B: Test 7, $p''(7) = \dfrac{78(7) - 480}{75,000} = 0.00088 > 0$

Then p is concave down on $(0, 6.15)$ and concave up on $(6.15, 40)$ and the point $(6.15, 7.52)$ is a point of inflection.

Sketch the graph for $0 \le x \le 40$ using the preceding information. Additional function values may be calculated if necessary.

x	$p(x)$
0	7.8
8	7.42
12	7.25
20	7.25
24	7.57
32	9.15
40	12.46

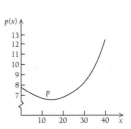

105. $V(r) = k(20r^2 - r^3), \quad 0 \le r \le 20$

$V'(r) = k(40r - 3r^2)$

$V''(r) = k(40 - 6r)$

$V'(r)$ exists for all r in $[0, 20]$, so the only critical points occur where $V'(r) = 0$.

$k(40r - 3r^2) = 0$

$40r - 3r^2 = 0$

$r(40 - 3r) = 0$

$r = 0 \quad$ or $\quad 40 - 3r = 0$

$r = 0 \quad$ or $\quad 40 = 3r$

$r = 0 \quad$ or $\quad \dfrac{40}{3} = r$

Using the Second Derivative Test:

$V''(0) = k(40 - 6(0)) = 40k > 0 \qquad [k > 0]$

$V''\left(\dfrac{40}{3}\right) = k\left(40 - 6\left(\dfrac{40}{3}\right)\right) = -40k < 0$

Since $V''\left(\dfrac{40}{3}\right) < 0$, we know that there is a

relative maximum at $x = \dfrac{40}{3}$. Thus, for an

object whose radius is $\dfrac{40}{3}$mm or 13.33mm, the

maximum velocity is needed to remove the object.

107. $\boxed{tw}$

109. $\boxed{tw}$

111. $f(x) = ax^2 + bx + c, \quad a \ne 0$

$f'(x) = 2ax + b$

$f''(x) = 2a$

Since $f'(x)$ exists for all real numbers, the only critical points occur when $f'(x) = 0$. We solve:

$2ax + b = 0$

$2ax = -b$

$x = \dfrac{-b}{2a}$

So the critical value will occur at $x = \dfrac{-b}{2a}$.

Applying the second derivative test, we see that

$f''(x) = 2a > 0, \quad$ for $a > 0$

$f''(x) = 2a < 0, \quad$ for $a < 0$

Therefore, a relative maximum occurs at

$x = \dfrac{-b}{2a}$ when $a < 0$ and a relative minimum

occurs at $x = \dfrac{-b}{2a}$ when $a > 0$.

113. $f(x) = 4x - 6x^{\frac{2}{3}}$

Graphing the function on the calculator we have:

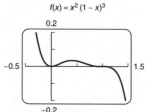

$f(x) = 4x - 6x^{2/3}$

Using the minimum/maximum feature on the calculator, we estimate a relative maximum at $(0, 0)$ and a relative minimum at $(1, -2)$.

115. $f(x) = x^2(1 - x)^3$

Graphing the function on the calculator we have:

$f(x) = x^2(1 - x)^3$

Using the minimum/maximum feature on the calculator, we estimate a relative maximum at $(0.4, 0.035)$ and a relative minimum at $(0, 0)$.

117. $f(x) = (x - 1)^{\frac{2}{3}} - (x + 1)^{\frac{2}{3}}$

Graphing the function on the calculator we have:

$f(x) = (x - 1)^{2/3} - (x + 1)^{2/3}$

Using the minimum/maximum feature on the calculator, we estimate a relative maximum at $(-1, 1.587)$ and a relative minimum at $(1, -1.587)$.

Exercise Set 2.3

1. $f(x) = \dfrac{2x-3}{x-5}$

The expression is in simplified form. We set the denominator equal to zero and solve.
$$x - 5 = 0$$
$$x = 5$$
The vertical asymptote is the line $x = 5$.

3. $f(x) = \dfrac{3x}{x^2 - 9}$

First, we write the function in simplified form.
$$f(x) = \dfrac{3x}{(x-3)(x+3)}.$$
Once the expression is in simplified form, we set the denominator equal to zero and solve.
$$(x-3)(x+3) = 0$$
$$x - 3 = 0 \quad \text{or} \quad x + 3 = 0$$
$$x = 3 \quad \text{or} \quad x = -3$$
The vertical asymptotes are the lines
$x = -3$ and $x = 3$.

5. $f(x) = \dfrac{x+2}{x^3 - 6x^2 + 8x}$

First, we write the function in simplified form.

$f(x) = \dfrac{x+2}{x(x^2 - 6x + 8)}$ Factor out x.

$ = \dfrac{x+2}{x(x-4)(x-2)}$

Once the expression is in simplified form, we set the denominator equal to zero and solve.
$$x(x-2)(x-4) = 0$$
$$x = 0 \quad \text{or} \quad x - 2 = 0 \quad \text{or} \quad x - 4 = 0$$
$$x = 0 \quad \text{or} \quad x = 2 \quad \text{or} \quad x = 4$$
The vertical asymptotes are the lines
$x = 0$, $x = 2$, and $x = 4$.

7. $f(x) = \dfrac{x+6}{x^2 + 7x + 6}$

First, we write the function in simplified form.

$f(x) = \dfrac{x+6}{(x+6)(x+1)}$

$ = \dfrac{1}{x+1}$ Dividing common terms

Once the expression is in simplified form, we set the denominator equal to zero and solve.
$$x + 1 = 0$$
$$x = -1$$
The vertical asymptote is the line $x = -1$.

9. $f(x) = \dfrac{6}{x^2 + 36}$

The function is in simplified form. The equation $x^2 + 36 = 0$ has no real solution; therefore, the function does not have any vertical asymptotes.

11. $f(x) = \dfrac{6x}{8x+3}$

To find the horizontal asymptote, we consider $\lim\limits_{x \to \infty} f(x)$. To find the limit, we will use some algebra and the fact that as $x \to \infty$, $\dfrac{b}{ax^n} \to 0$ for any positive integer n.

$\lim\limits_{x \to \infty} f(x) = \lim\limits_{x \to \infty} \dfrac{6x}{8x+3}$

$= \lim\limits_{x \to \infty} \dfrac{6x}{8x+3} \cdot \dfrac{\frac{1}{x}}{\frac{1}{x}}$ Multiplying by a form of 1

$= \lim\limits_{x \to \infty} \dfrac{\frac{6x}{x}}{\frac{8x}{x} + \frac{3}{x}}$

$= \lim\limits_{x \to \infty} \dfrac{6}{8 + \frac{3}{x}}$

$= \dfrac{6}{8+0}$ $\left[\text{as } x \to \infty, \dfrac{b}{ax^n} \to 0\right]$

$= \dfrac{6}{8} = \dfrac{3}{4}.$

In a similar manner, it can be shown that
$$\lim_{x \to -\infty} f(x) = \frac{3}{4}.$$

The horizontal asymptote is the line $y = \dfrac{3}{4}$.

13. $f(x) = \dfrac{4x}{x^2 - 3x}$

To find the horizontal asymptote, we consider $\lim\limits_{x \to \infty} f(x)$. To find the limit, we will use some algebra and the fact that as $x \to \infty$, $\dfrac{b}{ax^n} \to 0$ for any positive integer n.

$$\lim_{x \to \infty} f(x) = \lim_{x \to \infty} \frac{4x}{x^2 - 3x}$$

$$= \lim_{x \to \infty} \frac{4x}{x^2 - 3x} \cdot \frac{\frac{1}{x^2}}{\frac{1}{x^2}}$$ Multiplying by a form of 1

$$= \lim_{x \to \infty} \frac{\frac{4x}{x^2}}{\frac{x^2}{x^2} - \frac{3}{x^2}}$$

$$= \lim_{x \to \infty} \frac{\frac{1}{x}}{1 + \frac{3}{x^2}}$$

$$= \frac{0}{1 + 0} \qquad \left[\text{as } x \to \infty, \frac{b}{ax^n} \to 0 \right]$$

$$= 0.$$

In a similar manner, it can be shown that
$$\lim_{x \to -\infty} f(x) = 0.$$

The horizontal asymptote is the line $y = 0$.

15. $f(x) = 5 - \dfrac{3}{x}$

To find the horizontal asymptote, we consider $\lim_{x \to \infty} f(x)$. To find the limit, we will use some algebra and the fact that as $x \to \infty, \dfrac{b}{ax^n} \to 0$ for any positive integer n.

$$\lim_{x \to \infty} f(x) = \lim_{x \to \infty} 5 - \frac{3}{x}$$

$$= 5 - 0 \qquad \left[\text{as } x \to \infty, \frac{b}{ax^n} \to 0 \right]$$

$$= 5.$$

In a similar manner, it can be shown that
$$\lim_{x \to -\infty} f(x) = 5.$$

The horizontal asymptote is the line $y = 5$.

17. $f(x) = \dfrac{8x^4 - 5x^2}{2x^3 + x^2}$

To find the horizontal asymptote, we consider $\lim_{x \to \infty} f(x)$. To find the limit, we will use some algebra and the fact that as $x \to \infty, \dfrac{b}{ax^n} \to 0$ for any positive integer .

$$\lim_{x \to \infty} f(x) = \lim_{x \to \infty} \frac{8x^4 - 5x^2}{2x^3 + x^2}$$

$$= \lim_{x \to \infty} \frac{8x^4 - 5x^2}{2x^3 + x^2} \cdot \frac{\frac{1}{x^3}}{\frac{1}{x^3}}$$ Multiplying by a form of 1

$$= \lim_{x \to \infty} \frac{8x - \frac{5}{x}}{2 - \frac{1}{x^2}}$$

$$= \frac{\lim_{x \to \infty} 8x - \frac{5}{x}}{2 - 0}$$

$$= \infty$$

In a similar manner, it can be shown that
$$\lim_{x \to -\infty} f(x) = -\infty.$$

The function increases without bound as $x \to \infty$ and decreases without bound as $x \to -\infty$. Therefore, the function does not have a horizontal asymptote.

19. $f(x) = \dfrac{6x^4 + 4x^2 - 7}{2x^5 - x + 3}$

To find the horizontal asymptote, we consider $\lim_{x \to \infty} f(x)$. To find the limit, we will use some algebra and the fact that as $x \to \infty, \dfrac{b}{ax^n} \to 0$ for any positive integer n.

$$\lim_{x \to \infty} f(x) = \lim_{x \to \infty} \frac{6x^4 + 4x^2 - 7}{2x^5 - x + 3}$$

$$= \lim_{x \to \infty} \frac{6x^4 + 4x^2 - 7}{2x^5 - x + 3} \cdot \frac{\frac{1}{x^5}}{\frac{1}{x^5}}$$

$$= \lim_{x \to \infty} \frac{\frac{6}{x} + \frac{4}{x^3} - \frac{7}{x^5}}{2 - \frac{1}{x^4} + \frac{3}{x^5}}$$

$$= \frac{0}{2 + 0} \qquad \left[\text{as } x \to \infty, \frac{b}{ax^n} \to 0 \right]$$

$$= 0$$

In a similar manner, it can be shown that
$$\lim_{x \to -\infty} f(x) = 0.$$

The horizontal asymptote is the line $y = 0$.

21. $f(x) = \dfrac{2x^3 - 4x + 1}{4x^3 + 2x - 3}$

To find the horizontal asymptote, we consider $\lim\limits_{x \to \infty} f(x)$. To find the limit, we will use some

algebra and the fact that as $x \to \infty$, $\dfrac{b}{ax^n} \to 0$ for

any positive integer n.

$$\lim\limits_{x \to \infty} f(x) = \lim\limits_{x \to \infty} \frac{2x^3 - 4x + 1}{4x^3 + 2x - 3}$$

$$= \lim\limits_{x \to \infty} \frac{2x^3 - 4x + 1}{4x^3 + 2x - 3} \cdot \frac{\dfrac{1}{x^3}}{\dfrac{1}{x^3}}$$

$$= \lim\limits_{x \to \infty} \frac{2 - \dfrac{4}{x^2} + \dfrac{1}{x^3}}{4 + \dfrac{2}{x^2} - \dfrac{3}{x^3}}$$

$$= \frac{2 - 0 + 0}{4 + 0 - 0} \qquad \left[\text{as } x \to \infty, \frac{b}{ax^n} \to 0 \right]$$

$$= \frac{2}{4} = \frac{1}{2}$$

In a similar manner, it can be shown that

$$\lim\limits_{x \to -\infty} f(x) = \frac{1}{2}.$$

The horizontal asymptote is the line $y = \dfrac{1}{2}$.

23. $f(x) = -\dfrac{5}{x} = -5x^{-1}$

a) *Intercepts.* Since the numerator is the constant -5, there are no x-intercepts. The number 0 is not in the domain of the function, so there are no y-intercepts.

b) *Asymptotes.*
  Vertical. The denominator is 0 for $x = 0$, so the line $x = 0$ is a vertical asymptote.
  Horizontal. The degree of the numerator is less than the degree of the denominator, so $y = 0$ is the horizontal asymptote.
  Slant. There is no slant asymptote since the degree of the numerator is not one more than the degree of the denominator.

c) *Derivatives and Domain.*

$$f'(x) = 5x^{-2} = \frac{5}{x^2}$$

$$f''(x) = -10x^{-3} = -\frac{10}{x^3}$$

The domain of f is $(-\infty, 0) \cup (0, \infty)$ as determined in step (b).

d) *Critical Points.* $f'(x)$ exists for all values of x except 0, but 0 is not in the domain of the function, so $x = 0$ is not a critical value. The equation $f'(x) = 0$ has no solution, so there are no critical points.

e) *Increasing, decreasing, relative extrema.* We use 0 to divide the real number line into two intervals A: $(-\infty, 0)$ and B: $(0, \infty)$, and we test a point in each interval.

  A: Test -1, $f'(-1) = \dfrac{5}{(-1)^2} = 5 > 0$

  B: Test 1, $f'(1) = \dfrac{5}{(1)^2} = 5 > 0$

  Then $f(x)$ is increasing on both intervals. Since there are no critical points, there are no relative extrema.

f) *Inflection points.* $f''(x)$ does not exist at 0, but because 0 is not in the domain of the function, there cannot be an inflection point at 0. The equation $f''(x) = 0$ has no solution; therefore, there are no inflection points.

g) *Concavity.* We use 0 to divide the real number line into two intervals
  A: $(-\infty, 0)$ and B: $(0, \infty)$, and we test a point in each interval.

  A: Test -1, $f''(-1) = -\dfrac{10}{(-1)^3} = 10 > 0$

  B: Test 1, $f''(1) = -\dfrac{10}{(1)^3} = -10 < 0$

  Therefore, $f(x)$ is concave up on $(-\infty, 0)$ and concave down on $(0, \infty)$.

h) *Sketch.* Use the preceding information to sketch the graph. Compute additional function values as needed.

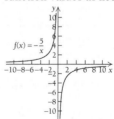

25. $f(x) = \dfrac{1}{x-5} = (x-5)^{-1}$

a) *Intercepts.* Since the numerator is the constant 1, there are no *x*-intercepts. To find the *y*-intercepts we compute $f(0)$

$$f(0) = \frac{1}{(0)-5} = -\frac{1}{5}$$

The point $\left(0, -\dfrac{1}{5}\right)$ is the *y*-intercept.

b) *Asymptotes.*
Vertical. The denominator is 0 for $x = 5$, so the line $x = 5$ is a vertical asymptote.
Horizontal. The degree of the numerator is less than the degree of the denominator, so $y = 0$ is the horizontal asymptote.

Slant. There is no slant asymptote since the degree of the numerator is not one more than the degree of the denominator.

c) *Derivatives and Domain.*

$$f'(x) = -(x-5)^{-2} = \frac{-1}{(x-5)^2}$$

$$f''(x) = 2(x-5)^{-3} = \frac{2}{(x-5)^3}$$

The domain of f is $(-\infty, 5) \cup (5, \infty)$ as determined in step (b).

d) *Critical Points.* $f'(x)$ exists for all values of *x* except 5, but 5 is not in the domain of the function, so $x = 5$ is not a critical value. The equation $f'(x) = 0$ has no solution, so there are no critical points.

e) *Increasing, decreasing, relative extrema.* We use 5 to divide the real number line into two intervals A: $(-\infty, 5)$ and B: $(5, \infty)$, and we test a point in each interval.

A: Test 4, $f'(4) = \dfrac{-1}{(4-5)^2} = -\dfrac{1}{1} = -1 < 0$

B: Test 6, $f'(6) = \dfrac{-1}{(6-5)^2} = -\dfrac{1}{1} = -1 < 0$

Then $f(x)$ is decreasing on both intervals. Since there are no critical points, there are no relative extrema.

f) *Inflection points.* $f''(x)$ does not exist at 5, but because 5 is not in the domain of the function, there cannot be an inflection point at 5. The equation $f''(x) = 0$ has no solution; therefore, there are no inflection points.

g) *Concavity.* We use 5 to divide the real number line into two intervals A: $(-\infty, 5)$ and B: $(5, \infty)$, and we test a point in each interval.

A: Test 4, $f''(4) = \dfrac{2}{(4-5)^3} = \dfrac{2}{-1} = -2 < 0$

B: Test 6, $f''(6) = \dfrac{2}{(6-5)^3} = \dfrac{2}{1} = 2 > 0$

Therefore, $f(x)$ is concave down on $(-\infty, 5)$ and concave up on $(5, \infty)$.

h) *Sketch.* Use the preceding information to sketch the graph. Compute additional function values as needed.

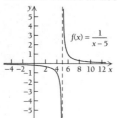

27. $f(x) = \dfrac{1}{x+2} = (x+2)^{-1}$

a) *Intercepts.* Since the numerator is the constant 1, there are no *x*-intercepts. To find the *y*-intercepts we compute $f(0)$

$$f(0) = \frac{1}{(0)+2} = \frac{1}{2}$$

The point $\left(0, \dfrac{1}{2}\right)$ is the *y*-intercept.

b) *Asymptotes.*
Vertical. The denominator is 0 for $x = -2$, so the line $x = -2$ is a vertical asymptote.
Horizontal. The degree of the numerator is less than the degree of the denominator, so $y = 0$ is the horizontal asymptote.

Slant. There is no slant asymptote since the degree of the numerator is not one more than the degree of the denominator.

c) *Derivatives and Domain.*

$$f'(x) = -(x+2)^{-2} = \frac{-1}{(x+2)^2}$$

$$f''(x) = 2(x+2)^{-3} = \frac{2}{(x+2)^3}$$

The domain of f is $(-\infty, -2) \cup (2, \infty)$ as determined in step (b).

d) *Critical Points.* $f'(x)$ exists for all values of x except -2, but -2 is not in the domain of the function, so $x = -2$ is not a critical value. The equation $f'(x) = 0$ has no solution, so there are no critical points.

e) *Increasing, decreasing, relative extrema.* We use -2 to divide the real number line into two intervals

A: $(-\infty, -2)$ and B: $(-2, \infty)$, and we test a point in each interval.

A: Test -3, $f'(-3) = \dfrac{-1}{\left((-3)+2\right)^2} = -1 < 0$

B: Test -1, $f'(-1) = \dfrac{-1}{\left((-1)+2\right)^2} = -1 < 0$

Then $f(x)$ is decreasing on both intervals. Since there are no critical points, there are no relative extrema.

f) *Inflection points.* $f''(x)$ does not exist at -2, but because -2 is not in the domain of the function, there cannot be an inflection point at -2. The equation $f''(x) = 0$ has no solution; therefore, there are no inflection points.

g) *Concavity.* We use -2 to divide the real number line into two intervals

A: $(-\infty, -2)$ and B: $(-2, \infty)$, and we test a point in each interval.

A: Test -3, $f''(-3) = \dfrac{2}{\left((-3)+2\right)^3} = -2 < 0$

B: Test -1, $f''(-1) = \dfrac{2}{\left((-1)+2\right)^3} = 2 > 0$

Therefore, $f(x)$ is concave down on $(-\infty, -2)$ and concave up on $(-2, \infty)$.

h) *Sketch.* Use the preceding information to sketch the graph. Compute additional function values as needed.

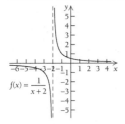

$f(x) = \dfrac{1}{x+2}$

29. $f(x) = \dfrac{-3}{x-3} = -3(x-3)^{-1}$

a) *Intercepts.* Since the numerator is the constant -3, there are no x-intercepts. To find the y-intercepts we compute $f(0)$

$$f(0) = \frac{-3}{(0)-3} = \frac{3}{3} = 1$$

The point $(0, 1)$ is the y-intercept.

b) *Asymptotes.*
Vertical. The denominator is 0 for $x = 3$, so the line $x = 3$ is a vertical asymptote.
Horizontal. The degree of the numerator is less than the degree of the denominator, so $y = 0$ is the horizontal asymptote.

Slant. There is no slant asymptote since the degree of the numerator is not one more than the degree of the denominator.

c) *Derivatives and Domain.*

$$f'(x) = 3(x-3)^{-2} = \frac{3}{(x-3)^2}$$

$$f''(x) = -6(x-3)^{-3} = \frac{-6}{(x-3)^3}$$

The domain of f is $(-\infty, 3) \cup (3, \infty)$ as determined in step (b).

d) *Critical Points.* $f'(x)$ exists for all values of x except 3, but 3 is not in the domain of the function, so $x = 3$ is not a critical value. The equation $f'(x) = 0$ has no solution, so there are no critical points.

e) *Increasing, decreasing, relative extrema.* We use 3 to divide the real number line into two intervals A: $(-\infty, 3)$ and B: $(3, \infty)$, and we test a point in each interval.

A: Test 2, $f'(2) = \dfrac{3}{\left((2)-3\right)^2} = 3 > 0$

B: Test 4, $f'(4) = \dfrac{3}{\left((4)-3\right)^2} = 3 > 0$

Then $f(x)$ is increasing on both intervals. Since there are no critical points, there are no relative extrema.

f) *Inflection points.* $f''(x)$ does not exist at 3, but because 3 is not in the domain of the function, there cannot be an inflection point at 3. The equation $f''(x) = 0$ has no solution; therefore, there are no inflection points.

g) *Concavity.* We use 3 to divide the real number line into two intervals
A: $(-\infty, 3)$ and B: $(3, \infty)$, and we test a point in each interval.

A: Test 2, $f''(2) = \dfrac{-6}{\left((2)-3\right)^3} = 6 > 0$

B: Test 4, $f''(4) = \dfrac{-6}{\left((4)-3\right)^3} = -6 < 0$

Therefore, $f(x)$ is concave up on $(-\infty, 3)$ and concave down on $(3, \infty)$.

h) *Sketch.* Use the preceding information to sketch the graph. Compute additional function values as needed.

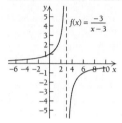

31. $f(x) = \dfrac{3x-1}{x}$

a) *Intercepts.* To find the *x*-intercepts, solve $f(x) = 0$.

$$\frac{3x-1}{x} = 0$$
$$3x - 1 = 0$$
$$3x = 1$$
$$x = \frac{1}{3}$$

Since $x = \dfrac{1}{3}$ does not make the denominator 0, the *x*-intercept is $\left(\dfrac{1}{3}, 0\right)$

The number 0 is not in the domain of $f(x)$ so there are no *y*-intercepts.

b) *Asymptotes.*
Vertical. The denominator is 0 for $x = 0$, so the line $x = 0$ is a vertical asymptote.
Horizontal. The numerator and the denominator have the same degree, so

$y = \dfrac{3}{1}$, or $y = 3$ is the horizontal asymptote.

Slant. There is no slant asymptote since the degree of the numerator is not one more than the degree of the denominator.

c) *Derivatives and Domain.*

$$f'(x) = \frac{1}{x^2}$$

$$f''(x) = -2x^{-3} = -\frac{2}{x^3}$$

The domain of f is $(-\infty, 0) \cup (0, \infty)$ as determined in step (b).

d) *Critical Points.* $f'(x)$ exists for all values of x except 0, but 0 is not in the domain of the function, so $x = 0$ is not a critical value. The equation $f'(x) = 0$ has no solution, so there are no critical points.

e) *Increasing, decreasing, relative extrema.* We use 0 to divide the real number line into two intervals A: $(-\infty, 0)$ and B: $(0, \infty)$, and we test a point in each interval.

A: Test -1, $f'(-1) = \dfrac{1}{(-1)^2} = 1 > 0$

B: Test 1, $f'(1) = \dfrac{1}{(-1)^2} = 1 > 0$

Then $f(x)$ is increasing on both intervals. Since there are no critical points, there are no relative extrema.

f) *Inflection points.* $f''(x)$ does not exist at 0, but because 0 is not in the domain of the function, there cannot be an inflection point at 0. The equation $f''(x) = 0$ has no solution; therefore, there are no inflection points.

g) *Concavity.* We use 0 to divide the real number line into two intervals
A: $(-\infty, 0)$ and B: $(0, \infty)$, and we test a point in each interval.

A: Test -1, $f''(-1) = \dfrac{-2}{(-1)^3} = 2 > 0$

B: Test 1, $f''(1) = \dfrac{-2}{(1)^3} = -2 < 0$

Therefore, $f(x)$ is concave up on $(-\infty, 0)$ and concave down on $(0, \infty)$.

h) *Sketch.* Use the preceding information to sketch the graph. Compute additional function values as needed.

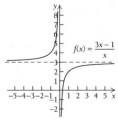

$f(x) = \dfrac{3x-1}{x}$

33. $f(x) = x + \dfrac{2}{x} = \dfrac{x^2 + 2}{x}$

a) *Intercepts.* The equation $f(x) = 0$ has no real solutions, so there are no x-intercepts. The number 0 is not in the domain of $f(x)$ so there are no y-intercepts.

b) *Asymptotes.*
   Vertical. The denominator is 0 for $x = 0$, so the line $x = 0$ is a vertical asymptote.
   Horizontal. The degree of the numerator is greater than the degree of the denominator, so there are no horizontal asymptotes.
   Slant. The degree of the numerator is exactly one greater than the degree of the denominator. As $|x|$ gets very large,

   $f(x) = x + \dfrac{2}{x}$ approaches x. Therefore,

   $y = x$ is the slant asymptote.

c) *Derivatives and Domain.*

   $f'(x) = 1 - 2x^{-2} = 1 - \dfrac{2}{x^2}$

   $f''(x) = 4x^{-3} = \dfrac{4}{x^3}$

   The domain of f is $(-\infty, 0) \cup (0, \infty)$ as determined in step (b).

d) *Critical Points.* $f'(x)$ exists for all values of x except 0, but 0 is not in the domain of the function, so $x = 0$ is not a critical value. The critical points will occur when $f'(x) = 0$.

   $1 - \dfrac{2}{x^2} = 0$

   $1 = \dfrac{2}{x^2}$

   $x^2 = 2$

   $x = \pm\sqrt{2}$

Thus, $-\sqrt{2}$ and $\sqrt{2}$ are critical values. $f\left(-\sqrt{2}\right) = -2\sqrt{2}$ and $f\left(\sqrt{2}\right) = 2\sqrt{2}$, so the critical points $\left(-\sqrt{2}, -2\sqrt{2}\right)$ and $\left(\sqrt{2}, 2\sqrt{2}\right)$ are on the graph.

e) *Increasing, decreasing, relative extrema.* We use $-\sqrt{2}, 0$, and $\sqrt{2}$ to divide the real number line into four intervals
   A: $\left(-\infty, -\sqrt{2}\right)$ B: $\left(-\sqrt{2}, 0\right)$, C: $\left(0, \sqrt{2}\right)$,
   and D: $\left(\sqrt{2}, \infty\right)$.

   A: Test -2, $f'(-2) = 1 - \dfrac{2}{(-2)^2} = \dfrac{1}{2} > 0$

   B: Test -1, $f'(-1) = 1 - \dfrac{2}{(-1)^2} = -1 < 0$

   C: Test 1, $f'(1) = 1 - \dfrac{2}{(1)^2} = -1 < 0$

   D: Test 2, $f'(2) = 1 - \dfrac{2}{(2)^2} = \dfrac{1}{2} > 0$

   Then $f(x)$ is increasing on
   $\left(-\infty, -\sqrt{2}\right]$ and $\left[\sqrt{2}, \infty\right)$ and is decreasing on $\left[-\sqrt{2}, 0\right)$ and $\left(0, \sqrt{2}\right]$. Therefore,
   $\left(-\sqrt{2}, -2\sqrt{2}\right)$ is a relative maximum, and
   $\left(\sqrt{2}, 2\sqrt{2}\right)$ is a relative minimum.

f) *Inflection points.* $f''(x)$ does not exist at 0, but because 0 is not in the domain of the function, there cannot be an inflection point at 0. The equation $f''(x) = 0$ has no solution; therefore, there are no inflection points.

g) *Concavity.* We use 0 to divide the real number line into two intervals
   A: $(-\infty, 0)$ and B: $(0, \infty)$, and we test a point in each interval.

   A: Test -1, $f''(-1) = \dfrac{4}{(-1)^3} = -4 < 0$

   B: Test 1, $f''(1) = \dfrac{4}{(1)^3} = 4 > 0$

   Therefore, $f(x)$ is concave down on $(-\infty, 0)$ and concave up on $(0, \infty)$.

h) *Sketch*. Use the preceding information to sketch the graph. Compute additional function values as needed.

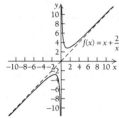

35. $f(x) = \dfrac{-1}{x^2} = -x^{-2}$

a) *Intercepts*. Since the numerator is the constant -1, there are no x-intercepts. The number 0 is not in the domain of the function, so there are no y-intercepts.

b) *Asymptotes*.
Vertical. The denominator is 0 for $x = 0$, so the line $x = 0$ is a vertical asymptote.
Horizontal. The degree of the numerator is less than the degree of the denominator, so $y = 0$ is the horizontal asymptote.

Slant. There is no slant asymptote since the degree of the numerator is not one more than the degree of the denominator.

c) *Derivatives and Domain*.

$$f'(x) = 2x^{-3} = \frac{2}{x^3}$$

$$f''(x) = -6x^{-4} = -\frac{6}{x^4}$$

The domain of f is $(-\infty, 0) \cup (0, \infty)$ as determined in step (b).

d) *Critical Points*. $f'(x)$ exists for all values of x except 0, but 0 is not in the domain of the function, so $x = 0$ is not a critical value. The equation $f'(x) = 0$ has no solution, so there are no critical points.

e) *Increasing, decreasing, relative extrema*. We use 0 to divide the real number line into two intervals A: $(-\infty, 0)$ and B: $(0, \infty)$, and we test a point in each interval.

A: Test -1, $f'(-1) = \dfrac{2}{(-1)^3} = -2 < 0$

B: Test 1, $f'(1) = \dfrac{2}{(1)^3} = 2 > 0$

Then $f(x)$ is decreasing on $(-\infty, 0)$ and is increasing on $(0, \infty)$. Since there are no critical points, there are no relative extrema.

f) *Inflection points*. $f''(x)$ does not exist at 0, but because 0 is not in the domain of the function, there cannot be an inflection point at 0. The equation $f''(x) = 0$ has no solution; therefore, there are no inflection points.

g) *Concavity*. We use 0 to divide the real number line into two intervals A: $(-\infty, 0)$ and B: $(0, \infty)$, and we test a point in each interval.

A: Test -1, $f''(-1) = -\dfrac{6}{(-1)^4} = -6 < 0$

B: Test 1, $f''(1) = -\dfrac{6}{(1)^4} = -6 < 0$

Therefore, $f(x)$ is concave down on both intervals.

h) *Sketch*. Use the preceding information to sketch the graph. Compute additional function values as needed.

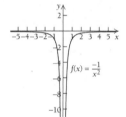

37. $f(x) = \dfrac{x}{x+2}$

a) *Intercepts*. To find the x-intercepts, solve $f(x) = 0$.

$$\frac{x}{x+2} = 0$$

$$x = 0$$

Since $x = 0$ does not make the denominator 0, the x-intercept is $(0, 0)$. $f(0) = 0$, so the y-intercept is $(0, 0)$ also.

b) *Asymptotes.*

Vertical. The denominator is 0 for $x = -2$, so the line $x = -2$ is a vertical asymptote.

Horizontal. The numerator and the denominator have the same degree, so

$$y = \frac{1}{1}, \text{ or } y = 1 \text{ is the horizontal asymptote.}$$

Slant. There is no slant asymptote since the degree of the numerator is not one more than the degree of the denominator.

c) *Derivatives and Domain.*

$$f'(x) = \frac{2}{(x+2)^2}$$

$$f''(x) = -4(x+2)^{-3} = -\frac{4}{(x+2)^3}$$

The domain of f is $(-\infty, -2) \cup (2, \infty)$ as determined in step (b).

d) *Critical Points.* $f'(x)$ exists for all values of x except -2, but -2 is not in the domain of the function, so $x = -2$ is not a critical value. The equation $f'(x) = 0$ has no solution, so there are no critical points.

e) *Increasing, decreasing, relative extrema.*
We use -2 to divide the real number line into two intervals A: $(-\infty, -2)$ and

B: $(-2, \infty)$, and we test a point in each interval.

A: Test -3, $f'(-3) = \frac{2}{((-3)+2)^2} = 2 > 0$

B: Test -1, $f'(-1) = \frac{2}{((-1)+2)^2} = 2 > 0$

Then $f(x)$ is increasing on both intervals. Since there are no critical points, there are no relative extrema.

f) *Inflection points.* $f''(x)$ does not exist at -2, but because -2 is not in the domain of the function, there cannot be an inflection point at -2. The equation $f''(x) = 0$ has no solution; therefore, there are no inflection points.

g) *Concavity.* We use -2 to divide the real number line into two intervals A: $(-\infty, -2)$ and B: $(-2, \infty)$, and we test a point in each interval.

A: Test -3, $f''(-3) = -\frac{4}{((-3)+2)^3} = 4 > 0$

B: Test -1, $f''(-1) = -\frac{4}{((-1)+2)^3} = -4 < 0$

Therefore, $f(x)$ is concave up on $(-\infty, -2)$ and concave down on $(-2, \infty)$.

h) *Sketch.* Use the preceding information to sketch the graph. Compute additional function values as needed.

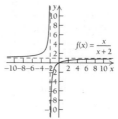

$$f(x) = \frac{x}{x+2}$$

39. $f(x) = \frac{-1}{x^2 + 2} = -(x^2 + 2)^{-1}$

a) *Intercepts.* Since the numerator is the constant -1, there are no x-intercepts.

$$f(0) = \frac{-1}{(0)^2 + 2} = -\frac{1}{2}, \text{ so the } y\text{-intercept is}$$

$$\left(0, -\frac{1}{2}\right).$$

b) *Asymptotes.*

Vertical. $x^2 + 2 = 0$ has no real solution, so there are no vertical asymptotes.

Horizontal. The degree of the numerator is less than the degree of the denominator, so $y = 0$ is the horizontal asymptote.

Slant. There is no slant asymptote since the degree of the numerator is not one more than the degree of the denominator.

c) *Derivatives and Domain.*

$$f'(x) = 2x(x^2 + 2)^{-2} = \frac{2x}{(x^2 + 2)^2}$$

$$f''(x) = \frac{-6x^2 + 4}{(x^2 + 2)^3}$$

The domain of f is $\mathbb{R}$ as determined in step (b).

d) *Critical Points.* $f'(x)$ exists for all real numbers. Solve $f'(x) = 0$

$$\frac{2x}{\left(x^2 + 2\right)^2} = 0$$

$$2x = 0$$

$$x = 0$$

The critical value is 0. From step (a) we found $\left(0, -\frac{1}{2}\right)$ is on the graph.

e) *Increasing, decreasing, relative extrema.* We use 0 to divide the real number line into two intervals A: $(-\infty, 0)$ and B: $(0, \infty)$, and we test a point in each interval.

A: Test -1, $f'(-1) = -\frac{2}{9} < 0$

B: Test 1, $f'(1) = \frac{2}{9} > 0$

Then $f(x)$ is decreasing on $(-\infty, 0]$ and is increasing on $[0, \infty)$. Thus $\left(0, -\frac{1}{2}\right)$ is a relative minimum.

f) *Inflection points.* $f''(x)$ exist for all real numbers. Solve $f''(x) = 0$.

$$\frac{-6x^2 + 4}{\left(x^2 + 2\right)^2} = 0$$

$$-6x^2 + 4 = 0$$

$$-6x^2 = -4$$

$$x^2 = \frac{2}{3}$$

$$x = \pm\sqrt{\frac{2}{3}}$$

$$f\left(-\sqrt{\frac{2}{3}}\right) = -\frac{3}{8} \text{ and } f\left(\sqrt{\frac{2}{3}}\right) = -\frac{3}{8}$$

So, $\left(-\sqrt{\frac{2}{3}}, -\frac{3}{8}\right)$ and $\left(\sqrt{\frac{2}{3}}, -\frac{3}{8}\right)$ are possible points of inflection.

g) *Concavity.* We use $-\sqrt{\frac{2}{3}}$ and $\sqrt{\frac{2}{3}}$ to divide the real number line into three intervals

A: $\left(-\infty, -\sqrt{\frac{2}{3}}\right)$ B: $\left(-\sqrt{\frac{2}{3}}, \sqrt{\frac{2}{3}}\right)$,

and C: $\left(\sqrt{\frac{2}{3}}, \infty\right)$

A: Test -1, $f''(-1) = -\frac{2}{27} < 0$

B: Test 0, $f''(0) = \frac{1}{2} > 0$

C: Test 1, $f''(1) = -\frac{2}{27} < 0$

Therefore, $f(x)$ is concave down on

$\left(-\infty, -\sqrt{\frac{2}{3}}\right)$ and $\left(\sqrt{\frac{2}{3}}, \infty\right)$ and concave up

on $\left(-\sqrt{\frac{2}{3}}, \sqrt{\frac{2}{3}}\right)$. Therefore the points

$\left(-\sqrt{\frac{2}{3}}, -\frac{3}{8}\right)$ and $\left(\sqrt{\frac{2}{3}}, -\frac{3}{8}\right)$ are points of

inflection.

h) *Sketch.* Use the preceding information to sketch the graph. Compute additional function values as needed.

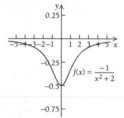

$$f(x) = \frac{-1}{x^2 + 2}$$

41. $f(x) = \frac{x+3}{x^2 - 9} = \frac{x+3}{(x+3)(x-3)} = \frac{1}{x-3}, \; x \neq \pm 3$

We write the expression in simplified form noting that the domain is restricted to all real numbers except for $x = \pm 3$.

a) *Intercepts.* $f(x) = 0$ has no solution.

$x = -3$ is not in the domain of the function. Therefore, there are no x-intercepts. To find the y-intercepts we compute $f(0)$

$$f(0) = \frac{1}{(0) - 3} = -\frac{1}{3}$$

The point $\left(0, -\frac{1}{3}\right)$ is the y-intercept.

b) *Asymptotes.*
Vertical. In the original function, the denominator is 0 for $x = -3$ or $x = 3$, however, $x = -3$ also made the numerator equal to 0. We look at the limits to determine if there are vertical asymptotes at these points.

$$\lim_{x \to -3} \frac{x+3}{x^2 - 9} = \lim_{x \to -3} \frac{1}{x - 3} = \frac{1}{-3 - 3} = -\frac{1}{6}.$$

Because the limit exists, the line $x = -3$ is not a vertical asymptote. Instead, we have a removable discontinuity, or a "hole" at the point $\left(-3, -\frac{1}{6}\right)$.

An open circle is drawn at $\left(-3, -\frac{1}{6}\right)$ to show that it is not part of the graph. The denominator is 0 for $x = 3$ and the numerator is not 0 at this value, so the line $x = 3$ is a vertical asymptote.
Horizontal. The degree of the numerator is less than the degree of the denominator, so $y = 0$ is the horizontal asymptote.

Slant. There is no slant asymptote since the degree of the numerator is not one more than the degree of the denominator.

c) *Derivatives and Domain.*

$$f'(x) = -(x-3)^{-2} = \frac{-1}{(x-3)^2}$$

$$f''(x) = 2(x-3)^{-3} = \frac{2}{(x-3)^3}$$

The domain of f is $(-\infty, -3) \cup (-3, \infty)$ as determined in step (b).

d) *Critical Points.* $f'(x)$ exists for all values of x except 3, but 3 is not in the domain of the function, so $x = 3$ is not a critical value. The equation $f'(x) = 0$ has no solution, so there are no critical points.

e) *Increasing, decreasing, relative extrema.*
We use -3 and 3 to divide the real number line into three intervals
A: $(-\infty, -3)$ B: $(-3, 3)$ and C: $(3, \infty)$.
We notice that $f'(x) < 0$ for all real numbers, $f(x)$ is decreasing on all three intervals $(-\infty, -3)$, $(-3, 3)$, and $(3, \infty)$.
Since there are no critical points, there are no relative extrema.

f) *Inflection points.* $f''(x)$ does not exist at 3, but because 3 is not in the domain of the function, there cannot be an inflection point at 3. The equation $f''(x) = 0$ has no solution; therefore, there are no inflection points.

g) *Concavity.* We use -3 and 3 to divide the real number line into three intervals
A: $(-\infty, -3)$ B: $(-3, 3)$ and C: $(3, \infty)$ and we test a point in each interval.

A: Test -4, $f''(-4) = \dfrac{2}{((-4)-3)^3} = -\dfrac{2}{343} < 0$

B: Test 2, $f''(2) = \dfrac{2}{((2)-3)^3} = -2 < 0$

C: Test 4, $f''(4) = \dfrac{2}{((4)-3)^3} = 2 > 0$

Therefore, $f(x)$ is concave down on $(-\infty, -3)$ and $(-3, 3)$ and concave up on $(3, \infty)$.

h) *Sketch.* Use the preceding information to sketch the graph. Compute additional function values as needed.

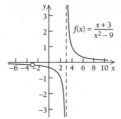

$$f(x) = \frac{x+3}{x^2 - 9}$$

43. $f(x) = \dfrac{x-1}{x+2}$

a) *Intercepts.* To find the x-intercepts, solve $f(x) = 0$.

$$\frac{x-1}{x+2} = 0$$

$$x = 1$$

Since $x = 1$ does not make the denominator 0, the x-intercept is $(1, 0)$.

$$f(0) = \frac{0-1}{0+2} = -\frac{1}{2}, \text{ so the } y\text{-intercept is}$$

$$\left(0, -\frac{1}{2}\right).$$

b) *Asymptotes.*
Vertical. The denominator is 0 for $x = -2$, so the line $x = -2$ is a vertical asymptote.
Horizontal. The numerator and the denominator have the same degree, so

$y = \dfrac{1}{1}$, or $y = 1$ is the horizontal asymptote.

Slant. There is no slant asymptote since the degree of the numerator is not one more than the degree of the denominator.

c) *Derivatives and Domain.*

$$f'(x) = \frac{3}{(x+2)^2}$$

$$f''(x) = -\frac{6}{(x+2)^3}$$

The domain of f is $(-\infty, -2) \cup (-2, \infty)$ as determined in part (b).

d) *Critical Points.* $f'(x)$ exists for all values of x except -2, but -2 is not in the domain of the function, so $x = -2$ is not a critical value. The equation $f'(x) = 0$ has no solution, so there are no critical points.

e) *Increasing, decreasing, relative extrema.* We use -2 to divide the real number line into two intervals
A: $(-\infty, -2)$ and B: $(-2, \infty)$, and we test a point in each interval.
A: Test -3, $f'(-3) = 3 > 0$
B: Test -1, $f'(-1) = 3 > 0$
Then $f(x)$ is increasing on both intervals.
Since there are no critical points, there are no relative extrema.

f) *Inflection points.* $f''(x)$ does not exist at -2, but because -2 is not in the domain of the function, there cannot be an inflection point at -2. The equation $f''(x) = 0$ has no solution; therefore, there are no inflection points.

g) *Concavity.* We use -2 to divide the real number line into two intervals
A: $(-\infty, -2)$ and B: $(-2, \infty)$, and we test a point in each interval.
A: Test -3, $f''(-3) = 6 > 0$
B: Test -1, $f''(-1) = -6 < 0$
Therefore, $f(x)$ is concave up on $(-\infty, -2)$ and concave down on $(-2, \infty)$.

h) *Sketch.* Use the preceding information to sketch the graph. Compute additional function values as needed.

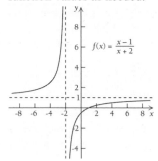

45. $f(x) = \dfrac{x^2 - 4}{x + 3}$

a) *Intercepts.* To find the x-intercepts, solve $f(x) = 0$.

$$\frac{x^2 - 4}{x + 3} = 0$$

$$x^2 - 4 = 0$$

$$x = \pm 2$$

Neither of these values make the denominator 0, so the x-intercepts are $(-2, 0)$ and $(2, 0)$. $f(0) = \dfrac{0^2 - 4}{0 + 3} = -\dfrac{4}{3}$, so the y-intercept is $\left(0, -\dfrac{4}{3}\right)$.

b) *Asymptotes.*
Vertical. The denominator is 0 for $x = -3$, so the line $x = -3$ is a vertical asymptote.
Horizontal. The degree of the numerator is greater than the degree of the denominator, so there are no horizontal asymptotes.
Slant. Divide the numerator by the denominator.

$$
\begin{array}{r}
x \quad\;\; -3 \\
x+3\overline{)x^2 \qquad\; -4} \\
\underline{x^2 + 3x} \\
-3x - 4 \\
\underline{-3x - 9} \\
5
\end{array}
$$

$$f(x) = x - 3 + \frac{5}{x + 3}$$

As $|x|$ gets very large, $f(x)$ approaches $x - 3$, so $y = x - 3$ is the slant asymptote.

c) *Derivatives and Domain.*

$$f'(x) = \frac{x^2 + 6x + 4}{(x+3)^2}$$

$$f''(x) = \frac{10}{(x+3)^3}$$

The domain of f is $(-\infty, -3) \cup (-3, \infty)$ as determined in part (b).

d) *Critical Points.* $f'(x)$ exists for all values of x except -3, but -3 is not in the domain of the function, so $x = -3$ is not a critical value. Solve $f'(x) = 0$.

$$\frac{x^2 + 6x + 4}{(x+3)^2} = 0$$

$$x^2 + 6x + 4 = 0$$

$$x = -3 \pm \sqrt{5} \quad \text{\small Using the Quadratic Formula}$$

$$x \approx -5.236 \text{ or } x \approx -0.764$$

$$f(-5.236) \approx -10.472 \text{ and}$$

$f(-0.764) \approx -1.528$, so $(-5.236, -10.472)$ and $(-0.764, -1.528)$ are on the graph.

e) *Increasing, decreasing, relative extrema.*
We use -5.236, -3, and -0.764 to divide the real number line into four intervals
A: $(-\infty, -5.236)$, B: $(-5.236, -3)$,
C: $(-3, -0.764)$, and D: $(-0.764, \infty)$
We test a point in each interval.

A: Test -6, $f'(-6) = \dfrac{4}{9} > 0$

B: Test -4, $f'(-4) = -4 < 0$

C: Test -2, $f'(-2) = -4 < 0$

D: Test 0, $f'(0) = \dfrac{4}{9} > 0$

Then $f(x)$ is increasing on the intervals $(-\infty, -5.236]$ and $[-0.764, \infty)$, and is decreasing on the intervals $[-5.226, -3)$ and $(-3, -0.764]$. Therefore, $(-5.236, -10.472)$ is a relative maximum and $(-0.764, -1.528)$ is a relative minimum.

f) *Inflection points.* $f''(x)$ does not exist at -3, but because -3 is not in the domain of the function, there cannot be an inflection point at -3. The equation $f''(x) = 0$ has no solution; therefore, there are no inflection points.

g) *Concavity.* We use -3 to divide the real number line into two intervals
A: $(-\infty, -3)$ and B: $(-3, \infty)$, and we test a point in each interval.
A: Test -4, $f''(-4) = -10 < 0$
B: Test -2, $f''(-2) = 10 > 0$
Therefore, $f(x)$ is concave down on $(-\infty, -3)$ and concave up on $(-3, \infty)$.

h) *Sketch.* Use the preceding information to sketch the graph. Compute additional function values as needed.

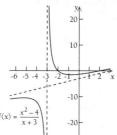

$f(x) = \dfrac{x^2 - 4}{x + 3}$

47. $f(x) = \dfrac{x+1}{x^2 - 2x - 3} = \dfrac{x+1}{(x-3)(x+1)} = \dfrac{1}{x-3}$

We write the expression in simplified form noting that the domain is restricted to all real numbers except for $x = -1$ and $x = 3$

a) *Intercepts.* $f(x) = 0$ has no solution.
$x = -1$ is not in the domain of the function. Therefore, there are no x-intercepts. To find the y-intercepts we compute $f(0)$

$$f(0) = \frac{0+1}{(0)^2 - 2(0) - 3} = -\frac{1}{3}$$

The point $\left(0, -\dfrac{1}{3}\right)$ is the y-intercept.

b) *Asymptotes.*
Vertical. In the original function, the denominator is 0 for $x = -1$ or $x = 3$, however, $x = -1$ also made the numerator equal to 0. We look at the limits to determine if there are vertical asymptotes at these points.

$$\lim_{x \to -1} \frac{x+1}{x^2 - 2x - 3} = \lim_{x \to -1} \frac{1}{x - 3} = \frac{1}{-1 - 3} = -\frac{1}{4}$$

Because the limit exists, the line $x = -1$ is not a vertical asymptote. Instead, we have a removable discontinuity, or a "hole" at the point $\left(-1, -\dfrac{1}{4}\right)$.

An open circle is drawn at $\left(-1, -\frac{1}{4}\right)$ to show that it is not part of the graph. The denominator is 0 for $x = 3$ and the numerator is not 0 at this value, so the line $x = 3$ is a vertical asymptote.

Horizontal. The degree of the numerator is less than the degree of the denominator, so $y = 0$ is the horizontal asymptote.

Slant. There is no slant asymptote since the degree of the numerator is not one more than the degree of the denominator.

c) *Derivatives and Domain.*

$$f'(x) = -(x-3)^{-2} = \frac{-1}{(x-3)^2}$$

$$f''(x) = 2(x-3)^{-3} = \frac{2}{(x-3)^3}$$

The domain of f is $(-\infty, 3) \cup (3, \infty)$ as determined in part (b).

d) *Critical Points.* $f'(x)$ exists for all values of x except 3, but 3 is not in the domain of the function, so $x = 3$ is not a critical value. The equation $f'(x) = 0$ has no solution, so there are no critical points.

e) *Increasing, decreasing, relative extrema.* We use -1 and 3 to divide the real number line into three intervals

   A: $(-\infty, -1)$ B: $(-1, 3)$ and C: $(3, \infty)$.

   We notice that $f'(x) < 0$ for all real numbers, $f(x)$ is decreasing on all three intervals $(-\infty, -1)$, $(-1, 3)$, and $(3, \infty)$.

   Since there are no critical points, there are no relative extrema.

f) *Inflection points.* $f''(x)$ does not exist at 3, but because 3 is not in the domain of the function, there cannot be an inflection point at 3. The equation $f''(x) = 0$ has no solution; therefore, there are no inflection points.

g) *Concavity.* We use -1 and 3 to divide the real number line into three intervals

   A: $(-\infty, -1)$ B: $(-1, 3)$ and C: $(3, \infty)$, and we test a point in each interval.

   A: Test -2, $f''(-2) = -\frac{2}{125} < 0$

   B: Test 2, $f''(2) = -2 < 0$

   C: Test 4, $f''(4) = 2 > 0$

Therefore, $f(x)$ is concave down on $(-\infty, -1)$ and $(-1, 3)$ and concave up on $(3, \infty)$.

h) *Sketch.* Use the preceding information to sketch the graph. Compute additional function values as needed.

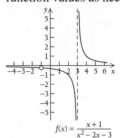

$$f(x) = \frac{x+1}{x^2 - 2x - 3}$$

49. $f(x) = \frac{2x^2}{x^2 - 16}$

a) *Intercepts.* The numerator is 0 for $x = 0$ and this value does not make the denominator 0, the x-intercept is $(0, 0)$

$f(0) = 0$, so the y-intercept is $(0, 0)$ also.

b) *Asymptotes.*
   Vertical. The denominator is 0 when

   $$x^2 - 16 = 0$$
   $$x^2 = 16$$
   $$x = \pm 4$$

   So the lines $x = -4$ and $x = 4$ are vertical asymptotes.
   Horizontal. The numerator and the denominator have the same degree, so

   $y = \frac{2}{1}$, or $y = 2$ is the horizontal asymptote.

   Slant. There is no slant asymptote since the degree of the numerator is not one more than the degree of the denominator.

c) *Derivatives and Domain.*

$$f'(x) = -\frac{64x}{(x^2 - 16)^2}$$

$$f''(x) = \frac{192x^2 + 1024}{(x^2 - 16)^3}$$

The domain of f is $(-\infty, -4) \cup (-4, 4) \cup (4, \infty)$ as determined in part (b).

d) *Critical Points.* $f'(x)$ exists for all values of x except $x = -4$ and $x = 4$, but -4 and 4 are not in the domain of the function, so $x = -4$ and $x = 4$ are not critical values. $f'(x) = 0$ for $x = 0$ so, $(0,0)$ is the only critical point.

e) *Increasing, decreasing, relative extrema.* We use -4, 0, and 4 to divide the real number line into four intervals A: $(-\infty, -4)$ B: $(-4, 0)$, C: $(0, 4)$, and D: $(4, \infty)$ We test a point in each interval.

A: Test -5, $f'(-5) = \dfrac{320}{81} > 0$

B: Test -1, $f'(-1) = \dfrac{64}{225} > 0$

C: Test 1, $f'(1) = -\dfrac{64}{225} < 0$

D: Test 5, $f'(5) = -\dfrac{320}{81} < 0$

Then $f(x)$ is increasing on the intervals $(-\infty, -4)$ and $(-4, 0]$, and is decreasing on the intervals $[0, 4)$ and $(4, \infty)$. Thus, there is a relative maximum at $(0, 0)$.

f) *Inflection points.* $f''(x)$ does not exist at -4 and 4, but because -4 and 4 are not in the domain of the function, there cannot be an inflection point at -4 or 4. The equation $f''(x) = 0$ has no real solution; therefore, there are no inflection points.

g) *Concavity.* We use -4 and 4 to divide the real number line into three intervals A: $(-\infty, -4)$ B: $(-4, 4)$ and C: $(4, \infty)$, and we test a point in each interval.

A: Test -5, $f''(-5) = \dfrac{5824}{729} > 0$

B: Test 0, $f''(0) = -\dfrac{1}{4} < 0$

C: Test 5, $f''(5) = \dfrac{5824}{729} > 0$

Therefore, $f(x)$ is concave up on the intervals $(-\infty, -4)$ and $(4, \infty)$ and concave down on the interval $(-4, 4)$.

h) *Sketch.* Use the preceding information to sketch the graph. Compute additional function values as needed.

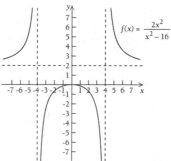

$$f(x) = \frac{2x^2}{x^2 - 16}$$

51. $f(x) = \dfrac{1}{x^2 - 1}$

a) *Intercepts.* Since the numerator is a constant 1, there are no x-intercepts.

$$f(0) = \frac{1}{(0)^2 - 1} = -1$$

The point $(0, -1)$ is the y-intercept.

b) *Asymptotes.*
Vertical. The denominator $x^2 - 1 = (x - 1)(x + 1)$ is 0 for $x = -1$ or $x = 1$, so the lines $x = -1$ and $x = 1$ are vertical asymptotes.
Horizontal. The degree of the numerator is less than the degree of the denominator, so $y = 0$ is the horizontal asymptote.
Slant. There is no slant asymptote since the degree of the numerator is not one more than the degree of the denominator.

c) *Derivatives and Domain.*

$$f'(x) = \frac{-2x}{(x^2 - 1)^2}$$

$$f''(x) = \frac{2(3x^2 + 1)}{(x^2 - 1)^3}$$

The domain of f is $(-\infty, -1) \cup (-1, 1) \cup (1, \infty)$ as determined in part (b).

d) *Critical Points.* $f'(x)$ exists for all values of x except -1 and 1, but these values are not in the domain of the function, so $x = -1$ and $x = 1$ are not a critical value. $f'(x) = 0$ for $x = 0$. From step (a) we know $f(0) = -1$, so the critical point is $(0, -1)$

e) *Increasing, decreasing, relative extrema.*
We use -1, 0, and 1 to divide the real number line into four intervals
A: $(-\infty, -1)$, B: $(-1, 0)$, C: $(0, 1)$, and
D: $(1, \infty)$, and we test a point in each interval.

A: Test -2, $f'(-2) = \dfrac{4}{9} > 0$

B: Test $-\dfrac{1}{2}$, $f'\left(-\dfrac{1}{2}\right) = \dfrac{16}{9} > 0$

C: Test $\dfrac{1}{2}$, $f'\left(\dfrac{1}{2}\right) = -\dfrac{16}{9} < 0$

D: Test 2, $f'(2) = -\dfrac{4}{9} < 0$

We see that $f(x)$ is increasing on the intervals $(-\infty, -1)$ and $(-1, 0]$, and is decreasing on the intervals $[0, 1)$ and $(1, \infty)$. Therefore, $(0, -1)$ is a relative maximum.

f) *Inflection points.* $f''(x)$ does not exist at -1 and 1, but because these values are not in the domain of the function, there cannot be an inflection point at -1 or 1. The equation $f''(x) = 0$ has no real solution; therefore, there are no inflection points.

g) *Concavity.* We use -1 and 1 to divide the real number line into three intervals
A: $(-\infty, -1)$ B: $(-1, 1)$ and C: $(1, \infty)$.
and we test a point in each interval.

A: Test -2, $f''(-2) = \dfrac{26}{27} > 0$

B: Test 0, $f''(0) = -2 < 0$

C: Test 2, $f''(2) = \dfrac{26}{27} > 0$

Therefore, $f(x)$ is concave up on $(-\infty, -1)$ and $(1, \infty)$, and concave down on $(-1, 1)$.

h) *Sketch.* Use the preceding information to sketch the graph. Compute additional function values as needed.

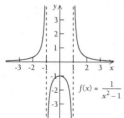

53. $f(x) = \dfrac{x^2 + 1}{x}$

a) *Intercepts.* The equation $f(x) = 0$ has no real solutions, so there are no x-intercepts. The number 0 is not in the domain of $f(x)$ so there are no y-intercepts.

b) *Asymptotes.*
Vertical. The denominator is 0 for $x = 0$, so the line $x = 0$ is a vertical asymptote.
Horizontal. The degree of the numerator is greater than the degree of the denominator, so there are no horizontal asymptotes.
Slant. The degree of the numerator is exactly one greater than the degree of the denominator. When we divide the numerator by the denominator we have
$f(x) = \dfrac{x^2 + 1}{x} = x + \dfrac{1}{x}$. As $|x|$ gets very large, $f(x) = x + \dfrac{1}{x}$ approaches x. Therefore, $y = x$ is the slant asymptote.

c) *Derivatives and Domain.*
$$f'(x) = \frac{x^2 - 1}{x^2}$$
$$f''(x) = \frac{2}{x^3}$$
The domain of f is $(-\infty, -0) \cup (0, \infty)$ as determined in part (b).

d) *Critical Points.* $f'(x)$ exists for all values of x except 0, but 0 is not in the domain of the function, so $x = 0$ is not a critical value. The critical points will occur when $f'(x) = 0$.
$$\frac{x^2 - 1}{x^2} = 0$$
$$x^2 - 1 = 0$$
$$x^2 = 1$$
$$x = \pm 1$$
Thus, -1 and 1 are critical values.
$f(-1) = -2$ and $f(1) = 2$, so the critical points $(-1, -2)$ and $(1, 2)$ are on the graph.

e) *Increasing, decreasing, relative extrema.*
We use $-1, 0$, and 1 to divide the real number line into four intervals
A: $(-\infty, -1)$ B: $(-1, 0)$, C: $(0, 1)$, and
D: $(1, \infty)$.

We test a point in each interval.

A: Test -2, $f'(-2) = \frac{3}{4} > 0$

B: Test $-\frac{1}{2}$, $f'\left(-\frac{1}{2}\right) = -3 < 0$

C: Test $\frac{1}{2}$, $f'\left(\frac{1}{2}\right) = -3 < 0$

D: Test 2, $f'(2) = \frac{3}{4} > 0$

Then $f(x)$ is increasing on $(-\infty, -1]$ and $[1, \infty)$ and is decreasing on $[-1, 0)$ and $(0, 1]$. Therefore, $(-1, -2)$ is a relative maximum, and $(1, 2)$ is a relative minimum.

f) *Inflection points.* $f''(x)$ does not exist at 0, but because 0 is not in the domain of the function, there cannot be an inflection point at 0. The equation $f''(x) = 0$ has no solution; therefore, there are no inflection points.

g) *Concavity.* We use 0 to divide the real number line into two intervals

A: $(-\infty, 0)$ and B: $(0, \infty)$, and we test a point in each interval.

A: Test -1, $f''(-1) = -2 < 0$

B: Test 1, $f''(1) = 2 > 0$

Therefore, $f(x)$ is concave down on $(-\infty, 0)$ and concave up on $(0, \infty)$.

h) *Sketch.* Use the preceding information to sketch the graph. Compute additional function values as needed.

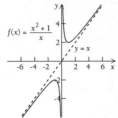

$f(x) = \dfrac{x^2+1}{x}$

55. $f(x) = \dfrac{x^2 - 9}{x - 3}$

We write the expression in simplified form:

$$f(x) = \frac{(x-3)(x+3)}{x-3} = x + 3, \quad x \neq 3$$

Note that the domain is restricted to all real numbers except for $x = 3$.

a) *Intercepts.* The numerator is 0 when $x = -3$ or $x = 3$ however, $x = 3$ is not in the domain of $f(x)$, so the x-intercept is $(-3, 0)$.

$$f(0) = \frac{0^2 - 9}{(0) - 3} = \frac{-9}{-3} = 3$$

The point $(0, 3)$ is the y-intercept.

b) *Asymptotes.*

In simplified form $f(x) = x + 3$, a linear function everywhere except $x = 3$. So there are no asymptotes of any kind.

In the original function, the denominator is 0 for $x = 3$, however, $x = 3$ also made the numerator equal to 0. We look at the limits to determine if there are vertical asymptotes at these points.

$\lim\limits_{x \to 3} \dfrac{x^2 - 9}{x - 3} = \lim\limits_{x \to 3} x + 3 = 6$. Because the limit exists, the line $x = 3$ is not a vertical asymptote. Instead, we have a removable discontinuity, or a "hole" at the point $(3, 6)$.

An open circle is drawn at this point to show that it is not part of the graph.

c) *Derivatives and Domain.*

$f'(x) = 1, \quad x \neq 3$

$f''(x) = 0$

The domain of f is $(-\infty, 3) \cup (3, \infty)$ as determined in part (b).

d) *Critical Points.* $f'(x)$ exists for all values of x except 3, but 3 is not in the domain of the function, so $x = 3$ is not a critical value. The equation $f'(x) = 0$ has no solution, so there are no critical points.

e) *Increasing, decreasing, relative extrema.* We use 3 to divide the real number line into two intervals A: $(-\infty, 3)$ and B: $(3, \infty)$.

We notice that $f'(x) > 0$ for all real numbers, $f(x)$ is increasing on both intervals. Since there are no critical points, there are no relative extrema.

f) *Inflection points.* $f''(x)$ is constant; therefore, there are no points of inflection.

g) *Concavity.* $f''(x)$ is 0; therefore, there is no concavity.

h) *Sketch.* Use the preceding information to sketch the graph.

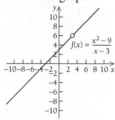

Note: In the preceding problem, we would have noticed that the graph of

$$f(x) = \frac{x^2 - 9}{x - 3} \text{ is the graph of } f(x) = x + 3$$

with the exception of the point $(3,6)$ which is a removable discontinuity. We simply need to graph $f(x) = x + 3$ with a hole at the point $(3,6)$. And determine all other aspects of the graph of $f(x)$ from the linear graph.

57. $V(t) = 50 - \dfrac{25t^2}{(t+2)^2}$

a) $V(0) = 50 - \dfrac{25(0)^2}{((0)+2)^2} = 50 - 0 = 50$

The inventory's value after 0 months is \$50.

$V(5) = 50 - \dfrac{25(5)^2}{((5)+2)^2} = 50 - \dfrac{625}{49} \approx 37.24$

The inventory's value after 5 months is \$37.24.

$V(10) = 50 - \dfrac{25(10)^2}{((10)+2)^2} = 50 - \dfrac{2500}{144} \approx 32.64$

The inventory's value after 10 months is \$32.64.

$V(70) = 50 - \dfrac{25(70)^2}{((70)+2)^2} = 50 - \dfrac{122,500}{5184} \approx 26.37$

The inventory's value after 70 months is \$26.37.

b) Find $V'(t)$ and $V''(t)$.

$$V'(t) = -\frac{(t+2)^2 (50t) - 25t^2 \left(2(t+2)(1)\right)}{\left((t+2)^2\right)^2}$$

$$= -\frac{(t+2)\left[(t+2)(50t) - 25t^2(2)\right]}{(t+2)^4}$$

$$= -\frac{50t^2 - 100t - 50t^2}{(t+2)^3}$$

$$= -\frac{100t}{(t+2)^3}$$

$$V''(t) = -\frac{(t+2)^3 (100) - 100t\left[3(t+2)^2(1)\right]}{\left((t+2)^3\right)^2}$$

$$= -\frac{(t+2)^2\left[100(t+2) - 100t(3)\right]}{(t+2)^6}$$

$$= -\frac{100t + 200 - 300t}{(t+2)^4}$$

$$= -\frac{-200t + 200}{(t+2)^4}$$

$$= \frac{200t - 200}{(t+2)^4}$$

$V'(t)$ exists for all values of t in $[0,\infty)$.

Solve $V'(t) = 0$.

$$-\frac{100t}{(t+2)^3} = 0$$

$$-100t = 0$$

$$t = 0$$

Since $t = 0$ is an endpoint of the domain, there cannot be a relative extrema at $t = 0$. We notice that $V'(t) < 0$ for all x in the domain, therefore, $V(t)$ is decreasing over the interval $[0,\infty)$. Since $V(t)$ is decreasing, the *absolute* maximum value of the inventory will be \$50 when $t = 0$.

c) Using the techniques of this section, we find the following additional information: *Intercepts.* There are no t-intercepts in $[0,\infty)$. The V-intercept is the point $(0,50)$

Asymptotes. There are no vertical asymptotes in $[0,\infty)$.

The line $V = 25$ is a horizontal asymptote. There are no slant asymptotes.

Increasing, decreasing, relative extrema. We have already seen that $V(t)$ is decreasing over the interval $[0,\infty)$. There are no relative extrema.

Inflection points, concavity. $V''(t)$ exist for all values of t in $[0,\infty)$. $V''(t) = 0$, when $t = 1$. We use this to split the domain into two intervals. A: $(0,1)$ and B: $(1,\infty)$. Testing points in each interval, we see $V(t)$ is concave down on $(0,1)$ and concave up on $(1,\infty)$.

We use this information to sketch the graph. Additional values may be computed as necessary.

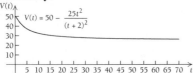

d) $\boxed{tw}$

59. $C(p) = \dfrac{48,000}{100 - p}$

a) $C(0) = \dfrac{48,000}{100 - 0} = \dfrac{48,000}{100} = 480$

The cost of removing 0% of the pollutants from a chemical spill is $480.

$C(20) = \dfrac{48,000}{100 - (20)} = \dfrac{48,000}{80} = 600$

The cost of removing 20% of the pollutants from a chemical spill is $600.

$C(80) = \dfrac{48,000}{100 - (80)} = \dfrac{48,000}{20} = 2400$

The cost of removing 80% of the pollutants from a chemical spill is $2400.

$C(90) = \dfrac{48,000}{100 - (90)} = \dfrac{48,000}{10} = 4800$

The cost of removing 90% of the pollutants from a chemical spill is $4800.

b) The domain of C is $0 \le p < 100$ since it is not possible to remove less than 0% or more than 100% of the pollutants, and $C(p)$ is not defined for $p = 100$.

c) Using the techniques of this section we find the following additional information.

Intercepts. No p-intercepts. The point $(0,480)$ is the C-intercept.

Asymptotes.   Vertical. $p = 100$
                 Horizontal. $C = 0$
                 Slant.   None

Increasing, decreasing, relative extrema. $C(p)$ is increasing over the interval $[0,100)$. There are no relative extrema.

Inflection points, concavity. $C(p)$ is concave up on the interval $(0,100)$. there are no inflection points.

We use this information and compute other function values as necessary to sketch the graph.

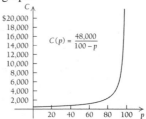

d) $\boxed{tw}$

61. $P(x) = \dfrac{2.632}{1 + 0.116x}$

a) $P(10) = \dfrac{2.632}{1 + 0.116(10)} = 1.21851852$

In 1980, the purchasing power of a dollar was $1.22.

$P(20) = \dfrac{2.632}{1 + 0.116(20)} = 0.79277108$

In 1990, the purchasing power of a dollar was $0.79.

$P(40) = \dfrac{2.632}{1 + 0.116(40)} = 0.4666667$

In 2010, the purchasing power of a dollar was $0.47.

b) Solve $P(x) = 0.50$

$$\frac{2.632}{1+0.116x} = 0.50$$

$$2.632 = 0.50(1+0.116x)$$

$$2.632 = 0.50 + 0.058x$$

$$2.132 = 0.058x$$

$$36.7586 = x$$

36.8 years after 1970, or in 2006, the purchasing power of a dollar will be $0.50.

c) Find $\lim\limits_{x\to\infty} P(x)$.

$$\lim_{x\to\infty} P(x) = \lim_{x\to\infty} \frac{2.632}{1+0.116x}$$

$$= \lim_{x\to\infty} \frac{2.632}{1+0.116x} \cdot \frac{\frac{1}{x}}{\frac{1}{x}}$$

$$= \lim_{x\to\infty} \frac{\frac{2.632}{x}}{\frac{1}{x}+0.116}$$

$$= \frac{0}{0+0.116} = 0$$

$$\lim_{x\to\infty} P(x) = 0.$$

63. $E(n) = 9 \cdot \dfrac{4}{n}$

a) Calculate each value for the given n.

$$E(9) = 9 \cdot \frac{4}{9} = 4.00$$

$$E(8) = 9 \cdot \frac{4}{8} = 4.50$$

$$E(7) = 9 \cdot \frac{4}{7} \approx 5.14$$

$$E(6) = 9 \cdot \frac{4}{6} = 6.00$$

$$E(5) = 9 \cdot \frac{4}{5} = 7.20$$

$$E(4) = 9 \cdot \frac{4}{4} = 9.00$$

$$E(3) = 9 \cdot \frac{4}{3} = 12.00$$

$$E(2) = 9 \cdot \frac{4}{2} = 18.00$$

$$E(1) = 9 \cdot \frac{4}{1} = 36.00$$

$$E\left(\frac{2}{3}\right) = 9 \cdot \frac{4}{\frac{2}{3}} = 9\left(4 \cdot \frac{3}{2}\right) = 54.00$$

$$E\left(\frac{1}{3}\right) = 9 \cdot \frac{4}{\frac{1}{3}} = 9\left(4 \cdot \frac{3}{1}\right) = 108.00$$

We complete the table.

Innings Pitched (n)	Earned-Run average (E)
9	4.00
8	4.50
7	5.14
6	6.00
5	7.20
4	9.00
3	12.00
2	18.00
1	36.00
$\frac{2}{3}$	54.00
$\frac{1}{3}$	108.00

b) $\lim\limits_{n\to 0} E(n) = \lim\limits_{n\to 0} 9 \cdot \dfrac{4}{n} = \lim\limits_{n\to 0} \dfrac{36}{n} = \infty$

65. $\boxed{tw}$

67. $\lim\limits_{x\to 0} \dfrac{|x|}{x}$

Using $|x| = \begin{cases} -x, & \text{for } x < 0 \\ x, & \text{for } x \geq 0 \end{cases}$, we have

$\lim\limits_{x\to 0^-} \dfrac{|x|}{x} = -1$ and $\lim\limits_{x\to 0^+} \dfrac{|x|}{x} = 1$.

Therefore, $\lim\limits_{x\to 0} \dfrac{|x|}{x}$ does not exist.

69. We divide the numerator and denominator by x^2 the highest power of x in the denominator.

$$\lim_{x\to\infty} \frac{-6x^3+7x}{2x^2-3x-10} = \lim_{x\to\infty} \frac{-6x+\frac{7}{x}}{2-\frac{3}{x}-\frac{10}{x^2}}$$

$$= \frac{\lim\limits_{x\to\infty}(-6x)+0}{2-0-0}$$

$$= -\infty$$

71. $\lim\limits_{x\to 1}\dfrac{x^3-1}{x^2-1}=\lim\limits_{x\to -2}\dfrac{(x-1)(x^2+x+1)}{(x-1)(x+1)}$

$$=\lim\limits_{x\to 1}\dfrac{(x^2+x+1)}{(x+1)}$$

$$=\dfrac{(1)^2+(1)+1}{(1)+1}$$

$$=\dfrac{3}{2}$$

73. $\lim\limits_{x\to -\infty}\dfrac{2x^4+x}{x+1}=\lim\limits_{x\to -\infty}\dfrac{2x^3+1}{1+\dfrac{1}{x}}$

$$=\dfrac{\lim\limits_{x\to -\infty}2x^3+1}{1+0}$$

$$=-\infty$$

75. $f(x)=\dfrac{x}{\sqrt{x^2+1}}$

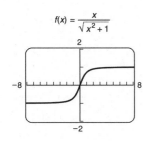

$f(x)=\dfrac{x}{\sqrt{x^2+1}}$

77. $f(x)=\dfrac{x^3+2x^2-15x}{x^2-5x-14}$

$f(x)=\dfrac{x^3+2x^2-15x}{x^2-5x-14}$

79. $f(x)=\left|\dfrac{1}{x}-2\right|$

$f(x)=\left|\dfrac{1}{x}-2\right|$

81. $\boxed{tw}$

Exercise Set 2.4

1. a) The absolute maximum gasoline mileage is obtained at a speed of 55 mph.
   b) The absolute minimum gasoline mileage is obtained at a speed of 5 mph.
   c) At 70 mph, the fuel economy is 25 mpg.

3. $f(x)=5+x-x^2;\ \ [0,2]$
   a) Find $f'(x)$
   $$f'(x)=1-2x$$
   b) Find the Critical Values. The derivative exists for all real numbers. Thus, we solve
   $$f'(x)=0$$
   $$1-2x=0$$
   $$1=2x$$
   $$\dfrac{1}{2}=x$$
   c) List the critical values and endpoints. These values are 0, $\dfrac{1}{2}$, and 2.
   d) Evaluate $f(x)$ at each value in step (c).
   $$f(0)=5+(0)-(0)^2=5$$
   $$f\left(\dfrac{1}{2}\right)=5+\left(\dfrac{1}{2}\right)-\left(\dfrac{1}{2}\right)^2=\dfrac{21}{4}=5.25$$
   $$f(2)=5+(2)-(2)^2=3$$
   The largest of these values, $\dfrac{21}{4}$, is the absolute maximum, it occurs at $x=\dfrac{1}{2}$. The smallest of these values, 3, is the absolute minimum, it occurs at $x=2$.

5. $f(x)=x^3-x^2-x+2;\ \ [-1,2]$
   a) Find $f'(x)$
   $$f'(x)=3x^2-2x-1$$
   b) Find the Critical Values. The derivative exists for all real numbers. Thus, we solve
   $$f'(x)=0$$
   $$3x^2-2x-1=0$$
   $$(3x+1)(x-1)=0$$
   $$3x+1=0\quad\text{or}\quad x-1=0$$
   $$3x=-1\quad\text{or}\quad x=1$$
   $$x=-\dfrac{1}{3}\quad\text{or}\quad x=1$$

c) List the critical values and endpoints. These
values are -1, $-\dfrac{1}{3}$, 1, and 2 .

d) Evaluate $f(x)$ at each value in step (c).

$f(-1) = (-1)^3 - (-1)^2 - (-1) + 2 = 1$

$f\left(\dfrac{-1}{3}\right) = \left(\dfrac{-1}{3}\right)^3 - \left(\dfrac{-1}{3}\right)^2 - \left(\dfrac{-1}{3}\right) + 2 = \dfrac{59}{27} \approx 2.2$

$f(1) = (1)^3 - (1)^2 - (1) + 2 = 1$

$f(2) = (2)^3 - (2)^2 - (2) + 2 = 4$

The largest of these values, 4,is the absolute
maximum, it occurs at $x = 2$. The smallest
of these values, 1, is the absolute minimum,
it occurs at $x = -1$ and $x = 1$.

7. $f(x) = x^3 - x^2 - x + 3$; $[-1,0]$

a) Find $f'(x)$

$f'(x) = 3x^2 - 2x - 1$

b) Find the Critical Values. The derivative
exists for all real numbers. Thus, we solve

$f'(x) = 0$

$3x^2 - 2x - 1 = 0$

$(3x + 1)(x - 1) = 0$

$3x + 1 = 0 \qquad \text{or} \quad x - 1 = 0$

$3x = -1 \qquad \text{or} \qquad x = 1$

$x = -\dfrac{1}{3} \qquad \text{or} \qquad x = 1$

c) List the critical values and endpoints. The
critical value $x = 1$ is not in the interval, so
we exclude it. We will test the values

-1, $-\dfrac{1}{3}$, and 0 .

d) Evaluate $f(x)$ at each value in step (c).

$f(-1) = (-1)^3 - (-1)^2 - (-1) + 3 = 2$

$f\left(\dfrac{-1}{3}\right) = \left(\dfrac{-1}{3}\right)^3 - \left(\dfrac{-1}{3}\right)^2 - \left(\dfrac{-1}{3}\right) + 3 = \dfrac{86}{27} \approx 3.2$

$f(0) = (0)^3 - (0)^2 - (0) + 3 = 3$

On the interval $[-1,0]$, the absolute

maximum is $\dfrac{86}{27}$, which occurs at $x = -\dfrac{1}{3}$.

The absolute minimum is 2, which occurs at
$x = -1$.

9. $f(x) = 5x - 7$; $[-2,3]$

a) Find $f'(x)$

$f'(x) = 5$

b)   and c)
The derivative exists and is 5 for all real
numbers. Note that the derivative is never 0.
Thus, there are no critical values for $f(x)$,
and the absolute maximum and absolute
minimum will occur at the endpoints of the
interval.

d) Evaluate $f(x)$ at the endpoints.

$f(-2) = 5(-2) - 7 = -17$

$f(3) = 5(3) - 7 = 8$

On the interval $[-2,3]$, the absolute
maximum is 8, which occurs at $x = 3$. The
absolute minimum is -17, which occurs at
$x = -2$.

11. $f(x) = 7 - 4x$; $[-2,5]$

a) Find $f'(x)$

$f'(x) = -4$

b)   and c)
The derivative exists and is -4 for all real
numbers. Note that the derivative is never 0.
Thus, there are no critical values for $f(x)$,
and the absolute maximum and absolute
minimum will occur at the endpoints of the
interval.

d) Evaluate $f(x)$ at the endpoints.

$f(-2) = 7 - 4(-2) = 15$

$f(5) = 7 - 4(5) = -13$

On the interval $[-2,5]$, the absolute
maximum is 15, which occurs at $x = -2$.
The absolute minimum is -13, which
occurs at $x = 5$.

13. $f(x) = -5$; $[-1,1]$

Note for all values of x, $f(x) = -5$. Thus, the
absolute maximum is -5 for $-1 \le x \le 1$ and the
absolute minimum is -5 for $-1 \le x \le 1$.

15. $f(x) = x^2 - 6x - 3;$ $[-1,5]$

a) $f'(x) = 2x - 6$

b) $f'(x)$ exists for all real numbers. Solve:

$2x - 6 = 0$

$2x = 6$

$x = 3$

c) The critical value and the endpoints are $-1, 3,$ and 5.

d) Evaluate $f(x)$ for each value in (c)

$f(-1) = (-1)^2 - 6(-1) - 3 = 4$

$f(3) = (3)^2 - 6(3) - 3 = -12$

$f(5) = (5)^2 - 6(5) - 3 = -8$

On the interval $[-1,5]$, the absolute maximum is 4, which occurs at $x = -1$. The absolute minimum is -12, which occurs at $x = 3$.

17. $f(x) = 3 - 2x - 5x^2;$ $[-3,3]$

a) $f'(x) = -2 - 10x$

b) $f'(x)$ exists for all real numbers. Solve:

$-2 - 10x = 0$

$-10x = 2$

$x = -\dfrac{1}{5}$

c) The critical value and the endpoints are $-3, -\dfrac{1}{5},$ and 3.

d) Evaluate $f(x)$ for each value in (c)

$f(-3) = 3 - 2(-3) - 5(-3)^2 = -36$

$f\left(-\dfrac{1}{5}\right) = 3 - 2\left(-\dfrac{1}{5}\right) - 5\left(-\dfrac{1}{5}\right)^2 = \dfrac{16}{5} = 3.2$

$f(3) = 3 - 2(3) - 5(3)^2 = -48$

On the interval $[-3,3]$, the absolute maximum is $\dfrac{16}{5}$, which occurs at $x = -\dfrac{1}{5}$. The absolute minimum is -48, which occurs at $x = 3$.

19. $f(x) = x^3 - 3x^2;$ $[0,5]$

a) $f'(x) = 3x^2 - 6x$

b) $f'(x)$ exists for all real numbers. Solve:

$3x^2 - 6x = 0$

$3x(x - 2) = 0$

$3x = 0$ or $x - 2 = 0$

$x = 0$ or $x = 2$

c) The critical value and the endpoints are $0, 2,$ and 5. Note, since 0 is an endpoint of the interval, $x = 0$ is included in this list as an endpoint, not a critical value.

d) Evaluate $f(x)$ for each value in (c)

$f(0) = (0)^3 - 3(0)^2 = 0$

$f(2) = (2)^3 - 3(2)^2 = -4$

$f(5) = (5)^3 - 3(5)^2 = 50$

On the interval $[0,5]$, the absolute maximum is 50, which occurs at $x = 5$. The absolute minimum is -4, which occurs at $x = 2$.

21. $f(x) = x^3 - 3x;$ $[-5,1]$

a) $f'(x) = 3x^2 - 3$

b) $f'(x)$ exists for all real numbers. Solve:

$3x^2 - 3 = 0$

$x^2 - 1 = 0$

$x = \pm 1$

c) The critical value and the endpoints are $-5, -1,$ and 1. Note, since 1 is an endpoint of the interval, $x = 1$ is included in this list as an endpoint, not a critical value.

d) Evaluate $f(x)$ for each value in (c)

$f(-5) = (-5)^3 - 3(-5) = -110$

$f(-1) = (-1)^3 - 3(-1) = 2$

$f(1) = (1)^3 - 3(1) = -2$

On the interval $[-5,1]$, the absolute maximum is 2, which occurs at $x = -1$. The absolute minimum is -110, which occurs at $x = -5$.

23. $f(x) = 1 - x^3$; $[-8, 8]$

a) $f'(x) = -3x^2$

b) $f'(x)$ exists for all real numbers. Solve:

$$-3x^2 = 0$$
$$x = 0$$

c) The critical value and the endpoints are -8, 0, and 8.

d) Evaluate $f(x)$ for each value in (c)

$$f(-8) = 1 - (-8)^3 = 513$$
$$f(0) = 1 - (0)^3 = 1$$
$$f(8) = 1 - (8)^3 = -511$$

On the interval $[-8, 8]$, the absolute maximum is 513, which occurs at $x = -8$. The absolute minimum is -511, which occurs at $x = 8$.

25. $f(x) = 12 + 9x - 3x^2 - x^3$; $[-3, 1]$

a) $f'(x) = 9 - 6x - 3x^2$

b) $f'(x)$ exists for all real numbers. Solve:

$$9 - 6x - 3x^2 = 0$$
$$x^2 + 2x - 3 = 0 \qquad \text{Divide by } -3.$$
$$(x + 3)(x - 1) = 0$$
$$x + 3 = 0 \quad \text{or} \quad x - 1 = 0$$
$$x = -3 \quad \text{or} \quad x = 1$$

c) The critical values and the endpoints are -3 and 1. Note, since the possible critical values are the endpoints of the interval, they are included in this list as endpoints, not as critical values.

d) Evaluate $f(x)$ for each value in (c)

$$f(-3) = 12 + 9(-3) - 3(-3)^2 - (-3)^3 = -15$$
$$f(1) = 12 + 9(1) - 3(1)^2 - (1)^3 = 17$$

On the interval $[-3, 1]$, the absolute maximum is 17, which occurs at $x = 1$. The absolute minimum is -15, which occurs at $x = -3$.

27. $f(x) = x^4 - 2x^3$; $[-2, 2]$

a) $f'(x) = 4x^3 - 6x^2$

b) $f'(x)$ exists for all real numbers. Solve:

$$4x^3 - 6x^2 = 0$$
$$2x^2(2x - 3) = 0$$
$$2x^2 = 0 \quad \text{or} \quad 2x - 3 = 0$$
$$x = 0 \quad \text{or} \quad x = \frac{3}{2}$$

c) The critical values and the endpoints are $-2, 0, \dfrac{3}{2}$, and 2.

d) Evaluate $f(x)$ for each value in (c)

$$f(-2) = (-2)^4 - 2(-2)^3 = 32$$
$$f(0) = (0)^4 - 2(0)^3 = 0$$
$$f\left(\frac{3}{2}\right) = \left(\frac{3}{2}\right)^4 - 2\left(\frac{3}{2}\right)^3 = -\frac{27}{16} = -1.6875$$
$$f(2) = (2)^4 - 2(2)^3 = 0$$

On the interval $[-2, 2]$, the absolute maximum is 32, which occurs at $x = -2$. The absolute minimum is $-\dfrac{27}{16}$, which occurs at $x = \dfrac{3}{2}$.

29. $f(x) = x^4 - 2x^2 + 5$; $[-2, 2]$

a) $f'(x) = 4x^3 - 4x$

b) $f'(x)$ exists for all real numbers. Solve:

$$4x^3 - 4x = 0$$
$$4x(x^2 - 1) = 0$$
$$4x = 0 \quad \text{or} \quad x^2 - 1 = 0$$
$$x = 0 \quad \text{or} \quad x = \pm 1$$

c) The critical values and the endpoints are $-2, -1, 0, 1$, and 2.

d) Evaluate $f(x)$ for each value in (c)

$$f(-2) = (-2)^4 - 2(-2)^2 + 5 = 13$$
$$f(-1) = (-1)^4 - 2(-1)^2 + 5 = 4$$
$$f(0) = (0)^4 - 2(0)^2 + 5 = 5$$
$$f(1) = (1)^4 - 2(1)^2 + 5 = 4$$
$$f(2) = (2)^4 - 2(2)^2 + 5 = 13$$

On the interval $[-2,2]$, the absolute maximum is 13, which occurs at $x = -2$ and $x = 2$. The absolute minimum is 4, which occurs at $x = -1$ and $x = 1$.

On the interval $[1,20]$, the absolute maximum is 20.05, which occurs at $x = 20$. The absolute minimum is 2, which occurs at $x = 1$.

31. $f(x) = (x+3)^{\frac{2}{3}} - 5;$ $\qquad [-4,5]$

a) $f'(x) = \dfrac{2}{3}(x+3)^{-\frac{1}{3}} = \dfrac{2}{3(x+3)^{\frac{1}{3}}}$

b) $f'(x)$ does not exist for $x = -3$. The equation $f'(x) = 0$ has no solution, so $x = -3$ is the only critical value.

c) The critical values and the endpoints are -4, -3, and 5.

d) Evaluate $f(x)$ for each value in (c)

$f(-4) = \left((-4)+3\right)^{\frac{2}{3}} - 5 = -4$

$f(-3) = \left((-3)+3\right)^{\frac{2}{3}} - 5 = -5$

$f(5) = \left((5)+3\right)^{\frac{2}{3}} - 5 = -1$

On the interval $[-4,5]$, the absolute maximum is -1, which occurs at $x = 5$. The absolute minimum is -5, which occurs at $x = -3$.

33. $f(x) = x + \dfrac{1}{x};$ $\qquad [1,20]$

a) $f'(x) = 1 - x^{-2} = 1 - \dfrac{1}{x^2}$

b) $f'(x)$ does not exist for $x = 0$. However, $x = 0$ is not in the interval. Solve $f'(x) = 0$.

$1 - \dfrac{1}{x^2} = 0$

$1 = \dfrac{1}{x^2}$

$x^2 = 1$

$x = \pm 1$

The critical value $x = -1$ is not in the interval, and the other critical value is an endpoint.

c) The critical values and the endpoints are 1 and 20.

d) Evaluate $f(x)$ for each value in (c)

$f(1) = 1 + \dfrac{1}{1} = 2$

$f(20) = 20 + \dfrac{1}{20} = \dfrac{401}{20} = 20.05$

35. $f(x) = \dfrac{x^2}{x^2+1};$ $\qquad [-2,2]$

a) $f'(x) = \dfrac{(x^2+1)(2x) - x^2(2x)}{(x^2+1)^2}$ $\qquad$ Quotient Rule

$\quad = \dfrac{2x^3 + 2x - 2x^3}{(x^2+1)^2}$

$\quad = \dfrac{2x}{(x^2+1)^2}$

b) $f'(x)$ exists for all real numbers. Solve:

$f'(x) = 0$

$\dfrac{2x}{(x^2+1)^2} = 0$

$2x = 0$

$x = 0$

c) The critical values and the endpoints are -2, 0, and 2.

d) Evaluate $f(x)$ for each value in (c)

$f(-2) = \dfrac{(-2)^2}{(-2)^2+1} = \dfrac{4}{5}$

$f(0) = \dfrac{(0)^2}{(0)^2+1} = 0$

$f(2) = \dfrac{(2)^2}{(2)^2+1} = \dfrac{4}{5}$

On the interval $[-2,2]$, the absolute maximum is $\dfrac{4}{5}$, which occurs at $x = -2$ and $x = 2$. The absolute minimum is 0, which occurs at $x = 0$.

37. $f(x) = (x+1)^{\frac{1}{3}};$ $\qquad [-2,26]$

a) $f'(x) = \dfrac{1}{3}(x+1)^{-\frac{2}{3}} = \dfrac{1}{3(x+1)^{\frac{2}{3}}}$

b) $f'(x)$ does not exist for $x = -1$. The equation $f'(x) = 0$ has no solution, so $x = -1$ is the only critical value.

c) The critical values and the endpoints are -2, -1, and 26.

d) Evaluate $f(x)$ for each value in (c)

$$f(-2) = ((-2)+1)^{\frac{1}{3}} = -1$$

$$f(-1) = ((-1)+1)^{\frac{1}{3}} = 0$$

$$f(26) = ((26)+1)^{\frac{1}{3}} = 3$$

On the interval $[-2, 26]$, the absolute minimum is -1, which occurs at $x = -2$. The absolute maximum is 3, which occurs at $x = 26$.

39. – 47.   Left to the student.

49. $f(x) = 12x - x^2$

When no interval is specified , we use the real line $(-\infty, \infty)$.

a) Find $f'(x)$

$$f'(x) = 12 - 2x$$

b) Find the critical values. The derivative exists for all real numbers. Thus, we solve
$$f'(x) = 0.$$
$$12 - 2x = 0$$
$$12 = 2x$$
$$6 = x$$
The only critical value is $x = 6$.

c) Since there is only one critical value, we can apply Max-Min Principle 2. First we find $f''(x)$.
$$f''(x) = -2.$$
The second derivative is constant, so $f''(6) = -2$. Since the second derivative is negative at 6, we have a maximum at $x = 6$. Next, we find the function value at $x = 6$.
$$f(6) = 12(6) - (6)^2 = 36$$
Therefore, the absolute maximum is 36, which occurs at $x = 6$. The function has no minimum value.

51. $f(x) = 2x^2 - 40x + 270$

When no interval is specified , we use the real line $(-\infty, \infty)$.

a) Find $f'(x)$

$$f'(x) = 4x - 40$$

b) Find the critical values. The derivative exists for all real numbers. Thus, we solve
$$f'(x) = 0.$$
$$4x - 40 = 0$$
$$4x = 40$$
$$x = 10$$
The only critical value is $x = 10$. .

c) Since there is only one critical value, we can apply Max-Min Principle 2. First we find $f''(x)$.
$$f''(x) = 4.$$
The second derivative is constant, so $f''(10) = 4$. Since the second derivative is positive at 10, we have a minimum at $x = 10$. Next, we find the function value at $x = 10$.
$$f(10) = 2(10)^2 - 40(10) + 270 = 70$$
Therefore, the absolute minimum is 70, which occurs at $x = 10$. The function has no maximum value.

53. $f(x) = x - \dfrac{4}{3}x^3$; $(0, \infty)$

a) Find $f'(x)$
$$f'(x) = 1 - 4x^2$$

b) Find the critical values. The derivative exists for all real numbers. Thus, we solve
$$f'(x) = 0.$$
$$1 - 4x^2 = 0$$
$$-4x^2 = -1$$
$$x^2 = \frac{1}{4}$$
$$x = \pm\sqrt{\frac{1}{4}}$$
$$x = \pm\frac{1}{2}$$
There are two critical values; however, the only critical value in $(0, \infty)$ is $x = \dfrac{1}{2}$.

c) Since there is only one critical value in the interval, we can apply Max-Min Principle 2. First we find $f''(x)$.
$$f''(x) = -8x$$
$$f''\left(\frac{1}{2}\right) = -4 \quad .$$

Since the second derivative is negative at $\frac{1}{2}$,

we have a maximum at $x = \frac{1}{2}$. Next, we

find the function value at $x = \frac{1}{2}$.

$$f\left(\frac{1}{2}\right) = \left(\frac{1}{2}\right) - \frac{4}{3}\left(\frac{1}{2}\right)^3 = \frac{1}{3}$$

Therefore, the absolute maximum is $\frac{1}{3}$,

which occurs at $x = \frac{1}{2}$. The function has no

minimum value.

55. $f(x) = x(60 - x) = 60x - x^2$

When no interval is specified , we use the real

line $(-\infty, \infty)$.

a) Find $f'(x)$

$f'(x) = 60 - 2x$

b) Find the critical values. The derivative exists
for all real numbers. Thus, we solve

$f'(x) = 0$.

$60 - 2x = 0$

$60 = 2x$

$30 = x$

The only critical value is $x = 30$.

c) Since there is only one critical value, we can
apply Max-Min Principle 2. First we find
$f''(x)$.

$f''(x) = -2$.

The second derivative is constant, so

$f''(30) = -2$. Since the second derivative is

negative at 30, we have a maximum at
$x = 30$. Next, we find the function value at
$x = 30$.

$f(30) = 30(60 - 30) = 900$

Therefore, the absolute maximum is 900,
which occurs at $x = 30$. The function has no
minimum value.

57. $f(x) = \frac{1}{3}x^3 - 3x;$ $[-2,2]$

a) Find $f'(x)$

$f'(x) = x^2 - 3$

b) Find the critical values. The derivative exists
for all real numbers. Thus, we solve

$f'(x) = 0$.

$x^2 - 3 = 0$

$x^2 = 3$

$x = \pm\sqrt{3} \approx \pm 1.732$

Both critical values are in the interval
$[-2,2]$.

c) The interval is closed and there is more than
one critical value, so we use Max-Min
Principle 1.
The critical points and the endpoints are
$-2, -\sqrt{3}, \sqrt{3},$ and 2.
Next, we find the function values at these
points.

$$f(-2) = \frac{1}{3}(-2)^3 - 3(-2) = \frac{10}{3} = 3.3\overline{3}$$

$$f(-\sqrt{3}) = \frac{1}{3}(-\sqrt{3})^3 - 3(-\sqrt{3})$$
$$= -\sqrt{3} + 3\sqrt{3} = 2\sqrt{3} \approx 3.464$$

$$f(\sqrt{3}) = \frac{1}{3}(\sqrt{3})^3 - 3(\sqrt{3})$$
$$= \sqrt{3} - 3\sqrt{3} = -2\sqrt{3} \approx -3.464$$

$$f(2) = \frac{1}{3}(2)^3 - 3(2) = -\frac{10}{3} = -3.3\overline{3}$$

The largest of these values, $2\sqrt{3}$, is the
maximum. It occurs at $x = -\sqrt{3}$. The
smallest of these values, $-2\sqrt{3}$, is the
minimum. It occurs at $x = \sqrt{3}$.
Thus, the absolute maximum over the
interval $[-2,2]$, is $2\sqrt{3}$, which occurs at
$x = -\sqrt{3}$, and the absolute minimum over
$[-2,2]$ is $-2\sqrt{3}$, which occurs at $x = \sqrt{3}$.

59. $f(x) = -0.001x^2 + 4.8x - 60$

When no interval is specified , we use the real
line $(-\infty, \infty)$.

a) Find $f'(x)$

$f'(x) = -0.002x + 4.8$

b) Find the critical values. The derivative exists for all real numbers. Thus, we solve $f'(x) = 0$.

$$-0.002x + 4.8 = 0$$
$$-0.002x = -4.8$$
$$x = 2400$$

The only critical value is $x = 2400$.

c) Since there is only one critical value, we can apply Max-Min Principle 2. First we find $f''(x)$.

$$f''(x) = -0.002.$$

The second derivative is constant, so $f''(2400) = -0.002$. Since the second derivative is negative at 2400, we have a maximum at $x = 2400$. Next, we find the function value at $x = 2400$.

$$f(2400) = -0.001(2400)^2 + 4.8(2400) - 60$$
$$= 5700$$

Therefore, the absolute maximum is 5700, which occurs at $x = 2400$. The function has no minimum value.

61. $f(x) = -\dfrac{1}{3}x^3 + 6x^2 - 11x - 50;$ $(0,3)$

a) Find $f'(x)$

$$f'(x) = -x^2 + 12x - 11$$

b) Find the critical values. The derivative exists for all real numbers. Thus, we solve $f'(x) = 0$.

$$-x^2 + 12x - 11 = 0$$
$$x^2 - 12 + 11 = 0$$
$$(x - 11)(x - 1) = 0$$
$$x - 11 = 0 \quad \text{or} \quad x - 1 = 0$$
$$x = 11 \quad \text{or} \quad x = 1$$

c) The interval $(0,3)$ is not closed. The only critical value in the interval is $x = 1$. Therefore, we can apply Max-Min Principle 2. First, we find the second derivative.

$$f''(x) = -2x + 12$$

Evaluating the second derivative at $x = 1$, we have:

$$f''(x) = -2(1) + 12 = 10 > 0.$$

Since the second derivative is positive when $x = 1$, there is a minimum at $x = 1$.

Next, find the function value at $x = 1$.

$$f(1) = -\frac{1}{3}(1)^3 + 6(1)^2 - 11(1) - 50$$
$$= -\frac{166}{3} = -55.33\overline{3}$$

Thus, the absolute minimum over the interval $(0,3)$ is $-\dfrac{166}{3}$, which occurs at $x = 1$.

63. $f(x) = 15x^2 - \dfrac{1}{2}x^3;$ $[0,30]$

a) Find $f'(x)$

$$f'(x) = 30x - \frac{3}{2}x^2$$

b) Find the critical values. The derivative exists for all real numbers. Thus, we solve $f'(x) = 0$.

$$30x - \frac{3}{2}x^2 = 0$$
$$60x - 3x^2 = 0$$
$$3x(20 - x) = 0$$
$$3x = 0 \quad \text{or} \quad 20 - x = 0$$
$$x = 0 \quad \text{or} \qquad x = 20$$

Both critical values are in the interval $[0,30]$.

c) Since the interval is closed and there is more than one critical value, we apply the Max-Min Principle 1.

The critical values and the endpoints are 0, 20, and 30.

Next, we find the function values.

$$f(0) = 15(0)^2 - \frac{1}{2}(0)^3 = 0$$

$$f(20) = 15(20)^2 - \frac{1}{2}(20)^3 = 2000$$

$$f(30) = 15(30)^2 - \frac{1}{2}(30)^3 = 0$$

The largest of these values, 2000, is the maximum. It occurs at $x = 20$. The smallest of these values, 0, is the minimum. It occurs at $x = 0$ and $x = 30$. .

Thus, the absolute maximum over the interval $[0,30]$, is 2000, which occurs at $x = 20$. , and the absolute minimum over $[0,30]$ is 0, which occurs at $x = 0$ and $x = 30$.

65. $f(x) = 2x + \dfrac{72}{x};$ $(0, \infty)$

$f(x) = 2x + 72x^{-1}$

a) Find $f'(x)$

$$f'(x) = 2 - 72x^{-2} = 2 - \frac{72}{x^2}$$

b) Find the critical values. $f'(x)$ does not exists for $x = 0$; however, 0 is not in the interval $(0, \infty)$. Therefore, we solve

$$f'(x) = 0$$

$$2 - \frac{72}{x^2} = 0$$

$$2 = \frac{72}{x^2}$$

$$2x^2 = 72 \quad \text{Multiplying by } x^2, \text{ since } x \neq 0.$$

$$x^2 = 36$$

$$x = \pm 6$$

c) The interval $(0, \infty)$ is not closed. The only critical value in the interval is $x = 6$. Therefore, we can apply Max-Min Principle 2. First, we find the second derivative.

$$f''(x) = 144x^{-3} = \frac{144}{x^3}$$

Evaluating the second derivative at $x = 6$, we have:

$$f''(6) = \frac{144}{(6)^3} = \frac{2}{3} > 0.$$

Since the second derivative is positive when $x = 6$, there is a minimum at $x = 6$. Next, find the function value at $x = 6$.

$$f(6) = 2(6) + \frac{72}{6} = 24$$

Thus, the absolute minimum over the interval $(0, \infty)$ is 24, which occurs at $x = 6$. The function has no maximum value over the interval $(0, \infty)$.

67. $f(x) = x^2 + \dfrac{432}{x};$ $(0, \infty)$

$f(x) = x^2 + 432x^{-1}$

a) Find $f'(x)$

$$f'(x) = 2x - 432x^{-2} = 2x - \frac{432}{x^2}$$

b) Find the critical values. $f'(x)$ does not exists for $x = 0$; however, 0 is not in the interval $(0, \infty)$. Therefore, we solve

$$f'(x) = 0$$

$$2x - \frac{432}{x^2} = 0$$

$$2x = \frac{432}{x^2}$$

$$2x^3 = 432 \quad \text{Multiplying by } x^2, \text{ since } x \neq 0.$$

$$x^3 = 216$$

$$x = 6$$

c) The interval $(0, \infty)$ is not closed. The only critical value in the interval is $x = 6$. Therefore, we can apply Max-Min Principle 2. First, we find the second derivative.

$$f''(x) = 2 + 864x^{-3} = 2 + \frac{864}{x^3}$$

Evaluating the second derivative at $x = 6$, we have:

$$f''(6) = 2 + \frac{864}{(6)^3} = 6 > 0.$$

Since the second derivative is positive when $x = 6$, there is a minimum at $x = 6$. Next, find the function value at $x = 6$.

$$f(6) = (6)^2 + \frac{432}{6} = 108$$

Thus, the absolute minimum over the interval $(0, \infty)$ is 108, which occurs at $x = 6$. The function has no maximum value over the interval $(0, \infty)$.

69. $f(x) = 2x^4 - x;$ $[-1, 1]$

a) Find $f'(x)$

$$f'(x) = 8x^3 - 1$$

b) Find the critical values. The derivative exists for all real numbers. Thus, we solve

$$f'(x) = 0.$$

$$8x^3 - 1 = 0$$

$$8x^3 = 1$$

$$x^3 = \frac{1}{8}$$

$$x = \frac{1}{2}$$

The only critical value $x = \dfrac{1}{2}$ is in the interval $[-1,1]$.

c) The interval is closed, and we are looking for both the absolute maximum and absolute minimum values, so we use Max-Min Principle 1.

The critical values and the endpoints are

-1, $\dfrac{1}{2}$, and 1.

Next, we find the function values at these points.

$$f(-1) = 2(-1)^4 - (-1) = 3$$

$$f\left(\dfrac{1}{2}\right) = 2\left(\dfrac{1}{2}\right)^4 - \left(\dfrac{1}{2}\right) = -\dfrac{3}{8}$$

$$f(1) = 2(1)^4 - (1) = 1$$

The largest of these values, 3, is the maximum. It occurs at $x = -1$. The smallest of these values, $-\dfrac{3}{8}$, is the minimum. It occurs at $x = \dfrac{1}{2}$.

Thus, the absolute maximum over the interval $[-1,1]$, is 3, which occurs at $x = -1$, and the absolute minimum over $[-1,1]$ is $-\dfrac{3}{8}$, which occurs at $x = \dfrac{1}{2}$.

71. $f(x) = \sqrt[3]{x} = x^{1/3}$; $[0,8]$

a) Find $f'(x)$

$$f'(x) = \dfrac{1}{3} x^{-2/3} = \dfrac{1}{3 \cdot \sqrt[3]{x^2}}$$

b) Find the critical values. $f'(x)$ does not exist for $x = 0$. The equation $f'(x) = 0$ has no solution, so the only critical value is 0, which is also an endpoint.

c) The interval is closed, and we are looking for both the absolute maximum and absolute minimum values, so we use Max-Min Principle 1.

The only critical value is an endpoint. The endpoints are 0 and 8. Next, we find the function values at these points.

$$f(0) = \sqrt[3]{0} = 0$$

$$f(8) = \sqrt[3]{8} = 2$$

The largest of these values, 2, is the maximum. It occurs at $x = 8$. The smallest of these values, 0, is the minimum. It occurs at $x = 0$.

Thus, the absolute maximum over the interval $[0,8]$, is 2, which occurs at $x = 8$, and the absolute minimum over $[0,8]$ is 0, which occurs at $x = 0$.

73. $f(x) = (x+1)^3$

When no interval is specified, we use the real line $(-\infty, \infty)$.

a) Find $f'(x)$

$$f'(x) = 3(x+1)^2$$

b) Find the critical values. The derivative exists for all real numbers. Thus, we solve $f'(x) = 0$.

$$3(x+1)^2 = 0$$

$$x + 1 = 0$$

$$x = -1$$

The only critical value is $x = -1$.

c) Since there is only one critical value, we can apply Max-Min Principle 2. First we find $f''(x)$.

$$f''(x) = 6(x+1).$$

Now,

$f''(-1) = 6\big((-1)+1\big) = 0$. So the Max-Min Principle 2 fails. We cannot use Max-Min Principle 1, because there are no endpoints.

We note that $f'(x) = 3(x+1)^2$ is never negative. Thus, $f(x)$ is increasing everywhere except at $x = -1$. Therefore, the function has no maximum or minimum over the interval $(-\infty, \infty)$.

Notice

$$f''(-2) = -6 < 0$$

$$f''(0) = 6 > 0$$

and

$$f(-1) = 0$$

Therefore, there is a point of inflection at $(-1,0)$.

75. $f(x) = 2x - 3;$ $[-1,1]$

   a) Find $f'(x)$

   $f'(x) = 2$

   b) and c)

   The derivative exists and is 2 for all real
   numbers. Therefore, $f'(x)$ is never 0. Thus,
   there are no critical values. We apply the
   Max-Min Principle 1. The endpoints are
   -1 and 1. We find the function values at the
   endpoints.

   $f(-1) = 2(-1) - 3 = -5$

   $f(1) = 2(1) - 3 = -1$

   Therefore, the absolute maximum over the
   interval $[-1,1]$ is -1, which occurs at $x = 1$,
   and the absolute minimum over the interval
   $[-1,1]$ is -5, which occurs at $x = -1$.

77. $f(x) = 2x - 3;$ $[-1,5)$

   a) Find $f'(x)$

   $f'(x) = 2$

   b) and c)

   The derivative exists and is 2 for all real
   numbers. Therefore, $f'(x)$ is never 0. Thus,
   there are no critical values. We apply the
   Max-Min Principle 1. There is only one
   endpoint, $x = -1$ We find the function value
   at the endpoint.

   $f(-1) = 2(-1) - 3 = -5$

   We know $f'(x) > 0$ over the interval, so the
   function is increasing over the interval
   $[-1,5)$. Therefore, the minimum value will
   be the left hand endpoint. The absolute
   minimum over the interval $[-1,5)$ is -5,
   which occurs at $x = -1$. Since the right
   endpoint is not included in the interval, the
   function has no maximum value over the
   interval $[-1,5)$.

79. $f(x) = x^{\frac{2}{3}};$ $[-1,1]$

   a) Find $f'(x)$

   $f'(x) = \frac{2}{3} x^{-\frac{1}{3}} = \frac{2}{3 \cdot \sqrt[3]{x}}$

   b) Find the critical values. $f'(x)$ does not exist
   for $x = 0$. The equation $f'(x) = 0$ has no
   solution, so the only critical value is 0.

   c) The interval is closed, and we are looking
   for both the absolute maximum and absolute
   minimum values, so we use Max-Min
   Principle 1.
   The critical value and the endpoints
   are -1, 0, and 1.
   Next, we find the function values at these
   points.

   $f(-1) = (-1)^{\frac{2}{3}} = 1$

   $f(0) = (0)^{\frac{2}{3}} = 0$

   $f(1) = (1)^{\frac{2}{3}} = 1$

   The largest of these values, 1, is the
   maximum. It occurs at $x = -1$ and $x = 1$. The
   smallest of these values, 0, is the minimum.
   It occurs at $x = 0$.
   Thus, the absolute maximum over the
   interval $[-1,1]$, is 1, which occurs at $x = -1$
   and $x = 1$, and the absolute minimum over
   $[-1,1]$ is 0, which occurs at $x = 0$.

81. $f(x) = \frac{1}{3} x^3 - x + \frac{2}{3}$

   When no interval is specified , we use the real
   line $(-\infty, \infty)$.

   a) Find $f'(x)$

   $f'(x) = x^2 - 1$

   b) Find the critical values. The derivative exists
   for all real numbers. Thus, we solve
   $f'(x) = 0$.

   $x^2 - 1 = 0$

   $x^2 = 1$

   $x = \pm 1$

   There are two critical values -1 and 1.

   c) The interval $(-\infty, \infty)$ is not closed, so the
   Max-Min Principle 1 does not apply. Since
   there is more than one critical value, the
   Max-Min Principle 2 does not apply.

A quick sketch of the graph will help us determine absolute or relative extrema occur at the critical values.

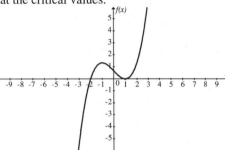

We determine that the function has no absolute extrema over the interval $(-\infty, \infty)$.

83. $f(x) = \dfrac{1}{3}x^3 - 2x^2 + x;$ $\quad [0, 4]$

a) Find $f'(x)$

$f'(x) = x^2 - 4x + 1$

b) Find the critical values. The derivative exists for all real numbers. Thus, we solve $f'(x) = 0$.

$x^2 - 4x + 1 = 0$

We use the quadratic formula to solve the equation.

$x = \dfrac{-(-4) \pm \sqrt{(-4)^2 - 4(1)(1)}}{2(1)}$

$= \dfrac{4 \pm \sqrt{12}}{2}$

$= \dfrac{4 \pm 2\sqrt{3}}{2}$

$= 2 \pm \sqrt{3}$

Both critical values $x = 2 - \sqrt{3} \approx 0.27$ and $x = 2 + \sqrt{3} \approx 3.73$ are in the closed interval $[0, 4]$.

c) The interval is closed, and there is more than one critical value in the interval, so we use Max-Min Principle 1.
The critical values and the endpoints are 0, $2 - \sqrt{3}$, $2 + \sqrt{3}$, and 4.
Next, we find the function values at these points.

$f(0) = \dfrac{1}{3}(0)^3 - 2(0)^2 + (0) = 0$

$f(2 - \sqrt{3}) = \dfrac{1}{3}(2 - \sqrt{3})^3 - 2(2 - \sqrt{3})^2 + (2 - \sqrt{3})$

$= -\dfrac{10}{3} + 2\sqrt{3} \approx 0.131$

$f(2 + \sqrt{3}) = \dfrac{1}{3}(2 + \sqrt{3})^3 - 2(2 + \sqrt{3})^2 + (2 + \sqrt{3})$

$= -\dfrac{10}{3} - 2\sqrt{3} \approx -6.797$

$f(4) = \dfrac{1}{3}(4)^3 - 2(4)^2 + (4)$

$= -\dfrac{20}{3} \approx -6.66\overline{6}$

The largest of these values, $-\dfrac{10}{3} + 2\sqrt{3}$, is the maximum. It occurs at $x = 2 - \sqrt{3}$ The smallest of these values, $-\dfrac{10}{3} - 2\sqrt{3}$, is the minimum. It occurs at $x = 2 + \sqrt{3}$
Thus, the absolute maximum over the interval $[0, 4]$, is $-\dfrac{10}{3} + 2\sqrt{3}$, which occurs at $x = 2 - \sqrt{3}$, and the absolute minimum over $[0, 4]$ is $-\dfrac{10}{3} - 2\sqrt{3}$, which occurs at $x = 2 + \sqrt{3}$.

85. $t(x) = x^4 - 2x^2$

When no interval is specified, we use the real line $(-\infty, \infty)$.

a) Find $t'(x)$

$t'(x) = 4x^3 - 4x$

b) Find the critical values. The derivative exists for all real numbers. Thus, we solve $t'(x) = 0$.

$4x^3 - 4x = 0$

$4x(x^2 - 1) = 0$

$x = 0 \quad$ or $\quad x^2 - 1 = 0$

$x = 0 \quad$ or $\quad x = \pm 1$

There are three critical values -1, 0, and 1.

c) The interval $(-\infty, \infty)$ is not closed, so the Max-Min Principle 1 does not apply. Since there is more than one critical value, the Max-Min Principle 2 does not apply.

A quick sketch of the graph will help us determine absolute or relative extrema occur at the critical values.

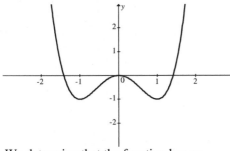

We determine that the function has no absolute maximum over the interval $(-\infty, \infty)$. The function's absolute minimum is -1, which occurs at $x = -1$ and $x = 1$.

87. – 95. Left to the Student.

97. $M(t) = -2t^2 + 100t + 180, \quad 0 \le t \le 40$

a) $M'(t) = -4t + 100$

b) $M'(t)$ exists for all real numbers. We solve $M'(t) = 0$.
$$-4t + 100 = 0$$
$$4t = 100$$
$$t = 25$$

c) Since there is only one critical value, we apply the Max-Min Principle 2. First, we find the second derivative.

$f''(t) = -4$. The second derivative is negative for all values of t in the interval, therefore, a maximum occurs at $t = 25$.

$$f(25) = -2(25)^2 + 100(25) + 180 = 1430$$

The maximum productivity for $0 \le t \le 40$ is 1430 units per month, which occurs at $t = 25$ years of service.

99. $p(x) = \dfrac{13x^3 - 240x^2 - 2460x + 585,000}{75,000}$

$$p(x) = \frac{1}{75,000}\left(13x^3 - 240x^2 - 2460x + 585,000\right)$$

We restrict our attention to the years 1970 to 2000. That is, we will look at the x-values $0 \le x \le 30$.

a) Find $p'(x)$

$$p'(x) = \frac{1}{75,000}\left(39x^2 - 480x - 2460\right)$$

b) Find the critical values. $p'(x)$ exists for all real numbers. Therefore, we solve $p'(x) = 0$.

$$\frac{1}{75,000}\left(39x^2 - 480x - 2460\right) = 0$$

$$39x^2 - 480x - 2460 = 0$$

Using the quadratic formula, we have:

$$x = \frac{-(-480) \pm \sqrt{(-480)^2 - 4(39)(-2460)}}{2(39)}$$

$$= \frac{480 \pm \sqrt{614,160}}{78}$$

$$x \approx -3.89 \quad \text{or} \quad x \approx 16.20$$

Only $x \approx 16.20$ is in the interval $[0, 30]$.

c) The critical values and the endpoints are 0, 16.20, and 30.

d) Using a calculator, we find the function values.

$$p(0) \approx 7.8$$
$$p(16.20) \approx 7.17$$
$$p(30) \approx 8.616$$

From 1970 to 2000 the minimum percentage of U.S. national income generated by nonfarm proprietors was 7.17 percent. This occurred about 16.20 years after 1970, or in 1986.

101. $P(t) = 0.000008533t^4 - 0.001685t^3 +$
$$0.090t^2 - 0.687t + 4.00, \quad 0 \le t \le 90$$

a) Find $P'(t)$.

$$P'(t) = 0.00003413t^3 - 0.005055t^2 +$$
$$0.18t - 0.687$$

b) $P'(t)$ exists for all real numbers. Solve

$$P'(t) = 0$$

$$0.00003413t^3 - 0.005055t^2 + 0.18t - 0.687 = 0$$

Using a graphing calculator, we approximate the zeros of $P'(t)$. We find the solutions:

$t \approx 4.3$

$t \approx 49.2$

$t \approx 94.6$

Two of the critical values are in the interval.

c) The critical values and the endpoints are: 0, 4.33, 49.16, and 90.

d) Find the function values.

$$P(0) = 4.0$$

$$P(4.3) \approx 2.6$$

$$P(49.2) \approx 37.4$$

$$P(90) \approx 2.7$$

The absolute maximum production of world wide oil was 37.4 billion barrels. The world achieved this production 49.2 years after 1950, or in the year 1999.

103. $C(x) = 5000 + 600x$

$$R(x) = -\frac{1}{2}x^2 + 1000x, \ 0 \le x \le 600$$

a) $P(x) = R(x) - C(x)$

$$= -\frac{1}{2}x^2 + 1000x - (5000 + 600x)$$

$$= -\frac{1}{2}x^2 + 400x - 5000$$

b) First, we find the critical values.

$$P'(x) = -x + 400$$

$P'(x)$ exists for all real numbers. Solve:

$$P'(x) = 0$$

$$-x + 400 = 0$$

$$x = 400$$

The critical value is 400 and the endpoints are 0 and 600. Using the Max-Min Principle 1, we have:

$$P(0) = -\frac{1}{2}(0)^2 + 400(0) - 5000$$

$$= -5000$$

$$P(400) = -\frac{1}{2}(400)^2 + 400(400) - 5000$$

$$= 75{,}000$$

$$P(600) = -\frac{1}{2}(600)^2 + 400(600) - 5000$$

$$= 55{,}000$$

The total profit is maximized when 400 items are produced.

105. $B(x) = 305x^2 - 1830x^3, \quad 0 \le x \le 0.16$

a) $B'(x) = 610x - 5490x^2$

b) $B'(x)$ exists for all real numbers. Solve:

$$B'(x) = 0$$

$$610x - 5490x^2 = 0$$

$$610x(1 - 9x) = 0$$

$$x = 0 \quad \text{or} \quad 1 - 9x = 0$$

$$x = 0 \quad \text{or} \quad x = \frac{1}{9} \approx 0.11$$

c) The critical points and the endpoints are 0, $\frac{1}{9}$, and 0.16.

d) We find the function values.

$$B(0) = 305(0)^2 - 1830(0)^3 = 0$$

$$B\left(\frac{1}{9}\right) = 305\left(\frac{1}{9}\right)^2 - 1830\left(\frac{1}{9}\right)^3$$

$$= \frac{305}{243} \approx 1.255$$

$$B(0.16) = 305(0.16)^2 - 1830(0.16)^3$$

$$\approx 0.312$$

The maximum blood pressure is approximately 1.255, which occurs at a dose of $x = \frac{1}{9}$ cc, or about 0.11cc of the drug.

107. $g(x) = x\sqrt{x+3}; \qquad [-3,3]$

a) Find $g'(x)$.

$$g'(x) = x\left[\frac{1}{2}(x+3)^{-\frac{1}{2}}(1)\right] + (1)(x+3)^{\frac{1}{2}}$$

$$= \frac{x}{2(x+3)^{\frac{1}{2}}} + (x+3)^{\frac{1}{2}}$$

$$= \frac{x}{2(x+3)^{\frac{1}{2}}} + \frac{(x+3)^{\frac{1}{2}}}{1} \cdot \frac{2(x+3)^{\frac{1}{2}}}{2(x+3)^{\frac{1}{2}}}$$

Multiplying by a form of 1

$$= \frac{x}{2(x+3)^{\frac{1}{2}}} + \frac{2(x+3)}{2(x+3)^{\frac{1}{2}}}$$

$$= \frac{3x+6}{2(x+3)^{\frac{1}{2}}}, \text{ or } \frac{3x+6}{2\sqrt{x+3}}$$

b) Find the critical values. $g'(x)$ exists for all values except -3. This is a critical value. To find the other critical values, we solve

$$g'(x) = 0$$

$$\frac{3x+6}{2\sqrt{x+3}} = 0$$

$$3x + 6 = 0$$

$$x = -2$$

The second critical value on the interval is -2.

c) On a closed interval, the Max-Min Principle 1 can always be used. The critical values and the endpoints are -3, -2, and 3.

d) Find the function value at each value in (c).

$$g(-3) = (-3)\sqrt{(-3)+3} = 0$$

$$g(-2) = (-2)\sqrt{(-2)+3} = -2$$

$$g(3) = (3)\sqrt{(3)+3} = 3\sqrt{6}$$

Thus, the absolute maximum over the interval $[-3,3]$ is $3\sqrt{6}$, which occurs at $x = 3$, and the absolute minimum is -2, which occurs at $x = -2$.

109. $C(x) = (2x+4) + \left(\dfrac{2}{x-6}\right), \qquad x > 6$

$$= 2x + 4 + 2(x-6)^{-1}$$

a) Find $C'(x)$.

$$C'(x) = 2 - 2(x-6)^{-2}(1)$$

$$= 2 - \frac{2}{(x-6)^2}$$

b) Find the critical values.

$C'(x)$ does not exist for $x = 6$; however, this value is not in the domain interval, so it is not a critical value. Solve $C'(x) = 0$.

$$2 - \frac{2}{(x-6)^2} = 0$$

$$2 = \frac{2}{(x-6)^2}$$

$$2(x-6)^2 = 2 \qquad \text{Multiplying by } (x-6)^2 \\ \text{Since } x \neq 6.$$

$$(x-6)^2 = 1$$

$$x - 6 = \pm 1 \qquad \text{Taking the square root} \\ \text{of both sides.}$$

$$x = 6 \pm 1$$

$$x = 5 \quad \text{or} \quad x = 7$$

The only critical value in $(6, \infty)$ is 7.

c) Since there is only one critical value, we apply the Max-Min Principle 2.

$$C''(x) = 4(x-6)^{-3} = \frac{4}{(x-6)^3}$$

$$C''(7) = \frac{4}{(7-6)^3} = 4 > 0$$

Therefore, since $C''(7) > 0$, there is a minimum at $x = 7$.

The firm should use 7 "quality units" to minimize its total cost of service.

111. $\boxed{tw}$

113. $P(t) = 0.0000000219t^4 - 0.0000167t^3 +$

$0.00155t^2 + 0.002t + 0.22, \quad 0 \le t \le 110$

a) Find $P'(t)$.

$$P'(t) = 0.0000000876t^3 - 0.0000501t^2 +$$

$$0.0031t + 0.002$$

$P'(t)$ exists for all real numbers. Solve $P'(t) = 0$. We use a calculator to find the zeros of $P'(t)$. We estimate the solutions to be:

$$x \approx -0.639$$

$$x \approx 71.333$$

$$x \approx 501.223$$

Only one of the critical values, $x \approx 71.333$, is in the interval $[0,110]$. We apply the Max-Min Principle 1, to find the absolute maximum.

The critical values and the endpoints are 0, 71.333, and 110.

The function values at these points are

$$P(0) = 0.22$$

$$P(71.333) \approx 2.755$$

$$P(110) \approx 0.173679$$

Thus, the absolute maximum oil production for the U.S. after 1910 was 2.755 billion barrels per year. This production level occurred 71.333 year after 1910, or in 1981.

115. $f(x) = \dfrac{3}{4}(x^2 - 1)^{2/3}; \qquad [\frac{1}{2}, \infty)$

Using a calculator, we enter the equation into the graphing editor:

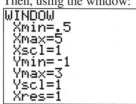

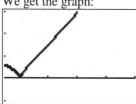

Then, using the window:

We get the graph:

Using the table feature, we locate the extrema. We estimate the absolute minimum to be 0, which occurs at $x = 1$. There is no absolute maximum.

117. a) Using a graphing calculator, we fit the linear equation $y = x + 8.857$. This corresponds to the model $P(t) = t + 8.857$. Where P is the pressure of the contractions and t is the time in minutes. We substitute 7 for t to find the pressure at 7 minutes.

$P(7) = 7 + 8.857 = 15.857$.

The pressure at 7 minutes is 15.857 mm of Hg.

b) Rounding the coefficients to 3 decimal places, we find the quartic regression

$y = 0.117x^4 - 1.520x^3 + 6.193x^2 - 7.018x + 10.009$

Changing the variables we get the model

$P(t) = 0.117t^4 - 1.520t^3 + 6.193t^2 - 7.018t + 10.009$

Using the table feature, when $x = 7$, $y = 24.857$. So the pressure at 7 minutes is 24.86 mm of mercury. (If we use the rounded coefficients above, we get 23.897 mm of Hg.)

Using the trace feature, we estimate the smallest contraction on the interval $[0,10]$ was about 7.62 mm of Hg. This occurred when $x \approx 0.765$.

Exercise Set 2.5

1. Express $Q = xy$ as a function of one variable.

First, we solve $x + y = 50$ for y.

$x + y = 50$

$y = 50 - x$

Next, we substitute $50 - x$ for y in $Q = xy$.

$Q = xy$

$Q = x(50 - x)$ Substituting

$= 50x - x^2$

Now that Q is a function of one variable we can find the maximum. First, we find the critical values.

$Q'(x) = 50 - 2x$. Since $Q'(x)$ exists for all real numbers, the only critical value will occur when $Q'(x) = 0$. We solve;

$50 - 2x = 0$

$50 = 2x$

$25 = x$

There is only one critical value. We use the second derivative to determine if the critical value is a maximum. Note that:

$Q''(x) = -2 < 0$. The second derivative is negative for all values of x. Therefore, a maximum occurs at $x = 25$.

Now,

$Q(25) = 50(25) - (25)^2 = 625$

Therefore, the maximum product is 625, which occurs when $x = 25$. If $x = 25$, then $y = 50 - 25 = 25$. The two numbers are 25 and 25.

3. $\boxed{tw}$

5. Let x be one number and y be the other number. Since the difference of the two numbers must be 4, we have $x - y = 4$.

The product, Q, of the two numbers is given by $Q = xy$, so our task is to minimize $Q = xy$, where $x - y = 4$.

First, we express $Q = xy$ as a function of one variable.

Solving $x - y = 4$ for y, we have:

$x - y = 4$

$-y = 4 - x$

$y = x - 4$

Next, we substitute $x - 4$ for y in $Q = xy$.

$$Q = x(x-4) = x^2 - 4x$$

$$Q(x) = x^2 - 4x$$

Find $Q'(x) = 2x - 4$

The derivative exists for all values of x; thus, the only critical values are where $Q'(x) = 0$.

$$2x - 4 = 0$$
$$2x = 4$$
$$x = 2$$

There is only one critical value. We can use the second derivative to determine whether we have a maximum.

$Q''(x) = 2 > 0$ for all values of x. Therefore, a minimum occurs at $x = 2$.

$$Q(2) = (2)^2 - 4(2) = -4$$

Thus, the minimum product is -4 when $x = 2$. Substitute 2 for x in $y = x - 4$ to find y.

$$y = 2 - 4 = -2.$$

The two numbers which have the minimum product are 2 and -2.

7. Maximize $Q = xy^2$, where x and y are positive numbers such that $x + y^2 = 1$.

Express $Q = xy^2$ as a function of one variable. First, we solve $x + y^2 = 1$ for y^2.

$$x + y^2 = 1$$
$$y^2 = 1 - x$$

Next, we substitute $1 - x$ for y^2 in $Q = xy^2$.

$$Q = xy^2$$
$$Q = x(1-x) \quad \text{Substituting}$$
$$= x - x^2$$

Now that Q is a function of one variable we can find the maximum. First, we find the critical values.

$Q'(x) = 1 - 2x$. Since $Q'(x)$ exists for all real numbers, the only critical value will occur when $Q'(x) = 0$. We solve;

$$1 - 2x = 0$$
$$1 = 2x$$
$$\frac{1}{2} = x$$

There is only one critical value. We use the second derivative to determine if the critical value is a maximum.

Note that: $Q''(x) = -2 < 0$. The second derivative is negative for all values of x.

Therefore, a maximum occurs at $x = \frac{1}{2}$.

Now,

$$Q\left(\frac{1}{2}\right) = \left(\frac{1}{2}\right) - \left(\frac{1}{2}\right)^2 = \frac{1}{4}$$

Substitute $\frac{1}{2}$ in for x in $x + y^2 = 1$ and solve for y.

$$\frac{1}{2} + y^2 = 1$$
$$y^2 = \frac{1}{2}$$
$$y = \pm\sqrt{\frac{1}{2}} = \pm\frac{1}{\sqrt{2}}$$
$$y = \frac{1}{\sqrt{2}}, \quad x \text{ and } y \text{ must be positive}$$

Therefore, the maximum value of Q is $\frac{1}{4}$ when

$$x = \frac{1}{2} \text{ and } y = \frac{1}{\sqrt{2}}.$$

9. Minimize $Q = 2x^2 + 3y^2$, where $x + y = 5$.

Express Q as a function of one variable. First, solve $x + y = 5$ for y.

$$x + y = 5$$
$$y = 5 - x$$

Then substitute $5 - x$ for y in $Q = 2x^2 + 3y^2$.

$$Q = 2x^2 + 3(5-x)^2$$
$$= 2x^2 + 3(25 - 10x + x^2)$$
$$= 2x^2 + 75 - 30x + 3x^2$$
$$= 5x^2 - 30x + 75$$

Find $Q'(x)$, where $Q(x) = 5x^2 - 30x + 75$.

$$Q'(x) = 10x - 30$$

This derivative exists for all values of x; thus the only critical values are where

$$Q'(x) = 0$$
$$10x - 30 = 0$$
$$10x = 30$$
$$x = 3$$

Since there is only one critical value, we can use the second derivative to determine whether we have a minimum.

Note that: $Q''(x) = 10$, which is positive for all real numbers. Thus $Q''(3) > 0$, so a minimum occurs when $x = 3$. The value of Q is

$$Q(3) = 2(3)^2 + 3(5-3)^2$$
$$= 18 + 12$$
$$= 30$$

Substitute 3 for x in $y = 5 - x$ to find y.

$$y = 5 - x$$
$$y = 5 - 3$$
$$y = 2$$

Thus, the minimum value of Q is 30 when $x = 3$ and $y = 2$.

11. Maximize $Q = xy$, where x and y are positive numbers such that $\frac{4}{3}x^2 + y = 16$.

Express Q as a function of one variable. First, solve $\frac{4}{3}x^2 + y = 16$ for y.

$$\frac{4}{3}x^2 + y = 16$$
$$y = 16 - \frac{4}{3}x^2$$

Then substitute $16 - \frac{4}{3}x^2$ for y in $Q = xy$.

$$Q = x\left(16 - \frac{4}{3}x^2\right)$$
$$= 16x - \frac{4}{3}x^3$$

Find $Q'(x)$, where $Q(x) = 16x - \frac{4}{3}x^3$.

$$Q'(x) = 16 - 4x^2$$

This derivative exists for all values of x; thus the only critical values are where

$$Q'(x) = 0$$
$$16 - 4x^2 = 0$$
$$-4x^2 = -16$$
$$x^2 = 4$$
$$x = \pm 2$$
$$x = 2 \qquad x \text{ must be positive}$$

Since there is only one critical value, we can use the second derivative to determine whether we have a maximum.

Note that: $Q''(x) = -8x$ and

$$Q''(2) = -8(2) = -16 < 0.$$

Since $Q''(2)$ is negative, a maximum occurs at $x = 2$.

$$Q(2) = 16(2) - \frac{4}{3}(2)^3$$
$$= 32 - \frac{32}{3}$$
$$= \frac{64}{3}$$

Substitute 2 for x in $y = 16 - \frac{4}{3}x^2$ to find y.

$$y = 16 - \frac{4}{3}x^2$$
$$y = 16 - \frac{4}{3}(2)^2$$
$$y = 16 - \frac{16}{3}$$
$$y = \frac{32}{3}$$

Thus, the maximum value of Q is $\frac{64}{3}$ when $x = 2$ and $y = \frac{32}{3}$.

13. Let x represent the length and y represent the width of the swimming area. It is helpful to draw a picture.

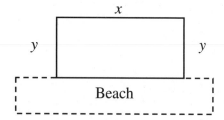

Since the life guard has 180 yd of rope and floats, the perimeter of the swimming area is $x + 2y = 180$. Solving this equation for x, we have $x = 180 - 2y$

The objective is to maximize area, which is given by

$$A = l \cdot w$$

Substituting $x = 180 - 2y$ for the length and y for the width, we have:

$$A = (180 - 2y)y = 180y - 2y^2.$$

We will maximize the area over the interval $0 < y < 90$, because y is the length of one side, and cannot be negative. Furthermore, since there is only 180 yards of rope, and we need two sides, y cannot be greater than 90 yards. If $y = 90$, then the length of the swimming area would be 0.

We now must find $A'(y)$, where

$A(y) = 180y - 2y^2$.

$A'(y) = 180 - 4y$

This derivative exists for all values of y in $(0,90)$. Thus the only critical values are where

$A'(y) = 0$

$180 - 4y = 0$

$-4y = -180$

$y = 45$

Since there is only one critical value in the interval, we use the second derivative to determine whether we have a maximum. Note, $A''(y) = -4 < 0$ for all values of y, so there is a maximum at $y = 45$.

Next, find the dimensions and the area. When $y = 45$,

we have $x = 180 - 2(45) = 90$

and

$A(45) = 180(45) - 2(45)^2 = 4050$.

Therefore, the maximum area is 4050 yd^2 when the overall dimensions are 45 yd by 90 yd.

15. Let x represent the length and y represent the width. It is helpful to draw a picture.

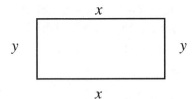

The perimeter is found by adding up the length of the sides. Since it is fixed at 54 feet, the equation of the perimeter is $2x + 2y = 54$.

The area is given by $A = xy$.

First, we solve the perimeter equation for y.

$2x + 2y = 54$

$2y = 54 - 2x$

$y = 27 - x$

Then we substitute for y into the area formula.

$A = xy$

$= x(27 - x) = 27x - x^2$

We want to maximize the area on the interval $(0, 27)$. We consider this interval because x is the length of the shed and cannot be negative. Since the perimeter cannot exceed 54 feet, x cannot be greater than 27, also if x is 27 feet, the width of the shed would be 0 feet. We begin by finding $A'(x)$.

$A'(x) = 27 - 2x$.

This derivative exists for all values of x in $(0, 27)$. Thus, the only critical values occur where $A'(x) = 0$. We solve the equation.

$27 - 2x = 0$

$-2x = -27$

$x = \dfrac{27}{2} = 13.5$

Since there is only one critical value in the interval, we can use the second derivative to determine whether we have a maximum. Note that

$A''(x) = -2 < 0$ for all values of x. Thus,

$A''(13.5) < 0$, so a maximum occurs at $x = 13.5$.

Now,

$A(x) = 27x - x^2$

$A(13.5) = 27(13.5) - (13.5)^2$

$= 182.25$

The maximum area is 182.25 ft^2.

Note: when $x = 13.5$, $y = 27 - 13.5 = 13.5$, so the overall dimensions that will achieve the maximum area are 13.5 ft by 13.5 ft.

17. When squares of length h on a side are cut out of the corners, we are left with a square base of length x. A picture will help.

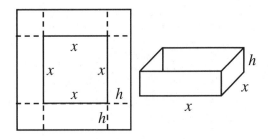

The resulting volume of the box is
$V = lwh = x \cdot x \cdot h = x^2 h$.
We want to express V in terms of one variable. Note that the overall length of a side of the aluminum is 50 cm. We see from the drawing, that $h + x + h = 50$, or $x + 2h = 50$. Solving for h we get:
$2h = 50 - x$

$$h = \frac{1}{2}(50 - x) = 25 - \frac{1}{2}x.$$

Substituting h into the volume equation, we have:

$$V = x^2\left(25 - \frac{1}{2}x\right) = 25x^2 - \frac{1}{2}x^3.$$ The objective is to maximize $V(x)$ on the interval $(0, 50)$.
First, we find the derivative.

$$V'(x) = 50x - \frac{3}{2}x^2$$

This derivative exists for all x in the interval $(0, 50)$, so the critical values will occur when $V'(x) = 0$. Solving this equation, we have:

$$50x - \frac{3}{2}x^2 = 0$$

$$x\left(50 - \frac{3}{2}x\right) = 0$$

$$x = 0 \quad \text{or} \quad 50 - \frac{3}{2}x = 0$$

$$x = 0 \quad \text{or} \quad -\frac{3}{2}x = -50$$

$$x = 0 \quad \text{or} \quad x = \frac{100}{3} \approx 33\frac{1}{3}$$

The only critical value in $(0, 50)$ is $\frac{100}{3}$, or about 33.33. Therefore, we can use the second derivative $V''(x) = 50 - 3x$ to determine if we have a maximum. We have

$$V''\left(\frac{100}{3}\right) = 50 - 3\left(\frac{100}{3}\right) = -50 < 0.$$

Therefore, there is a maximum at $\frac{100}{3}$.

$$V\left(\frac{100}{3}\right) = 25\left(\frac{100}{3}\right)^2 - \frac{1}{2}\left(\frac{100}{3}\right)^3$$

$$= \frac{250,000}{27} = 9259\frac{7}{27}$$

Now, we find the height of the box.

$$h = 25 - \frac{1}{2}\left(\frac{100}{3}\right) = \frac{25}{3} = 8\frac{1}{3}.$$

Therefore, a box with dimensions $33\frac{1}{3}$ cm. by $33\frac{1}{3}$ cm. by $8\frac{1}{3}$ cm. will yield a maximum volume of $9259\frac{7}{27}$ cm^3.

19. First, we make a drawing.

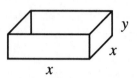

The surface area of the open-top, square-based, rectangular box is found by adding the area of the base and the four sides. x^2 is the area of the base, xy is the area of one of the sides and there are four sides, therefore the surface area is given by $S = x^2 + 4xy$.
The volume must by 62.5 cubic inches, and is given by $V = l \cdot w \cdot h = x^2 y = 62.5$.
To express S in terms of one variable, we solve $x^2 y = 62.5$ for y:

$$y = \frac{62.5}{x^2}$$

Then

$$S(x) = x^2 + 4x\left(\frac{62.5}{x^2}\right)$$

$$= x^2 + \frac{250}{x} = x^2 + 250x^{-1}$$

Now S is defined only for positive numbers, so we minimize S on the interval $(0, \infty)$.

First, we find $S'(x)$.

$$S'(x) = 2x - 250x^{-2}$$

$$= 2x - \frac{250}{x^2}$$

Since $S'(x)$ exists for all x in $(0, \infty)$, the only critical values are where $S'(x) = 0$. We solve the following equation:

$$S'(x) = 0$$

$$2x - \frac{250}{x^2} = 0$$

$$2x = \frac{250}{x^2}$$

$$x^3 = 125$$

$$x = 5$$

Since there is only one critical value, we use the second derivative to determine whether we have a minimum. Note that

$$S''(x) = 2 + 500x^{-3} = 2 + \frac{500}{x^3}.$$

$S''(5) = 2 + \frac{500}{5^3} = 6 > 0$. Since the second

derivative is positive, we have a minimum at $x = 5$. We find y when $x = 5$.

$$y = \frac{62.5}{x^2}$$

$$= \frac{62.5}{5^2}$$

$$= 2.5$$

The surface area is minimized when $x = 5$ in. and $y = 2.5$ in. We find the minimum surface area by substituting these values into the surface area equation.

$$S = x^2 + 4xy$$

$$= (5)^2 + 4(5)(2.5)$$

$$= 25 + 50$$

$$= 75$$

The minimum surface area is 75 in^2 when the dimensions are 5 in. by 5 in. by 2.5 in.

21. First, we make a drawing.

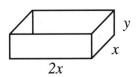

The surface area of the open-top, square-based, rectangular dumpster is found by adding the area of the base and the four sides. $2x^2$ is the area of the base, xy is the area of two of the sides, while $2xy$ is the area of the other two sides. Therefore the surface area is given by $S = 2x^2 + 2xy + 2(2xy) = 2x^2 + 6xy$.

The volume must by 12 cubic yards, and is given by $V = l \cdot w \cdot h = 2x \cdot x \cdot y = 2x^2 y = 12$.

To express S in terms of one variable, we solve $2x^2 y = 12$ for y:

$$y = \frac{6}{x^2}$$

Then

$$S(x) = 2x^2 + 6x\left(\frac{6}{x^2}\right)$$

$$= 2x^2 + \frac{36}{x} = 2x^2 + 36x^{-1}$$

Now S is defined only for positive numbers, so we minimize S on the interval $(0, \infty)$.

First, we find $S'(x)$.

$$S'(x) = 4x - 36x^{-2}$$

$$= 4x - \frac{36}{x^2}$$

Since $S'(x)$ exists for all x in $(0, \infty)$, the only critical values are where $S'(x) = 0$. We solve the following equation:

$$4x - \frac{36}{x^2} = 0$$

$$4x = \frac{36}{x^2}$$

$$x^3 = 9$$

$$x = \sqrt[3]{9} \approx 2.08$$

Since there is only one critical value, we use the second derivative to determine whether we have a minimum. Note that

$$S''(x) = 4 + 72x^{-3} = 4 + \frac{72}{x^3}.$$

$S''(\sqrt[3]{9}) = 4 + \frac{72}{\left(\sqrt[3]{9}\right)^3} = 12 > 0$. Since the second

derivative is positive, we have a minimum at $x = \sqrt[3]{9} \approx 2.08$. The width is 2.08 yd.; therefore, the length is $2(2.08) \approx 4.16$. We find the height y

$$y = \frac{6}{x^2}$$

$$= \frac{6}{(2.08)^2}$$

$$\approx 1.387$$

The overall dimensions of the dumpster that will minimize surface area are 2.08 yd. by 4.16 yd. by 1.387 yd.

23. $R(x) = 50x - 0.5x^2;\ C(x) = 4x + 10$

Profit is equal to revenue minus cost.

$$P(x) = R(x) - C(x)$$
$$= 50x - 0.5x^2 - (4x + 10)$$
$$= -0.5x^2 + 46x - 10$$

Because x is the number of units produced and sold, we are only concerned with the non-negative values of x. Therefore, we will find the maximum of $P(x)$ on the interval $[0, \infty)$.

First, we find $P'(x)$.

$$P'(x) = -x + 46$$

The derivative exists for all values of x in $[0, \infty)$. Thus, we solve $P'(x) = 0$.

$$-x + 46 = 0$$
$$-x = -46$$
$$x = 46$$

There is only one critical value. We can use the second derivative to determine whether we have a maximum.

$$P''(x) = -1 < 0$$

The second derivative is less than zero for all values of x. Thus, a maximum occurs at $x = 46$.

$$P(46) = -0.5(46)^2 + 46(46) - 10$$
$$= -1058 + 2116 - 10$$
$$= 1048$$

The maximum profit is $1048 when 46 units are produced and sold.

25. $R(x) = 2x;\ C(x) = 0.01x^2 + 0.6x + 30$

Profit is equal to revenue minus cost.

$$P(x) = R(x) - C(x)$$
$$= 2x - (0.01x^2 + 0.6x + 30)$$
$$= -0.01x^2 + 1.4x - 30$$

Because x is the number of units produced and sold, we are only concerned with the non-negative values of x. Therefore, we will find the maximum of $P(x)$ on the interval $[0, \infty)$.

First, we find $P'(x)$.

$$P'(x) = -0.02x + 1.4$$

The derivative exists for all values of x in $[0, \infty)$. Thus, we solve $P'(x) = 0$.

$$-0.02x + 1.4 = 0$$
$$-0.02x = -1.4$$
$$x = 70$$

There is only one critical value. We can use the second derivative to determine whether we have a maximum.

$$P''(x) = -0.02 < 0$$

The second derivative is less than zero for all values of x. Thus, a maximum occurs at $x = 70$.

$$P(70) = -0.01(70)^2 + 1.4(70) - 30$$
$$= -49 + 98 - 30$$
$$= 19$$

The maximum profit is $19 when 70 units are produced and sold.

27. $R(x) = 9x - 2x^2$

$$C(x) = x^3 - 3x^2 + 4x + 1$$

$R(x)$ and $C(x)$ are in thousands of dollars and x is in thousands of units.

Profit is equal to revenue minus cost.

$$P(x) = R(x) - C(x)$$
$$= 9x - 2x^2 - (x^3 - 3x^2 + 4x + 1)$$
$$= -x^3 + x^2 + 5x - 1$$

Because x is the number of units produced and sold, we are only concerned with the non-negative values of x. Therefore, we will find the maximum of $P(x)$ on the interval $[0, \infty)$.

First, we find $P'(x)$.

$$P'(x) = -3x^2 + 2x + 5$$

The derivative exists for all values of x in $[0, \infty)$. Thus, we solve $P'(x) = 0$.

$$-3x^2 + 2x + 5 = 0$$
$$3x^2 - 2x - 5 = 0$$
$$(3x - 5)(x + 1) = 0$$

$3x - 5 = 0$ or $x + 1 = 0$

$3x = 5$ or $x = -1$

$x = \dfrac{5}{3}$ or $x = -1$

There is only one critical value in the interval $[0, \infty)$. We can use the second derivative to determine whether we have a maximum.

$$P''(x) = -6x + 2$$

Therefore,

$$P''\left(\frac{5}{3}\right) = -6\left(\frac{5}{3}\right) + 2 = -10 + 2 = -8 < 0$$

The second derivative is less than zero for $x = \dfrac{5}{3}$. Thus, a maximum occurs at $x = \dfrac{5}{3}$.

$$P\left(\frac{5}{3}\right) = -\left(\frac{5}{3}\right)^3 + \left(\frac{5}{3}\right)^2 + 5\left(\frac{5}{3}\right) - 1$$

$$= -\frac{125}{27} + \frac{25}{9} + \frac{25}{3} - 1$$

$$= -\frac{125}{27} + \frac{75}{27} + \frac{225}{27} - \frac{27}{27}$$

$$= \frac{148}{27}$$

Note that $x = \dfrac{5}{3}$ thousand is approximately 1.667 thousand or 1667 units, and that $\dfrac{148}{27}$ thousand is approximately 5.481 thousand or 5481.

Thus, the maximum profit is approximately $5481 when approximately 1667 units are produced and sold.

29. $p = 150 - 0.5x$ Price per suit.

$C(x) = 4000 + 0.25x^2$ Cost per suit.

a) Revenue is price times quantity. Therefore, revenue can be found by multiplying the number of suits sold, x, by the price of the suit, p. Substituting $150 - 0.5x$ for p, we have:

$$R(x) = x \cdot p$$
$$= x(150 - 0.5x)$$
$$R(x) = 150x - 0.5x^2$$

b) Profit is revenue minus cost. Therefore,

$$P(x) = R(x) - C(x)$$
$$= 150x - 0.5x^2 - \left(4000 + 0.25x^2\right)$$
$$= -0.75x^2 + 150x - 4000$$

Since x is the number of suits produced and sold, we will restrict the domain to the interval $0 \le x < \infty$.

c) To determine the number of suits required to maximize profit, we first find $P'(x)$.

$$P'(x) = -1.5x + 150.$$

The derivative exists for all real numbers in the interval $[0, \infty)$. Thus, we solve

$$P'(x) = 0$$
$$-1.5x + 150 = 0$$
$$-1.5x = -150$$
$$x = 100$$

Since there is only one critical value, we can use the second derivative to determine whether we have a maximum.

$$P''(x) = -1.5 < 0$$

The second derivative is negative for all values of x; therefore, a maximum occurs at $x = 100$.

Raggs, Ltd. must sell 100 suits to maximize profit.

d) The maximum profit is found by substituting 100 for x in the profit function.

$$P(100) = -0.75(100)^2 + 150(100) - 4000$$
$$= -7500 + 15,000 - 4000$$
$$= 3500$$

The maximum profit is $3500.

e) The price per suit is given by:

$$p = 150 - 0.5x$$

Substituting 100 for x, we have:

$$p = 150 - 0.5(100) = 150 - 50 = 100.$$

The price per suit will be $100.

31.   Let x be the amount by which the price of $18 should be decreased (if x is negative, the price would be increased to maximize revenue). First, we express total revenue R as a function of x. There are two sources of revenue, revenue from tickets and revenue from concessions.

$$R(x) = \left(\begin{array}{c}\text{Revenue from} \\ \text{tickets}\end{array}\right) + \left(\begin{array}{c}\text{Revenue from} \\ \text{concessions}\end{array}\right)$$

$$= \left(\begin{array}{c}\text{Number of} \\ \text{People}\end{array}\right) \cdot \left(\begin{array}{c}\text{Ticket} \\ \text{Price}\end{array}\right) + \left(\begin{array}{c}\text{Number of} \\ \text{People}\end{array}\right) \cdot 4.50$$

Note, the increase in ticket sales is $10,000\,x$, when price drops $3x$ dollars. Therefore, the increase in ticket sales is $\dfrac{10,000}{3}x$ when price drops x dollars.

$$R(x) = \left(40,000 + \frac{10,000}{3}x\right)(18 - x) +$$
$$\left(40,000 + \frac{10,000}{3}x\right)(4.50)$$

$$= -\frac{10,000}{3}x^2 + 20,000x + 720,000 +$$
$$180,000 + 15,000x$$

$$= -\frac{10,000}{3}x^2 + 35,000x + 900,000$$

Therefore, the total revenue function is

$$R(x) = -\frac{10,000}{3}x^2 + 35,000x + 900,000$$

To find x such that $R(x)$ is a maximum, we first find $R'(x)$:

$$R'(x) = -\frac{20,000}{3}x + 35,000$$

This derivative exists for all real numbers x. thus, the only critical values are where $R'(x) = 0$; so we solve that equation:

$$-\frac{20,000}{3}x + 35,000 = 0$$

$$-\frac{20,000}{3}x = -35,000$$

$$x = 5.25$$

Since this is the only critical value, we can use the second derivative,

$$R''(x) = -\frac{20,000}{3} < 0$$

to determine whether we have a maximum. since $R''(5.25)$ is negative, $R(5.25)$ is a maximum. Therefore, in order to maximize revenue, the university should charge $18 - \$5.25$, or \$12.75. Since \$12.75 is \$5.25 less than \$18, we can find the attendance using

$$40,000 + \frac{10,000}{3}(5.25) = 57,500$$

The average attendance when ticket price is \$12.75 is 57,500 people.

33. Let x equal the number of additional trees per acre which should be planted. Then the number of trees planted per acre is represented by $(20 + x)$ and the yield per tree by $(30 - x)$. The total yield per acre is equal to the yield per tree times the number of trees so, we have:

$$Y(x) = (30 - x)(20 + x)$$
$$= 600 + 10x - x^2$$

To find x such that $Y(x)$ is a maximum, we first find $Y'(x)$:

$$Y'(x) = 10 - 2x.$$

This derivative exists for all real numbers x. Thus, the only critical values are where $Y'(x) = 0$; so we solve that equation:

$$10 - 2x = 0$$
$$-2x = -10$$
$$x = 5$$

This correspond to planting 5 trees.

Since this is the only critical value, we can use the second derivative,

$$R''(x) = -2 < 0,$$

to determine whether we have a maximum. Since $R''(5)$ is negative, $R(5)$ is a maximum.

Therefore, in order to maximize yield, the apple farm should plant $20 + 5$, or 25 trees per acre.

35. a) When $x = 25$, $q = 2.13$. When $x = 25 + 1$, or 26, then $q = 2.13 - 0.04 = 2.09$. We use the points $(25, 2.13)$ and $(26, 2.09)$ to find the linear demand function $q(x)$. First, we find the slope:

$$m = \frac{2.13 - 2.09}{25 - 26} = \frac{0.04}{-1} = -0.04.$$

Next, we use the point-slope equation:

$$q - 2.13 = -0.04(x - 25)$$
$$q - 2.13 = -0.04x + 1$$
$$q = -0.04x + 3.13$$

Therefore, the linear demand function is:

$$q(x) = -0.04x + 3.13.$$

b) Revenue is price times quantity; therefore, the revenue function is:

$$R(x) = x \cdot q(x)$$
$$= x(-0.04x + 3.13)$$
$$= -0.04x^2 + 3.13x$$

To find x such that $R(x)$ is a maximum, we first find $R'(x)$:

$$R'(x) = -0.08x + 3.13.$$

This derivative exists for all values of x. So the only critical values occur when $R'(x) = 0$; so we solve that equation:

$$-0.08x + 3.13 = 0$$
$$-0.08x = -3.13$$
$$x = 39.125$$

Since this is the only critical value, we can use the second derivative,

$$R''(x) = -0.08 < 0,$$

to determine whether we have a maximum. Since $R''(39.125)$ is negative, $R(39.125)$ is a maximum.

In order to maximize revenue, the State of Maryland should charge \$39.125 or rounding up to \$39.13 per license plate.

37. The volume of the box is given by
$$V = x \cdot x \cdot y = x^2 y = 320.$$

The area of the base is x^2. The cost of the base is $15x^2$ cents.

The area of the top is x^2. The cost of the top is $10x^2$ cents.

The area of each side is xy. The total area for the four sides is $4xy$. The cost of the four sides is $2.5(4xy)$ cents.

The total costs in cents is given by
$$C = 15x^2 + 10x^2 + 2.5(4xy) = 25x^2 + 10xy.$$

To express C in terms of one variable, we solve $x^2 y = 320$ for y:
$$y = \frac{320}{x^2}.$$

Then,
$$C(x) = 25x^2 + 10x\left(\frac{320}{x^2}\right)$$
$$= 25x^2 + \frac{3200}{x}$$

The function is defined only for positives numbers, and the problem dictates that the length x must be positive, so we are minimizing C on the interval $(0,\infty)$.

First, we find $C'(x)$.
$$C'(x) = 50x - 3200x^{-2} = 50x - \frac{3200}{x^2}$$

Since $C'(x)$ exists for all x in $(0,\infty)$, the only critical values are where $C'(x) = 0$. Thus, we solve the following equation:
$$50x - \frac{3200}{x^2} = 0$$
$$50x = \frac{3200}{x^2}$$
$$50x^3 = 3200$$
$$x^3 = 64$$
$$x = 4$$

This is the only critical value, so we can use the second derivative to determine whether we have a minimum.
$$C''(x) = 50 + 6400x^{-3} = 50 + \frac{6400}{x^3}$$

Note that the second derivative is positive for all positive values of x, therefore we have a minimum at $x = 4$.

We find y when $x = 4$.
$$y = \frac{320}{x^2} = \frac{320}{(4)^2} = \frac{320}{16} = 20$$

The cost is minimized when the dimensions are 4 ft by 4 ft by 20 ft.

39. Let x equal the lot size. Now the inventory costs are given by:
$$C(x) = \overset{\text{Yearly Carrying}}{\underset{\text{Cost}}{}} + \overset{\text{Yearly reorder}}{\underset{\text{Cost}}{}}$$

We consider each cost separately.

Yearly carrying costs, $C_c(x)$: Can be found by multiplying the cost to store the items by the number of items in storage. The average amount held in stock is $\frac{x}{2}$, and it cost \$20 per pool table for storage. Thus:
$$C_c(x) = 20 \cdot \frac{x}{2}$$
$$= 10x$$

Yearly reorder costs, $C_r(x)$: Can be found by multiplying the cost of each order by the number of reorders. The cost of each order is $40 + 16x$, and the number of orders per year is $\frac{100}{x}$. Therefore,
$$C_r(x) = (40 + 16x)\left(\frac{100}{x}\right)$$
$$= \frac{4000}{x} + 1600$$

Hence, the total inventory cost is:
$$C(x) = C_c(x) + C_r(x)$$
$$= 10x + \frac{4000}{x} + 1600.$$

We want to find the minimum value of C on the interval $[1,100]$. First, we find $C'(x)$:
$$C'(x) = 10 - 4000x^{-2} = 10 - \frac{4000}{x^2}.$$

The derivative exists for all x in $[1,100]$, so the only critical values are where $C'(x) = 0$. We solve that equation:

$$C'(x) = 0$$

$$10 - \frac{4000}{x^2} = 0$$

$$10 = \frac{4000}{x^2}$$

$$10x^2 = 4000$$

$$x^2 = 400$$

$$x = \pm 20$$

The only critical value in the interval is $x = 20$, so we can use the second derivative to determine whether we have a minimum.

$$C''(x) = 8000x^{-3} = \frac{8000}{x^3}$$

Notice that $C''(x)$ is positive for all values of x in $[1,100]$, we have a minimum at $x = 20$. Thus, to minimize inventory costs, the store should order pool tables $\dfrac{100}{20} = 5$ times per year. The lot size will be 20 tables.

41. Let x equal the lot size. Now the inventory costs are given by:

$$C(x) = \underset{\text{Cost}}{\text{Yearly Carrying}} + \underset{\text{Cost}}{\text{Yearly reorder}}$$

We consider each cost separately.

Yearly carrying costs, $C_c(x)$: Can be found by multiplying the cost to store the items by the number of items in storage. The average amount held in stock is $\dfrac{x}{2}$, and it cost \$2 per calculator for storage. Thus:

$$C_c(x) = 2 \cdot \frac{x}{2}$$

$$= x$$

Yearly reorder costs, $C_r(x)$: Can be found by multiplying the cost of each order by the number of reorders. The cost of each order is $5 + 2.50x$, and the number of orders per year is $\dfrac{720}{x}$. Therefore,

$$C_r(x) = (5 + 2.50x)\left(\frac{720}{x}\right)$$

$$= \frac{3600}{x} + 1800$$

Hence, the total inventory cost is:

$$C(x) = C_c(x) + C_r(x)$$

$$= x + \frac{3600}{x} + 1800.$$

We want to find the minimum value of C on the interval $[1,720]$. First, we find $C'(x)$:

$$C'(x) = 1 - 3600x^{-2} = 1 - \frac{3600}{x^2}.$$

The derivative exists for all x in $[1,720]$, so the only critical values are where $C'(x) = 0$. We solve that equation:

$$1 - \frac{3600}{x^2} = 0$$

$$1 = \frac{3600}{x^2}$$

$$x^2 = 3600$$

$$x = \pm 60$$

The only critical value in the interval is $x = 60$, so we can use the second derivative to determine whether we have a minimum.

$$C''(x) = 7200x^{-3} = \frac{7200}{x^3}$$

Notice that $C''(x)$ is positive for all values of x in $[1,720]$, we have a minimum at $x = 60$. Thus, to minimize inventory costs, the store should order calculators $\dfrac{720}{60} = 12$ times per year. The lot size will be 60 calculators.

43. Let x equal the lot size.
Yearly carrying costs:

$$C_c(x) = 2 \cdot \frac{x}{2}$$

$$= x$$

Yearly reorder costs:

$$C_r(x) = (4 + 2.50x)\left(\frac{256}{x}\right)$$

$$= \frac{1024}{x} + 640$$

Hence, the total inventory cost is:

$$C(x) = C_c(x) + C_r(x)$$

$$= x + \frac{1024}{x} + 640.$$

We want to find the minimum value of C on the interval $[1,256]$. First, we find $C'(x)$:

$$C'(x) = 1 - 1024x^{-2} = 1 - \frac{1024}{x^2}.$$

The derivative exists for all x in $[1, 256]$, so the only critical values are where $C'(x) = 0$. We solve that equation:

$$1 - \frac{1024}{x^2} = 0$$

$$1 = \frac{1024}{x^2}$$

$$x^2 = 1024$$

$$x = \pm 32$$

The only critical value in the interval is $x = 32$, so we can use the second derivative to determine whether we have a minimum.

$$C''(x) = 2048x^{-3} = \frac{2048}{x^3}$$

Notice that $C''(x)$ is positive for all values of x in $[1, 256]$, we have a minimum at $x = 32$. Thus, to minimize inventory costs, the store should order calculators $\dfrac{256}{32} = 8$ times per year. The lot size will be 32 calculators.

45. Case I.
If y is the length, the girth is
$x + x + x + x$, or $4x$.
Case II.
If x is the length, the girth is
$x + y + x + y$, or $2x + 2y$.
Case I.
The combine length and girth is
$y + 4x = 84$.

The volume is $V = x \cdot x \cdot y = x^2 y$.

We want express V in terms of one variable.
We solve $y + 4x = 84$, for y.

$$y + 4x = 84$$

$$y = 84 - 4x$$

Thus,

$$V = x^2(84 - 4x) = 84x^2 - 4x^3.$$

To maximize $V(x)$ we first find $V'(x)$.

$$V'(x) = 168x - 12x^2$$

This derivative exists for all x, so the critical values will occur when $V'(x) = 0$; therefore, we solve that equation.

$$168x - 12x^2 = 0$$

$$12x(14 - x) = 0$$

$$12x = 0 \quad \text{or} \quad 14 - x = 0$$

$$x = 0 \quad \text{or} \qquad x = 14$$

Since $x \neq 0$, the only critical value is $x = 14$.
We can use the second derivative,

$$V'(x) = 168 - 24x,$$

to determine whether we have a maximum.

$$V'(14) = 168 - 24(14) = -168 < 0$$

Therefore, we have a maximum at $x = 14$.
If $x = 14$, then $y = 84 - 4(14) = 28$.

Therefore, the dimensions that will maximize the volume of the package are 14 in. by 14 in. by 28 in. The volume is
$14 \times 14 \times 28 = 5488 \text{ in}^3$.

Case II.
The combine length and girth is
$x + 2x + 2y = 3x + 2y = 84$.

The volume is $V = x \cdot x \cdot y = x^2 y$.

We want express V in terms of one variable.
We solve $3x + 2y = 84$, for y.

$$3x + 2y = 84$$

$$2y = 84 - 3x$$

$$y = 42 - \frac{3}{2}x$$

Thus,

$$V = x^2\left(42 - \frac{3}{2}x\right) = 42x^2 - \frac{3}{2}x^3.$$

To maximize $V(x)$ we first find $V'(x)$.

$$V'(x) = 84x - \frac{9}{2}x^2$$

This derivative exists for all x, so the critical values will occur when $V'(x) = 0$; therefore, we solve that equation.

$$84x - \frac{9}{2}x^2 = 0$$

$$3x\left(28 - \frac{3}{2}x\right) = 0$$

$$3x = 0 \quad \text{or} \quad 28 - \frac{3}{2}x = 0$$

$$x = 0 \quad \text{or} \qquad -\frac{3}{2}x = -28$$

$$x = 0 \quad \text{or} \qquad x = \frac{56}{3}$$

Since $x \neq 0$, the only critical value is $x = \dfrac{56}{3}$

We can use the second derivative,

$$V''(x) = 84 - 9x,$$

to determine whether we have a maximum.

$$V''\left(\frac{56}{3}\right) = 84 - 9\left(\frac{56}{3}\right) = -84 < 0$$

Therefore, we have a maximum at $x = \frac{56}{3}$.

If $x = \frac{56}{3} \approx 18.67$, then $y = 42 - \frac{3}{2}\left(\frac{56}{3}\right) = 14$.

Therefore, the dimensions that will maximize the volume of the package are 18.67 in. by 18.67 in. by 14 in. The volume is

$$\frac{56}{3} \times \frac{56}{3} \times 14 \approx 4878.2 \text{ in}^3.$$

Comparing Case I and Case II, we see that the maximum volume is 5488 in^3 when the dimensions are 14 in. by 14 in. by 28 in.

47. Use the figure in the text book. Since the radius of the window is x, the diameter of the window is $2x$, which is also the length of the base of the window.
The circumference of a circle whose radius is x is given by:
$$C = 2\pi x. \qquad (C = 2\pi r)$$
Therefore, the perimeter of the semicircle is
$$\frac{1}{2}C = \frac{2\pi x}{2} = \pi x.$$
The perimeter of the three sides of the rectangle which form the remaining part of the total perimeter of the window is given by:
$$2x + y + y = 2x + 2y.$$
The total perimeter of the window is:
$$\pi x + 2x + 2y = 24.$$
Maximizing the amount of light is the same as maximizing the area of the window. The area of the circle with radius x is:
$$A = \pi x^2, \qquad (A = \pi r^2).$$
Therefore, the area of the semicircle is:
$$\frac{1}{2}A = \frac{1}{2}\pi x^2.$$
The area of the rectangle is $2x \cdot y$.
The total area of the Norman window is
$$A = \frac{1}{2}\pi x^2 + 2xy.$$
To express A in terms of one variable, we solve $\pi x + 2x + 2y = 24$ for y:
$$\pi x + 2x + 2y = 24$$
$$2y = 24 - 2x - \pi x$$
$$y = 12 - x - \frac{\pi}{2}x$$

Then,
$$A(x) = \frac{1}{2}\pi x^2 + 2x\left(12 - x - \frac{\pi}{2}x\right)$$
$$= \frac{1}{2}\pi x^2 + 24x - 2x^2 - \pi x^2$$
$$= \left(-\frac{1}{2}\pi - 2\right)x^2 + 24x$$

We maximize A on the interval $(0, 24)$. We first find $A'(x)$.
$$A'(x) = (-\pi - 4)x + 24.$$
Since $A'(x)$ exists for all x in $(0, 24)$, the only critical points are where $A'(x) = 0$. Thus, we solve the following equation:
$$A'(x) = 0$$
$$(-\pi - 4)x + 24 = 0$$
$$(-\pi - 4)x = -24$$
$$x = \frac{-24}{(-\pi - 4)}$$
$$x = \frac{-24}{-(\pi + 4)}$$
$$x = \frac{24}{\pi + 4} \approx 3.36$$

This is the only critical value, so we can use the second derivative to determine whether we have a maximum.
$$A''(x) = -\pi - 4 < 0$$
Since $A''(x)$ is negative for all values of x, we have a maximum at $x = \frac{24}{\pi + 4}$. We find y when
$$x = \frac{24}{\pi + 4}:$$
$$y = 12 - \frac{\pi}{2}x - x$$
$$= 12 - \frac{\pi}{2}\left(\frac{24}{\pi + 4}\right) - \frac{24}{\pi + 4}$$
$$= 12\left(\frac{\pi + 4}{\pi + 4}\right) - \frac{12\pi}{\pi + 4} - \frac{24}{\pi + 4}$$
$$= \frac{12\pi + 48 - 12\pi - 24}{\pi + 4}$$
$$= \frac{24}{\pi + 4} \approx 3.36$$

To maximize the amount of light through the window, the dimensions must be

$x = \dfrac{24}{\pi + 4}$ ft and $y = \dfrac{24}{\pi + 4}$ ft, or approximately

$x \approx 3.36$ ft and $y \approx 3.36$ ft.

49. Let x represent a positive number. Then, $\dfrac{1}{x}$ is

the reciprocal of the number, and x^2 is the square of the number. The sum, S, of the reciprocal and five times the square is given by:

$$S(x) = \frac{1}{x} + 5x^2.$$

We want to minimize $S(x)$ on the interval $(0, \infty)$. First, we find $S'(x)$

$$S'(x) = -x^{-2} + 10x = -\frac{1}{x^2} + 10x$$

Since $S'(x)$ exists for all values of x in $(0, \infty)$, the only critical values occur when $S'(x) = 0$. We solve the following equation:

$$-\frac{1}{x^2} + 10x = 0$$

$$10x = \frac{1}{x^2}$$

$$10x^3 = 1$$

$$x^3 = \frac{1}{10}$$

$$x = \sqrt[3]{\frac{1}{10}} = \frac{1}{\sqrt[3]{10}}$$

Since there is only one critical value, we use the second derivative,

$$S''(x) = 2x^{-3} + 10 = \frac{2}{x^3} + 10,$$

to determine whether it is a minimum. The second derivative is positive for all x in $(0, \infty)$; therefore, the sum is a minimum when

$$x = \frac{1}{\sqrt[3]{10}}.$$

51. Let A represent the amount deposited in savings accounts and i represent the interest rate paid on the money deposited. If A is directly proportional to i, then there is some positive constant k such that $A = ki$. The interest earned by the bank is represented by $18\%A$, or $0.18A$. The interest paid by the bank is represented by iA. Thus the profit received by the bank is given by

$P = 0.18A - iA.$

We express P as a function of the interest the bank pays on the money deposited, i, by substituting ki for A.

$P = 0.18(ki) - i(ki)$

$\quad = 0.18ki - ki^2$

We maximize P on the interval $(0, \infty)$. First, we find $P'(i)$.

$P'(i) = 0.18k - 2ki$

Since $P'(i)$ exists for all i in $(0, \infty)$, the only critical values are where $P'(i) = 0$. We solve the following equation:

$0.18k - 2ki = 0$

$\quad -2ki = -0.18k$

$\quad i = \dfrac{-0.18k}{-2k}$

$\quad i = 0.09$

Since there is only one critical point, we can use the second derivative to determine whether we have a minimum. Notice that $P''(i) = -2k$, which is a negative constant $(k > 0)$. Thus, $P''(0.09)$ is negative, so $P(0.09)$ is a maximum. To maximize profit, the bank should pay 9% on its savings accounts.

53. Using the drawing in the text, we write a function that gives the cost of the power line. The length of the power line on the land is given by $4 - x$, so the cost of laying the power line underground is given by:

$C_L(x) = 3000(4 - x) = 12,000 - 3000x.$

The length of the power line that will be under water is $\sqrt{1 + x^2}$, so the cost of laying the power line underwater is given by:

$C_W(x) = 5000\sqrt{1 + x^2}$

Therefore, the total cost of laying the power line is:

$$C(x) = C_L(x) + C_W(x)$$

$$= 12,000 - 3000x + 5000\sqrt{1+x^2}.$$

We want to minimize $C(x)$ over the interval $0 \le x \le 4$. First, we find the derivative.

$$C'(x) = -3000 + 5000\left(\frac{1}{2}\right)\left(1+x^2\right)^{-\frac{1}{2}}(2x)$$

$$= -3000 + 5000x\left(1+x^2\right)^{-\frac{1}{2}}$$

$$= -3000 + \frac{5000x}{\sqrt{1+x^2}}$$

Since the derivative exists for all x, we find the critical values by solving the equation:

$$C'(x) = 0$$

$$-3000 + \frac{5000x}{\sqrt{1+x^2}} = 0$$

$$-3000\sqrt{1+x^2} + 5000x = 0$$

$$5000x = 3000\sqrt{1+x^2}$$

$$\frac{5}{3}x = \sqrt{1+x^2}$$

$$\left(\frac{5}{3}x\right)^2 = \left(\sqrt{1+x^2}\right)^2$$

$$\frac{25}{9}x^2 = 1+x^2$$

$$\frac{16}{9}x^2 = 1$$

$$x^2 = \frac{9}{16}$$

$$x = \pm\sqrt{\frac{9}{16}}$$

$$x = \pm\frac{3}{4}$$

The only critical value in the interval $[0,4]$ is $x = \frac{3}{4}$, so we can use the second derivative to determine if we have a minimum.

$$C''(x) = \frac{\left(1+x^2\right)^{\frac{1}{2}}(5000) - 5000x\left[\frac{1}{2}\left(1+x^2\right)^{-\frac{1}{2}}(2x)\right]}{\left[\left(1+x^2\right)^{\frac{1}{2}}\right]^2}$$

$$= \frac{5000\sqrt{1+x^2} - \dfrac{5000x^2}{\sqrt{1+x^2}}}{\left(1+x^2\right)}$$

$$= \frac{\dfrac{5000}{\sqrt{1+x^2}} - \dfrac{5000x^2}{\left(1+x^2\right)^{\frac{3}{2}}}}{}$$

$$= \frac{5000\left(1+x^2\right) - 5000x^2}{\left(1+x^2\right)^{\frac{3}{2}}}$$

$$= \frac{5000}{\left(1+x^2\right)^{\frac{3}{2}}}$$

$C''(x)$ is positive for all x in $[0,4]$; therefore, a minimum occurs at $x = \frac{3}{4}$. When $x = \frac{3}{4}$,

$$4 - \frac{3}{4} = \frac{13}{4} = 3.25.$$

Therefore, S should be 3.25 miles down shore from the power station.

Note: since we are minimizing cost over a closed interval, we could have used Max-Min Principle 1 to determine the minimum, and avoided finding the second derivative. The critical value and the endpoints are $0, \frac{3}{4}$, and 4. The function values at these three points are:

$$C(0) = 12,000 - 3000(0) + 5000\left(\sqrt{1+(0)^2}\right)$$

$$= 17,000$$

$$C\left(\frac{3}{4}\right) = 12,000 - 3000\left(\frac{3}{4}\right) + 5000\left(\sqrt{1+\left(\frac{3}{4}\right)^2}\right)$$

$$= 16,000$$

$$C(4) = 12,000 - 3000(4) + 5000\left(\sqrt{1+(4)^2}\right)$$

$$\approx 20,615.53$$

Therefore, the minimum occurs when $x = \frac{3}{4}$, or when S is 3.25 miles down shore from the power station.

55. Using the drawing in the text, we write a function which gives the total distance between the cities.

The distance from C_1 to the bridge can be given by $\sqrt{a^2 + (p-x)^2}$. The distance over the bridge is r. The distance from the bridge to C_2 can be given by $\sqrt{b^2 + x^2}$. Therefore, the total distance between the two cities is given by:

$$D(x) = \sqrt{a^2 + (p-x)^2} + r + \sqrt{x^2 + b^2}$$

To minimize the distance, we find the derivative of the function first.

$$D'(x) = \frac{1}{2}\left[a^2 + (p-x)^2\right]^{-\frac{1}{2}} \cdot 2(p-x)(-1) +$$

$$\frac{1}{2}\left[b^2 + x^2\right]^{-\frac{1}{2}}(2x)$$

$$= \frac{x-p}{\sqrt{a^2 + (p-x)^2}} + \frac{x}{\sqrt{b^2 + x^2}}$$

The derivative exists for all values of x in the interval $[0, p]$. Therefore, the only critical values occur when $D'(x) = 0$. We solve this equation.

$$\frac{x-p}{\sqrt{a^2 + (p-x)^2}} + \frac{x}{\sqrt{b^2 + x^2}} = 0.$$

The solution to this equation is

$$x = \frac{bp}{b-a} \text{ or } x = \frac{bp}{b+a}.$$

Only $x = \dfrac{bp}{b+a}$ is in $[0, p]$.

Since there is only one critical value, we can use the second derivative to determine if there is a minimum.

$$D''(x) = \frac{a^2}{\left[a^2 + (p-x)^2\right]^{\frac{3}{2}}} + \frac{b^2}{\left[x^2 + b^2\right]^{\frac{3}{2}}}.$$

$D''(x) > 0$ for all values of x; therefore, a minimum occurs at $x = \dfrac{bp}{b+a}$.

The bridge should be located such that the distance x is $\dfrac{bp}{b+a}$ units.

57. $A(x) = \dfrac{C(x)}{x}$

a) Taking the derivative of $A(x)$ we have:

$$A'(x) = \frac{d}{dx}\left[\frac{C(x)}{x}\right]$$

$$= \frac{x \cdot C'(x) - C(x) \cdot 1}{x^2} \quad \text{Quotient Rule}$$

$$= \frac{x \cdot C'(x) - C(x)}{x^2}$$

b) The derivative exists for all x in $(0, \infty)$, therefore, the critical values will occur when $A'(x_0) = 0$. We solve the equation:

$$\frac{x_0 \cdot C'(x_0) - C(x_0)}{x_0^2} = 0$$

$$x_0 \cdot C'(x_0) - C(x_0) = 0 \quad \text{Multiplying by } x_0^2 \neq 0.$$

$$x_0 \cdot C'(x_0) = C(x_0)$$

$$C'(x_0) = \frac{C(x_0)}{x_0} = A(x_0)$$

59. Express Q as a function of one variable. First, solve $x^2 + y^2 = 2$ for y. We have:

$$y^2 = 2 - x^2$$

$$y = \pm\sqrt{2 - x^2}$$

y is a real number of x in the interval $\left[-\sqrt{2}, \sqrt{2}\right]$.

If $y = -\sqrt{2 - x^2}$, we substitute for y to get:

$$Q = 3x + y^3$$

$$Q = 3x + \left(-\sqrt{2 - x^2}\right)^3$$

$$= 3x - \left(2 - x^2\right)^{\frac{3}{2}}$$

Next, we find $Q'(x)$.

$$Q'(x) = 3 - \left(\frac{3}{2}\right)\left(2 - x^2\right)^{\frac{1}{2}}(-2x)$$

$$= 3 + 3x\left(2 - x^2\right)^{\frac{1}{2}}$$

$$= 3 + 3x\sqrt{2 - x^2}$$

The derivative exists for all values of x in the interval $\left[-\sqrt{2}, \sqrt{2}\right]$; thus, the only critical values are where $Q'(x) = 0$. We solve the equation:

$$Q'(x) = 0$$

$$3 + 3x\sqrt{2 - x^2} = 0$$

$$3 = -3x\sqrt{2 - x^2}$$

$$1 = -x\sqrt{2 - x^2}$$

$$1^2 = \left(-x\sqrt{2 - x^2}\right)^2$$

$$1 = x^2\left(2 - x^2\right)$$

$$1 = 2x^2 - x^4$$

$$x^4 - 2x^2 + 1 = 0$$

$$\left(x^2 - 1\right)^2 = 0$$

$$x^2 - 1 = 0$$

$$x^2 = 1$$

$$x = \pm 1$$

We notice that $x = 1$ is an extraneous solution which does not work.

$$3 + 3(1)\sqrt{2 - (1)^2} = 3 + 3 = 6 \neq 0.$$

Therefore, the only critical value is $x = -1$. The critical point and the endpoints are $-\sqrt{2}$, -1, and $\sqrt{2}$.

$$Q\left(-\sqrt{2}\right) = 3\left(-\sqrt{2}\right) - \left(2 - \left(-\sqrt{2}\right)^2\right)^{3/2} = -3\sqrt{2}$$

$$Q(-1) = 3(-1) - \left(2 - (-1)^2\right)^{3/2} = -4$$

$$Q\left(\sqrt{2}\right) = 3\left(\sqrt{2}\right) - \left(2 - \left(\sqrt{2}\right)^2\right)^{3/2} = 3\sqrt{2}$$

The minimum value of Q is $-3\sqrt{2}$ and occurs when $x = -\sqrt{2}$ and $y = -\sqrt{2 - \left(-\sqrt{2}\right)^2} = 0$

Next, we repeat the process for $y = \sqrt{2 - x^2}$. We notice that:

$$Q = 3x + y^3$$

$$Q = 3x + \left(\sqrt{2 - x^2}\right)^3$$

$$= 3x + \left(2 - x^2\right)^{3/2}$$

Next, we find $Q'(x)$.

$$Q'(x) = 3 + \left(\frac{3}{2}\right)\left(2 - x^2\right)^{1/2}(-2x)$$

$$= 3 - 3x\left(2 - x^2\right)^{1/2}$$

$$= 3 - 3x\sqrt{2 - x^2}$$

The derivative exists for all values of x in the interval $\left[-\sqrt{2}, \sqrt{2}\right]$; thus, the only critical values are where $Q'(x) = 0$. We solve the equation:

$$3 - 3x\sqrt{2 - x^2} = 0$$

$$3 = 3x\sqrt{2 - x^2}$$

$$1 = x\sqrt{2 - x^2}$$

$$1^2 = \left(x\sqrt{2 - x^2}\right)^2$$

$$1 = x^2\left(2 - x^2\right)$$

$$1 = 2x^2 - x^4$$

$$x^4 - 2x^2 + 1 = 0$$

$$\left(x^2 - 1\right)^2 = 0$$

$$x^2 - 1 = 0$$

$$x^2 = 1$$

$$x = \pm 1$$

We notice that $x = -1$ is an extraneous solution which does not work.

$$3 - 3(-1)\sqrt{2 - (-1)^2} = 3 + 3 = 6 \neq 0.$$

Therefore, the only critical value is $x = 1$. The critical point and the endpoints are $-\sqrt{2}$, 1, and $\sqrt{2}$.

$$Q\left(-\sqrt{2}\right) = 3\left(-\sqrt{2}\right) + \left(2 - \left(-\sqrt{2}\right)^2\right)^{3/2} = -3\sqrt{2}$$

$$Q(1) = 3(1) + \left(2 - (1)^2\right)^{3/2} = 4$$

$$Q\left(\sqrt{2}\right) = 3\left(\sqrt{2}\right) + \left(2 - \left(\sqrt{2}\right)^2\right)^{3/2} = 3\sqrt{2}$$

The minimum value of Q is $-3\sqrt{2}$ occurs when

$$x = -\sqrt{2} \text{ and } y = -\sqrt{2 - \left(-\sqrt{2}\right)^2} = 0.$$

Regardless of what value of y we chose, we see that the minimum of Q, is $-3\sqrt{2}$, when $x = -\sqrt{2}$ and $y = 0$.

61. From Exercise 60, we know that the store should order a lot size of $\sqrt{\dfrac{2bQ}{a}}$ units,

$\sqrt{\dfrac{aQ}{2b}}$ times per year.

When $Q = 2500$, $a = 10$, $b = 20$, $c = 9$, the store should order:

$\sqrt{\dfrac{aQ}{2b}} = \sqrt{\dfrac{10(2500)}{2(20)}} = 25$ times per year.

The lot size of each order should be:

$\sqrt{\dfrac{2(20)(2500)}{10}} = 100$ units.

Exercise Set 2.6

---

1. $R(x) = 5x;\quad C(x) = 0.001x^2 + 1.2x + 60$

a) Total profit is revenue minus cost.

$P(x) = R(x) - C(x)$

$\quad = 5x - \left(0.001x^2 + 1.2x + 60\right)$

$\quad = 5x - 0.001x^2 - 1.2x - 60$

$\quad = -0.001x^2 + 3.8x - 60$

b) Substituting 100 for x into the three functions, we have:

$R(100) = 5(100) = 500$

The total revenue from the sale of the first 50 units is $500.

$C(100) = 0.001(100)^2 + 1.2(100) + 60 = 190$

The total cost of producing the first 100 units is $190.

$P(100) = R(100) - C(100)$

$\quad = 500 - 190$

$\quad = 310$

The total profit is $310 when the first 100 units are produced and sold.

Note, we could have also used the profit function, $P(x)$, from part (a) to find the profit.

$P(100) = -0.001(100)^2 + 3.8(100) - 60 = 310$

c) Finding the derivative for each of the functions, we have:

$R'(x) = 5$

$C'(x) = 0.002x + 1.2$

$P'(x) = -0.002x + 3.8$

d) Substituting 100 for x in each of the three marginal functions, we have:

$R'(100) = 5$

Once 100 units have been sold, the approximate revenue for the 51st unit is $5.

$C'(100) = 1.4$

Once 100 units have been produced, the approximate cost for the 101st unit is $1.40.

$P'(100) = -0.002(100) + 3.8 = 3.6$

Once 100 units have been produced and sold, the approximate profit from the sale of the 101st unit is $3.60.

e) $\boxed{tw}$ In part (b), we are observing the total revenue, cost and profit from the production and sale of the first 100 items. In part (d), we are observing the approximate revenue, cost and profit from the production and sale of the 101st unit only. These quantities are also known as the marginal revenue, marginal cost and marginal profit.

3. $C(x) = 0.001x^3 + 0.07x^2 + 19x + 700$

a) Substituting 25 for x into the cost function, we have:

$C(25) = 0.001(25)^3 + 0.07(25)^2 + 19(25) + 700$

$\quad = 1234.375$

The current monthly cost of producing 25 chairs is $1234.38.

b) In order to find the additional cost of producing 26 chairs monthly, we first find the total cost of producing 26 chairs in a month.

$C(26) = 0.001(26)^3 + 0.07(26)^2 + 19(26) + 700$

$\quad = 1258.896$

Next, we subtract the cost of producing 25 chairs monthly found in part (a) from the cost of producing 26 chairs monthly.

$C(26) - C(25) = 1258.896 - 1234.375$

$\quad = 24.521$

The additional cost of increasing production to 26 chairs monthly is $24.52.

c) First, we find the marginal cost function,

$C'(x) = 0.003x^2 + 0.14x + 19$.

Next, we substituting 25 for x, we have:

$C'(25) = 0.003(25)^2 + 0.14(25) + 19$

$\quad = 24.375$

The marginal cost when 25 chairs have been produced is $24.38.

d) Using the marginal cost from part (c), the additional cost required to produce 2 additional chairs monthly is:
$$2(24.375) = 48.75.$$
Therefore, the difference in cost between producing 25 and 27 chairs per month is approximately $48.75.

e) In part (a) we found that it cost 1234.38 to produce 25 chairs per month. In part (d) we found that the difference in cost between 25 chairs and 27 chairs per month was $48.75. Therefore, the approximate total cost of producing 27 chairs per month is
$$C(27) \approx 1234.38 + 48.75 = 1283.13.$$
We predict the cost of producing 27 chairs monthly will be $1283.13.

5. $R(x) = 0.005x^3 + 0.01x^2 + 0.5x$

a) Substituting 70 for x, we have:
$$R(70) = 0.005(70)^3 + 0.01(70)^2 + 0.5(70)$$
$$= 1715 + 49 + 35$$
$$= 1799$$
The currently daily revenue from selling 70 lawn chairs per day is $1799.

b) Substituting 73 for x, we have:
$$R(73) = 0.005(73)^3 + 0.01(73)^2 + 0.5(73)$$
$$= 2034.875$$
$$= 2034.88$$
Therefore, the increase in revenue from increasing sales to 73 chairs per day is:
$$R(73) - R(70) = 2034.88 - 1799$$
$$= 235.88$$
Revenue will increase $235.88 per day if the number of chairs sold increases to 73 per day.

c) First we find the marginal revenue function by finding the derivative of the revenue function.
$$R'(x) = 0.015x^2 + 0.02x + 0.5$$
Substituting 70 for x, we have:
$$R'(70) = 0.015(70)^2 + 0.02(70) + 0.5$$
$$= 75.40$$
The marginal revenue when 70 lawn chairs are sold daily is $75.40.

d) In part (a) we found that selling 70 lawn chairs per day resulted in a revenue of $1799. In part (c) we found that the marginal revenue when 70 chairs were sold is $75.40. Using these two numbers, we estimate the daily revenue generated by selling 71 chairs is
$$R(71) \approx R(70) + R'(70)$$
$$= \$1799 + \$75.40 = \$1874.40.$$
Similarly, the daily revenue generated by selling 72 chairs, or 2 additional chairs, daily is approximately
$$R(72) \approx R(70) + 2 \cdot R'(70)$$
$$\approx \$1799 + 2(\$75.40) \approx \$1949.80.$$
The daily revenue generated by selling 73 chairs, or 3 additional chairs, daily is approximately
$$R(73) \approx R(70) + 3 \cdot R'(70)$$
$$\approx \$1799 + 3(\$75.40) \approx \$2025.20.$$

7. $P(x) = -0.004x^3 - 0.3x^2 + 600x - 800$

a) Substituting 9 for x, we have:
$$P(9) = -0.004(9)^3 - 0.3(9)^2 + 600(9) - 800$$
$$= -2.916 - 24.30 + 5400 - 800$$
$$= 4572.784$$
The currently weekly profit is $4572.78.

b) First, we find the total weekly profit of selling 8 laptops per week.
$$P(8) = -0.004(8)^3 - 0.3(8)^2 + 600(8) - 800$$
$$= 3978.752$$
The difference in weekly profit from selling 8 laptops and 9 laptops per week is
$$P(9) - P(8) = 4572.78 - 3978.75$$
$$= 594.03$$
Therefore, Crawford Computing would lose $594.03 each week if 8 laptops were sold each week instead of 9.

c) First, we find the marginal profit function by taking the derivative of the profit function.
$$P'(x) = -0.012x^2 - 0.6x + 600$$
Substituting 9 for x, we have:
$$P'(9) = -0.012(9)^2 - 0.6(9) + 600$$
$$= 593.628$$
The marginal profit is $593.63 when 9 laptops are sold weekly.

d) From part (a), we know that when 9 laptops are built and sold, total weekly profit is $4572.78. From part (c), we know that when 9 laptops are built and sold, marginal profit is $593.63. Therefore, we estimate:

$$P(10) \approx P(9) + P'(9)$$
$$\approx 4572.78 + 593.63$$
$$\approx 5166.41$$

The total weekly profit is approximately $5166.41 when 10 laptops are built and sold weekly.

9. $N(1000) = 500{,}000$ means that 500,000 computers will be sold annually when the price of the computer is $1000.

$N'(1000) = -100$ means that when the price is increased $1 to $1001, sales will decrease by 100 computers per year.

11. $C(x) = 0.01x^2 + 0.6x + 30$

$$\Delta C = C(x + \Delta x) - C(x)$$

Substituting $x = 70$, and $\Delta x = 1$ we have

$$\Delta C = C(70 + 1) - C(70)$$
$$= C(71) - C(70)$$
$$= 0.01(71)^2 + 0.6(71) + 30 - $$
$$\left[0.01(70)^2 + 0.6(70) + 30 \right]$$
$$= 2.01$$

The additional cost of producing the 71st unit is $2.01.

Finding the derivative of $C(x)$ we have:

$$C'(x) = 0.02x + 0.6$$

Substituting 70 for x, we have:

$$C'(70) = 0.02(70) + 0.6 = 2.00$$

The marginal cost when 70 units are produced is $2.00.

13. $R(x) = 2x$

$$\Delta R = R(x + \Delta x) - R(x)$$

Substituting $x = 70$, and $\Delta x = 1$ we have

$$\Delta R = R(70 + 1) - R(70)$$
$$= R(71) - R(70)$$
$$= 2(71) - \left[2(70) \right]$$
$$= 2$$

The additional cost of producing the 71st unit is $2.00.

Finding the derivative of $R(x)$ we have:

$$R'(x) = 2.$$

The derivative is constant; therefore,

$$R'(70) = 2$$

The marginal cost when 70 units are produced is $2.00.

15. $C(x) = 0.01x^2 + 0.6x + 30; \ R(x) = 2x$

a) Finding the profit function we have:

$$P(x) = R(x) - C(x)$$
$$= 2x - (0.01x^2 + 0.6x + 30)$$
$$= -0.01x^2 + 1.4x - 30$$

b) $\Delta P = P(x + \Delta x) - P(x)$

Substituting $x = 70$, and $\Delta x = 1$ we have

$$\Delta P = P(70 + 1) - P(70)$$
$$= P(71) - P(70)$$
$$= -0.01(71)^2 + 1.4(71) - 30 - $$
$$\left[-0.01(70)^2 + 1.4(70) - 30 \right]$$
$$= -0.01$$

The additional profit of producing and selling the 71st unit is −$0.01.

Finding the derivative of $P(x)$ we have:

$$P'(x) = -0.02x + 1.4$$

Substituting 70 for x, we have:

$$P'(70) = -0.02(70) + 1.4 = 0.00$$

The marginal profit when 70 units are produced and sold is $0.00.

17. $D = 0.007p^3 - 0.5p^2 + 150p$

a) We take the derivative of the demand function with respect to price.

$$\frac{dD}{dp} = 0.021p^2 - p + 150$$

b) Substituting 25 for p in the demand function we have:

$$D = 0.007(25)^3 - 0.5(25)^2 + 150(25)$$
$$= 109.375 - 312.50 + 3750$$
$$= 3546.875$$

Consumers will want to buy 3547 units when price is $25 per unit.

c) $\boxed{tw}$

d) $\boxed{tw}$

Me
Rachel
Kaley
Danielle
Ashley

19. $A(x) = \dfrac{13x + 100}{x}$

To estimate the change in average cost as production goes from 100 to 101 units, we establish that $x = 100$ and $\Delta x = 1$. Next, we find the derivative of $A(x)$.

$A'(x) = \dfrac{x(13) - (13x + 100)(1)}{x^2}$

$= \dfrac{13x - 13x - 100}{x^2}$

$= -\dfrac{100}{x^2}$

Therefore,

$\Delta A \approx A'(x)\Delta x$

$\approx A'(100)\Delta x \qquad (x = 100)$

$\approx -\dfrac{100}{(100)^2}\Delta x$

$\approx -\dfrac{100}{(100)^2}(1) \qquad (\Delta x = 1)$

≈ -0.01

The average cost changes by about $-\$0.01$. (We see an approximate decrease in average cost of one cent.)

21. $P(x) = 567 + x(36x^{0.6} - 104)$

$= 567 + 36x^{1.6} - 104x$

x is the number of years since 1960; therefore, the year 2009 corresponds to $x = 2009 - 1960 = 49$, and the year 2010 corresponds to $x = 2010 - 1960 = 50$. To estimate the increase in gross domestic product from 2009 to 2010, we establish that $x = 49$ and $\Delta x = 1$. Next, we find the derivative of $P(x)$:

$P'(x) = 36(1.6)x^{0.6} - 104 = 57.6x^{0.6} - 104$.

Therefore,

$\Delta P \approx P'(x)\Delta x$

$\approx P'(49)\Delta x \qquad [x = 49]$

$\approx \left(57.6(49)^{0.6} - 104\right)\Delta x$

$\approx (491.03173875)(1) \qquad [\Delta x = 1]$

≈ 491.03

The gross domestic product should increase about $491.03 billion between 2009 and 2010.

23. $\boxed{tw}$

25. Alan's marginal tax rate is 21%, therefore for each additional dollar he earns, he will have to pay $0.21 in taxes. If he earns another $2000, dollars, we will pay an additional $2000(0.21) = \$420$ in taxes.

27. $y = f(x) = x^2$, $x = 2$, and $\Delta x = 0.01$

$\Delta y = f(x + \Delta x) - f(x)$

$= f(2 + 0.01) - f(2)$ Substituting 2 for x and

0.01 for Δx.

$= f(2.01) - f(2)$

$= (2.01)^2 - (2)^2$

$= 0.0401$

$f'(x)\Delta x = 2x \cdot \Delta x \qquad \left[f(x) = x^2; f'(x) = 2x\right]$

$f'(2)\Delta x = 2(2) \cdot (0.01)$ Substituting 2 for x and

0.01 for Δx.

$= 4(0.01)$

$= 0.04$

29. $y = f(x) = x + x^2$, $x = 3$, and $\Delta x = 0.04$

$\Delta y = f(x + \Delta x) - f(x)$

$= f(3 + 0.04) - f(3)$ Substituting 3 for x and

0.04 for Δx.

$= f(3.04) - f(3)$

$= \left[(3.04) + (3.04)^2\right] - \left[(3) + (3)^2\right]$

$= [12.2816] - [12]$

$= 0.2816$

$f'(x)\Delta x = (1 + 2x) \cdot \Delta x \qquad \begin{bmatrix} f(x) = x + x^2; \\ f'(x) = 1 + 2x \end{bmatrix}$

$f'(3)\Delta x = \left[1 + 2(3)\right] \cdot (0.04)$ Substituting 3 for x

and 0.04 for Δx.

$= [7](0.04)$

$= 0.28$

31. $y = f(x) = \dfrac{1}{x^2} = x^{-2}$, $x = 1$, and $\Delta x = 0.5$

$\Delta y = f(x + \Delta x) - f(x)$

$ = f(1 + 0.5) - f(1)$ Substituting 1 for x and

 0.5 for Δx.

$ = f(1.5) - f(1)$

$ = \left[\dfrac{1}{(1.5)^2}\right] - \left[\dfrac{1}{(1)^2}\right]$

$ = \left[\dfrac{1}{2.25}\right] - [1]$

$ = -0.5556$

$f'(x)\Delta x = -2x^{-3} \cdot \Delta x$ $\begin{bmatrix} f(x) = x^{-2}; \\ f'(x) = -2x^{-3} \end{bmatrix}$

$f'(1)\Delta x = \left[-2(1)^{-3}\right] \cdot (0.5)$ Substituting 1 for x

$\phantom{f'(1)\Delta x = [-2(1)^{-3}]}$ and 0.5 for Δx.

$ = [-2](0.5)$

$ = -1$

33. $y = f(x) = 3x - 1$, $x = 4$, and $\Delta x = 2$

$\Delta y = f(x + \Delta x) - f(x)$

$ = f(4 + 2) - f(4)$ Substituting 4 for x and

 2 for Δx.

$ = f(6) - f(4)$

$ = [3(6) - 1] - [3(4) - 1]$

$ = [17] - [11]$

$ = 6$

$f'(x)\Delta x = (3) \cdot \Delta x$ $[f(x) = 3x - 1; f'(x) = 3]$

$f'(4)\Delta x = (3) \cdot (2)$ Substituting 4 for x

 and 2 for Δx.

$ = 6$

35. We first think of the number closest to 26 that is a perfect square. This is 25. What we will do is approximate how $y = \sqrt{x}$, changes when x changes from 25 to 26. Let

$y = f(x) = \sqrt{x} = x^{1/2}$

Then $f'(x) = \dfrac{1}{2} x^{-1/2} = \dfrac{1}{2\sqrt{x}}$

Using, $\Delta y \approx f'(x)\Delta x$, we have

$\Delta y \approx f'(x)\Delta x$

$ \approx \dfrac{1}{2\sqrt{x}} \cdot \Delta x$

We are interested in Δy as x changes from 25 to 26, so

$\Delta y \approx \dfrac{1}{2\sqrt{x}} \cdot \Delta x$

$ \approx \dfrac{1}{2\sqrt{25}} \cdot 1$ Replacing x with 25 and Δx with 1

$ \approx \dfrac{1}{2 \cdot 5}$

$ \approx \dfrac{1}{10} = 0.1$

We can now approximate $\sqrt{26}$;

$\sqrt{26} = \sqrt{25} + \Delta y$

$\phantom{\sqrt{26}} = 5 + \Delta y$

$\phantom{\sqrt{26}} \approx 5 + 0.1$

$\phantom{\sqrt{26}} \approx 5.1$

To five decimal places $\sqrt{26} = 5.09902$. Thus, our approximation is fairly accurate.

37. We first think of the number closest to 102 that is a perfect square. This is 100. What we will do is approximate how $y = \sqrt{x}$, changes when x changes from 100 to 102. Let

$y = f(x) = \sqrt{x} = x^{1/2}$

Then $f'(x) = \dfrac{1}{2} x^{-1/2} = \dfrac{1}{2\sqrt{x}}$

Using, $\Delta y \approx f'(x)\Delta x$, we have

$\Delta y \approx f'(x)\Delta x$

$ \approx \dfrac{1}{2\sqrt{x}} \cdot \Delta x$

We are interested in Δy as x changes from 100 to 102, so

$\Delta y \approx \dfrac{1}{2\sqrt{x}} \cdot \Delta x$

$ \approx \dfrac{1}{2\sqrt{100}} \cdot 2$ Replacing x with 100 and Δx with 2

$ \approx \dfrac{1}{2 \cdot 10} \cdot 2$

$ \approx \dfrac{1}{10} = 0.1$

We can now approximate $\sqrt{102}$;
$$\sqrt{102} = \sqrt{100} + \Delta y$$
$$= 10 + \Delta y$$
$$\approx 10 + 0.1$$
$$\approx 10.1$$

To five decimal places $\sqrt{26} = 10.09950$. Thus, our approximation is fairly accurate.

39. We first think of the number closest to 1005 that is a perfect cube. This is 1000. What we will do is approximate how $y = \sqrt[3]{x}$, changes when x changes from 1000 to 1005. Let

$$y = f(x) = \sqrt[3]{x} = x^{\frac{1}{3}}$$

Then $f'(x) = \dfrac{1}{3}x^{-\frac{2}{3}} = \dfrac{1}{2\sqrt[3]{x^2}}$

Using, $\Delta y \approx f'(x)\Delta x$, we have

$$\Delta y \approx f'(x)\Delta x$$

$$\approx \frac{1}{2\sqrt[3]{x^2}} \cdot \Delta x$$

We are interested in Δy as x changes from 1000 to 1005, so

$$\Delta y \approx \frac{1}{3\sqrt[3]{x^2}} \cdot \Delta x$$

Replacing x with 1000 and Δx with 5, we have

$$\approx \frac{1}{3 \cdot \sqrt[3]{(1000)^2}} \cdot 5$$

$$\approx \frac{1}{3 \cdot 100} \cdot 5$$

$$\approx \frac{1}{60} = 0.017$$

We can now approximate $\sqrt[3]{1005}$;
$$\sqrt[3]{1005} = \sqrt[3]{1000} + \Delta y$$
$$= 10 + \Delta y$$
$$\approx 10 + 0.017$$
$$\approx 10.017$$

To five decimal places $\sqrt[3]{1000} = 10.01664$ Thus, our approximation is fairly accurate.

41. $y = \sqrt{x+1} = (x+1)^{\frac{1}{2}}$

First, we find $\dfrac{dy}{dx}$:

$$\frac{dy}{dx} = \frac{1}{2}(x+1)^{-\frac{1}{2}}(1) = \frac{1}{2\sqrt{x+1}} .$$

Then
$$dy = \frac{1}{2\sqrt{x+1}} dx .$$

Note that the expression for dy contains two variables x and dx.

43. $y = (2x^3 + 1)^{\frac{3}{2}}$

First, we find $\dfrac{dy}{dx}$:

$$\frac{dy}{dx} = \frac{3}{2}(2x^3 + 1)^{\frac{1}{2}}(6x^2) \text{ By the extended power rule}$$

$$= 9x^2(2x^3 + 1)^{\frac{1}{2}}$$

$$= 9x^2\sqrt{2x^3 + 1}.$$

Then
$$dy = 9x^2\sqrt{2x^3 + 1}\, dx .$$

Note that the expression for dy contains two variables x and dx.

45. $y = \sqrt[5]{x+27} = (x+27)^{\frac{1}{5}}$

First, we find $\dfrac{dy}{dx}$. By the extended power rule we have:

$$\frac{dy}{dx} = \frac{1}{5}(x+27)^{-\frac{4}{5}}(1) = \frac{1}{5 \cdot \sqrt[5]{(x+27)^4}} .$$

Then
$$dy = \frac{1}{5 \cdot \sqrt[5]{(x+27)^4}} dx .$$

Note that the expression for dy contains two variables x and dx.

47. $y = x^4 - 2x^3 + 5x^2 + 3x - 4$

First, we find $\dfrac{dy}{dx}$:

$$\frac{dy}{dx} = 4x^3 - 6x^2 + 10x + 3 .$$

Then
$$dy = (4x^3 - 6x^2 + 10x + 3)dx .$$

Note that the expression for dy contains two variables x and dx.

49. From Exercise 47, we know:

$dy = \left(4x^3 - 6x^2 + 10x + 3\right) dx$.

When $x = 2$ and $dx = 0.1$ we have:

$dy = \left(4(2)^3 - 6(2)^2 + 10(2) + 3\right)(0.1)$

$= (32 - 24 + 20 + 3)(0.1)$

$= (31)(0.1)$

$= 3.1$

51. $y = (3x - 10)^5$

First, we find $\dfrac{dy}{dx}$:

$\dfrac{dy}{dx} = 5(3x - 10)^4 (3)$ By the extended power rule

$= 15(3x - 10)^4$

Then

$dy = 15(3x - 10)^4 \, dx$.

When $x = 4$ and $dx = 0.03$ we have:

$dy = 15(3(4) - 10)^4 (0.03)$

$= 15(2)^4 (0.03)$

$= 7.2$

53. Let $y = f(x) = x^4 - x^2 + 8$

First we find $f'(x)$:

$f'(x) = 4x^3 - 2x$.

Then

$dy = f'(x) dx$

$= \left(4x^3 - 2x\right) dx$

To approximate $f(5.1)$, we will use

$x = 5$ and $dx = 0.1$ to determine the differential dy.

Substituting 5 for x and 0.1 for dx we have:

$dy = f'(5) dx$

$= \left(4(5)^3 - 2(5)\right)(0.1)$

$= (4(125) - 10)(0.1)$

$= (500 - 10)(0.1)$

$= (490)(0.1)$

$= 49$

Next, we find

$f(5) = (5)^4 - (5)^2 + 8$

$= 625 - 25 + 8$

$= 608$

Now,

$f(5.1) \approx f(5) + f'(5) dx$

$\approx 608. + 49.00$

≈ 657

55. $S = 0.02235 h^{0.42246} w^{0.51456}$

We begin by noticing that we are wanting to estimate the change in surface area due to a change in weight w; therefore, we will first find $\dfrac{dS}{dw}$. Since $h = 160$, we have:

$S = 0.02235 (160)^{0.42246} w^{0.51456}$

$= 0.02235 (8.53399783) w^{0.51456}$

$= 0.19073485 w^{0.51456}$

Now we can take the derivative of S with respect to w.

$\dfrac{dS}{dw} = 0.19073485 (0.51456) w^{-0.48544}$

$= 0.09814452 w^{-0.48544}$

Therefore,

$dS = \left(0.09814452 w^{-0.48544}\right) dw$

Now that we have the differential, we can use her weight of 60 kg to approximate how much her surface area changes when her weight drops 1 kg. We substitute 60 for w and -1 for dw to get:

$dS \approx \left(0.09814452 (60)^{-0.48544}\right)(-1)$

≈ -0.01345

The patients surface area will change by -0.01345 m^2.

57. $N(t) = \dfrac{0.8t + 1000}{5t + 4}$

First we find $N'(t)$ by the quotient rule.

$N'(t) = \dfrac{(5t + 4)(0.8) - (0.8t + 1000)(5)}{(5t + 4)^2}$

$= \dfrac{4t + 3.2 - 4t - 5000}{(5t + 4)^2}$

$= -\dfrac{4996.8}{(5t + 4)^2}$

The differential is:

$dN = N'(t) dt$

$= -\dfrac{4996.8}{(5t + 4)^2} \cdot dt$

We approximate the change in bodily concentration from 1.0 hr to 1.1 hr by using 1.0 for t and 0.1 for dt.

$$dN = -\frac{4996.8}{\left(5(1.0)+4\right)^2} \cdot (0.1)$$

$$= -\frac{4996.8}{(9)^2}(0.1)$$

$$= -\frac{4996.8}{81}(0.1)$$

$$\approx -61.6889(0.1)$$

$$\approx -6.16889$$

Next, we approximate the change in bodily concentration from 2.8 hr to 2.9 hr by using 2.8 for t and 0.1 for dt.

$$dN = -\frac{4996.8}{\left(5(2.8)+4\right)^2} \cdot (0.1)$$

$$= -\frac{4996.8}{(18)^2}(0.1)$$

$$= -\frac{4996.8}{324}(0.1)$$

$$\approx -15.4222(0.1)$$

$$\approx -1.54222$$

The concentration changes more from 1.0 hr to 1.1 hr.

59. The circumference of the earth, which is the original length of the rope, is given by $C(r) = 2\pi r$, where r is the radius of the earth.

We need to find the change in the length of the radius, Δr, when the length of the rope is increased 10 feet.

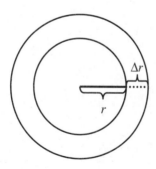

Using differentials, $\Delta C \approx C'(r)\Delta r$ represents the change in the length of the rope. Therefore, $\Delta C = 10$, and we have:

$$10 = C'(r)\Delta r.$$

Noticing that $C'(r) = 2\pi$, we have:

$$10 = 2\pi \cdot \Delta r$$

$$\frac{10}{2\pi} = \Delta r$$

Therefore, the rope is raised approximately

$$\Delta r = \frac{5}{\pi} \approx 1.59 \text{ feet above the earth.}$$

61. $p = 100 - \sqrt{x}$

Since revenue is price times quantity, the revenue function is given by:

$$R(x) = p \cdot x$$

$$= \left(100 - \sqrt{x}\right)x$$

$$= 100x - x^{3/2}.$$

To find the marginal revenue, we take the derivative of the revenue function. Thus:

$$R'(x) = 100 - \frac{3}{2}x^{1/2}$$

$$= 100 - \frac{3\sqrt{x}}{2}.$$

63. $p = 500 - x$

Since revenue is price times quantity, the revenue function is given by:

$$R(x) = p \cdot x$$

$$= (500 - x)x$$

$$= 500x - x^2$$

To find the marginal revenue, we take the derivative of the revenue function. Thus:

$$R'(x) = 500 - 2x.$$

65. $p = \frac{3000}{x} + 5$

Since revenue is price times quantity, the revenue function is given by:

$$R(x) = p \cdot x$$

$$= \left(\frac{3000}{x} + 5\right)x$$

$$= 3000 + 5x.$$

To find the marginal revenue, we take the derivative of the revenue function. Thus:

$$R'(x) = 5.$$

67. $\boxed{tw}$

Exercise Set 2.7

1. Differentiate implicitly to find $\dfrac{dy}{dx}$.

We have

$x^3 + 2y^3 = 6$

Differentiating both sides with respect to x yields:

$$\frac{d}{dx}\left(x^3 + 2y^3\right) = \frac{d}{dx}(6)$$

$$\frac{d}{dx}x^3 + 2\frac{d}{dx}y^3 = \frac{d}{dx}6$$

$$3x^2 + 2\cdot 3y^2 \cdot \frac{dy}{dx} = 0$$

Next, we isolate $\dfrac{dy}{dx}$

$$6y^2 \cdot \frac{dy}{dx} = -3x^2$$

$$\frac{dy}{dx} = \frac{-3x^2}{6y^2}$$

$$\frac{dy}{dx} = \frac{-x^2}{2y^2}$$

Find the slope of the tangent line to the curve at $(2,-1)$.

$$\frac{dy}{dx} = \frac{-x^2}{2y^2}$$

Replacing x with 2 and y with -1, we have:

$$\frac{dy}{dx} = \frac{-(2)^2}{2(-1)^2} = \frac{-4}{2} = -2 \ .$$

The slope of the tangent line to the curve at $(2,-1)$ is -2.

3. Differentiate implicitly to find $\dfrac{dy}{dx}$.

We have

$2x^2 - 3y^3 = 5$

Differentiating both sides with respect to x yields:

$$\frac{d}{dx}\left(2x^2 - 3y^3\right) = \frac{d}{dx}(5)$$

$$\frac{d}{dx}2x^2 - 3\frac{d}{dx}y^3 = \frac{d}{dx}5$$

$$2\cdot 2x - 3\cdot 3y^2 \cdot \frac{dy}{dx} = 0$$

Next, we isolate $\dfrac{dy}{dx}$

$$-9y^2 \cdot \frac{dy}{dx} = -4x$$

$$\frac{dy}{dx} = \frac{-4x}{-9y^2}$$

$$\frac{dy}{dx} = \frac{4x}{9y^2}$$

Find the slope of the tangent line to the curve at $(-2,1)$.

$$\frac{dy}{dx} = \frac{4x}{9y^2}$$

Replacing x with -2 and y with 1, we have:

$$\frac{dy}{dx} = \frac{4(-2)}{9(1)^2} = \frac{-8}{9} = -\frac{8}{9} \ .$$

The slope of the tangent line to the curve at $(-2,1)$ is $-\dfrac{8}{9}$.

5. Differentiate implicitly to find $\dfrac{dy}{dx}$.

We have

$x^2 - y^2 = 1$

Differentiating both sides with respect to x yields:

$$\frac{d}{dx}\left(x^2 - y^2\right) = \frac{d}{dx}(1)$$

$$\frac{d}{dx}x^2 - \frac{d}{dx}y^2 = \frac{d}{dx}1$$

$$2x - 2y\cdot\frac{dy}{dx} = 0$$

Next, we isolate $\dfrac{dy}{dx}$

$$-2y\cdot\frac{dy}{dx} = -2x$$

$$\frac{dy}{dx} = \frac{-2x}{-2y}$$

$$\frac{dy}{dx} = \frac{x}{y}$$

Find the slope of the tangent line to the curve at $\left(\sqrt{3},\sqrt{2}\right)$.

$$\frac{dy}{dx} = \frac{x}{y}$$

Replacing x with $\sqrt{3}$ and y with $\sqrt{2}$, we have:

$$\frac{dy}{dx} = \frac{\sqrt{3}}{\sqrt{2}} = \sqrt{\frac{3}{2}}.$$

The slope of the tangent line to the curve at $\left(\sqrt{3}, \sqrt{2}\right)$ is $\sqrt{\frac{3}{2}}$.

7. Differentiate implicitly to find $\dfrac{dy}{dx}$.

We have
$$3x^2 y^4 = 12$$

Differentiating both sides with respect to x yields:

$$\frac{d}{dx}\left(3x^2 y^4\right) = \frac{d}{dx}(12)$$

$$3x^2 \frac{d}{dx} y^4 + y^4 \frac{d}{dx} 3x^2 = \frac{d}{dx} 12 \quad \text{Product Rule}$$

$$3x^2\left(4y^3 \cdot \frac{dy}{dx}\right) + y^4 \cdot (3 \cdot 2x) = 0$$

$$12x^2 y^3 \cdot \frac{dy}{dx} + 6xy^4 = 0$$

$$12x^2 y^3 \cdot \frac{dy}{dx} = -6xy^4$$

$$\frac{dy}{dx} = \frac{-6xy^4}{12x^2 y^3}$$

$$\frac{dy}{dx} = -\frac{y}{2x}$$

Find the slope of the tangent line to the curve at $(2, -1)$.

$$\frac{dy}{dx} = -\frac{y}{2x}$$

Replacing x with 2 and y with -1, we have:

$$\frac{dy}{dx} = -\frac{(-1)}{2(2)} = \frac{1}{4}.$$

The slope of the tangent line to the curve at $(2, -1)$ is $\frac{1}{4}$.

9. Differentiate implicitly to find $\dfrac{dy}{dx}$.

We have
$$x^3 - x^2 y^2 = -9$$

Differentiating both sides with respect to x yields:

$$\frac{d}{dx}\left(x^3 - x^2 y^2\right) = \frac{d}{dx}(-9)$$

$$\frac{d}{dx} x^3 - \frac{d}{dx} x^2 y^2 = \frac{d}{dx}(-9)$$

$$3x^2 - \left[x^2\left(2y \cdot \frac{dy}{dx}\right) + y^2 (2x)\right] = 0$$

$$3x^2 - 2x^2 y \cdot \frac{dy}{dx} - 2xy^2 = 0$$

$$-2x^2 y \cdot \frac{dy}{dx} = 2xy^2 - 3x^2$$

$$\frac{dy}{dx} = \frac{2xy^2 - 3x^2}{-2x^2 y}$$

$$\frac{dy}{dx} = \frac{-x\left(3x - 2y^2\right)}{-x(2xy)}$$

$$\frac{dy}{dx} = \frac{3x - 2y^2}{2xy}$$

Find the slope of the tangent line to the curve at $(3, -2)$.

$$\frac{dy}{dx} = \frac{3x - 2y^2}{2xy}$$

Replacing x with 3 and y with -2, we have:

$$\frac{dy}{dx} = \frac{3(3) - 2(-2)^2}{2(3)(-2)} = \frac{9 - 8}{-12} = -\frac{1}{12}.$$

The slope of the tangent line to the curve at $(3, -2)$ is $-\frac{1}{12}$.

11. Differentiate implicitly to find $\dfrac{dy}{dx}$.

We have
$$xy - x + 2y = 3$$

Differentiating both sides with respect to x yields:

$$\frac{d}{dx}(xy - x + 2y) = \frac{d}{dx}(3)$$

$$\frac{d}{dx}xy - \frac{d}{dx}x + 2\frac{d}{dx}y = \frac{d}{dx}(3)$$

$$\left[x\left(\frac{dy}{dx}\right) + y(1)\right] - 1 + 2 \cdot \frac{dy}{dx} = 0$$

$$x \cdot \frac{dy}{dx} + y - 1 + 2 \cdot \frac{dy}{dx} = 0$$

$$x \cdot \frac{dy}{dx} + 2 \cdot \frac{dy}{dx} = 1 - y$$

$$(x+2) \cdot \frac{dy}{dx} = 1 - y$$

$$\frac{dy}{dx} = \frac{1-y}{x+2}$$

Find the slope of the tangent line to the curve at $\left(-5, \frac{2}{3}\right)$.

$$\frac{dy}{dx} = \frac{1-y}{x+2}$$

Replacing x with -5 and y with $\frac{2}{3}$, we have:

$$\frac{dy}{dx} = \frac{1 - \left(\frac{2}{3}\right)}{(-5) + 2} = \frac{\frac{1}{3}}{-3} = -\frac{1}{9}.$$

The slope of the tangent line to the curve at $\left(-5, \frac{2}{3}\right)$ is $-\frac{1}{9}$.

13. Differentiate implicitly to find $\frac{dy}{dx}$.

We have

$$x^2 y - 2x^3 - y^3 + 1 = 0$$

Differentiating both sides with respect to x yields:

$$\frac{d}{dx}(x^2 y - 2x^3 - y^3 + 1) = \frac{d}{dx}(0)$$

$$\frac{d}{dx}x^2 y - \frac{d}{dx}2x^3 - \frac{d}{dx}y^3 + \frac{d}{dx}1 = 0$$

$$\left[x^2\left(\frac{dy}{dx}\right) + y(2x)\right] - 2(3x^2) - (3y^2) \cdot \frac{dy}{dx} = 0$$

$$x^2 \cdot \frac{dy}{dx} + 2xy - 6x^2 - 3y^2 \cdot \frac{dy}{dx} = 0$$

$$x^2 \cdot \frac{dy}{dx} - 3y^2 \cdot \frac{dy}{dx} = 6x^2 - 2xy$$

$$(x^2 - 3y^2) \cdot \frac{dy}{dx} = 6x^2 - 2xy$$

$$\frac{dy}{dx} = \frac{6x^2 - 2xy}{x^2 - 3y^2}$$

Find the slope of the tangent line to the curve at $(2, -3)$.

$$\frac{dy}{dx} = \frac{6x^2 - 2xy}{x^2 - 3y^2}$$

Replacing x with 2 and y with -3, we have:

$$\frac{dy}{dx} = \frac{6(2)^2 - 2(2)(-3)}{(2)^2 - 3(-3)^2} = \frac{24 + 12}{4 - 27} = \frac{36}{-23} = -\frac{36}{23}$$

The slope of the tangent line to the curve at $(2, -3)$ is $-\frac{36}{23}$.

15. Differentiate implicitly to find $\frac{dy}{dx}$.

We have
$$2xy + 3 = 0$$

Differentiating both sides with respect to x yields:

$$\frac{d}{dx}(2xy + 3) = \frac{d}{dx}(0)$$

$$2 \cdot \frac{d}{dx}xy + \frac{d}{dx}3 = \frac{d}{dx}(0)$$

$$2 \cdot \left[x\left(\frac{dy}{dx}\right) + y(1)\right] + 0 = 0$$

$$2x \cdot \frac{dy}{dx} + 2y = 0$$

$$2x \cdot \frac{dy}{dx} = -2y$$

$$\frac{dy}{dx} = \frac{-2y}{2x}$$

$$\frac{dy}{dx} = -\frac{y}{x}$$

17. Differentiate implicitly to find $\dfrac{dy}{dx}$.

We have

$$x^2 - y^2 = 16$$

Differentiating both sides with respect to x yields:

$$\frac{d}{dx}(x^2 - y^2) = \frac{d}{dx}(16)$$

$$\frac{d}{dx}x^2 - \frac{d}{dx}y^2 = \frac{d}{dx}(16)$$

$$2x - 2y \cdot \frac{dy}{dx} = 0$$

$$-2y \cdot \frac{dy}{dx} = -2x$$

$$\frac{dy}{dx} = \frac{-2x}{-2y}$$

$$\frac{dy}{dx} = \frac{x}{y}$$

19. Differentiate implicitly to find $\dfrac{dy}{dx}$.

We have

$$y^5 = x^3$$

Differentiating both sides with respect to x yields:

$$\frac{d}{dx}(y^5) = \frac{d}{dx}(x^3)$$

$$5y^4 \cdot \frac{dy}{dx} = 3x^2$$

$$\frac{dy}{dx} = \frac{3x^2}{5y^4}$$

21. Differentiate implicitly to find $\dfrac{dy}{dx}$.

We have

$$x^2 y^3 + x^3 y^4 = 11$$

Differentiating both sides with respect to x yields:

$$\frac{d}{dx}(x^2 y^3 + x^3 y^4) = \frac{d}{dx}(11)$$

$$\frac{d}{dx}(x^2 y^3) + \frac{d}{dx}(x^3 y^4) = 0$$

Notice:

$$\frac{d}{dx}(x^2 y^3) = x^2\left(3y^2 \cdot \frac{dy}{dx}\right) + y^3(2x)$$

$$= 3x^2 y^2 \cdot \frac{dy}{dx} + 2xy^3$$

and

$$\frac{d}{dx}(x^3 y^4) = x^3\left(4y^3 \cdot \frac{dy}{dx}\right) + y^4(3x^2)$$

$$= 4x^3 y^3 \cdot \frac{dy}{dx} + 3x^2 y^4$$

Therefore,

$$\frac{d}{dx}(x^2 y^3) + \frac{d}{dx}(x^3 y^4) = 0$$

$$3x^2 y^2 \cdot \frac{dy}{dx} + 2xy^3 + 4x^3 y^3 \cdot \frac{dy}{dx} + 3x^2 y^4 = 0$$

Isolating $\dfrac{dy}{dx}$, we have:

$$\left(4x^3 y^3 + 3x^2 y^2\right) \cdot \frac{dy}{dx} = -3x^2 y^4 - 2xy^3$$

$$\frac{dy}{dx} = \frac{-3x^2 y^4 - 2xy^3}{4x^3 y^3 + 3x^2 y^2}$$

$$\frac{dy}{dx} = \frac{xy^2\left(-3xy^2 - 2y\right)}{xy^2\left(4x^2 y + 3x\right)}$$

$$\frac{dy}{dx} = -\frac{2y + 3xy^2}{4x^2 y + 3x}$$

23. Differentiate implicitly to find $\dfrac{dp}{dx}$.

$$p^3 + p - 3x = 50$$

Differentiating both sides with respect to x yields:

$$\frac{d}{dx}(p^3 + p - 3x) = \frac{d}{dx}(50)$$

$$3p^2 \cdot \frac{dp}{dx} + \frac{dp}{dx} - 3 \cdot 1 = 0$$

$$\left(3p^2 + 1\right) \cdot \frac{dp}{dx} = 3$$

$$\frac{dp}{dx} = \frac{3}{3p^2 + 1}$$

25. Differentiate implicitly to find $\dfrac{dp}{dx}$.

$$xp^3 = 24$$

Differentiating both sides with respect to x yields:

$$\frac{d}{dx}\left(xp^3\right) = \frac{d}{dx}(24)$$

$$x\left(3p^2 \cdot \frac{dp}{dx}\right) + p^3\left(1\right) = 0 \qquad \text{Product Rule}$$

$$3xp^2 \cdot \frac{dp}{dx} = -p^3$$

$$\frac{dp}{dx} = \frac{-p^3}{3xp^2}$$

$$\frac{dp}{dx} = -\frac{p}{3x}$$

27. Differentiate implicitly to find $\dfrac{dp}{dx}$.

$$\frac{xp}{x+p} = 2$$

First, we multiply both sides by $x + p$ to clear the fraction.

$$\left(x+p\right)\left(\frac{xp}{x+p}\right) = (2)(x+p)$$

$$xp = 2x + 2p$$

Next, differentiating both sides with respect to x yields:

$$\frac{d}{dx}\left(xp\right) = \frac{d}{dx}\left(2x+2p\right)$$

$$x \cdot \frac{dp}{dx} + p \cdot 1 = 2 \cdot 1 + 2 \cdot \frac{dp}{dx}$$

$$x \cdot \frac{dp}{dx} - 2 \cdot \frac{dp}{dx} = 2 - p$$

$$\left(x-2\right) \cdot \frac{dp}{dx} = 2 - p$$

$$\frac{dp}{dx} = \frac{2-p}{x-2}$$

29. Differentiate implicitly to find $\dfrac{dp}{dx}$.

$$\left(p+4\right)\left(x+3\right) = 48$$

Expanding the left hand side of the equation we have:

$$px + 3p + 4x + 12 = 48$$

$$px + 3p + 4x = 36$$

Differentiating both sides with respect to x yields:

$$\frac{d}{dx}\left(px + 3p + 4x\right) = \frac{d}{dx}(36)$$

$$p \cdot 1 + x \cdot \frac{dp}{dx} + 3 \cdot \frac{dp}{dx} + 4 \cdot 1 = 0$$

$$\left(x+3\right) \cdot \frac{dp}{dx} = -p - 4$$

$$\frac{dp}{dx} = \frac{-p-4}{x+3}$$

31. $A^3 + B^3 = 9$

We differentiate both sides with respect to t.

$$\frac{d}{dt}\left(A^3 + B^3\right) = \frac{d}{dt}(9)$$

$$3A^2 \cdot \frac{dA}{dt} + 3B^2 \cdot \frac{dB}{dt} = 0$$

$$3A^2 \cdot \frac{dA}{dt} = -3B^2 \cdot \frac{dB}{dt}$$

$$\frac{dA}{dt} = \frac{-3B^2}{3A^2} \cdot \frac{dB}{dt}$$

We find B when $A = 2$:

$$A^3 + B^3 = 9$$

$$\left(2\right)^3 + B^3 = 9$$

$$8 + B^3 = 9$$

$$B^3 = 1$$

$$B = 1$$

Next, we substitute 2 for A, 1 for B, and 3 for $\dfrac{dB}{dt}$ into the formula for $\dfrac{dA}{dt}$:

$$\frac{dA}{dt} = \frac{-3B^2}{3A^2} \cdot \frac{dB}{dt}$$

$$= \frac{-3\left(1\right)^2}{3\left(2\right)^2} \cdot \left(3\right)$$

$$= \frac{-3}{12} \cdot 3$$

$$= -\frac{3}{4}$$

33. $R(x) = 50x - 0.5x^2$

Differentiating with respect to time we have:

$$\frac{d}{dt}R(x) = \frac{d}{dt}\left(50x - 0.5x^2\right)$$

$$\frac{dR}{dt} = 50 \cdot \frac{dx}{dt} - x \cdot \frac{dx}{dt}$$

$$\frac{dR}{dt} = \left(50 - x\right) \cdot \frac{dx}{dt}$$

Next, we substitute 30 for x and 20 for dx/dt.

$$\frac{dR}{dt} = (50 - 30) \cdot 20 = (20) \cdot 20 = 400$$

The rate of change of total revenue with respect to time is $400 per day.

$$C(x) = 4x + 10$$

Differentiating with respect to time we have:

$$\frac{d}{dt}C(x) = \frac{d}{dt}(4x + 10)$$

$$\frac{dC}{dt} = 4 \cdot \frac{dx}{dt}$$

Next, we substitute 30 for x and 20 for dx/dt.

$$\frac{dC}{dt} = 4 \cdot (20) = 80$$

The rate of change of total cost with respect to time is $80 per day.

Profit is revenue minus cost. Therefore;

$$P(x) = R(x) - C(x)$$

$$= 50x - 0.5x^2 - (4x + 10)$$

$$= -0.5x^2 + 46x - 10$$

Differentiating with respect to time we have:

$$\frac{d}{dt}P(x) = \frac{d}{dt}(-0.5x^2 + 46x - 10)$$

$$\frac{dP}{dt} = -x \cdot \frac{dx}{dt} + 46\frac{dx}{dt}$$

$$\frac{dP}{dt} = (46 - x) \cdot \frac{dx}{dt}$$

Next, we substitute 30 for x and 20 for dx/dt.

$$\frac{dP}{dt} = (46 - (30)) \cdot (20) = (16)(20) = 320$$

The rate of change of total profit with respect to time is $320 per day.

35. $R(x) = 2x$

Differentiating with respect to time we have:

$$\frac{d}{dt}R(x) = \frac{d}{dt}(2x)$$

$$\frac{dR}{dt} = 2 \cdot \frac{dx}{dt}$$

Next, we substitute 20 for x and 8 for dx/dt.

$$\frac{dR}{dt} = 2 \cdot 8 = 16$$

The rate of change of total revenue with respect to time is $16 per day.

$$C(x) = 0.01x^2 + 0.6x + 30$$

Differentiating with respect to time we have:

$$\frac{d}{dt}C(x) = \frac{d}{dt}(0.01x^2 + 0.6x + 30)$$

$$\frac{dC}{dt} = 0.02x \cdot \frac{dx}{dt} + 0.6 \cdot \frac{dx}{dt}$$

$$\frac{dC}{dt} = (0.02x + 0.6) \cdot \frac{dx}{dt}$$

Next, we substitute 20 for x and 8 for dx/dt.

$$\frac{dC}{dt} = (0.02(20) + 0.6) \cdot 8 = 8$$

The rate of change of total cost with respect to time is $8 per day.

Profit is revenue minus cost. Therefore;

$$P(x) = R(x) - C(x)$$

$$= 2x - (0.01x^2 + 0.6x + 30)$$

$$= -0.01x^2 + 1.4x - 30$$

Differentiating with respect to time we have:

$$\frac{d}{dt}P(x) = \frac{d}{dt}(-0.01x^2 + 1.4x - 30)$$

$$\frac{dP}{dt} = -0.02x \cdot \frac{dx}{dt} + 1.4\frac{dx}{dt}$$

$$\frac{dP}{dt} = (1.4 - 0.02x) \cdot \frac{dx}{dt}$$

Next, we substitute 20 for x and 8 for dx/dt.

$$\frac{dP}{dt} = (1.4 - 0.02(20)) \cdot (8) = 8$$

The rate of change of total profit with respect to time is $8 per day.

37. $5p + 4x + 2px = 60$

First, we take the derivative of both sides of the equation with respect to t.

$$\frac{d}{dt}[5p + 4x + 2px] = \frac{d}{dt}[60]$$

$$5\frac{dp}{dt} + 4\frac{dx}{dt} + 2\underbrace{\left(p \cdot \frac{dx}{dt} + \frac{dp}{dt} \cdot x\right)}_{\text{Product Rule}} = 0$$

$$5\frac{dp}{dt} + 4\frac{dx}{dt} + 2p \cdot \frac{dx}{dt} + 2x \cdot \frac{dp}{dt} = 0$$

Next, we solve for $\frac{dx}{dt}$.

$$4\frac{dx}{dt} + 2p \cdot \frac{dx}{dt} = -5\frac{dp}{dt} - 2x \cdot \frac{dp}{dt}$$

$$(4 + 2p)\frac{dx}{dt} = -(5 + 2x) \cdot \frac{dp}{dt}$$

$$\frac{dx}{dt} = \frac{-(5 + 2x)}{(4 + 2p)} \cdot \frac{dp}{dt}$$

Substituting 3 for x, 5 for p, and 1.5 for $\dfrac{dp}{dt}$, we have:

$$\frac{dx}{dt} = \frac{-(5+2(3))}{(4+2(5))} \cdot (1.5)$$

$$= \frac{-(11)}{14} \cdot (1.5)$$

$$= \frac{-16.5}{14}$$

$$\approx -1.18$$

Sales are changing at a rate of -1.18 sales per day.

39. $A = \pi r^2$

To find the rate of change of the area of the Arctic ice cap with respect to time, we take the derivative of both sides of the equation with respect to t.

$$\frac{d}{dt} A = \frac{d}{dt} \left[\pi r^2 \right]$$

$$\frac{dA}{dt} = \pi \frac{d}{dt} \left[r^2 \right] \qquad \text{Constant Multiple Rule}$$

$$\frac{dA}{dt} = \pi \left[2r \cdot \frac{dr}{dt} \right] \qquad \text{Chain Rule}$$

$$\frac{dA}{dt} = 2\pi r \cdot \frac{dr}{dt}$$

In 2005, the r was 808 miles, and $\dfrac{dr}{dt} = -4.3$ miles per year. Substituting these values into the derivative, we have:

$$\frac{dA}{dt} = 2\pi (808)(-4.3)$$

$$\approx -21,830.2990313$$

$$\approx -21,830$$

Therefore, in 2005 the Arctic ice cap was changing at a rate of $-21,830$ mi^2 per year.

Another way of stating this is to say that the Arctic ice cap was *shrinking* at a rate of 21,830 mi^2/yr.

41. $S = \dfrac{\sqrt{hw}}{60}$

First, we substitute 165 for h, and then we take the derivative of both sides with respect to t.

$$S = \frac{\sqrt{165w}}{60} = \frac{\sqrt{165}}{60} \cdot w^{\frac{1}{2}}$$

$$\frac{d}{dt}[S] = \frac{d}{dt}\left[\frac{\sqrt{165}}{60} \cdot w^{\frac{1}{2}} \right]$$

$$\frac{dS}{dt} = \frac{\sqrt{165}}{60} \cdot \frac{1}{2} w^{-\frac{1}{2}} \cdot \frac{dw}{dt}$$

$$= \frac{\sqrt{165}}{120} \cdot \frac{1}{w^{\frac{1}{2}}} \cdot \frac{dw}{dt}$$

$$= \frac{\sqrt{165}}{120\sqrt{w}} \cdot \frac{dw}{dt}$$

Now, we will substitute 70 for w and -2 for $\dfrac{dw}{dt}$.

$$\frac{dS}{dt} = \frac{\sqrt{165}}{120\sqrt{70}} \cdot (-2)$$

$$\approx -0.0256$$

Therefore, Tom's surface area is changing at a rate of -0.0256 m^2/month. We could also say that Tom's surface area is *decreasing* by 0.0256 m^2/month.

43. $V = \dfrac{p}{4Lv}\left(R^2 - r^2 \right)$

We assume that r, p, L and v are constants

a) Taking the derivative of both sides with respect to t, we have:

$$\frac{dV}{dt} = \frac{d}{dt}\left[\frac{p}{4Lv}\left(R^2 - r^2 \right) \right]$$

$$= \frac{p}{4Lv}\left[\frac{d}{dt} R^2 - \frac{d}{dt} r^2 \right]$$

$$= \frac{p}{4Lv}\left[2R \cdot \frac{dR}{dt} - 0 \right]$$

$$= \frac{pR}{2Lv} \cdot \frac{dR}{dt}$$

Substituting 70 for L, 400 for p and 0.003 for v, we have:

$$\frac{dV}{dt} = \frac{400R}{2(70)(0.003)} \cdot \frac{dR}{dt}$$

$$= \frac{400R}{0.42} \cdot \frac{dR}{dt}$$

$$\approx 952.38R \cdot \frac{dR}{dt}$$

b) Using the derivative in part (a), we substitute 0.00015 for dR/dt and 0.1 for R to get:

$$\frac{dV}{dt} = 952.38(0.1) \cdot (0.00015)$$

$$\approx 0.0143$$

The speed of the persons blood will be increasing at a rate of 0.0143 mm/sec^2.

45. Since the ladder forms a right triangle with the wall and the ground, we know that:

$$x^2 + y^2 = 26^2$$

$$x^2 + y^2 = 676$$

We are looking for $\dfrac{dy}{dt}$. Differentiating both sides with respect to t, we have:

$$\frac{d}{dt}\left[x^2 + y^2\right] = \frac{d}{dt}[676]$$

$$2x\frac{dx}{dt} + 2y\frac{dy}{dt} = 0$$

$$2y\frac{dy}{dt} = -2x\frac{dx}{dt} \qquad \text{Subtracting}$$

$$\frac{dy}{dt} = \frac{-2x}{2y} \cdot \frac{dx}{dt} \qquad \text{Dividing by } 2y$$

$$\frac{dy}{dt} = \frac{-x}{y} \cdot \frac{dx}{dt}$$

The lower end of the wall is being pulled away from the wall at a rate of 5 feet per second; therefore, $\dfrac{dx}{dt} = 5$. When the lower end is 10 feet away from the wall, $x = 10$, and

$$(10)^2 + y^2 = 676$$

$$100 + y^2 = 676$$

$$y^2 = 676 - 100$$

$$y^2 = 576$$

$$y = \pm\sqrt{576}$$

$$y = \pm 24 \qquad \text{Since } y \text{ must be positive.}$$

$$y = 24$$

We substitute 10 for x, 24 for y and 5 for $\dfrac{dx}{dt}$ into the derivative to get:

$$\frac{dy}{dt} = \frac{-x}{y} \cdot \frac{dx}{dt}$$

$$= -\frac{(10)}{(24)} \cdot (5)$$

$$= -\frac{25}{12}$$

$$= -2\tfrac{1}{12}$$

When the lower end of the ladder is 10 feet from the wall, the top of the ladder is moving down the wall at a rate of $-2\tfrac{1}{12}$ feet per second.

47. $V = \dfrac{4}{3}\pi r^3$

Differentiating both sides with respect to t, we have:

$$\frac{dV}{dt} = \frac{d}{dt}\left[\frac{4}{3}\pi r^3\right]$$

$$= \frac{4}{3}\pi \cdot \frac{d}{dt}\left[r^3\right]$$

$$= \frac{4}{3}\pi\left[3r^2\frac{dr}{dt}\right]$$

$$= 4\pi r^2 \cdot \frac{dr}{dt}$$

Next, substituting 0.7 for dr/dt and 7.5 for r, we have:

$$\frac{dV}{dt} = 4\pi(7.5)^2(0.7)$$

$$= 4\pi(56.25)(0.7)$$

$$= 157.5\pi$$

$$\approx 494.8$$

The cantaloupe's volume is changing at the rate of 494.8 cm^3/week.

49. $\dfrac{1}{x^2} + \dfrac{1}{y^2} = 5$

$$x^{-2} + y^{-2} = 5$$

Differentiating both sides with respect to x, we have:

$$\frac{d}{dx}\left[x^{-2}\right]+\frac{d}{dx}\left[y^{-2}\right]=\frac{d}{dx}[5]$$

$$-2x^{-3}-2y^{-3}\frac{dy}{dx}=0$$

$$-2y^{-3}\frac{dy}{dx}=2x^{-3}$$

$$\frac{dy}{dx}=-\frac{x^{-3}}{y^{-3}}$$

$$\frac{dy}{dx}=-\frac{y^{3}}{x^{3}}$$

51. $y^{2}=\dfrac{x^{2}-1}{x^{2}+1}$

Differentiating both sides with respect to x, we have:

$$\frac{d}{dx}\left[y^{2}\right]=\frac{d}{dx}\left[\frac{x^{2}-1}{x^{2}+1}\right]$$

$$2y\frac{dy}{dx}=\frac{\left(x^{2}+1\right)(2x)-\left(x^{2}-1\right)(2x)}{\left(x^{2}+1\right)^{2}}$$

$$2y\frac{dy}{dx}=\frac{2x^{3}+2x-2x^{3}+2x}{\left(x^{2}+1\right)^{2}}$$

$$2y\frac{dy}{dx}=\frac{4x}{\left(x^{2}+1\right)^{2}}$$

$$\frac{dy}{dx}=\frac{4x}{2y\left(x^{2}+1\right)^{2}}$$

$$\frac{dy}{dx}=\frac{2x}{\left(x^{2}+1\right)^{2}}$$

53. $(x-y)^{3}+(x+y)^{3}=x^{5}+y^{5}$

Differentiating both sides with respect to x, we have:

$$\frac{d}{dx}\left[(x-y)^{3}+(x+y)^{3}\right]=\frac{d}{dx}\left[x^{5}+y^{5}\right]$$

$$\frac{d}{dx}(x-y)^{3}+\frac{d}{dx}(x+y)^{3}=\frac{d}{dx}x^{5}+\frac{d}{dx}y^{5}$$

$$3(x-y)^{2}\cdot\frac{d}{dx}(x-y)+3(x+y)^{2}\frac{d}{dx}(x+y)=$$
$$5x^{4}+5y^{4}\frac{dy}{dx}$$

$$3(x-y)^{2}\left(1-\frac{dy}{dx}\right)+3(x+y)^{2}\left(1+\frac{dy}{dx}\right)=$$
$$5x^{4}+5y^{4}\frac{dy}{dx}$$

$$3(x-y)^{2}-3(x-y)^{2}\frac{dy}{dx}+3(x+y)^{2}+$$
$$3(x+y)^{2}\frac{dy}{dx}=5x^{4}+5y^{4}\frac{dy}{dx}$$

$$\left[3(x+y)^{2}-3(x-y)^{2}-5y^{4}\right]\frac{dy}{dx}=$$
$$5x^{4}-3(x-y)^{2}-3(x+y)^{2}$$

$$\frac{dy}{dx}=\frac{5x^{4}-3(x-y)^{2}-3(x+y)^{2}}{3(x+y)^{2}-3(x-y)^{2}-5y^{4}}$$

Simplification will yield:
$$\frac{dy}{dx}=\frac{5x^{4}-6x^{2}-6y^{2}}{12xy-5y^{4}}$$

55. $y^{2}-xy+x^{2}=5$

Differentiate implicitly to find $\dfrac{dy}{dx}$

$$\frac{d}{dx}\left[y^{2}-xy+x^{2}\right]=\frac{d}{dx}[5]$$

$$2y\frac{dy}{dx}-\left[x\frac{dy}{dx}+y\cdot1\right]+2x=0$$

$$(2y-x)\frac{dy}{dx}-y+2x=0$$

$$(2y-x)\frac{dy}{dx}=y-2x$$

$$\frac{dy}{dx}=\frac{y-2x}{2y-x}$$

Differentiate $\dfrac{dy}{dx}$ implicitly to find $\dfrac{d^{2}y}{dx^{2}}$

$$\frac{d}{dx}\left[\frac{dy}{dx}\right] = \frac{d}{dx}\left[\frac{y-2x}{2y-x}\right]$$

$$\frac{d^2y}{dx^2} = \frac{(2y-x)\left(\dfrac{dy}{dx}-2\right) - (y-2x)\left(2\dfrac{dy}{dx}-1\right)}{(2y-x)^2}$$

Simplifying the numerator we have:

$$\frac{d^2y}{dx^2} = \left[\left(2y\frac{dy}{dx} - 4y - x\frac{dy}{dx} + 2x\right) - \right.$$

$$\left.\left(2y\frac{dy}{dx} - y - 4x\frac{dy}{dx} + 2x\right)\right] \div (2y-x)^2$$

$$\frac{d^2y}{dx^2} = \frac{-3y + 3x\dfrac{dy}{dx}}{(2y-x)^2}$$

Substituting $\dfrac{y-2x}{2y-x}$ for $\dfrac{dy}{dx}$

$$\frac{d^2y}{dx^2} = \frac{-3y + 3x\cdot\dfrac{y-2x}{2y-x}}{(2y-x)^2}$$

$$= \frac{-3y\dfrac{2y-x}{2y-x} + 3x\cdot\dfrac{y-2x}{2y-x}}{(2y-x)^2}$$

$$= \frac{-6y^2 + 3xy + 3xy - 6x^2}{(2y-x)^3}$$

$$= \frac{-6y^2 + 6xy - 6x^2}{(2y-x)^3}$$

$$= \frac{-6\left(y^2 - xy + x^2\right)}{(2y-x)^3}$$

57. $x^3 - y^3 = 8$

Differentiate implicitly to find $\dfrac{dy}{dx}$

$$\frac{d}{dx}\left[x^3 - y^3\right] = \frac{d}{dx}[8]$$

$$3x^2 - 3y^2\frac{dy}{dx} = 0$$

$$-3y^2\frac{dy}{dx} = -3x^2$$

$$\frac{dy}{dx} = \frac{-3x^2}{-3y^2}$$

$$\frac{dy}{dx} = \frac{x^2}{y^2}$$

Differentiate $\dfrac{dy}{dx}$ implicitly to find $\dfrac{d^2y}{dx^2}$

$$\frac{d^2y}{dx^2} = \frac{d}{dx}\left[\frac{x^2}{y^2}\right]$$

$$= \frac{y^2\cdot(2x) - x^2\cdot 2y\dfrac{dy}{dx}}{\left(y^2\right)^2}$$

$$= \frac{2xy^2 - 2x^2 y\left[\dfrac{x^2}{y^2}\right]}{y^4} \quad \text{Substituting for } \frac{dy}{dx}$$

$$= \frac{2xy^2 - \dfrac{2x^4}{y}}{y^4}$$

$$= \frac{\dfrac{2xy^2}{1}\cdot\dfrac{y}{y} - \dfrac{2x^4}{y}}{y^4}$$

$$= \frac{\dfrac{2xy^3 - 2x^4}{y}}{y^4}$$

$$= \frac{2xy^3 - 2x^4}{y^5}$$

$$= \frac{2x\left(y^3 - x^3\right)}{y^5}$$

59. $\boxed{tw}$

61.   Using the calculator, we have:

$x^4 = y^2 + x^6$

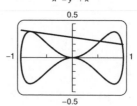

63.   Using the calculator, we have:

$x^3 = y^2(2-x)$

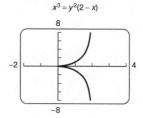

Chapter 3

Exponential and Logarithmic Functions

Exercise Set 3.1

1. Graph: $y = 4^x$

   First, we find some function values.

   $x = -2, y = 4^{-2} = \dfrac{1}{4^2} = \dfrac{1}{16} = 0.0625$

   $x = -1, y = 4^{-1} = \dfrac{1}{4} = 0.25$

   $x = 0, \quad y = 4^0 = 1$

   $x = 1, \quad y = 4^1 = 4$

   $x = 2, \quad y = 4^2 = 16$

x	y
-2	0.0625
-1	0.25
0	1
1	4
2	16

   Next, we plot the points and connect them with a smooth curve.

   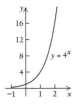

3. Graph: $y = (0.25)^x$

   First note that:

   $y = (0.25)^x$

   $\quad = \left(\dfrac{1}{4}\right)^x$

   $\quad = \left(4^{-1}\right)^x$

   $\quad = 4^{-x}$

   This will ease our work in calculating function values.

$x = -2, y = 4^{-(-2)} = 4^2 = 16$

$x = -1, y = 4^{-(-1)} = 4^1 = 4$

$x = 0, \quad y = 4^{-(0)} = 1$

$x = 1, \quad y = 4^{-(1)} = \dfrac{1}{4} = 0.25$

$x = 2, \quad y = 4^{-(2)} = \dfrac{1}{4^2} = \dfrac{1}{16} = 0.0625$

x	y
-2	16
-1	4
0	1
1	0.25
2	0.0625

Next, we plot the points and connect them with a smooth curve.

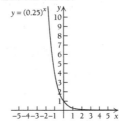

5. Graph: $f(x) = \left(\dfrac{3}{2}\right)^x$

   First, we find some function values.

   $x = -2, f(-2) = \left(\dfrac{3}{2}\right)^{-2} = \dfrac{1}{\left(\dfrac{3}{2}\right)^2} = \dfrac{1}{\left(\dfrac{9}{4}\right)} = \dfrac{4}{9} = .44\overline{4}$

   $x = -1, f(-1) = \left(\dfrac{3}{2}\right)^{-1} = \dfrac{1}{\left(\dfrac{3}{2}\right)} = \dfrac{2}{3} = .66\overline{6}$

   $x = 0, \quad f(0) = \left(\dfrac{3}{2}\right)^0 = 1$

   $x = 1, \quad f(1) = \left(\dfrac{3}{2}\right)^1 = \dfrac{3}{2} = 1.5$

   $x = 2, \quad f(2) = \left(\dfrac{3}{2}\right)^2 = \dfrac{9}{4} = 2.25$

   We organize these values in an input-output table.

x	$f(x)$
-2	0.444
-1	0.666
0	1
1	1.5
2	2.25

Next, we plot the points and connect them with a smooth curve.

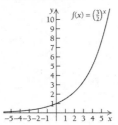

7. Graph: $g(x) = \left(\dfrac{2}{3}\right)^x$

First, we find some function values.

$x = -2, g(-2) = \left(\dfrac{2}{3}\right)^{-2} = \dfrac{1}{\left(\dfrac{2}{3}\right)^2} = \dfrac{1}{\dfrac{4}{9}} = \dfrac{9}{4} = 2.25$

$x = -1, g(-1) = \left(\dfrac{2}{3}\right)^{-1} = \dfrac{1}{\left(\dfrac{2}{3}\right)} = \dfrac{3}{2} = 1.5$

$x = 0, \; g(0) = \left(\dfrac{2}{3}\right)^0 = 1$

$x = 1, \; g(1) = \left(\dfrac{2}{3}\right)^1 = 0.66\overline{6}$

$x = 2, \; g(2) = \left(\dfrac{2}{3}\right)^2 = \dfrac{4}{9} = 0.44\overline{4}$

x	$g(x)$
-2	2.25
-1	1.5
0	1
1	0.666
2	0.444

Next, we plot the points and connect them with a smooth curve.

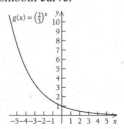

9. Graph: $f(x) = (2.5)^x$

First, we find some function values.

$x = -2, f(-2) = (2.5)^{-2} = \dfrac{1}{(2.5)^2} = \dfrac{1}{6.25} = 0.16$

$x = -1, f(-1) = (2.5)^{-1} = \dfrac{1}{2.5} = 0.4$

$x = 0, \; f(0) = (2.5)^0 = 1$

$x = 1, \; f(1) = (2.5)^1 = 2.5$

$x = 2, \; f(2) = (2.5)^2 = 6.25$

x	$f(x)$
-2	0.16
-1	0.4
0	1
1	2.5
2	6.25

Next, we plot the points and connect them with a smooth curve.

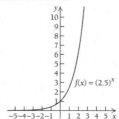

11. $f(x) = e^x$

By Theorem 1, $\dfrac{d}{dx} e^x = e^x$. Therefore,

$f'(x) = e^x$

13. $g(x) = e^{2x}$

Using the chain rule $\dfrac{d}{dx} e^{f(x)} = f'(x)e^{f(x)}$, we have $g'(x) = e^{2x} \cdot 2 = 2e^{2x}$.

15. $f(x) = 6e^x$

Using $\dfrac{d}{dx}\left[c \cdot f(x) \right] = c \cdot f'(x)$, we have:

$f'(x) = 6e^x$.

17. $F(x) = e^{-7x}$

$F'(x) = -7e^{-7x}$

19. $G(x) = 2e^{4x}$

We use $\dfrac{d}{dx}[c \cdot f(x)] = c \cdot f'(x)$:

$G'(x) = \dfrac{d}{dx}\left[2 \cdot e^{4x}\right] = 2 \cdot \dfrac{d}{dx}e^{4x}$

Next, we use the Chain Rule:

$\dfrac{d}{dx}e^{f(x)} = f'(x)e^{f(x)}$; therefore,

$G'(x) = 2 \cdot 4e^{4x}$

$G'(x) = 8e^{4x}$.

21. $f(x) = -3e^{-x}$

$f'(x) = \dfrac{d}{dx}\left[-3 \cdot e^{-x}\right] = -3 \cdot \dfrac{d}{dx}e^{-x}$

$= -3 \cdot e^{-x} \cdot (-1)$

$= 3e^{-x}$

23. $g(x) = \dfrac{1}{2}e^{-5x}$

$g'(x) = \dfrac{d}{dx}\left[\dfrac{1}{2} \cdot e^{-5x}\right] = \dfrac{1}{2} \cdot \dfrac{d}{dx}e^{-5x}$

$= \dfrac{1}{2}e^{-5x} \cdot (-5)$

$= -\dfrac{5}{2}e^{-5x}$

25. $F(x) = -\dfrac{2}{3}e^{x^2}$

$F'(x) = \dfrac{d}{dx}\left[-\dfrac{2}{3} \cdot e^{x^2}\right] = -\dfrac{2}{3} \cdot \dfrac{d}{dx}e^{x^2}$

$= -\dfrac{2}{3} \cdot e^{x^2} \cdot (2x)$

$= -\dfrac{4x}{3}e^{x^2}$

27. $G(x) = 7 + 3e^{5x}$

Using the Sum Rule, we have

$G'(x) = \dfrac{d}{dx}\left[7 + 3e^{5x}\right]$

$= \dfrac{d}{dx}7 + 3\dfrac{d}{dx}e^{5x}$.

To differentiate the two terms remember that the derivative of a constant is zero. All that is left is to apply the Chain Rule.

$G'(x) = 0 + 3 \cdot e^{5x} \cdot (5)$

$= 15e^{5x}$

29. $f(x) = x^5 - 2e^{6x}$

First, we use the Difference Rule

$f'(x) = \dfrac{d}{dx}x^5 - 2\dfrac{d}{dx}e^{6x}$.

Next, we use the Power Rule on the first term and the Chain Rule on the second term

$f'(x) = 5x^4 - 2 \cdot e^{6x} \cdot (6)$

$= 5x^4 - 12e^{6x}$

31. $g(x) = x^5 e^{2x}$

First, we use the Product Rule.

$g'(x) = x^5 \cdot \left[\dfrac{d}{dx}e^{2x}\right] + \left[\dfrac{d}{dx}x^5\right] \cdot e^{2x}$

$= x^5 \cdot e^{2x} \cdot (2) + 5x^4 \cdot e^{2x}$

$= 2x^5 e^{2x} + 5x^4 e^{2x}$

$= (2x + 5)x^4 e^{2x}$ Factoring

33. $F(x) = \dfrac{e^{2x}}{x^4}$

First, we use the Quotient Rule.

$F'(x) = \dfrac{x^4 \cdot (2)e^{2x} - e^{2x} \cdot 4x^3}{x^8}$

$= \dfrac{2x^3 e^{2x}(x - 2)}{x^3 \cdot x^5}$ Factoring

$= \dfrac{x^3}{x^3} \cdot \dfrac{2e^{2x}(x - 2)}{x^5}$ Remove the factor equal to 1.

$= \dfrac{2e^{2x}(x - 2)}{x^5}$

35. $f(x) = (x^2 + 3x - 9)e^x$

$f'(x) = (2x + 3)e^x + (x^2 + 3x - 9)e^x$ Product Rule

$= (2x + 3 + x^2 + 3x - 9)e^x$ Factoring

$= (x^2 + 5x - 6)e^x$

37. $f(x) = \dfrac{e^x}{x^4}$

$f'(x) = \dfrac{x^4 \cdot e^x - 4x^3 \cdot e^x}{x^8}$ By the Quotient Rule

$= \dfrac{x^3 e^x (x-4)}{x^3 \cdot x^5}$ Factoring

$= \dfrac{x^3}{x^3} \cdot \dfrac{e^x (x-4)}{x^5}$ Remove the factor equal to 1.

$= \dfrac{e^x (x-4)}{x^5}$

39. $f(x) = e^{-x^2 + 7x}$

$f'(x) = e^{-x^2 + 7x} \cdot \left[\dfrac{d}{dx} \left(-x^2 + 7x \right) \right]$ By the Chain Rule

$= e^{-x^2 + 7x} \cdot [-2x + 7]$

$= (-2x + 7) \cdot e^{-x^2 + 7x}$

41. $f(x) = e^{-x^2/2}$

$f'(x) = e^{-x^2/2} \cdot \left[\dfrac{d}{dx} \left(-\dfrac{x^2}{2} \right) \right]$

$= e^{-x^2/2} \left[\dfrac{-1}{2}(2x) \right]$

$= -x \cdot e^{-x^2/2}$

43. $y = e^{\sqrt{x-7}}$

First note that

$y = e^{\sqrt{x-7}} = e^{(x-7)^{1/2}}$.

Using the Chain Rule, we have

$\dfrac{dy}{dx} = e^{\sqrt{x-7}} \cdot \left[\tfrac{d}{dx}(x-7)^{1/2} \right]$.

Using the Chain Rule again, we have

$\dfrac{dy}{dx} = e^{\sqrt{x-7}} \cdot \left[\dfrac{1}{2}(x-7)^{-1/2} \right]$

$= e^{\sqrt{x-7}} \cdot \dfrac{1}{2(x-7)^{1/2}}$ Properties of exponents

$= \dfrac{e^{\sqrt{x-7}}}{2\sqrt{x-7}}$

45. $y = \sqrt{e^x - 1}$

$y = \sqrt{e^x - 1} = \left(e^x - 1 \right)^{1/2}$

$\dfrac{dy}{dx} = \dfrac{1}{2} \left(e^x - 1 \right)^{-1/2} \cdot \dfrac{d}{dx} \left(e^x - 1 \right)$ By the Chain Rule

$= \dfrac{1}{2} \left(e^x - 1 \right)^{-1/2} \cdot \left(e^x - 0 \right)$

$= \dfrac{e^x}{2 \left(e^x - 1 \right)^{1/2}}$

$= \dfrac{e^x}{2\sqrt{e^x - 1}}$

47. $y = xe^{-2x} + e^{-x} + x^3$

Differentiate the function term by term. We apply the Product Rule to the first term, the Chain Rule to the second term, and the Power Rule to the third term.

$\dfrac{dy}{dx} = x(-2)e^{-2x} + 1 \cdot e^{-2x} + (-1)e^{-x} + 3x^2$

    Product Rule   Chain Rule   Power Rule

Now we simplify:

$\dfrac{dy}{dx} = -2xe^{-2x} + e^{-2x} - e^{-x} + 3x^2$

$= (1 - 2x)e^{-2x} - e^{-x} + 3x^2$.

49. $y = 1 - e^{-x}$

$\dfrac{dy}{dx} = 0 - e^{-x} \cdot (-1)$

$= e^{-x}$

51. $y = 1 - e^{-kx}$

We use the Chain Rule, remembering that k is a constant:

$\dfrac{dy}{dx} = 0 - e^{-kx} \cdot (-k)$

$= ke^{-kx}$

53. $g(x) = \left(4x^2 + 3x \right)e^{x^2 - 7x}$

We use the Product Rule first, and apply the Chain Rule when taking the derivative of the exponential term.

$g'(x) = \left(4x^2 + 3x \right)(2x - 7)e^{x^2 - 7x} +$

                               Chain Rule

$\left(8x + 3 \right)e^{x^2 - 7x}$

Factoring out the common term, we have

$$g'(x) = \left[\left(4x^2 + 3x\right)\left(2x - 7\right) + \left(8x + 3\right)\right]e^{x^2 - 7x}.$$

Simplifying inside the bracket, yields

$$g'(x) = \left[\left(8x^3 - 22x^2 - 21x\right) + \left(8x + 3\right)\right]e^{x^2 - 7x}$$

$$= \left[8x^3 - 22x^2 - 13x + 3\right]e^{x^2 - 7x}.$$

55. Graph: $f(x) = e^{2x}$

Using a calculator, we first find some function values.

$$f(-2) = e^{2(-2)} = e^{-4} \approx 0.0183$$

$$f(-1) = e^{2(-1)} = e^{-2} \approx 0.1353$$

$$f(0) = e^{2(0)} = e^0 = 1$$

$$f(1) = e^{2(1)} = e^2 \approx 7.3891$$

$$f(2) = e^{2(2)} = e^4 \approx 54.598$$

x	$f(x)$
-2	0.0183
-1	0.1353
0	1
1	7.3891
2	54.598

Next, we plot the points and connect them with a smooth curve.

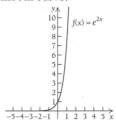

Derivatives. $f'(x) = 2e^{2x}$ and $f''(x) = 4e^{2x}$.

Critical values of f. Since $f'(x) > 0$ for all real numbers x, we know that the derivative exists for all real numbers and there is no solution to the equation $f'(x) = 0$. There are no critical values and therefore no maximum or minimum values.

Increasing. Since $f'(x) > 0$ for all real numbers x, the function f is increasing over the entire real line.

Inflection points. Since $f''(x) > 0$ for all real numbers x, the equation $f''(x) = 0$ has no solution and there are no points of inflection.

Concavity. Since $f''(x) > 0$ for all real numbers x, the function f' is increasing over the entire real line and the graph is concave up over the entire real line.

57. Graph: $g(x) = e^{(1/2)x}$

First, we find some function values.

$$g(-2) = e^{(1/2)(-2)} = e^{-1} \approx 0.3679$$

$$g(-1) = e^{(1/2)(-1)} = e^{-1/2} \approx 0.6065$$

$$g(0) = e^{(1/2)(0)} = e^0 = 1$$

$$g(1) = e^{(1/2)(1)} = e^{1/2} \approx 1.6487$$

$$g(2) = e^{(1/2)(2)} = e^1 \approx 2.7183$$

x	$g(x)$
-2	0.3679
-1	0.6065
0	1
1	1.6487
2	2.7183

Next, we plot the points and connect them with a smooth curve.

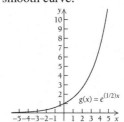

Derivatives. $g'(x) = \frac{1}{2}e^{(1/2)x}$ and

$$g''(x) = \frac{1}{4}e^{(1/2)x}.$$

Critical values of g. Since $g'(x) > 0$ for all real numbers x, we know that the derivative exists for all real numbers and there is no solution to the equation $g'(x) = 0$. There are no critical values and therefore no maximum or minimum values.

Increasing. Since $g'(x) > 0$ for all real numbers x, the function g is increasing over the entire real line.

Inflection points. Since $g''(x) > 0$ for all real numbers x, the equation $g''(x) = 0$ has no solution and there are no points of inflection.

Concavity. Since $g''(x) > 0$ for all real numbers x, the function g' is increasing over the entire real line and the graph is concave up over the entire real line.

59. Graph: $f(x) = \dfrac{1}{2}e^{-x}$

First, we find some function values.

$f(-2) = \dfrac{1}{2}e^{-(-2)} = \dfrac{1}{2}e^2 \approx 3.6945$

$f(-1) = \dfrac{1}{2}e^{-(-1)} = \dfrac{1}{2}e^1 \approx 1.3591$

$f(0) = \dfrac{1}{2}e^{-(0)} = \dfrac{1}{2}e^0 = 0.5$

$f(1) = \dfrac{1}{2}e^{-(1)} = \dfrac{1}{2}e^{-1} \approx 0.1839$

$f(2) = \dfrac{1}{2}e^{-(2)} = \dfrac{1}{2}e^{-2} \approx 0.6767$

x	$f(x)$
-2	3.6945
-1	1.3591
0	0.5
1	0.1839
2	0.6767

Next, we plot the points and connect them with a smooth curve.

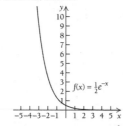

Derivatives. $f'(x) = -\dfrac{1}{2}e^{-x}$ and

$$f''(x) = \dfrac{1}{2}e^{-x}.$$

Critical values of f. Since $f'(x) < 0$ for all real numbers x, we know that the derivative exists for all real numbers and there is no solution to the equation $f'(x) = 0$. There are no critical values and therefore no maximum or minimum values.

Decreasing. Since $f'(x) < 0$ for all real numbers x, the function f is decreasing over the entire real line.

Inflection points. Since $f''(x) > 0$ for all real numbers x, the equation $f''(x) = 0$ has no solution and there are no points of inflection.

Concavity. Since $f''(x) > 0$ for all real numbers x, the function f' is increasing over the entire real line and the graph is concave up over the entire real line.

61. Graph: $F(x) = -e^{\left(\frac{1}{3}\right)x}$

First, we find some function values.

$F(-2) = -e^{\left(\frac{1}{3}\right)(-2)} = -e^{-\frac{2}{3}} \approx -0.5134$

$F(-1) = -e^{\left(\frac{1}{3}\right)(-1)} = -e^{-\frac{1}{3}} \approx -0.7165$

$F(0) = -e^{\left(\frac{1}{3}\right)(0)} = -e^0 = -1$

$F(1) = -e^{\left(\frac{1}{3}\right)(1)} = -e^{\frac{1}{3}} \approx -1.3956$

$F(2) = -e^{\left(\frac{1}{3}\right)(2)} = -e^{\frac{2}{3}} \approx -1.9477$

x	$F(x)$
-2	- 0.5134
-1	- 0.7165
0	- 1
1	- 1.3956
2	- 1.9477

Next, we plot the points and connect them with a smooth curve.

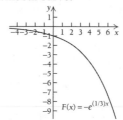

Derivatives $F'(x) = -\dfrac{1}{3}e^{\left(\frac{1}{3}\right)x}$ and

$$F''(x) = -\dfrac{1}{9}e^{\left(\frac{1}{3}\right)x}.$$

Critical values of F. Since $F'(x) < 0$ for all real numbers x, we know that the derivative exists for all real numbers and there is no solution to the equation $F'(x) = 0$. There are no critical values and therefore no maximum or minimum values.

Decreasing. Since $F'(x) < 0$ for all real numbers x, the function F is decreasing over the entire real line.

Inflection points. Since $F''(x) < 0$ for all real numbers x, the equation $F''(x) = 0$ has no solution and there are no points of inflection.

Concavity. Since $F''(x) < 0$ for all real numbers x, the function F' is decreasing over the entire real line and the graph is concave down over the entire real line.

63. Graph: $g(x) = 2(1 - e^{-x})$, for $x \geq 0$

First, we find some function values:

$g(0) = 2(1 - e^{-0}) = 0$

$g(1) = 2(1 - e^{-1}) \approx 1.2642$

$g(2) = 2(1 - e^{-2}) \approx 1.7293$

$g(3) = 2(1 - e^{-3}) \approx 1.9004$

$g(4) = 2(1 - e^{-4}) \approx 1.9634$

$g(5) = 2(1 - e^{-5}) \approx 1.9865$

x	$G(x)$
0	0
1	1.2642
2	1.7293
3	1.9004
4	1.9634
5	1.9865

Next, we plot the points and connect them with a smooth curve.

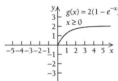

Derivatives. $g'(x) = 2e^{-x}$ and

$$g''(x) = -2e^{-x}.$$

Critical values of g. Since $g'(x) > 0$ for all real numbers x, we know that the derivative exists for all real numbers and there is no solution to the equation $g'(x) = 0$. There are no critical values and therefore no maximum or minimum values.

Increasing. Since $g'(x) > 0$ for all real numbers x, the function g is increasing over the entire real line.

Inflection points. Since $g''(x) < 0$ for all real numbers x, the equation $g''(x) = 0$ has no solution and there are no points of inflection.

Concavity. Since $g''(x) < 0$ for all real numbers x, the function g' is decreasing over the entire real line and the graph is concave down over the entire real line.

65. From exercise 55 we know that:

$f(x) = e^{2x}$

$f'(x) = 2e^{2x}$

$f''(x) = 4e^{2x}$

Enter each function into your graphing calculator.

Set the window on the calculator by pressing the Window button and changing the dimensions of the window.

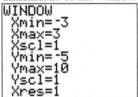

Now press the graph key. The graphs are shown in the screen shot.

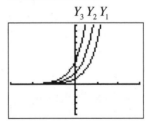

67. From Exercise 57 we know that:

$g(x) = e^{(1/2)x}$

$g'(x) = \frac{1}{2}e^{(1/2)x}$

$g''(x) = \frac{1}{4}e^{(1/2)x}$

Enter each function into your graphing calculator.

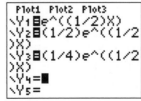

Set the window on the calculator by pressing the Window button and changing the dimensions of the window.

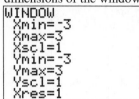

Now press the graph key. The graphs are shown in the screen shot.

69. From Exercise 59 we know that:

$$f(x) = \frac{1}{2}e^{-x}$$

$$f'(x) = -\frac{1}{2}e^{-x}$$

$$f''(x) = \frac{1}{2}e^{-x}$$

Enter each function into your graphing calculator.

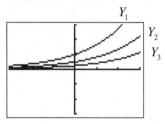

Set the window on the calculator by pressing the Window button and changing the dimensions of the window.

Now press the graph key. The graphs are shown in the screen shot.

$Y_1 = Y_3$

71. From Exercise 61 we know that:

$$F(x) = -e^{(1/3)x}$$

$$F'(x) = -\frac{1}{3}e^{(1/3)x}$$

$$F''(x) = -\frac{1}{9}e^{(1/3)x}$$

Enter each function into your graphing calculator.

Set the window on the calculator by pressing the Window button and changing the dimensions of the window.

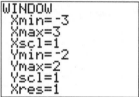

Now press the graph key. The graphs are shown in the screen shot.

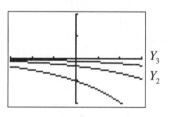

73. From Exercise 63 we know that:

$$g(x) = 2(1 - e^{-x}); \quad x \geq 0$$

$$g'(x) = 2e^{-x}; \qquad x \geq 0$$

$$g''(x) = -2e^{-x}; \qquad x \geq 0$$

Enter each function into your graphing calculator. Notice how you can specify the domain in your calculator. Consult your owners manual for a detailed explanation on how to do this.

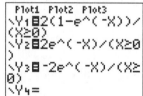

Set the window on the calculator by pressing the Window button and changing the dimensions of the window.

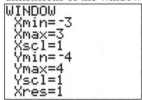

Now press the graph key. The graphs are shown in the screen shot.

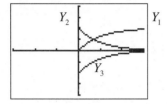

75. First find the derivative of the function.

$$f(x) = e^x$$

$$f'(x) = e^x$$

Now, evaluate the derivative at $x = 0$.

$$f'(0) = e^0 = 1$$

The slope of the tangent line at the point $(0,1)$ is 1.

77. First find the slope of the tangent line at $(0,1)$, by evaluating the derivative at $x = 0$.

$$g(x) = e^{-x}$$

$$g'(x) = -e^{-x}$$

$$g'(0) = -e^{-0} = -1$$

Then we find the equation of the line with slope -1 and containing the point $(0,1)$.

$$y - y_1 = m(x - x_1) \quad \text{Point-slope equation}$$

$$y - 1 = -1(x - 0)$$

$$y - 1 = -x$$

$$y = -x + 1$$

79. Enter the function and the tangent line found in problem 77 into the calculator.

Set your window in order to see the functions.

Press the graph button.

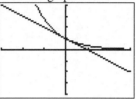

81. a) Evaluate $A(t)$ at $t = 14$. Note: 2009 is 14 years after 1995.

$$A(14) = 2.43e^{0.18(14)}$$

$$= 2.43e^{2.52}$$

$$= 30.201490$$

$$A(14) \approx 30.2$$

Americans will spend approximately $30.2 billion on organic food and beverages in 2009.

b) Find the derivative of the function.

$$A(t) = 2.43e^{0.18t}$$

$$A'(t) = 2.43(0.18)e^{0.18t}$$

$$A'(t) = 0.4374e^{0.18t}$$

Find the rate of change in 2006 by evaluating the derivative at $t = 11$.

$$A'(11) = 0.4374e^{0.18(11)}$$

$$= 0.4374e^{1.98}$$

$$= 3.167976$$

$$A'(t) \approx 3.17$$

American expenditure on organic food and beverages is growing at a rate of $3.17 billion per year in 2006.

83. $T(x) = \dfrac{5200}{1 + 16.3e^{-0.7x}}$

a) In $2005, x = 2005 - 1994 = 11$.

$$T(11) = \frac{5200}{1 + 16.3e^{-0.7(11)}}$$

$$= \frac{5200}{1 + 16.3e^{-7.7}}$$

$$\approx 5162$$

In 2005 there will be approximately 5162 thousand, or 5.162 million, cellular phones in Israel.

In 2010, $x = 2010 - 1994 = 16$.

$$T(16) = \frac{5200}{1 + 16.3e^{-0.7(16)}}$$

$$= \frac{5200}{1 + 16.3e^{-11.2}}$$

$$\approx 5199$$

In 2005 there will be approximately 5199 thousand, or 5.199 million, cellular phones in Israel.

b)   Evaluate the limit of $T(x)$ as $x \to \infty$

$$\lim_{x \to \infty} T(x) = \lim_{x \to \infty} \frac{5200}{1 + 16.3e^{-0.7x}}$$

$$= \frac{\lim_{x \to \infty} 5200}{\lim_{x \to \infty} 1 + \lim_{x \to \infty} 16.3e^{-0.7x}}$$

$$= \frac{5200}{1 + 0}$$

$$= 5200$$

The limit is 5200 thousand cell phones or 5.2 million cell phones. This represents the saturation point for Israel's cellular phone market.

85. a) $C'(t) = 0 - 50(-1)e^{-t}$

$\qquad C'(t) = 50e^{-t}$

b) $C'(0) = 50e^{-0} = 50 \cdot 1 = 50$

Marginal cost when $t = 0$ is 50 million dollars per year.

c) $C'(4) = 50e^{-4} \approx 0.916$

Marginal cost when $t = 4$ is 0.916 million dollars per year or $916,000 per year.

d) $\boxed{tw}$

87. $q = 240e^{-0.003x}$

a) When $x = 250$

$q = 240e^{-0.003(250)}$

$= 240e^{-0.75}$

$q \approx 113$

At a price of $250 the demand for this MP3 player is 113 thousand, or 113,000 units.

b) Create a table of values as needed. Find the function values for q as shown in part 'a'.

x	q
0	240
100	177.796
200	131.715
250	113.368
300	97.577
400	72.287

Next plot the points and connect the lines with a smooth curve.

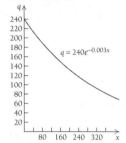

c) Take the derivative of the demand function with respect to x.

$q'(x) = 240(-0.003)e^{-0.003x}$

$\qquad = -0.72e^{-0.003x}$

d) $\boxed{tw}$

89. a) $C(0) = 10 \cdot 0^2 e^{-0} = 0$

The initial concentration is 0 parts per million (ppm).

$C(1) = 10 \cdot 1^2 e^{-1}$

$\qquad \approx 10(0.367879)$

$\qquad \approx 3.68$

After 1 hour, the concentration is approximately 3.7 ppm.

$C(2) = 10 \cdot 2^2 e^{-2}$

$\qquad \approx 40(0.135335)$

$\qquad \approx 5.41$

After 2 hours, the concentration is approximately 5.4 ppm.

$C(3) = 10 \cdot 3^2 e^{-3}$

$\qquad \approx 90(0.049787)$

$\qquad \approx 4.48$

After 3 hours, the concentration is approximately 4.5 ppm.

$C(10) = 10 \cdot 10^2 e^{-10}$

$\qquad \approx 1000(0.000045)$

$\qquad \approx 0.05$

After 10 hours, the concentration is approximately 0.05 ppm.

b) Plot the points $(0,0); (1,3.7); (2,5.4);$

$(3,4.5)$ and $(10,0.05)$ and other points as needed. Then we connect the points with a smooth curve.

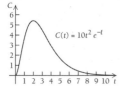

c) $C'(t) = 10t^2(-1)e^{-t} + 20te^{-t}$

$= (20t - 10t^2)e^{-t}$

d) Since $C'(x)$ exist for all values of $x \geq 0$, the only critical values are where $C'(x) = 0$. Solve:

$$C'(x) = 0$$
$$(20t - 10t^2)e^{-t} = 0$$
$$20t - 10t^2 = 0 \qquad (e^{-t} \neq 0)$$
$$10t(2 - t) = 0$$
$$t = 0 \ \ or \ \ t = 2$$

The are two critical values. The critical value $t = 2$ is the obvious maximum. Find $C(2)$.

$$C(2) = 10 \cdot 2^2 e^{-2}$$
$$\approx 40(0.135335)$$
$$\approx 5.41$$

The maximum value of the concentration will be 5.41 parts per million and will occur 2 hours after the drug has been administered.

e) $\boxed{tw}$

91. $D(t) = 34.4 - \dfrac{30.48}{1 + 29.44e^{-0.072t}}$

a) Take the derivative using the quotient rule.

$$D'(t) = 0 - \frac{d}{dt}\left[\frac{30.48}{1 + 29.44e^{-0.072t}}\right]$$

$$= -\frac{\left(1 + 29.44e^{-0.072t}\right)(0) - 29.44e^{-0.072}(-0.072) \cdot 30.48}{\left(1 + 29.44e^{-0.072t}\right)^2}$$

Simplifying the numerator the derivative is

$$D'(t) \approx -\frac{64.6e^{-0.072t}}{\left(1 + 29.44e^{-0.072}\right)^2}$$

The derivative represents the rate of change of the death rate per thousand people per year in Mexico t years after 1990.

b) In 2009, $t = 2009 - 1900 = 109$.

$$D'(109) \approx -\frac{64.6e^{-0.072(109)}}{\left(1 + 29.44e^{-0.072(109)}\right)^2}$$

$$\approx -\frac{64.6e^{-7.848}}{\left(1 + 29.44e^{-7.848}\right)^2}$$

$$\approx -0.025$$

Between 2009 and 2010, Mexico's death rate will decline approximately 0.025 deaths per thousand people.

93. $y = \left(e^{3x} + 1\right)^5$

Using the Extended Power Rule.

$$\frac{dy}{dx} = 5\left(e^{3x} + 1\right)^4 \cdot 3e^{3x}$$

$$= 15e^{3x}\left(e^{3x} + 1\right)^4$$

95. $y = \dfrac{e^{3t} - e^{7t}}{e^{4t}}$

Simplify the expression.

$$y = \frac{e^{3t}\left(1 - e^{4t}\right)}{e^{3t} \cdot e^{t}}$$

$$= \frac{1 - e^{4t}}{e^{t}}$$

Using the Quotient Rule.

$$\frac{dy}{dt} = \frac{e^{t}\left(-4e^{4t}\right) - e^{t}\left(1 - e^{4t}\right)}{\left(e^{t}\right)^2}$$

$$= \frac{e^{t}\left(-4e^{4t} - 1 + e^{4t}\right)}{e^{t} \cdot e^{t}}$$

$$= \frac{-1 - 3e^{4t}}{e^{t}}$$

$$\frac{dy}{dt} = -e^{-t} - 3e^{3t}$$

97. $y = \dfrac{e^x}{x^2+1}$

$\dfrac{dy}{dx} = \dfrac{\left(x^2+1\right)e^x - e^x\left(2x\right)}{\left(x^2+1\right)^2}$ Quotient Rule

$= \dfrac{e^x\left(x^2+1-2x\right)}{\left(x^2+1\right)^2}$ Factoring the Numerator

$= \dfrac{e^x\left(x^2-2x+1\right)}{\left(x^2+1\right)^2}$

$= \dfrac{e^x\left(x-1\right)^2}{\left(x^2+1\right)^2}$

99. $f(x) = e^{\sqrt{x}} + \sqrt{e^x}$

$= e^{x^{1/2}} - e^{x/2}$

$f'(x) = e^{x^{1/2}} \cdot \dfrac{1}{2}x^{-1/2} + \dfrac{1}{2}e^{x/2}$ Chain Rule

$= \dfrac{e^{\sqrt{x}}}{2\sqrt{x}} + \dfrac{\sqrt{e^x}}{2}$

101. $f(x) = e^{x/2} \cdot \sqrt{x-1}$

First, we note that

$f(x) = e^{x/2} \cdot (x-1)^{1/2}$.

Next, we use the Product Rule.

$f'(x) = e^{x/2} \cdot \dfrac{1}{2}(x-1)^{-1/2} + \dfrac{1}{2}e^{x/2}(x-1)^{1/2}$

$= \dfrac{1}{2}e^{x/2}\left[(x-1)^{-1/2} + (x-1)^{1/2}\right]$ Factoring

$= \dfrac{1}{2}e^{x/2}\left[\dfrac{1}{\sqrt{x-1}} + \sqrt{x-1}\right]$

$= \dfrac{1}{2}e^{x/2}\left[\dfrac{1}{\sqrt{x-1}} + \sqrt{x-1}\cdot\dfrac{\sqrt{x-1}}{\sqrt{x-1}}\right]$ Multiplying by 1

$= \dfrac{1}{2}e^{x/2}\left[\dfrac{1}{\sqrt{x-1}} + \dfrac{x-1}{\sqrt{x-1}}\right]$

$= \dfrac{1}{2}e^{x/2}\left[\dfrac{x}{\sqrt{x-1}}\right]$ Adding Fractions

$= e^{x/2}\left[\dfrac{x}{2\sqrt{x-1}}\right]$

103. $f(x) = \dfrac{e^x - e^{-x}}{e^x + e^{-x}}$

Using the Quotient Rule

$f'(x)$

$= \dfrac{\left(e^x + e^{-x}\right)\left(e^x + e^{-x}\right) - \left(e^x - e^{-x}\right)\left(e^x - e^{-x}\right)}{\left(e^x + e^{-x}\right)^2}$

$= \dfrac{\left(e^{2x} + e^0 + e^0 + e^{-2x}\right) - \left(e^{2x} - e^0 - e^0 + e^{-2x}\right)}{\left(e^x + e^{-x}\right)^2}$

$= \dfrac{e^{2x} + e^0 + e^0 + e^{-2x} - e^{2x} + e^0 + e^0 - e^{-2x}}{\left(e^x + e^{-x}\right)^2}$

$= \dfrac{4}{\left(e^x + e^{-x}\right)^2}$

105. $e = \lim\limits_{t\to 0} f(t);\ f(t) = (1+t)^{1/t}$

$f(1) = (1+1)^{1/1} = 2^1 = 2$

$f(0.5) = (1+0.5)^{1/0.5} = 1.5^2 = 2.25$

$f(0.2) = (1+0.2)^{1/0.2} = 1.2^5 = 2.48832$

$f(0.1) = (1+0.1)^{1/0.1} = 1.1^{10} \approx 2.59374$

$f(0.001) = (1+0.001)^{1/0.001} = 1.001^{1000} \approx 2.71692$

107. $f(x) = x^2 e^{-x};\ [0,4]$

$f'(x) = x^2 \cdot (-1)e^{-x} + 2xe^{-x}$

$= -x^2 e^{-x} + 2xe^{-x}$

Since $f'(x)$ exist for all values of x in $[0,4]$, the only critical values are where $f'(x) = 0$.

$f'(x) = 0$

$-x^2 e^{-x} + 2xe^{-x} = 0$

$-e^{-x}\left(x^2 - 2x\right) = 0$

$x^2 - 2x = 0 \quad \left(-e^{-x} \neq 0\right)$

$x(x-2) = 0$

$x = 0 \ \text{ or } \ x = 2$

Use the Max-Min Principle 1. Find the function values at $x = 0, x = 2,$ and $x = 4$.

$f(x) = x^2 e^{-x}$

$f(0) = 0^2 \cdot e^{-0} = 0 \cdot 1 = 0$

$f(2) = 2^2 \cdot e^{-2} = 4 \cdot (0.135335) \approx 0.5413$

$f(4) = 4^2 \cdot e^{-4} = 16(0.018315) \approx 0.2930$

The maximum value of $f(x)$ is $4e^{-2}$ or approximately 0.5413 when $x = 2$.

109. $\boxed{tw}$

111. Graph: $y = x^2 e^{-x}$

From Exercise 107 we see that here is a relative maximum at $\left(2, 4e^{-2}\right)$, or approximately $\left(2, 0.5413\right)$.

Using our work from Exercise 107 and the First-Derivative Test we find that there is a relative minimum at $\left(0, 0\right)$. Also observe that $y \geq 0$ for all x.

For:

$x = -3,\; y = \left(-3\right)^2 e^{-(-3)} = 9e^3 \approx 180.77$

$x = -2,\; y = \left(-2\right)^2 e^{-(-2)} = 4e^2 \approx 29.56$

$x = -1,\; y = \left(-1\right)^2 e^{-(-1)} = 1e^1 \approx 2.72$

$x = 0,\; y = \left(0\right)^2 e^{-(0)} = 0 \cdot 1 = 0$

$x = 1,\; y = \left(1\right)^2 e^{-(1)} = 1e^{-1} \approx 0.37$

$x = 2,\; y = \left(2\right)^2 e^{-(2)} = 4e^{-2} \approx 0.54$

$x = 3,\; y = \left(3\right)^2 e^{-(3)} = 9e^{-3} \approx 0.45$

$x = 4,\; y = \left(4\right)^2 e^{-(4)} = 16e^{-4} \approx 0.29$

$x = 5,\; y = \left(5\right)^2 e^{-(5)} = 25e^{-5} \approx 0.17$

$x = 10,\; y = \left(10\right)^2 e^{-(10)} = 100e^{-10} \approx 0.005$

Next, we plot these points and connect them with a smooth curve.

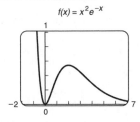

113. See page 310 in the text. The graphs of $f\left(x\right), f'\left(x\right),$ and $f''\left(x\right)$ are all the graph of $y = e^x$.

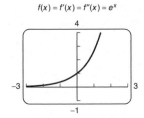

115. $f\left(x\right) = 2e^{0.3x}$

$f'\left(x\right) = 0.6e^{0.3x}$

$f''\left(x\right) = 1.8e^{0.3x}$

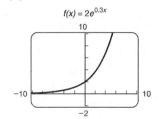

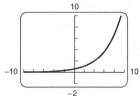

117.

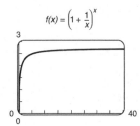

Exercise Set 3.2

---

1. $\log_2 8 = 3$ Logarithmic equation.

  $2^3 = 8$ Exponential equation; 2 is the base, 3 is the exponent.

3. $\log_8 2 = \dfrac{1}{3}$ Logarithmic equation.

  $8^{1/3} = 2$ Exponential equation; 8 is the base, 3 is the exponent.

5. $\log_a K = J$ Logarithmic equation.

     $a^J = K$ Exponential equation;
                          a is the base, J is the exponent.

7. $-\log_{10} h = p$ Logarithmic equation.

     $\log_{10} h = -p$ Multiply by -1.

       $10^{-p} = h$ Exponential equation;
                       10 is the base,
                       $-p$ is the exponent.

9. $e^M = b$ Exponential equation;
                   e is the base, M is the exponent.

     $\log_e b = M$ Logarithmic equation.

     or $\ln b = M$ $\ln b$ is the abbreviation for $\log_e b$.

11. $10^2 = 100$ Exponential equation;
                       10 is the base,
                       2is the exponent.

     $\log_{10} 100 = 2$ Logarithmic equation.

13. $10^{-1} = 0.1$ Exponential equation;
                       10 is the base,
                       -1is the exponent.

     $\log_{10} 0.1 = -1$ Logarithmic equation.

15. $M^p = V$ Exponential equation;
                       M is the base, p is the exponent.

     $\log_M V = p$ Logarithmic equation.

17. $\log_b\left(\dfrac{5}{3}\right) = \log_b 5 - \log_b 3$ (P2)

               $= 1.609 - 1.099$

               $= 0.51$

19. $\log_b 15 = \log_b (3 \cdot 5)$

            $= \log_b 3 + \log_b 5$ (P1)

            $= 1.099 + 1.609$

            $= 2.708$

21. $\log_b 5b = \log_b 5 + \log_b b$ (P1)

           $= 1.609 + 1$ (P5)

           $= 2.609$

23. $\ln 20 = \ln(4 \cdot 5)$

           $= \ln 4 + \ln 5$ (P1)

           $= 1.3863 + 1.6094$

           $= 2.9957$

25. $\ln\left(\dfrac{1}{5}\right) = \ln 1 - \ln 5$ (P2)

           $= 0 - 1.6094$ (P6)

           $= -1.6094$

27. $\ln\sqrt{e^8} = \ln e^{8/2}$

           $= \ln e^4$

           $= 4$ (P5)

29. Using a calculator and rounding to six decimal places, we have

     $\ln 5894 \approx 8.681690$.

31. $\ln 0.0182 \approx -4.006334$

33. $\ln 8100 \approx 8.999619$

35. $e^t = 80$

     $\ln e^t = \ln 80$ Taking the natural log on
                            both sides

         $t = \ln 80$ (P5)

         $t \approx 4.382027$ Using a calculator

         $t \approx 4.382$

37. $e^{2t} = 1000$

     $\ln e^{2t} = \ln 1000$ Taking the natural log on
                            both sides

         $2t = \ln 1000$ (P5)

         $t = \dfrac{\ln 1000}{2}$

         $t \approx \dfrac{6.907755}{2}$ Using a calculator

         $t \approx 3.453878$

         $t \approx 3.454$

39. $e^{-t} = 0.1$

     $\ln e^{-t} = \ln 0.1$ Taking the natural log on
                            both sides

       $-t = \ln 0.1$ (P5)

         $t = -\ln 0.1$

         $t \approx 2.302585$ Using a calculator

         $t \approx 2.303$

41. $e^{-0.02t} = 0.06$

$\ln e^{-0.02t} = \ln 0.06$ Taking the natural log on both sides

$-0.02t = \ln 0.06$ (P5)

$t = \dfrac{\ln 0.06}{-0.02}$

$t \approx \dfrac{-2.813411}{-0.02}$ Using a calculator

$t \approx 140.671$

43. $y = -8 \ln x$

$\dfrac{dy}{dx} = -8 \cdot \dfrac{1}{x}$ $\left[\frac{d}{dx} \big[c \cdot f(x) \big] = c \cdot f'(x) \right]$

$\dfrac{dy}{dx} = \dfrac{-8}{x}$

45. $y = x^4 \ln x - \frac{1}{2} x^2$

Differentiate this function term by term. We apply the Product Rule to the first term, and the Power Rule to the second term.

$\dfrac{dy}{dx} = \underbrace{x^4 \cdot \dfrac{1}{x} + 4x^3 \cdot \ln x}_{\text{Product Rule}} - \underbrace{\dfrac{1}{2} \cdot 2x}_{\text{Power Rule}}$

$\dfrac{dy}{dx} = x^3 + 4x^3 \ln x - x$

47. $f(x) = \ln(6x)$

Using Theorem 7, we have

$\dfrac{d}{dx} \ln f(x) = \dfrac{1}{f(x)} \cdot f'(x) = \dfrac{f'(x)}{f(x)}$

$f'(x) = \dfrac{1}{6x} \cdot 6$

$= \dfrac{1}{x}$

49. $g(x) = x^2 \ln(7x)$

We apply the Product Rule. Remember to use Theorem 7 when taking the derivative of the logarithm.

$g'(x) = x^2 \cdot \underbrace{\dfrac{1}{7x} \cdot 7}_{\text{Theorem 7}} + 2x \cdot \ln(7x)$

$\qquad\qquad \text{Product Rule}$

$g'(x) = x^2 \cdot \dfrac{1}{x} + 2x \ln(7x)$

$= x + 2x \ln(7x)$

51. $y = \dfrac{\ln x}{x^4}$

Using the Quotient Rule, we have

$\dfrac{dy}{dx} = \dfrac{x^4 \cdot \frac{1}{x} - 4x^3 \cdot \ln x}{x^8}$

$= \dfrac{x^3 - 4x^3 \ln x}{x^8}$

$= \dfrac{x^3(1 - 4\ln x)}{x^3 \cdot x^5}$ Factoring

$= \dfrac{x^3}{x^3} \cdot \dfrac{1 - 4\ln x}{x^5}$ Removing a factor equal to 1

$= \dfrac{1 - 4\ln x}{x^5}$

53. $y = \ln\left(\dfrac{x^2}{4}\right)$

$y = \ln x^2 - \ln 4$ (P2)

$\dfrac{dy}{dx} = \dfrac{1}{x^2} \cdot 2x - 0$ Theorem 7

$= \dfrac{2}{x}$

55. $y = \ln(3x^2 + 2x - 1)$

Using Theorem 7, we note:

$\dfrac{d}{dx} \ln g(x) = \dfrac{g'(x)}{g(x)}$

$g(x) = 3x^2 + 2x - 1$

$g'(x) = 6x + 2$

Therefore,

$\dfrac{dy}{dx} = \dfrac{6x + 2}{3x^2 + 2x - 1}$

$\dfrac{dy}{dx} = \dfrac{2(3x + 1)}{3x^2 + 2x - 1}$

57. $f(x) = \ln\left(\dfrac{x^2 - 7}{x}\right)$

Using the Theorem 7.

$f'(x) = \dfrac{\left(\dfrac{d}{dx}\left[\dfrac{x^2 - 7}{x} \right] \right)}{\dfrac{x^2 - 7}{x}}$

$= \left(\dfrac{d}{dx}\left[\dfrac{x^2 - 7}{x} \right] \right) \cdot \dfrac{x}{x^2 - 7}$ Dividing fractions

We apply the Quotient Rule to take the derivative of the inside function to get

$$\frac{d}{dx}\left[\frac{x^2-7}{x}\right] = \frac{x \cdot 2x - (x^2-7)\cdot 1}{x^2}$$

$$= \frac{2x^2 - x^2 + 7}{x^2}$$

$$= \frac{x^2 + 7}{x^2}$$

Now we substitute back into the original derivative.

$$f'(x) = \frac{x^2+7}{x^2} \cdot \frac{x}{x^2-7}$$

$$= \frac{x^2+7}{x(x^2-7)}$$

An alternate solution to this problem involves using property P2.

$$f(x) = \ln\left(\frac{x^2-7}{x}\right)$$

$$= \ln(x^2-7) - \ln x \qquad \text{(P2)}$$

$$= \frac{2x}{(x^2-7)} - \frac{1}{x} \qquad \text{Theorem 7}$$

$$= \frac{2x}{x^2-7} - \frac{1}{x}$$

Combine the fractions by finding a common denominator.

$$f'(x) = \frac{2x}{x^2-7} \cdot \frac{x}{x} - \frac{1}{x} \cdot \frac{x^2-7}{x^2-7} \qquad \text{Multiply by 1}$$

$$= \frac{2x^2}{x(x^2-7)} - \frac{x^2-7}{x(x^2-7)}$$

$$= \frac{x^2+7}{x(x^2-7)}$$

59. $g(x) = e^x \ln x^2$

$$g'(x) = e^x \cdot \frac{2x}{x^2} + e^x \cdot \ln x^2 \qquad \text{By the Product Rule}$$

$$\underset{\text{Theorem 7}}{}$$

$$= e^x \cdot \frac{2}{x} + e^x \ln x^2$$

$$= \frac{2e^x}{x} + 2e^x \ln x \qquad \text{(P3)}$$

61. $f(x) = \ln(e^x + 1)$

$$f'(x) = \frac{(e^x + 0)}{e^x + 1} \qquad \text{Theorem 7}$$

$$= \frac{e^x}{e^x + 1}$$

63. $g(x) = (\ln x)^4$

Using the Extended Power Rule, we have

$$g'(x) = 4(\ln x)^3 \cdot \frac{1}{x}$$

$$= \frac{4(\ln x)^3}{x}$$

65. $f(x) = \ln(\ln(8x))$

First we apply Theorem 7.

$$f'(x) = \frac{1}{\ln(8x)} \cdot \left(\frac{d}{dx}\ln(8x)\right)$$

Using Theorem 7 again, we have

$$f'(x) = \frac{1}{\ln(8x)} \cdot \frac{1}{8x} \cdot 8$$

$$= \frac{1}{x\ln(8x)}$$

67. $g(x) = \ln(5x) \cdot \ln(3x)$

Using the Product Rule along with Theorem 7, we have

$$g'(x) = \ln(5x)\cdot\left(\frac{3}{3x}\right) + \left(\frac{5}{5x}\right)\cdot \ln(3x)$$

$$\underset{\text{Theorem 7} \qquad \text{Theorem 7}}{}$$

$$= \ln(5x)\frac{1}{x} + \frac{1}{x}\ln(3x)$$

$$= \frac{\ln(5x) + \ln(3x)}{x}$$

If we wanted to simplify the expression, we could use Property 1 to combine the logarithms.

$$g'(x) = \frac{\ln(5x \cdot 3x)}{x} = \frac{\ln(15x^2)}{x}$$

69. First, we find the point of tangency by evaluating the function at $x = 2$.

$$x = 2;$$

$$y = (2^2 - 2)\ln(6\cdot 2) = 2\ln(12) \approx 4.9698$$

Point of tangency: $(2, 4.9698)$

Now, we find the slope of the function at $x = 2$ by taking the derivative.

$$\frac{dy}{dx} = \left(x^2 - x\right)\left(6 \cdot \frac{1}{6x}\right) + (2x - 1)\ln(6x)$$

$$= \left(x^2 - x\right)\left(\frac{1}{x}\right) + (2x - 1)\ln(6x)$$

$$= x - 1 + (2x - 1)\ln(6x)$$

Next, we evaluate the derivative at $x = 2$

$$\left.\frac{dy}{dx}\right|_{x=2} = 2 - 1 + (2 \cdot 2 - 1)\ln(6 \cdot 2)$$

$$= 1 + 3\ln(12) \approx 8.4547$$

Now we have the slope and a point on the tangent line. We use the Point-Slope formula to find the equation of the tangent line.

$$y - y_1 = m(x - x_1)$$

$$y - 4.970 = 8.455(x - 2)$$

$$y - 4.970 = 8.455x - 16.91$$

$$y = 8.455x - 11.94$$

71. First, we find the point of tangency by evaluating the function at $x = 3$.

$x = 3$;

$$y = (\ln 3)^2 \approx 1.207$$

Point of tangency: $(3, 1.207)$

Now, we find the slope of the function at $x = 3$ by taking the derivative of the function using the Extended Power Rule.

$$\frac{dy}{dx} = 2\left(\ln(x)\right) \cdot \frac{1}{x}$$

$$= \frac{2\ln x}{x}$$

Next, we evaluate the derivative at $x = 3$

$$\left.\frac{dy}{dx}\right|_{x=3} = \frac{2\ln(3)}{3} \approx 0.732$$

Now we have the slope and a point on the tangent line. We use the Point-Slope formula to find the equation of the tangent line.

$$y - y_1 = m(x - x_1)$$

$$y - 1.207 = 0.732(x - 3)$$

$$y - 1.207 = 0.732x - 2.196$$

$$y = 0.732x - 0.989$$

73. $N(a) = 2000 + 500\ln a, \ a \geq 1$

a) We substitute 1 in for a.

$$N(1) = 2000 + 500 \cdot \ln 1$$

$$= 2000 + 500 \cdot 0$$

$$= 2000$$

Thus, 2000 units were sold after spending $1000 on advertising.

b) Taking the derivative with respect to a, we have

$$N'(a) = 0 + 500 \cdot \frac{1}{a}$$

$$= \frac{500}{a} .$$

Therefore,

$$N'(10) = \frac{500}{10} = 50 .$$

c) $N'(a) > 0$ for all $a \geq 1$. Thus $N(a)$ is an increasing function and has a minimum value of 2000 when $a = 1$ thousand dollars. There is no maximum, because there is not an upper limit on the advertising budget.

d) $\boxed{tw}$

75. How long should the campaign last in order to maximize profit? We need to find the revenue and cost functions in order to find the profit. We find the revenue function first.

$$R(t) = \begin{pmatrix} \text{Price} \\ \text{Per} \\ \text{Unit} \end{pmatrix} \cdot \begin{pmatrix} \text{Target} \\ \text{Market} \end{pmatrix} \cdot \begin{pmatrix} \text{Percentage} \\ \text{Buying} \end{pmatrix}$$

$$R(t) = 0.50(1,000,000)\left(1 - e^{-0.04t}\right)$$

$$= 500,000 - 500,000e^{-0.04t}$$

Next, we find the cost function.

$$C(t) = \begin{pmatrix} \text{Advertising cost} \\ \text{per day} \end{pmatrix} \cdot \begin{pmatrix} \text{Number of} \\ \text{days} \end{pmatrix}$$

$$C(t) = 2000 \cdot t$$

Now, we find the profit function.

$$P(t) = R(t) - C(t)$$

$$= 500,000 - 500,000e^{-0.04t} - 2000t$$

We take the derivative of the profit function.

$$P'(t) = 0 - 500,000(-0.04)e^{-0.04t} - 2000$$

$$= 20,000e^{-0.04t} - 2000$$

Next, we set the derivative of the profit function equal to zero and solve for t to find the critical values.

$$P'(t) = 0$$

$$20,000e^{-0.04t} - 2000 = 0$$

$$20,000e^{-0.04t} = 2000$$

$$e^{-0.04t} = \frac{2000}{20,000}$$

$$e^{-0.04t} = 0.1$$

$$\ln\left(e^{-0.04t}\right) = \ln(0.1)$$

$$-0.04t = -2.30258$$

$$t \approx \frac{-2.30258}{-0.04} \approx 57.565$$

Rounding up, we see that the only critical value is 58 days. We will apply the second derivative test to see if we have a maximum.

The second derivative of the profit function is $P''(t) = -800e^{-0.04t}$.

Evaluating this at the critical value we get $P''(58) = -800e^{-0.04 \cdot 58} < 0$ so we have a maximum.

The length of the advertising campaign must be 58 days to result in maximum profit.

77. $V(t) = 58\left(1 - e^{-1.1t}\right) + 20$

Where $V(t)$ is the value of the stock after time t, in months.

a) $V(1) = 58\left(1 - e^{-1.1(1)}\right) + 20$

$$= 58(1 - 0.332871) + 20$$

$$= 58(0.667129) + 20$$

$$= 38.693482 + 20$$

$$= 58.693482$$

$$\approx 58.69$$

The value of the stock one month after purchase is \$58.69

$$V(12) = 58\left(1 - e^{-1.1(12)}\right) + 20$$

$$= 58(1 - 0.000002) + 20$$

$$= 58(0.999998) + 20$$

$$= 57.999884 + 20$$

$$= 77.999884$$

$$\approx 78.00$$

The value of the stock 12 months after purchase is \$78.00

b) $V'(t) = 58\left(0 - (-1.1)e^{-1.1t}\right) + 0$

$$= 58\left(1.1e^{-1.1t}\right)$$

$$V'(t) = 63.8e^{-1.1t}$$

c) Solve: $V(t) = 75$

$$58\left(1 - e^{-1.1t}\right) + 20 = 75$$

$$58\left(1 - e^{-1.1t}\right) = 55$$

$$\left(1 - e^{-1.1t}\right) = \frac{55}{58}$$

$$1 - e^{-1.1t} = 0.948276$$

$$-e^{-1.1t} = -0.051724$$

$$e^{-1.1t} = 0.051724$$

$$\ln e^{-1.1t} = \ln(0.051724)$$

$$-1.1t = -2.96183$$

$$t = \frac{-2.96183}{-1.1}$$

$$t \approx 2.6926$$

The stock will first reach \$75 approximately 2.7 months after the purchase.

d) $\boxed{tw}$

79. a) Taking the derivative of the profit function $P(x) = 2x - 0.3x \ln x$ will give us marginal profit.

$$P'(x) = 2 - \left[0.3x\left(\frac{1}{x}\right) + 0.3 \ln x\right]$$

$$= 2 - 0.3 - 0.3 \ln x$$

$$= 1.7 - 0.3 \ln x$$

b) $\boxed{tw}$

c) Since $P'(x)$ is defined for all values of $x > 0$. We find the critical values by setting $P'(x) = 0$.

$$1.7 - 0.3 \ln x = 0$$

$$-0.3 \ln x = -1.7$$

$$\ln x = \frac{-1.7}{-0.3}$$

$$\ln x = 5.666667$$

$$x = e^{5.66667}$$

$$x \approx 289.069$$

Notice that:

$$P''(x) = \frac{-0.3}{x} < 0 \text{ for all values of } x > 0$$

The critical value results in a maximum.

Therefore, 289,069 mechanical pencils should be sold to maximize profit.

81. $R = A \ln r - Br$

$$\frac{dR}{dr} = \frac{A}{r} - B$$

Since $\frac{dR}{dr}$ exists for all $r > 0$. Therefore we

find the critical values by solving $\frac{dR}{dr} = 0$.

$$\frac{dR}{dr} = 0$$

$$\frac{A}{r} - B = 0$$

$$\frac{A}{r} = B$$

$$\frac{A}{B} = r$$

The critical value occurs when $r = \frac{A}{B}$.

Using the Second-Derivative Test to see if it is a maximum.

$$\frac{d^2R}{dr^2} = -Ar^{-2} = \frac{-A}{r^2} < 0 \text{ for all } r > 0.$$

Thus, we have maximum at $r = \frac{A}{B}$. The

maximum function value will be:

$$R = A \ln\left(\frac{A}{B}\right) - B\left(\frac{A}{B}\right)$$

$$= A(\ln A - \ln B) - A$$

$$= A \ln\left(\frac{A}{B}\right) - A$$

or

$$= A \ln A - A \ln B - A$$

83. $S(t) = 78 - 15\ln(t+1), \quad t \geq 0$

a) $S(0) = 78 - 15\ln(0+1)$

$$= 78 - 15\ln(1)$$

$$= 78 - 15 \cdot 0$$

$$= 78$$

The average score when they initially took the exam was 78%.

b) $S(4) = 78 - 15\ln(4+1)$

$$= 78 - 15\ln(5)$$

$$= 78 - 15(1.609438)$$

$$= 78 - 24.141569$$

$$\approx 53.9$$

The average score 4 months after they took the exam was 53.9%.

c) $S(24) = 78 - 15\ln(24+1)$

$$= 78 - 15\ln(25)$$

$$= 78 - 15(3.218876)$$

$$= 78 - 48.283137$$

$$\approx 29.7$$

The average score 24 months after they took the exam was 29.7%.

d) First, we reword the question:
"29.7 (the average score after 24 months) is what percent of 78 (the average score on the initial exam)?
Next, we translate the sentence into an equation and solve:

$$29.7 = x \cdot 78$$

$$\frac{29.7}{78} = x$$

$$0.381 = x$$

$$38.1\% \approx x$$

The students retained approximately 38.1% of their original answers after 2 years.

e) $S(t) = 78 - 15\ln(t+1)$

$$S'(t) = 0 - 15(1)\left(\frac{1}{t+1}\right)$$

$$= -\frac{15}{t+1}$$

f) $S'(t) < 0$ for all values of $t \geq 0$. Therefore, $S(t)$ is a decreasing function and has a maximum value of 78% when $t = 0$. The function does not have a minimum value.

g) $\boxed{tw}$

85. $W(t) = 100\left(1 - e^{-0.3t}\right)$

a) $W(1) = 100\left(1 - e^{-0.3(1)}\right)$

   $= 100\left(1 - e^{-0.3}\right)$

   $= 100\left(1 - 0.740818\right)$

   $= 100\left(0.259182\right)$

   $= 25.91812$

   ≈ 26

   After 1 week of practice, a keyboarder will type approximately 26 words per minute.

   $W(8) = 100\left(1 - e^{-0.3(8)}\right)$

   $= 100\left(1 - e^{-2.4}\right)$

   $= 100\left(1 - 0.090718\right)$

   $= 100\left(0.909282\right)$

   $= 90.9282$

   ≈ 91

   After 8 weeks of practice, a keyboarder will type approximately 91 words per minute.

b) $W(t) = 100\left(1 - e^{-0.3t}\right)$

   $W'(t) = 100\left[-\left(-0.3\right)e^{-0.3t}\right]$

   $= 30e^{-0.3t}$

c) Solve:

$$W(t) = 95$$

$$100\left(1 - e^{-0.3t}\right) = 95$$

$$\left(1 - e^{-0.3t}\right) = 0.95 \qquad \text{Divide by 100}$$

$$-e^{-0.3t} = -0.05 \qquad \text{Subtract 1}$$

$$e^{-0.3t} = 0.05 \qquad \text{Multiply by -1}$$

$$\ln\left(e^{-0.3t}\right) = \ln\left(0.05\right) \qquad \text{Taking the}$$

Natural log on both sides

$$-0.3t = \ln\left(0.05\right)$$

$$t = \frac{\ln\left(0.05\right)}{-0.3} \qquad \text{Simplifying}$$

$$t \approx 9.985774$$

$$t \approx 10.0$$

It will take a keyboarder approximately 10 weeks of practice in order to type 95 words per minute.

d) $\boxed{tw}$

87. $P = P_0 e^{kt}$

$\dfrac{P}{P_0} = e^{kt} \qquad$ Divide by P_0

$\ln\left(\dfrac{P}{P_0}\right) = \ln\left(e^{kt}\right) \qquad$ Take the Natural Log on both sides

$\ln\left(\dfrac{P}{P_0}\right) = kt \qquad$ (P5)

$\dfrac{\ln\left(\dfrac{P}{P_0}\right)}{k} = t \qquad$ Divide by k.

89. $f(t) = \ln\left(t^2 - t\right)^7$

$f(t) = 7\ln\left(t^2 - t\right) \qquad$ (P3)

$f'(t) = 7\left(\dfrac{2t - 1}{t^2 - t}\right) \qquad$ By Theorem 7

$f'(t) = \dfrac{7(2t - 1)}{t^2 - t} \qquad$ Simplifying

91. $f(x) = \ln\left[\ln\left(\ln\left(3x\right)\right)\right]$

Differentiating this function will require several applications of the Chain Rule. We will start on the outside and work our way in to the middle.

$$f'(x) = \left(\frac{1}{\ln\left(\ln\left(3x\right)\right)}\right) \cdot \frac{d}{dx}\left(\ln\left(\ln\left(3x\right)\right)\right)$$

$$= \left(\frac{1}{\ln\left(\ln\left(3x\right)\right)}\right) \cdot \left[\left(\frac{1}{\ln\left(3x\right)}\right) \cdot \frac{d}{dx}\left(\ln\left(3x\right)\right)\right]$$

$$= \left(\frac{1}{\ln\left(\ln\left(3x\right)\right)}\right) \cdot \left(\frac{1}{\ln\left(3x\right)}\right)\left[\frac{1}{3x} \cdot 3\right]$$

$$= \frac{1}{x\ln\left(3x\right) \cdot \ln\left(\ln\left(3x\right)\right)}$$

93. $f(t) = \ln\dfrac{1-t}{1+t}$

$f(t) = \ln(1-t) - \ln(1+t)$ (P2)

$f'(t) = \dfrac{(-1)}{1-t} - \dfrac{(1)}{1+t}$ By Theorem 7

$= \dfrac{-1}{1-t} - \dfrac{1}{1+t}$

$= \dfrac{1}{(-1)(1-t)} - \dfrac{1}{1+t}$ Properties of fractions

$= \dfrac{1}{(t-1)} - \dfrac{1}{t+1}$ Distribution

Next, we find a common denominator.

$= \dfrac{t+1}{(t-1)(t+1)} - \dfrac{t-1}{(t-1)(t+1)}$

$f'(t) = \dfrac{2}{t^2 - 1}$

95. $f(x) = \log_5 x$

Using (P7), the change-of-base formula.

$f(x) = \dfrac{\ln x}{\ln 5}$

Remember that $\ln 5$ is a constant.

$f'(x) = \dfrac{1}{\ln 5} \cdot \dfrac{1}{x}$

$= \dfrac{1}{x \ln 5}$

97. $y = \ln\sqrt{5 + x^2}$

$y = \ln\left(5 + x^2\right)^{\frac{1}{2}}$

$y = \tfrac{1}{2}\ln\left(5 + x^2\right)$ (P3)

$\dfrac{dy}{dx} = \dfrac{1}{2} \cdot \dfrac{2x}{5 + x^2}$ Theorem 7

$\dfrac{dy}{dx} = \dfrac{x}{5 + x^2}$

99. $f(x) = \tfrac{1}{5}x^5\left(\ln x - \dfrac{1}{5}\right)$

Using the Product Rule.

$f'(x) = \dfrac{1}{5}x^5\left(\dfrac{1}{x} - 0\right) + \dfrac{1}{5}\left(5x^4\right)\left(\ln x - \dfrac{1}{5}\right)$

$= \dfrac{1}{5}x^4 + x^4\ln x - \dfrac{1}{5}x^4$

$= x^4\ln x$

101. $f(x) = \ln\dfrac{1+\sqrt{x}}{1-\sqrt{x}}$

$f(x) = \ln\left(1+\sqrt{x}\right) - \ln\left(1-\sqrt{x}\right)$ (P2)

We use theorem 7, and the Chain Rule to differentiate both logarithms.

Note: $\dfrac{d}{dx}\left(1+\sqrt{x}\right) = \dfrac{1}{2}x^{-\frac{1}{2}}$

$\dfrac{d}{dx}\left(1-\sqrt{x}\right) = -\dfrac{1}{2}x^{-\frac{1}{2}}$

$f'(x) = \dfrac{\left(\frac{1}{2}\right)x^{-\frac{1}{2}}}{1+\sqrt{x}} - \dfrac{\left(-\frac{1}{2}\right)x^{-\frac{1}{2}}}{1-\sqrt{x}}$

Simplifying

$f'(x) = \dfrac{1}{1+\sqrt{x}} \cdot \dfrac{1}{2x^{\frac{1}{2}}} + \dfrac{1}{1-\sqrt{x}} \cdot \dfrac{1}{2x^{\frac{1}{2}}}$

$= \dfrac{1}{2\sqrt{x}\left(1+\sqrt{x}\right)} + \dfrac{1}{2\sqrt{x}\left(1-\sqrt{x}\right)}$

Now, we combine the fractions.

$f'(x)$

$= \dfrac{1}{2\sqrt{x}\left(1+\sqrt{x}\right)} \cdot \dfrac{1-\sqrt{x}}{1-\sqrt{x}} + \dfrac{1}{2\sqrt{x}\left(1-\sqrt{x}\right)} \cdot \dfrac{1+\sqrt{x}}{1+\sqrt{x}}$

$= \dfrac{1-\sqrt{x}}{2\sqrt{x}\left(1+\sqrt{x}\right)\left(1-\sqrt{x}\right)} + \dfrac{1+\sqrt{x}}{2\sqrt{x}\left(1+\sqrt{x}\right)\left(1-\sqrt{x}\right)}$

$= \dfrac{2}{2\sqrt{x}\left(1+\sqrt{x}\right)\left(1-\sqrt{x}\right)}$

$= \dfrac{1}{\sqrt{x}\left(1+\sqrt{x}\right)\left(1-\sqrt{x}\right)}$

or, if we multiply the two binomials

$f'(x) = \dfrac{1}{\sqrt{x}\left(1-x\right)}$.

103. Let $X = \log_a M$ and $Y = \log_a N$

Proof of Property 1:

$M = a^X$ and $N = a^Y$ Definition of Logarithm

So,

$MN = a^X a^Y = a^{X+Y}$ Product Rule for Exponents

Thus,

$\log_a(MN) = X + Y$ Definition of Logarithm

$= \log_a M + \log_a N$ Substitution

105. Proof of Property 3:

$$M = a^X \qquad \text{Definition of Logarithm}$$

So,

$$M^k = \left(a^X\right)^k \quad \left[a = b \Rightarrow a^c = b^c\right]$$

$$= a^{X \cdot k} \qquad \text{Power Rule for Exponents}$$

Thus,

$$\log_a\left(M^k\right) = X \cdot k \qquad \text{Definition of Logarithm}$$

$$= k \cdot \log_a M \qquad \text{Substitution}$$

107. $\lim\limits_{h \to 0} \dfrac{\ln(1+h)}{h}$

Note: $\ln(1+h)$ exist only for $h > -1$.

Since the function is not continuous at $h = 0$, we will use input-output tables.

First, we look as h approaches 0 from the left.

h	$\dfrac{\ln(1+h)}{h}$
-0.9	2.56
-0.5	1.39
-0.1	1.05
-0.01	1.01
-0.001	1.001

From the table we observe that

$$\lim\limits_{h \to 0^-} \dfrac{\ln(1+h)}{h} = 1.$$

Now, we look as h approaches 0 from the right.

h	$\dfrac{\ln(1+h)}{h}$
0.9	0.71
0.5	0.81
0.1	0.95
0.01	0.995
0.001	0.9995

From the table we observe that

$$\lim\limits_{h \to 0^+} \dfrac{\ln(1+h)}{h} = 1.$$

Thus,

$$\lim\limits_{h \to 0} \dfrac{\ln(1+h)}{h} = 1.$$

109. We consider the function $y = \dfrac{\ln x}{x}$. It can be shown that this function has a maximum at $x = e$. Thus:

$$\dfrac{\ln e}{e} > \dfrac{\ln \pi}{\pi} \qquad \text{Definition of Maximum}$$

$$\pi \cdot \ln e > e \cdot \ln \pi$$

$$\ln e^\pi > \ln \pi^e \qquad \text{(P3)}$$

$$e^\pi > \pi^e.$$

Therefore; e^π is larger.

111. Find $\lim\limits_{x \to 1} \ln x$

First, we observe as x approaches 1 from the left.

x	$\ln x$
0.1	-2.30
0.4	-0.92
0.7	-0.36
0.9	-0.11
0.99	-0.01
0.999	-0.001

$$\lim\limits_{x \to 1^-} \ln x = 0$$

Now, we observe as x approaches 1 from the right.

x	$\ln x$
1.9	0.64
1.7	0.53
1.4	0.34
1.1	0.10
1.01	0.01
1.001	0.001

$$\lim\limits_{x \to 1^+} \ln x = 0$$

Thus, $\lim\limits_{x \to 1} \ln x = 0$

113. Using the window:

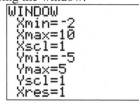

We graph the functions using a calculator.

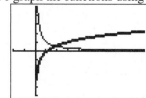

Notice the function is the thicker graph.

115. Using the window:

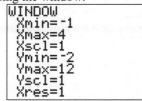

We graph the functions using a calculator.

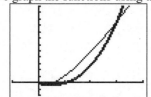

Notice, the function is the thicker graph.

117. $f(x) = x \ln x$

$$f'(x) = x \cdot \frac{1}{x} + \ln x$$
$$= 1 + \ln x$$

$f'(x)$ exists for all x in the domain of $f(x)$.

Solve:

$$f'(x) = 0$$
$$1 + \ln x = 0$$
$$\ln x = -1$$
$$x = e^{-1}$$

The critical value occurs when $x = e^{-1}$.

Use the Second-Derivative test to determine if the critical value represents a relative maximum or a relative minimum.

$$f''(x) = \frac{1}{x}$$

$$f''(e^{-1}) = \frac{1}{e^{-1}} = e > 0$$

Therefore the function is concave up and there is a relative minimum at the critical value $x = e^{-1}$.

$$f(e^{-1}) = e^{-1} \ln e^{-1}$$
$$= e^{-1} \cdot (-1)$$
$$= -\frac{1}{e} \approx -0.368$$

The minimum value that occurs at $x = e^{-1}$ is

$$-\frac{1}{e} \approx -0.368$$

Exercise Set 3.3

1. Using Theorem 8, the general form of f that satisfies the equation $f'(x) = 4 \cdot f(x)$ is

$f(x) = ce^{4x}$ for some constant c.

3. Using Theorem 8, the general form of A that satisfies the equation $\dfrac{dA}{dt} = -9 \cdot A$ is

$A = ce^{-9t}$, or $A(t) = ce^{-9t}$ for some constant c.

Note: If we were to let the initial population, the population when $t = 0$, be represented by $A(0) = A_0$, then we have

$A_0 = A(0) = ce^{-9 \cdot 0} = ce^0 = c$. Thus, $A_0 = c$, and we can express $A(t) = A_0 e^{-9t}$.

5. Using Theorem 8, the general form of Q that satisfies the equation $\dfrac{dQ}{dt} = k \cdot Q$ is

$Q = ce^{kt}$, or $Q(t) = ce^{kt}$ for some constant c.

Note: If we were to let the initial population, the population when $t = 0$, be represented by $Q(0) = Q_0$, then we have

$Q_0 = Q(0) = ce^{k \cdot 0} = ce^0 = c$. Thus, $Q_0 = c$, and we can express $Q(t) = Q_0 e^{kt}$.

7. a) Using Theorem 8, the general form of N that satisfies the equation

$N'(t) = 0.046 \cdot N(t)$ is

$N(t) = ce^{0.046t}$ for some constant c.

Allowing $t = 0$ to correspond to 1980 when approximately 112,000 patent applications were received, gives us the initial condition $N(0) = 112,000$. Therefore:

$$112,000 = ce^{0.046 \cdot 0}$$
$$112,000 = ce^0$$
$$112,000 = c$$

Substituting this value for c. We get:

$$N(t) = 112,000e^{0.046t}$$

b) In 2010, $t = 2010 - 1980 = 30$.

$$N(30) = 112,000e^{0.046(30)}$$

$$= 112,000e^{1.38}$$

$$= 112,000(3.974902)$$

$$= 445,188.98$$

$$\approx 445,189$$

There will be approximately 445,189 patent applications received in 2010.

c) From Theorem 9, the doubling time T is given by $T = \dfrac{\ln 2}{k}$.

$$T = \frac{\ln 2}{0.046} \approx 15.1$$

It will take approximately 15.1 years for the number of patent applications to double. This means that in 1995, the number of patents will have doubled.

9. a) Using Theorem 8, the general form of P that satisfies the equation

$$\frac{dP}{dt} = 0.065 \cdot P(t) \text{ is }$$

$P(t) = P_0 e^{0.065t}$ for some initial principal P_0.

b) If \$1000 is invested, then $P_0 = 1000$ and

$$P(t) = 1000e^{0.065t}$$

$$P(1) = 1000e^{0.065(1)}$$

$$= 1000e^{0.065}$$

$$= 1000(1.067159)$$

$$\approx 1067.16$$

The balance after 1 year is \$1067.16.

$$P(2) = 1000e^{0.065(2)}$$

$$= 1000e^{0.13}$$

$$= 1000(1.138828)$$

$$\approx 1138.83$$

The balance after 2 years is \$1138.83.

c) From Theorem 9, the doubling time T is given by $T = \dfrac{\ln 2}{k}$.

$$T = \frac{\ln 2}{0.065} \approx 10.7$$

It will take approximately 10.7 years for the balance to double.

11. a) Using Theorem 8, the general form of G that satisfies the equation

$$\frac{dG}{dt} = 0.093 \cdot G(t) \text{ is}$$

$G(t) = ce^{0.093t}$ for some constant c.

Allowing $t = 0$ to correspond to 2000 when approximately 4.7 billion gallons of bottled water were sold, gives us the initial condition $G(0) = 4.7$. Therefore:

$$4.7 = ce^{0.093 \cdot 0}$$

$$4.7 = ce^0$$

$$4.7 = c$$

Substituting this value for c. We get:

$$G(t) = 4.7e^{0.093t}$$

Where $G(t)$ is in billions of gallons sold and t is the number of years since 2000.

b) In 2010, $t = 2010 - 2000 = 10$.

$$N(10) = 4.7e^{0.093(10)}$$

$$= 4.7e^{0.93}$$

$$= 4.7(2.534509)$$

$$= 11.912193$$

$$\approx 11.91$$

There will be approximately 11.91 billion gallons of bottle water sold in 2010.

c) From Theorem 9, the doubling time T is given by $T = \dfrac{\ln 2}{k}$.

$$T = \frac{\ln 2}{0.093} \approx 7.5$$

It will take approximately 7.5 years for the amount of bottled water sold to double. Sometime in the middle of 2007 according to our model.

13. From Theorem 9, the doubling time T is given by $T = \dfrac{\ln 2}{k}$. Substituting $T = 15$ we get:

$$15 = \frac{\ln 2}{k}$$

Solve for k to find the interest rate.

$$15 \cdot k = \ln 2$$

$$k = \frac{\ln 2}{15}$$

$$k \approx 0.046210$$

The annual interest rate is 4.62%.

15. From Theorem 9, the doubling time T is

given by $T = \dfrac{\ln 2}{k}$. Substituting the growth rate

$k = 0.10$ into the formula.

$$T = \dfrac{\ln 2}{0.10}$$

$$\approx 6.9$$

The doubling time for the demand for oil is 6.9 years; therefore, at the end of the year 2012 the demand for oil will be double the demand for oil in 2006.

17. Find the doubling time:

$$T = \dfrac{\ln 2}{k} = \dfrac{\ln 2}{0.062} \approx 11.2$$

The doubling time is 11.2 years.

$$P(t) = P_0 e^{kt}$$

$$P(t) = 75,000 e^{0.062t}$$

We substitute $t = 5$ to find the amount after five years:

$$P(5) = 75,000 e^{0.062(5)}$$

$$= 75,000 e^{0.31}$$

$$\approx 102,256.88$$

In five years the account will have $102,256.88.

19. First, we find the initial investment:

$$P(t) = P_0 e^{kt}$$

Substituting, we have

$$11,414.71 = P_0 e^{0.084(5)}$$

$$11,414.71 = P_0 e^{0.42}$$

$$\dfrac{11,414.71}{e^{0.42}} = P_0$$

$$7499.9989 \approx P_0$$

$$7500.00 \approx P_0$$

The initial investment is $7500.
Next, we find the doubling time:

$$T = \dfrac{\ln 2}{k} = \dfrac{\ln 2}{0.084} \approx 8.3$$

The doubling time is 8.3 years.

21. a) The exponential growth function is

$V(t) = V_0 e^{kt}$. We will express $V(t)$ in

dollars and t as the number of
years since 1950. This set up gives us the
initial value of $V_0 = 30,000$ dollars.
Substituting this into the function, we have:

$$V(t) = 30,000 e^{kt}$$

In 2004, $t = 2004 - 1950 = 54$, and the
painting sold for 104,168,000 so we know
$V(54) = 104,168,000$ million dollars.

Substitute this information into the function
and solve for k.

$$104,168,000 = 30,000 e^{k(54)}$$

$$\dfrac{104,168,000}{30,000} = e^{54k}$$

$$\ln\left(\dfrac{104,168,000}{30,000}\right) = \ln e^{54k}$$

$$\ln\left(\dfrac{104,168,000}{30,000}\right) = 54k$$

$$\dfrac{\ln\left(\dfrac{104,168,000}{30,000}\right)}{54} = k$$

$$0.151 \approx k$$

The exponential growth rate is
approximately 15.1%. Substituting this
back into the function, we find the
exponential growth function is

$$V(t) = 30,000 e^{0.151t}.$$

b) In 2010, $t = 2010 - 1950 = 60$. We evaluate
the exponential growth function found in
part 'a' to get:

$$V(60) = 30,000 e^{0.151(60)}$$

$$= 30,000 e^{9.06}$$

$$= 30,000(8604.150654)$$

$$\approx 258,124,520$$

In the year 2010, the value of the painting
will be approximately $258,124,520.

c) From Theorem 9, the doubling time T is

given by $T = \dfrac{\ln 2}{k}$.

$$T = \dfrac{\ln 2}{0.151} \approx 4.59 \approx 4.6$$

The painting doubles in value approximately
every 4.6 years.

d) We set $V(t) = 1,000,000,000$ and solve for t.

$$30,000e^{0.151t} = 1,000,000,000$$

$$e^{0.151t} = \frac{1,000,000,000}{30,000}$$

$$\ln e^{0.151t} = \ln\left(\frac{1,000,000,000}{30,000}\right)$$

$$0.151t = \ln\left(\frac{1,000,000,000}{30,000}\right)$$

$$t = \frac{\ln\left(\dfrac{1,000,000,000}{30,000}\right)}{0.151}$$

$$t \approx 68.969$$

$$t \approx 69$$

It will take approximately 69 years for the value of the painting to reach 1 billion dollars. This will occur in the year 2019. $\left[1950 + 69 = 2019\right]$

23. a) Letting $t = 0$ correspond to 1960, means that $P_0 = 2277$. Therefore, the exponential growth function will be $P(t) = 2277e^{kt}$. In 2004, $t = 2004 - 1960 = 44$, the per capital income was $32,907. Using this information we find k.

$$32,907 = 2277e^{k(44)}$$

$$\frac{32,907}{2277} = e^{44k}$$

$$\ln\left(\frac{32,907}{2277}\right) = \ln e^{44k}$$

$$\ln\left(\frac{32,907}{2277}\right) = 44k$$

$$\frac{\ln\left(\dfrac{32,907}{2277}\right)}{44} = k$$

$$0.0607 \approx k$$

The exponential growth rate is 0.0607 or 6.07%.
Substituting the growth rate back into the function, we find the exponential growth function.
$$P(t) = 2277e^{0.0607t}.$$

b) In 2020, $t = 2020 - 1960 = 60$.
$$P(60) = 2277e^{0.0607(60)} \approx 86,908.76$$

In 2020, the U.S. per capita personal income will be approximately $86,908.76.

c) We need to solve the equation $P(t) = 100,000$ for t.

$$2277e^{0.0607t} = 100,000$$

$$e^{0.0607t} = \frac{100,000}{2277}$$

$$\ln e^{0.0607t} = \ln\left(\frac{100,000}{2277}\right)$$

$$0.0607t = \ln\left(\frac{100,000}{2277}\right)$$

$$t = \frac{\ln\left(\dfrac{100,000}{2277}\right)}{0.0607}$$

$$t \approx 62.3$$

The U.S. per capita income will be $100,000 approximately 62.3 years after 1960, or in 2022.

d) From Theorem 9, the doubling time T is given by $T = \dfrac{\ln 2}{k}$.

$$T = \frac{\ln 2}{0.0607} \approx 11.419 \approx 11.4$$

The U.S. per capita income doubles approximately every 11.4 years. Therefore, during 1971 the U.S. per capita income doubled that of the U.S. per capita income in 1960.

25. a) Enter the data into your calculator statistics editor.

Now use the regression command to find the exponential growth function.

```
EDIT CALC TESTS
8↑LinReg(a+bx)
9:LnReg
0:ExpReg
A:PwrReg
B:Logistic
C:SinReg
D:Manual-Fit
```

This gives us:

```
ExpReg
 y=a*b^x
 a=136.3939183
 b=1.071842825
```

The calculator models the function.

$y = 136.3939183(1.071842825)^x$, where y is in millions of dollars and x is the number of years after 1990.

From the technology connection section, we know that $b^x = e^{x(\ln b)}$. Apply this to the model our calculator found to get:

$1.071842825^x = e^{x(\ln 1.071842825)}$

$= e^{0.0693794334x}$

Rounding the values and substituting, we get the exponential growth function:

$P(t) = 136.4e^{0.0694t}$, where t is years since 1990, $k \approx 0.069$, or 6.9%, and $P_0 \approx 136.4$ million.

b) Using the model in part 'a'.
In 2007, $t = 2007 - 1990 = 17$

$P(17) = 136.4e^{0.0964(17)} \approx 443.8$

In 2007, paper shredder sales were approximately $444 million.

In 2012, $t = 2012 - 1990 = 22$

$P(22) = 136.4e^{0.0694(22)} \approx 627.9$

In 2012, paper shredder sales were approximately $628 million.

c) We need to solve the equation $P(t) = 500$ for t.

$136.4e^{0.0694t} = 500$

$e^{0.0694t} = \dfrac{500}{136.4}$

$\ln e^{0.0694t} = \ln\left(\dfrac{500}{136.4}\right)$

$0.0694t = \ln\left(\dfrac{500}{136.4}\right)$

$t = \dfrac{\ln\left(\dfrac{500}{136.4}\right)}{0.0694}$

$t \approx 18.7$

It will take approximately 18.7 years for the total sales of paper shredders to reach $500 million. This should occur in the year 2008.

d) From Theorem 9, the doubling time T is given by $T = \dfrac{\ln 2}{k}$.

$T = \dfrac{\ln 2}{0.0694} \approx 9.9877 \approx 10.0$

The doubling time for sales of paper shredders is approximately 10 years.

27. If we let $t = 0$ correspond to the year 1626 the initial value of Manhattan is $V_0 = 24$ Using the exponential growth function $V(t) = V_0e^{kt}$.

Assuming an exponential rate of inflation of 5% means that $k = 0.05$. Substituting these values into the growth function gives us:

$V(t) = 24e^{0.05t}$.

In 2010, $t = 2010 - 1626 = 384$

$V(384) = 24e^{0.05(384)}$

$= 24e^{19.2}$

$\approx 5,231,970,592$

Manhattan Island will be worth approximately $5,231,970,592 or $5.23 billion.

29. Let $S(t)$ be the average salary in dollars t years after 1970. Therefore $S_0 = 29,303$. Therefore the exponential growth model is $S(t) = 29,303e^{kt}$. Use the average salary in 2005 to find the growth rate.
In 2005, $t = 2005 - 1970 = 35$,
$S(35) = 2,632,655$, so:

$2,632,655 = 29,303e^{k(35)}$

$\dfrac{2,632,655}{29,303} = e^{35k}$

$\ln\left(\dfrac{2,632,655}{29,303}\right) = \ln e^{35k}$

$\ln\left(\dfrac{2,632,655}{29,303}\right) = 35k$

$\dfrac{\ln\left(\dfrac{2,632,655}{29,303}\right)}{35} = k$

$0.1285 \approx k$

The growth rate is 12.85% per year.
Thus, the exponential growth model is
$S(t) = 29,303e^{.1285t}$

In 2010, $t = 2010 - 1970 = 40$

$S(40) = 29,303e^{0.1285(40)} \approx 5,002,484.16$

The average salary will be $5,002,484 in 2010.
In 2020, $t = 2020 - 1970 = 50$

$S(50) = 29,303e^{0.1285(50)} \approx 18,082,319.20$

The average salary will be $18,082,319 in 2020.

31. $P(x) = \dfrac{100}{1 + 49e^{-0.13x}}$

a) $P(0) = \dfrac{100}{1 + 49e^{-0.13(0)}}$

$= \dfrac{100}{1 + 49e^{0}}$

$= \dfrac{100}{1 + 49}$

$= \dfrac{100}{50}$

$= 2$

About 2% purchased the game without seeing the advertisement.

b) $P(5) = \dfrac{100}{1 + 49e^{-0.13(5)}}$

$= \dfrac{100}{1 + 49e^{-0.65}}$

≈ 3.8

About 3.8% will purchase the game after the advertisement runs 5 times.

$P(10) = \dfrac{100}{1 + 49e^{-0.13(10)}}$

$= \dfrac{100}{1 + 49e^{-1.3}}$

≈ 7.0

About 7.0% will purchase the game after the advertisement runs 10 times.

$P(20) = \dfrac{100}{1 + 49e^{-0.13(20)}}$

$= \dfrac{100}{1 + 49e^{-2.6}}$

≈ 21.6

About 21.6% will purchase the game after the advertisement runs 20 times.

$P(30) = \dfrac{100}{1 + 49e^{-0.13(30)}}$

$= \dfrac{100}{1 + 49e^{-3.9}}$

≈ 50.2

About 50.2% will purchase the game after the advertisement runs 30 times.

$P(50) = \dfrac{100}{1 + 49e^{-0.13(50)}}$

$= \dfrac{100}{1 + 49e^{-6.5}}$

≈ 93.1

About 93.1% will purchase the game after the advertisement runs 50 times.

$P(60) = \dfrac{100}{1 + 49e^{-0.13(60)}}$

$= \dfrac{100}{1 + 49e^{-7.8}}$

≈ 98.0

About 98.0% will purchase the game after the advertisement runs 60 times.

c) We apply the Quotient Rule to take the derivative.

$P'(x)$

$= \dfrac{\left(1 + 49e^{-0.13x}\right)(0) - 49(-0.13)e^{-0.13x} \cdot 100}{\left(1 + 49e^{-0.13x}\right)^{2}}$

$= \dfrac{637e^{-0.13x}}{\left(1 + 49e^{-0.13x}\right)^{2}}$

d) The derivative $P'(x)$ exists for all real numbers. The equation $P'(x) = 0$ has no solution. Thus, the function has no critical points and hence, no relative extrema. $P'(x) > 0$ for all real numbers, so $P(x)$ is increasing on $[0, \infty)$. The second derivative can be used to show that the graph has an inflection point at $(29.9, 50)$. The function is concave up on the interval $(0, 29.9)$ and concave down on the interval $(29.9, \infty)$

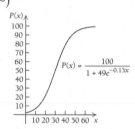

33. $T = \dfrac{\ln 2}{k} = \dfrac{\ln 2}{0.035} \approx 19.8$

The doubling time T is 19.8 years.

35. $k = \dfrac{\ln 2}{T} = \dfrac{\ln 2}{6.931} \approx 0.10$

The growth rate k is 0.10 or 10% per year.

37. $T = \dfrac{\ln 2}{k} = \dfrac{\ln 2}{0.02794} \approx 24.8$

The doubling time T is 24.8 years.

39. Let $t = 0$ correspond to 1972. Then the initial population of grizzly bears is $P_0 = 190$. The set up of the exponential growth function is $P(t) = 190e^{kt}$. In 2005, $t = 2005 - 1972 = 33$. The population of grizzly bears had grown to 610, hence $P(33) = 610$. Use this information to find the growth rate k.

$$610 = 190e^{k(33)}$$

$$\frac{610}{190} = e^{33k}$$

$$\ln\left(\frac{610}{190}\right) = 33k$$

$$\frac{\ln\left(\frac{610}{190}\right)}{33} = k$$

$$0.035 \approx k$$

The growth rate is approximately 3.5%. Now we can use the information to find the exponential growth function.

$$P(t) = 190e^{0.035t}$$

In 2012, $t = 2012 - 1972 = 40$

$$P(40) = 190e^{0.035(40)} \approx 770$$

Yellowstone National Park will be home to approximately 770 grizzly bears in 2012.

41. From Example 7 we know: $R(b) = e^{21.4b}$

For the risk of an accident to be 80% we have

$$80 = e^{21.4b}$$

$$\ln 80 = \ln e^{21.4b}$$

$$\ln 80 = 21.4b$$

$$\frac{\ln 80}{21.4} = b$$

$$0.205 \approx b$$

When the blood alcohol level is 0.205%, the risk of having an accident is 80%.

43. $P(t) = \dfrac{5780}{1+4.78e^{-0.4t}}$

a) $P(0) = \dfrac{5780}{1+4.78e^{-0.4(0)}}$

$$= \frac{5780}{1+4.78e^0}$$

$$= \frac{5780}{1+4.78}$$

$$= 1000$$

The island population in year zero is 1000.

$$P(1) = \frac{5780}{1+4.78e^{-0.4(1)}}$$

$$= \frac{5780}{1+4.78e^{-0.4}}$$

$$\approx 1375$$

The island population after 1 year is 1375.

$$P(2) = \frac{5780}{1+4.78e^{-0.4(2)}}$$

$$= \frac{5780}{1+4.78e^{-0.8}}$$

$$\approx 1836$$

The island population after 2 years is 1836.

$$P(5) = \frac{5780}{1+4.78e^{-0.4(5)}}$$

$$= \frac{5780}{1+4.78e^{-2}}$$

$$\approx 3510$$

The island population after 5 years is 3510.

$$P(10) = \frac{5780}{1+4.78e^{-0.4(10)}}$$

$$= \frac{5780}{1+4.78e^{-4}}$$

$$\approx 5315$$

The island population after 10 years is 5315.

$$P(20) = \frac{5780}{1+4.78e^{-0.4(20)}}$$

$$= \frac{5780}{1+4.78e^{-8}}$$

$$\approx 5771$$

The island population after 20 years is 5771.

b) We apply the Quotient Rule to take the derivative.

$$P'(t) = \frac{\left(1+4.78e^{-0.4t}\right)(0) - 4.78(-0.4)e^{-0.4t} \cdot 5780}{\left(1+4.78e^{-0.4t}\right)^2}$$

$$= \frac{11,051.36e^{-0.4t}}{\left(1+4.78e^{-0.4t}\right)^2}$$

c) The derivative $P'(t)$ exists for all real numbers. The equation $P'(t) = 0$ has no solution. Thus, the function has no critical points and hence, no relative extrema. $P'(t) > 0$ for all real numbers, so $P(t)$ is increasing on $[0,\infty)$. The second derivative can be used to show that the graph has an inflection point at $(3.911, 2890)$.

The function is concave up on the interval $(0, 3.911)$ and concave down on the interval $(3.911, \infty)$

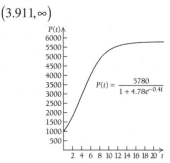

45. Let $t = 0$ correspond to 1930. Thus the initial number of women earning bachelor's degrees is $P_0 = 48,869$. So the exponential growth function is $P(t) = 48,869e^{kt}$ In 2005, $t = 2005 - 1930 = 75$, approximately 832,000 women received a degree. Using this information we can find the exponential growth rate k. Substitute the information into the growth function.

$$832,000 = 48,869e^{k(75)}$$

$$\frac{832,000}{48,869} = e^{75k}$$

$$\ln\left(\frac{832,000}{48,869}\right) = \ln e^{75k}$$

$$\ln\left(\frac{832,000}{48,869}\right) = 75k$$

$$\frac{\ln\left(\frac{832,000}{48,869}\right)}{75} = k$$

$$0.0378 \approx k$$

The exponential growth rate is 0.0378 or 3.78%. Therefore the exponential growth function is:
$$P(t) = 48,869e^{0.0378t}$$

Where t is time in years since 1930 and $P(t)$ is number of women earning bachelor's degrees. To determine the rate at which the number of women earning bachelor's degrees is growing, take the derivative of the exponential growth function.

$$P'(t) = 48,869(0.378)e^{0.0378t}$$

$$= 1847e^{0.378t}$$

The rate at which the number of women earning bachelor degrees is growing t years after 1930 is $1847e^{0.378t}$.

47. $P(t) = 100\left(1 - e^{-0.4t}\right)$

a) $P(0) = 100\left(1 - e^{-0.4(0)}\right)$
$$= 100(1 - 1)$$
$$= 0$$
0% of doctors are prescribing this medication after zero months.

$P(1) = 100\left(1 - e^{-0.4(1)}\right)$
$$= 100\left(1 - e^{-0.4}\right)$$
$$\approx 33.0$$
33% of doctors are prescribing this medication after 1 month.

$P(2) = 100\left(1 - e^{-0.4(2)}\right)$
$$= 100\left(1 - e^{-0.8}\right)$$
$$\approx 55.1$$
55% of doctors are prescribing this medication after 2 months.

$P(3) = 100\left(1 - e^{-0.4(3)}\right)$
$$= 100\left(1 - e^{-1.2}\right)$$
$$\approx 69.9$$
70% of doctors are prescribing this medication after 3 months.

$P(5) = 100\left(1 - e^{-0.4(5)}\right)$
$$= 100\left(1 - e^{-2}\right)$$
$$\approx 86.47$$
86% of doctors are prescribing this medication after 5 months.

$P(12) = 100\left(1 - e^{-0.4(12)}\right)$
$$= 100\left(1 - e^{-4.8}\right)$$
$$\approx 99.2$$
99.2% of doctors are prescribing this medication after 12 months.

$P(16) = 100\left(1 - e^{-0.4(16)}\right)$
$$= 100\left(1 - e^{-6.4}\right)$$
$$\approx 99.8$$
99.8% of doctors are prescribing this medication after 16 months.

b) $P'(t) = 100\left[0 - (-0.4)e^{-0.4t}\right]$
$$= 100\left(0.4e^{-0.4t}\right)$$
$$= 40e^{-0.4t}$$

Therefore,
$$P'(7) = 40e^{-0.4(7)}$$
$$= 40e^{-2.8}$$
$$\approx 2.432$$
At 7 months, the percentage of doctors who are prescribing the medicine are growing at a rate of 2.4% per month.

c) The derivative $P'(t)$ exists for all real numbers. The equation $P'(t) = 0$ has no solution. Thus, the function has no critical points and hence, no relative extrema. $P'(t) > 0$ for all real numbers, so $P(t)$ is increasing on $[0, \infty)$. $P''(t) = -0.16e^{-0.4t}$,so $P''(t) < 0$ for all real numbers and hence $P(t)$ is concave down for all on $[0, \infty)$.

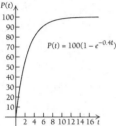

49. $\boxed{tw}$

51. $\boxed{tw}$

53. $\boxed{tw}$

55. $\boxed{tw}$

57. Let T_4 be the time it takes for a population to quadruple, then according to the exponential growth function:
$$4 \cdot P_0 = P_0 e^{kT_4}$$
Solve this equation for T_4.
$$\frac{4P_0}{P_0} = e^{kT_4}$$
$$\ln 4 = \ln e^{kT_4}$$
$$\ln 4 = kT_4$$
$$\frac{\ln 4}{k} = T_4$$
$$T_4 = \frac{\ln 4}{k}$$

59. Let k_1 and k_2 represent the growth rates of Q_1 and Q_2. Using the Theorem 9, we know that:
$$k_1 = \frac{\ln 2}{1} \quad \text{and } k_2 = \frac{\ln 2}{2}$$
Since the initial amounts of both quantities are the same we have:
$$Q_1 = Q_0 e^{\ln 2 \cdot t} \text{ and } Q_2 = Q_0 e^{\left(\frac{\ln 2}{2}\right)t}$$
Q_1 is twice the size of Q_2 when
$$Q_1 = 2Q_2$$
$$Q_0 e^{t \cdot \ln 2} = 2Q_0 e^{t\left(\frac{\ln 2}{2}\right)}$$
$$\frac{Q_0 e^{t \cdot \ln 2}}{Q_0 e^{t\left(\frac{\ln 2}{2}\right)}} = 2$$
$$e^{t\left(\ln 2 - \left(\frac{\ln 2}{2}\right)\right)} = 2$$
$$\ln\left[e^{t\left(\frac{\ln 2}{2}\right)}\right] = \ln 2$$
$$t\left(\frac{\ln 2}{2}\right) = \ln 2$$
$$t = \frac{\ln 2}{\frac{\ln 2}{2}}$$
$$t = \frac{1}{\frac{1}{2}} = 2$$
It will take 2 years for Q_1 to be twice the size of Q_2.

61. $i = e^k - 1$
Substitute 0.073 into the equation for k.
$$i = e^{0.073} - 1$$
$$\approx 1.075731 - 1$$
$$\approx 0.07573$$
The effective annual yield is 7.57%.

63. $i = e^k - 1$
Substitute 0.0942 into the equation for i.
$$0.0942 = e^k - 1$$
$$1.0942 = e^k$$
$$\ln 1.0942 = \ln e^k$$
$$0.09002 \approx k$$
The interest rate was 9.0%.

65. Consider the system of equations.

$$y_1 = Ce^{kt_1}$$

$$y_2 = Ce^{kt_2}$$

Divide the first equation by the second equation.

$$\frac{y_1}{y_2} = \frac{Ce^{kt_1}}{Ce^{kt_2}}$$

$$\frac{y_1}{y_2} = e^{k(t_1-t_2)}$$

$$\ln\left(\frac{y_1}{y_2}\right) = \ln e^{k(t_1-t_2)}$$

$$\ln\left(\frac{y_1}{y_2}\right) = k(t_1 - t_2) \qquad \text{(P5)}$$

Solving the equation for k we have:

$$\frac{\ln\left(\dfrac{y_1}{y_2}\right)}{t_1 - t_2} = k$$

$$\frac{\ln y_1 - \ln y_2}{t_1 - t_2} = k$$

Now that we have a value for k, substitute k back into either equation to find the value for C. We substitute into the equation involving y_1.

$$y_1 = Ce^{kt_1}$$

$$\frac{y_1}{e^{kt_1}} = C$$

$$y_1 \cdot e^{-kt} = C$$

$$y_1 \cdot e^{-\left(\frac{\ln y_1 - \ln y_2}{t_1 - t_2}\right)(t_1)} = C$$

67. ⬜ *tw*

Exercise Set 3.4

---

1. a) $N(t) = N_0 e^{-kt}$

We substitute 0.096 in for k.

$$N(t) = N_0 e^{-0.096t}$$

b) $N_0 = 500$

$$N(t) = 500 e^{-0.096t}$$

$$N(4) = 500 e^{-0.096(4)}$$

$$= 500 e^{-0.384}$$

$$\approx 340.565713$$

$$\approx 341$$

There will be approximately 341g of Iodine -131 present after 4 days.

c) From Theorem 10, we know that the half-life T and the decay rate k are related by:

$$T = \frac{\ln 2}{k}.$$

Substitute $k = .096$

$$T = \frac{\ln 2}{0.096} \approx 7.22028$$

It will take about 7.2 days for half of the 500g of Iodine - 131 to remain.

3. a) When A decomposes at a rate proportional to the amount of A present, we know that

$$\frac{dA}{dt} = -kA.$$

The solution to this equation is

$$A(t) = A_0 e^{-kt}.$$

b) First, we find k. The half-life of A is 3.3 hrs. From theorem 10 we know:

$$k = \frac{\ln 2}{T}$$

$$k = \frac{\ln 2}{3.3} \approx 0.21$$

The initial amount $A_0 = 10$, so the exponential decay function is:

$$A(t) = 10 e^{-0.21t}.$$

To determine how long it will take to reduce A to 1 lb we solve $A(t) = 1$ for t.

$$10 e^{-0.21t} = 1$$

$$e^{-0.21t} = 0.1$$

$$\ln\left(e^{-0.21t}\right) = \ln(0.1)$$

$$-0.21t = \ln(0.1)$$

$$t = \frac{\ln 0.1}{-0.21}$$

$$t \approx 10.96469$$

$$t \approx 11$$

It will take approximately 11 hours for 10 lbs of substance A to reduce to 1 lb.

5. From theorem 10 we know:
$$k = \frac{\ln 2}{T}$$
$$k = \frac{\ln 2}{3} \approx 0.231$$
The decay rate is 0.231 or 23.1% per minute.

7. From theorem 10 we know:
$$T = \frac{\ln 2}{k}$$
$$T = \frac{\ln 2}{.0315} \approx 22.0$$
The half-life is 22.0 years.

9. $P(t) = P_0 e^{-kt}$

We substitute 1000 for P_0, 0.0315 for k, and 100 for t.
$$P(100) = 1000 e^{-0.0315(100)}$$
$$= 1000 e^{-3.15}$$
$$\approx 42.9$$
Approximately 42.9 grams of lead-210 will remain after 100 years.

11. If an ivory tusk has lost 40% of its carbon-14 from its initial amount P_0, then 60% of the initial amount remains. To find the age of the tusk, we solve the following equation for t:
$$0.60 P_0 = P_0 e^{-0.0001205t}$$
$$0.60 = e^{-0.0001205t}$$
$$\ln 0.60 = \ln e^{-0.0001205t}$$
$$\ln 0.60 = -0.0001205t$$
$$\frac{\ln 0.60}{-0.0001205} = t$$
$$4239.216 \approx t$$
$$4239 \approx t$$
The ivory tusk is approximately 4239 years old.

13. First, we find k. The half-life is 60.1 days. From Theorem 10 we know.
$$k = \frac{\ln 2}{T}$$
$$k = \frac{\ln 2}{60.1}$$
$$k \approx 0.0115$$
The decay rate is approximately 0.0115 or 1.15% per day.

If the initial amount A_0 decreased by 25%, then 75% of A_0 remains. We solve the following equation for t.
$$0.75 A_0 = A_0 e^{-0.0115t}$$
$$0.75 = e^{-0.0115t}$$
$$\ln 0.75 = \ln e^{-0.0115t}$$
$$\ln 0.75 = -0.0115t$$
$$\frac{\ln 0.75}{-0.0115} = t$$
$$25.015 \approx t$$
$$25 \approx t$$
The sample was sitting on the shelf for approximately 25 days.

15. Since the corn pollen had lost 38.1% of its carbon-14, then 61.9% of the initial carbon-14 remains. The decay rate of carbon-14 is $k = 0.0001205$. In order to find out the age of the corn pollen, we must solve the following equation for t.
$$0.619 P_0 = P_0 e^{-0.0001205t}$$
$$0.619 = e^{-0.0001205t}$$
$$\ln 0.619 = -0.0001205t$$
$$\frac{\ln 0.619}{-0.0001205} = t$$
$$3980 \approx t$$
The corn pollen was approximately 3980 years old.

17. Use the exponential growth function $P(t) = P_0 e^{kt}$. The interest rate is 5.3% so $k = 0.053$. When $t = 20$ the parents wish the future value to be $40,000, substituting this information into the function gives us the equation:
$$40,000 = P_0 e^{0.053(20)}$$
$$\frac{40,000}{e^{1.06}} = P_0$$
$$13,858.23 \approx P_0$$
The parents should invest $13,858.23 in order to have $40,000 on their child's 20th birthday.

19. The interest rate is 5.7%, so $k = 0.057$. In 6 years, the athletes salary will be $9 million so $P(6) = 9$ million dollars. Substituting this information into the exponential growth function we get:

$$9 = P_0 e^{0.057(6)}$$

$$9 = P_0 e^{0.342}$$

$$\frac{9}{e^{0.342}} = P_0$$

$$6.393134 \approx P_0$$

The present value is $6.393134 million or $6,393,134.

21. Using the exponential growth function $P(t) = P_0 e^{kt}$. It is known that the interest rate is 8.3%, therefore $k = 0.083$. In 13 years the value of the trust fund will be $80,000, so $P(13) = 80,000$. Substitute this information into the exponential growth function and solve for the present value P_0.

$$80,000 = P_0 e^{0.083(13)}$$

$$80,000 = P_0 e^{1.079}$$

$$\frac{80,000}{e^{1.079}} = P_0$$

$$27,194.82 \approx P_0$$

The present value of the trust fund is $27,194.82.

23. a) $V(0) = 40,000 e^{-0} = 40,000$

The initial cost of the machinery was $40,000.

b) $V(2) = 40,000 e^{-2} \approx 5413.41$

The salvage value after 2 years is approximately $5413.41.

c) $\boxed{tw}$

25. a) Enter the data into your calculator.

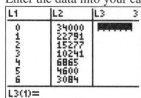

Then use the regression to fit the data.

We have:

```
ExpReg
 y=a*b^x
 a=34001.78697
 b=.6702977719
```

The calculator determined the exponential function to be

$$y = 34,001.78697 (0.6702977719)^x.$$

Using $b^x = e^{x(\ln b)}$, we have

$$0.6702977719^x = e^{x(\ln(0.6702977719))}$$

$$= e^{-0.4000332297 \cdot x}$$

Converting the calculator equation to $V(t) = V_0 e^{-kt}$ we have:

$$V(t) = 34,001.78697 e^{-0.4000332297 t}$$

b) $V(7) = 34,001.78697 e^{-0.4000332297(7)}$

$$\approx 2067.17$$

The salvage value of the copier is approximately $2067.17 after 7 years.

$$V(10) = 34,001.78697 e^{-0.4000332297(10)}$$

$$\approx 622.56$$

The salvage value of the copier is approximately $622.56 after 10 years.

c) Solve $V(t) = 1000$ for t.

$$34,001.78697 e^{-0.4000332297 t} = 1000$$

$$e^{-0.4000332297 t} = \frac{1000}{34,001.78697}$$

$$\ln e^{-0.4000332297 t} = \ln\left(\frac{1000}{34,001.78697}\right)$$

$$-0.4000332297 t = \ln\left(\frac{1000}{34,001.78697}\right)$$

$$t = \frac{\ln\left(\frac{1000}{34,001.78967}\right)}{-0.4000332297}$$

$$t \approx 8.82$$

The salvage value of the copier will be $1000 approximately 8.8 years after the purchase.

d) From Theorem 10 we know:

$$T = \frac{\ln 2}{k}$$

$$T = \frac{\ln 2}{0.4} \approx 1.73$$

The copier will be worth half of its original value approximately 1.7 years after it was purchased.

e) $\boxed{tw}$

27. a) Use the exponential-decay model
$B(t) = B_0 e^{-kt}$ Let t be years since 1985, this gives us the initial consumption of beef $B_0 = 80$. Using the fact that in 1996, $t = 1996 - 1985 = 11$, and the beef consumption was 67 lbs, we can substitute into the exponential-decay function to find the decay rate k.

$$67 = 80e^{-k(11)}$$

$$\frac{67}{80} = e^{-11k}$$

$$\ln\left(\frac{67}{80}\right) = \ln e^{-11k}$$

$$\ln\left(\frac{67}{80}\right) = -11k$$

$$\frac{\ln\left(\frac{67}{80}\right)}{-11} = k$$

$$0.016 \approx k$$

The decay rate is 0.016, or 1.6% per year. Substituting this into the exponential decay function we get:
$B(t) = 80e^{-0.016t}$.

b) In 2000, $t = 2000 - 1985 = 15$.

$$B(15) = 80e^{-0.016(15)} \approx 62.9$$

In the year 2000, annual consumption of beef will be approximately 62.9 pounds per person.

c) Set $B(t) = 20$ and solve for t.

$$80e^{-0.016t} = 20$$

$$e^{-0.016t} = \frac{20}{80}$$

$$\ln e^{-0.016t} = \ln\left(\tfrac{1}{4}\right)$$

$$-0.016t = \ln\left(\tfrac{1}{4}\right)$$

$$t = \frac{\ln\left(\tfrac{1}{4}\right)}{-0.016}$$

$$t \approx 86.6$$

Theoretically, consumption of beef will be 20lbs per person about 86.6 years after 1985, or in the year 2072.

29. a) Use the exponential decay model
$P(t) = P_0 e^{-kt}$. If we let $t = 0$ correspond to 1995, then $P_0 = 51.9$ million people. In 2006, the population of Ukraine was 46.7 million, so $P(11) = 46.7$. Substituting this information into the model we get:

$$46.7 = 51.9e^{-k(11)}$$

$$\frac{46.7}{51.9} = e^{-11k}$$

$$\ln\left(\frac{46.7}{51.9}\right) = \ln e^{-11k}$$

$$\ln\left(\frac{46.7}{51.9}\right) = -11k$$

$$\frac{\ln\left(\frac{46.7}{51.9}\right)}{-11} = k$$

$$0.0096 \approx k$$

Thus, $P(t) = 51.9e^{-0.0096t}$, where $P(t)$ is in millions of people and t is the number of years since 1995.

b) In 2015, $t = 2015 - 1995 = 20$.

$$P(20) = 51.9e^{-0.0096(20)} \approx 42.8$$

The population of the Ukraine will be approximately 42.8 million in 2015.

c) Note that 1 = 0.000001 million.
Set $P(t) = 0.000001$ and solve for t.

$$51.9e^{-0.0096t} = 0.000001$$

$$e^{-0.0096t} = \frac{0.000001}{51.9}$$

$$\ln e^{-0.0096t} = \ln\left(\frac{0.000001}{51.9}\right)$$

$$-0.0096t = \ln\left(\frac{0.000001}{51.9}\right)$$

$$t = \frac{\ln\left(\dfrac{0.000001}{51.9}\right)}{-0.0096}$$

$$t \approx 1851$$

According to the model, the population of the Ukraine will be 1 person 1851 years after 1995, or in the year 3846.

31. a) According to Newton's Law of Cooling $T(t) = ae^{-kt} + C$. The room temperature is 75 degrees so $C = 75$. At $t = 0$, $T = 102$ degrees. We substitute these values into Newton's law of Cooling.

$$102 = ae^{-k(0)} + 75$$

$$102 = ae^0 + 75$$

$$102 = a + 75$$

$$27 = a$$

b) From part 'a' we have $T(t) = 27e^{-kt} + 75$.
Using the fact that when $t = 10$, $T = 90$ we have:

$$90 = 27e^{-k(10)} + 75$$

$$15 = 27e^{-10k}$$

$$\frac{15}{27} = e^{-10k}$$

$$\ln\left(\frac{15}{27}\right) = \ln\left(e^{-10k}\right)$$

$$\ln\left(\frac{15}{27}\right) = -10k$$

$$\frac{\ln\left(\dfrac{15}{27}\right)}{-10} = k$$

$$0.05878 \approx k$$

c) From 'a' and 'b' parts we have

$$T(t) = 27e^{-0.05878t} + 75$$

$$T(20) = 27e^{-0.05878(20)} + 75$$

$$= 27e^{-1.1756} + 75$$

$$\approx 83.3$$

After 20 minutes the water temperature is approximately 83 degrees.

d) Solve $T(t) = 80$ for t.

$$27e^{-0.05878t} + 75 = 80$$

$$27e^{-0.05878t} = 5$$

$$e^{-0.05878t} = \frac{5}{27}$$

$$\ln e^{-0.05878t} = \ln\left(\frac{5}{27}\right)$$

$$-0.05878t = \ln\left(\frac{5}{27}\right)$$

$$t = \frac{\ln\left(\dfrac{5}{27}\right)}{-0.05878}$$

$$t \approx 28.7$$

It will take approximately 29 minutes for the water temperature to cool to 80 degrees.

e) $\boxed{tw}$

33. Newton's law of Cooling states $T(t) = ae^{-kt} + C$.

Assume the body had a normal temperature of 98.6 degrees at the time of death and the room remained a constant 60 degrees giving us the constant $C = 60$. We find the constant a first, using the fact that 98.6 degrees is normal body temperature.

$$98.6 = ae^{-k(0)} + 60$$

$$98.6 = a + 60$$

$$38.6 = a$$

So $T(t) = 38.6e^{-kt} + 60$

Next, we use the two temperature readings to find k. We want to find the number of hours since death, t. When the corner took the first temperature reading t hours since death, the body temperature was 85.9 degrees. One hour later at $t + 1$ hours since death, the body temperature was 83.4 degrees. Using these two pieces of information we get two equations.

$$85.9 = 38.6e^{-kt} + 60$$

$$83.4 = 38.6e^{-k(t+1)} + 60$$

Subtracting 60 from each equation we get:

$25.9 = 38.6e^{-kt}$

$23.4 = 38.6e^{-k(t+1)}$

A quick way to solve this system of equations is to divide the first equation by the second equation. This gives us:

$\dfrac{25.9}{23.4} = \dfrac{38.6e^{-kt}}{38.6e^{-k(t+1)}}$

$\dfrac{25.9}{23.4} = e^{-kt-(-kt-k)}$

$\dfrac{25.9}{23.4} = e^{k}$

$\ln\left(\dfrac{25.9}{23.4}\right) = \ln e^{k}$

$0.10 \approx k$

Now we substitute 0.10 in for k into the equation $25.9 = 38.6e^{-kt}$ and solve for t.

$25.9 = 38.6e^{-0.10t}$

$\dfrac{25.9}{38.6} = e^{-0.1t}$

$\ln\left(\dfrac{25.9}{38.6}\right) = \ln e^{-0.1t}$

$\ln\left(\dfrac{25.9}{38.6}\right) = -0.1t$

$\dfrac{\ln\left(\dfrac{25.9}{38.6}\right)}{-0.10} = t$

$4 \approx t$

Therefore, the body had been dead for 4 hours. Since the temperature was taken at 11 P.M, the time of death was 7 P.M.

35. $W = 170e^{-0.008t}$

a) We substitute 20 in for t.

$W(20) = 170e^{-0.008(20)}$

$= 170e^{-0.16}$

≈ 144.86

The monk weighs approximately 145 pounds after 20 days.

b) We take the derivative of the function to find the rate of change.

$W'(t) = 170(-0.008)e^{-0.008t}$

$= -1.36e^{-0.008t}$

Now, we substitute 20 in for t.

$W'(20) = -1.36e^{-0.009(20)} \approx -1.16$

The monk is losing approximately 1.2 pounds per day after 20 days.

37. $P(t) = 50e^{-0.004t}$

a) We substitute 375 for t.

$P(375) = 50e^{-0.004(375)}$

$= 50e^{-1.5}$

≈ 11.2

After 375 days, approximately 11.2 watts of power will be available.

b) From theorem 10 we know:

$T = \dfrac{\ln 2}{k}$

$T = \dfrac{\ln 2}{0.004} \approx 173$

The half-life of the power supply is approximately 173 days.

c) Set $P(t) = 10$ and solve for t.

$50e^{-0.004t} = 10$

$e^{-0.004t} = \dfrac{10}{50}$

$\ln e^{-0.004t} = \ln 0.2$

$-0.004t = \ln 0.2$

$t = \dfrac{\ln 0.2}{-0.004}$

$t \approx 402.36$

The satellite can stay in operation for 402 days.

d) When $t = 0$

$P(0) = 50e^{-0.004(0)}$

$= 50e^{0}$

$= 50$

At the beginning, the satellite had 50 watts of power.

e) $\boxed{tw}$

39. (a)

41. (c)

43. (d)

45. (f)

47. (b)

49. $\boxed{tw}$

51. $\boxed{tw}$

53. $\dfrac{dP}{dt} = 0.009P$ implies that

$P(t) = P_0 e^{0.009t}$

The doubling time for the U.S. population is

$T = \dfrac{\ln 2}{0.009} \approx 77.02$

Therefore the U.S. population doubles approximately every 77 years. This means that the 2007 population will double in the year 2084.

Exercise Set 3.5

1. $y = 7^x$

$\dfrac{dy}{dx} = (\ln 7)7^x$ Theorem 12: $\dfrac{dy}{dx}a^x = (\ln a)a^x$

3. $f(x) = 8^x$

$f'(x) = (\ln 8)8^x$ Theorem 12

5. $g(x) = x^3 (5.4)^x$

Using the Product Rule, we get

$g'(x) = x^3 \left(\dfrac{dy}{dx}(5.4)^x\right) + \left(\dfrac{dy}{dx}x^3\right)(5.4)^x$

$= x^3 \underbrace{\left(\ln(5.4)(5.4)^x\right)}_{\text{Theorem 12}} + 3x^2 (5.4)^x$

$= x^2 (5.4)^x \left(x\ln(5.4) + 3\right)$ Factoring

7. $y = 7^{x^4+2}$

Using the Chain Rule and Theorem 12, we get

$\dfrac{dy}{dx} = (\ln 7)7^{x^4+2}\left(\dfrac{d}{dx}(x^4+2)\right)$

$= (\ln 7)7^{x^4+2}(4x^3)$

$= 4x^3 (\ln 7)7^{x^4+2}$

9. $y = e^{8x}$

$\dfrac{dy}{dx} = 8e^{8x}$ Theorem 2

11. $f(x) = 3^{x^4+1}$

Using the Chain Rule and Theorem 12, we get

$\dfrac{dy}{dx} = (\ln 3)3^{x^4+1}\left(\dfrac{d}{dx}(x^4+1)\right)$

$= (\ln 3)3^{x^4+1}(4x^3)$

$= 4x^3 (\ln 3)3^{x^4+1}$

13. $y = \log_4 x$

$\dfrac{dy}{dx} = \dfrac{1}{\ln 4}\cdot\dfrac{1}{x}$ Theorem 14

$= \dfrac{1}{x\ln 4}$

15. $y = \log_{17} x$

$\dfrac{dy}{dx} = \dfrac{1}{\ln 17}\cdot\dfrac{1}{x}$ Theorem 14.

$= \dfrac{1}{x\ln 17}$

17. $g(x) = \log_6 (5x+1)$

Using the Chain Rule and Theorem 14, we get

$g'(x) = \dfrac{1}{\ln 6}\cdot\dfrac{1}{5x+1}\cdot\dfrac{d}{dx}(5x+1)$

$= \dfrac{1}{\ln 6}\cdot\dfrac{1}{5x+1}\cdot 5$

$= \dfrac{5}{(5x+1)\ln 6}$

19. $F(x) = \log(6x-7)$ $(\log x = \log_{10} x)$

Using the Chain Rule and Theorem 14 we get

$F'(x) = \dfrac{1}{\ln 10}\cdot\dfrac{1}{6x-7}\cdot\dfrac{d}{dx}(6x-7)$

$= \dfrac{1}{\ln 10}\cdot\dfrac{1}{6x-7}\cdot 6$

$= \dfrac{6}{(6x-7)\ln 10}$

21. $y = \log_8 (x^3+x)$

Using the Chain Rule and Theorem 14, we get

$\dfrac{dy}{dx} = \dfrac{1}{\ln 8}\cdot\dfrac{1}{x^3+x}\cdot\dfrac{d}{dx}(x^3+x)$

$= \dfrac{1}{\ln 8}\cdot\dfrac{1}{x^3+x}\cdot(3x^2+1)$

$= \dfrac{3x^2+1}{(x^3+x)\ln 8}$

23. $f(x) = 4\log_7\left(\sqrt{x} - 2\right)$

$f'(x) = 4\dfrac{d}{dx}\log_7\left(\sqrt{x} - 2\right)$

Using the Chain Rule and Theorem 14, we get:

$f'(x) = 4 \cdot \dfrac{1}{\ln 7} \cdot \dfrac{1}{\sqrt{x} - 2} \cdot \dfrac{d}{dx}\left(\sqrt{x} - 2\right)$

$= \dfrac{4}{\left(\sqrt{x} - 2\right)\ln 7} \cdot \dfrac{d}{dx}\left(x^{1/2} - 2\right) \quad \left[\sqrt[n]{x} = x^{1/n}\right]$

$= \dfrac{4}{\left(\sqrt{x} - 2\right)\ln 7}\left(\dfrac{1}{2}x^{-1/2}\right) \quad$ Power Rule

$= \dfrac{4}{\left(\sqrt{x} - 2\right)\ln 7} \cdot \dfrac{1}{2x^{1/2}} \quad$ Properties of Exponents

$= \dfrac{4}{2\sqrt{x}\left(\sqrt{x} - 2\right)\ln 7}$

$= \dfrac{2}{\left(x - 2\sqrt{x}\right)\ln 7}$

25. $y = 6^x \cdot \log_7 x$

Since y is of the form $f(x) \cdot g(x)$, we apply the Product Rule.

$\dfrac{dy}{dx} = 6^x \cdot \left(\dfrac{1}{\ln 7} \cdot \dfrac{1}{x}\right) + (\ln 6)6^x \cdot \log_7 x$

Next, we use the commutative property of multiplication to rearrange the derivative:

$\dfrac{dy}{dx} = \dfrac{6x}{x\ln 7} + 6^x \ln 6 \cdot \log_7 x$

27. $G(x) = \left(\log_{12} x\right)^5$

Using the Extended Power Rule, we have:

$G'(x) = 5 \cdot \left(\log_{12} x\right)^4 \cdot \dfrac{d}{dx}\left(\log_{12} x\right)$

$= 5 \cdot \left(\log_{12} x\right)^4 \cdot \left(\dfrac{1}{\ln 12} \cdot \dfrac{1}{x}\right) \quad$ Theorem 14

$= 5\left(\log_{12} x\right)^4\left(\dfrac{1}{x\ln 12}\right)$

29. $g(x) = \dfrac{7^x}{4x + 1}$

Since $g(x)$ is in the form $g(x) = \dfrac{f(x)}{h(x)}$ we apply the Quotient Rule.

$g'(x) = \dfrac{(4x+1)\left(\dfrac{d}{dx}7^x\right) - 7^x\left(\dfrac{d}{dx}(4x+1)\right)}{(4x+1)^2}$

$= \dfrac{(4x+1)(\ln 7)7^x - 7^x(4)}{(4x+1)^2}$

$= \dfrac{7^x\left((4x+1)\ln 7 - 4\right)}{(4x+1)^2} \quad$ Factor out 7^x

$= \dfrac{7^x\left(4x\ln 7 + \ln 7 - 4\right)}{(4x+1)^2} \quad$ Distribute $\ln 7$

31. $y = 5^{2x^3 - 1} \cdot \log(6x + 5)$

Using the Product Rule we have:

$\dfrac{dy}{dx} = 5^{2x^3 - 1}\left(\dfrac{d}{dx}\log(6x + 5)\right)$

$\quad + \left(\dfrac{d}{dx}5^{2x^3 - 1}\right)\log(6x + 5)$

Next, using the Chain Rule, we have:

$\dfrac{dy}{dx} = 5^{2x^3 - 1}\left(\dfrac{1}{\ln 10} \cdot \dfrac{1}{6x + 5} \cdot 6\right)$

$\quad + (\ln 5)5^{2x^3 - 1}\left(6x^2\right) \cdot \log(6x + 5)$

Next, using properties of multiplication, we have:

$\dfrac{dy}{dx}$

$= \dfrac{6 \cdot 5^{2x^3 - 1}}{(6x + 5)\ln 10} + (\ln 5)5^{2x^3 - 1} \cdot 6x^2 \log(6x + 5)$

33. $F(x) = 7^x\left(\log_4 x\right)^9$

Using the Product Rule, the Extended Power Rule and Theorem 12, we have:

$F'(x)$

$= 7^x \cdot \dfrac{d}{dx}\left[\left(\log_4 x\right)^9\right] + \left(\log_4 x\right)^9 \cdot \dfrac{d}{dx}\left[7^x\right]$

$= 7^x \cdot \left[9\left(\log_4 x\right)^8 \cdot \left(\dfrac{1}{x \cdot \ln 4}\right)\right] +$

$\quad \left(\log_4 x\right)^9 \cdot \left[(\ln 7)7^x\right]$

$= \dfrac{7^x \cdot 9 \cdot \left(\log_4 x\right)^8}{x \cdot \ln 4} + (\ln 7)7^x \cdot \left(\log_4 x\right)^9$

35. $f(x) = (3x^5 + x)^5 \log_3 x$

First, using the Product Rule, we have:

$$f'(x) = (3x^5 + x)^5 \frac{d}{dx} \log_3 x$$
$$+ \left(\frac{d}{dx}(3x^5 + x)^5 \right) \log_3 x$$

Next, we will apply the Chain Rule:

$$f'(x) = (3x^5 + x)^5 \left(\frac{1}{\ln 3} \cdot \frac{1}{x} \right)$$
$$+ 5(3x^5 + x)^4 \frac{d}{dx}(3x^5 + x) \cdot \log_3 x$$

Which gives us:

$$f'(x) = (3x^5 + x)^5 \left(\frac{1}{x \ln 3} \right)$$
$$+ 5(3x^5 + x)^4 (15x^4 + 1) \log_3 x$$

37. a) $V(t) = 5200(0.80)^t$

$$V'(t) = 5200 \frac{d}{dt}(0.80)^t$$
$$= 5200(\ln 0.80)(0.80)^t \quad \text{Theorem 12}$$

b) $\boxed{tw}$

39. a) $L(t) = 1547(1.083)^t$

Since t is the number of years since 1980, the year 2012 corresponds to $t = 2012 - 1980 = 32$ years.

$$L(32) = 1547(1.083)^{32}$$
$$= 1547(12.82657226)$$
$$= 19,842.7072800$$
$$\approx 19,842.71$$

In the year 2012. total financial liability of U.S. households will be approximately $19,842.71 billion.

b) First, we find the derivative using Theorem 12:

$$L'(t) = 1547(\ln 1.083)(1.083)^t$$

Next, we evaluate the derivative at $t = 25$

$$L'(25) = 1547(\ln 1.083)(1.083)^{25}$$
$$= 1547(\ln 1.083)(7.34025953)$$
$$= 905.42098032$$
$$\approx 905.42$$

c) $\boxed{tw}$

41. a) $P(t) = (0.98)^t$

$$P(10) = (0.98)^{10} = 0.81707281$$

Ten years after the neighboring farm begins to use the GMO seed, the GMO-free farm will be 81.7 percent GMO-free.

b) First, we find the derivative

$$P'(t) = \ln(0.98)(0.98)^t.$$

Next, we evaluate the derivative at $t = 15$:

$$P'(15) = (\ln 0.98)(0.98)^{15}$$
$$= (\ln 0.98) \cdot 0.73856910$$
$$= -0.01492110$$
$$\approx -0.0149$$

c) $\boxed{tw}$

43. $R = \log \frac{I}{I_0}$

We substitute $10^{6.9} \cdot I_0$ for I

$$R = \log \frac{10^{6.9} \cdot I_0}{I_0}$$
$$= \log 10^{6.9}$$
$$= 6.9 \qquad \qquad \text{(P5)}$$

45. $I = I_0 10^{0.1L}$

a) Substituting 100 for L we have:

$$I = I_0 10^{0.1(100)}$$
$$= I_0 10^{10}$$
$$= 10^{10} \cdot I_0$$

The intensity of a power mower is $10^{10} \cdot I_0$.

b) Substituting 10 for L we have:

$$I = I_0 10^{0.1(10)}$$
$$= I_0 10^1$$
$$= 10 \cdot I_0$$

The intensity of a just audible sound is $10 \cdot I_0$

c) Comparing the intensity in parts (a) and (b) we have $10^{10} \cdot I_0 = 10^9 (10 \cdot I_0)$. The intensity of (a) is 10^9 times more than the intensity of (b).

d) $I = I_0 10^{0.1L}$

$$\frac{dI}{dL} = I_0 \frac{d}{dL} 10^{0.1L} \qquad I_0 \text{ is a constant.}$$

Next, using Theorem 12 and the Chain Rule we have:

$$\frac{dI}{dL} = I_0 (\ln 10) 10^{0.1L} \cdot \left(\frac{d}{dL} 0.1L \right)$$

$$= I_0 (\ln 10) 10^{0.1L} (0.1)$$

$$= 0.1 \cdot \ln 10 \cdot I_0 10^{0.1L}$$

e) $\boxed{tw}$

47. a) $L = 10 \log \dfrac{I}{I_0}$

First, we rearrange the equation using Property 2.

$$L = 10 (\log I - \log I_0)$$

$$= 10 \log I - 10 \log I_0$$

Next, we take the derivative using the Difference Rule:

$$\frac{dL}{dI} = 10 \frac{d}{dI} \log I - \frac{d}{dI} 10 \log I_0$$

Using Theorem 14 we have:

$$\frac{dL}{dI} = 10 \frac{1}{\ln 10} \cdot \frac{1}{I} - 0 \qquad I_0 \text{ is a constant}$$

$$= \frac{10}{\ln 10} \cdot \frac{1}{I}$$

b) $\boxed{tw}$

49. By Theorem 13: $\displaystyle\lim_{h \to 0} \frac{3^h - 1}{h} = \ln 3$

51. $y = 2^{x^4}$

$$\frac{dy}{dx} = (\ln 2) 2^{x^4} \frac{d}{dx} x^4$$

$$= (\ln 2) 2^{x^4} \cdot 4x^3$$

53. $y = \log_3 (\log x)$

Using the Chain Rule and Theorem 14, we have:

$$\frac{dy}{dx} = \frac{1}{\ln 3} \cdot \frac{1}{\log x} \cdot \frac{d}{dx} \log x$$

$$= \frac{1}{\ln 3} \cdot \frac{1}{\log x} \cdot \frac{1}{\ln 10} \cdot \frac{1}{x}$$

$$= \frac{1}{\ln 3 \cdot \log x \cdot \ln 10 \cdot x}$$

55. $f(x) = a^{f(x)}$

$$f(x) = e^{f(x) \ln a} \qquad\qquad \left[a^x = e^{x \ln a} \right]$$

$$f'(x) = e^{f(x) \ln a} \frac{d}{dx} f(x) \ln a \qquad \textbf{Chain Rule}$$

$$f'(x) = e^{f(x) \ln a} \cdot f'(x) \ln a$$

$$f'(x) = \ln a \cdot a^{f(x)} \cdot f'(x) \qquad \left[a^x = e^{x \ln a} \right]$$

57. $y = [f(x)]^{g(x)}, \quad f(x > 0)$

$$y = e^{g(x) \cdot \ln f(x)} \qquad\qquad \left[a^x = e^{x \ln a} \right]$$

$$\frac{dy}{dx} = e^{g(x) \cdot \ln f(x)} \frac{d}{dx} g(x) \ln f(x) \quad \textbf{Chain Rule}$$

Next, using the Product Rule, we have:

$$\frac{dy}{dx} = e^{g(x) \cdot \ln f(x)} \left(g(x) \cdot \frac{1}{f(x)} \cdot f'(x) + g'(x) \ln f(x) \right)$$

Simplifying we get:

$$\frac{dy}{dx} = e^{g(x) \ln f(x)} \left(\frac{g(x) f'(x)}{f(x)} + g'(x) \ln f(x) \right)$$

$$= [f(x)]^{g(x)} \left(\frac{g(x) f'(x)}{f(x)} + g'(x) \ln f(x) \right)$$

59. $\boxed{tw}$

Exercise Set 3.6

---

1. a) The demand function is

$$q = D(x) = 400 - x.$$

The definition of the elasticity of demand is given by: $E(x) = -\dfrac{x \cdot D'(x)}{D(x)}$. In order to find the elasticity of demand, we need to find the derivative of the demand function first.

$$\frac{dq}{dx} = D'(x) = \frac{d}{dx}(400 - x) = -1.$$

Next, we substitute -1 for $D'(x)$, and $400 - x$ for $D(x)$ into the expression for elasticity.

$$E(x) = -\frac{x \cdot (-1)}{400 - x} = \frac{x}{400 - x}$$

b) Substituting $x = 125$ into the expression found in part (a) we have:

$$E(125) = \frac{(125)}{400 - (125)} = \frac{125}{275} = \frac{5}{11}$$

Since $E(125) = \frac{5}{11}$ is less than one, the demand is inelastic.

c) The values of x for which $E(x) = 1$ will maximize total revenue. We solve:

$$E(x) = 1$$

$$\frac{x}{400 - x} = 1$$

$$x = 400 - x$$

$$2x = 400$$

$$x = 200$$

A price of $200 will maximize total revenue.

3. $q = D(x) = 200 - 4x;\ x = 46$

a) $D'(x) = -4$

$$E(x) = -\frac{x \cdot D'(x)}{D(x)} = -\frac{x \cdot (-4)}{200 - 4x} = \frac{4x}{200 - 4x}$$

$$= \frac{x}{50 - x}$$

b) Substituting $x = 46$ into the expression found in part (a) we have:

$$E(46) = \frac{46}{50 - 46} = \frac{46}{4} = \frac{23}{2} = 11.5$$

Since $E(46) > 1$, demand is elastic.

c) We solve $E(x) = 1$

$$\frac{x}{50 - x} = 1$$

$$x = 50 - x$$

$$2x = 50$$

$$x = 25$$

A price of $25 will maximize total revenue.

5. $q = D(x) = \dfrac{400}{x};\ x = 50$

a) First, we rewrite the demand function.

$$D(x) = \frac{400}{x} = 400x^{-1}.$$

Next, we take the derivative of the demand function, using the Power Rule.

$$D'(x) = 400(-1)x^{-2} = -400x^{-2}$$

Making the appropriate substitutions into the elasticity function, we have

$$E(x) = -\frac{x \cdot D'(x)}{D(x)} = -\frac{x \cdot \left(-400x^{-2}\right)}{400x^{-1}} =$$

$$= \frac{400x^{-1}}{400x^{-1}} = 1$$

Therefore, $E(x) = 1$ for all values of x.

b) $E(50) = 1$, so demand is unit elastic.

c) $E(x) = 1$ for all values of x. Therefore, total revenue is maximized for all values of x. In other words, total revenue is the same regardless of the price.

7. $q = D(x) = \sqrt{600 - x};\ x = 100$

a) First rewrite the demand function:

$$D(x) = (600 - x)^{\frac{1}{2}}.$$

Next, we take the derivative of the demand function, using the Chain Rule:

$$D'(x) = \frac{1}{2}(600 - x)^{-\frac{1}{2}} \cdot \frac{d}{dx}(600 - x)$$

$$= \frac{-1}{2\sqrt{600 - x}}$$

Making the appropriate substitutions into the elasticity function, we have

$$E(x) = -\frac{x \cdot D'(x)}{D(x)} = -\frac{x \cdot \left(\dfrac{-1}{2\sqrt{600 - x}}\right)}{\sqrt{600 - x}} =$$

$$= \frac{\dfrac{x}{2\sqrt{600 - x}}}{\sqrt{600 - x}} = \frac{x}{2(600 - x)}$$

$$= \frac{x}{1200 - 2x}$$

b) Substituting $x = 100$ into the expression found in part (a) we have:

$$E(100) = \frac{100}{1200 - 2(100)} = \frac{100}{1000} = \frac{1}{10}$$

Since $E(100) < 1$, demand is inelastic.

c) Solve $E(x) = 1$

$$\frac{x}{1200 - 2x} = 1$$

$$x = 1200 - 2x$$

$$3x = 1200$$

$$x = 400$$

A price of $400 will maximize total revenue.

9. $q = D(x) = 100e^{-0.25x}$; $x = 10$

a) Using the Chain Rule we have:

$D'(x) = 100e^{-0.25x}(-0.25)$

$= -25e^{-0.25x}$

Making the appropriate substitutions into the elasticity function, we have

$E(x) = -\dfrac{x \cdot D'(x)}{D(x)} = -\dfrac{x \cdot \left(-25e^{-0.25x}\right)}{100e^{-0.25x}}$

$= \dfrac{25xe^{-0.25x}}{100e^{-0.25x}} = \dfrac{x}{4}$

b) Substituting $x = 10$ into the expression found in part (a) we have:

$E(10) = \dfrac{10}{4} = \dfrac{5}{2} = 2.5$

Since $E(10) > 1$, demand is elastic.

c) Solve $E(x) = 1$

$\dfrac{x}{4} = 1$

$x = 4$

A price of $4 will maximize total revenue.

11. $q = D(x) = \dfrac{100}{(x+3)^2}$; $x = 1$

a) First, we rewrite the demand function:

$D(x) = 100(x+3)^{-2}$.

Next, we take the derivative of the demand function, using the Chain Rule:

$D'(x) = 100(-2)(x+3)^{-3}\left(\dfrac{d}{dx}(x+3)\right)$

$= -200(x+3)^{-3}$

$= -\dfrac{200}{(x+3)^3}$

Making the appropriate substitutions into the elasticity function, we have:

$E(x) = -\dfrac{x \cdot D'(x)}{D(x)} = -\dfrac{x \cdot \left(-\dfrac{200}{(x+3)^3}\right)}{\dfrac{100}{(x+3)^2}}$

$= x \cdot \left(\dfrac{200}{(x+3)^3}\right)\dfrac{(x+3)^2}{100}$

$= \dfrac{2x}{x+3}$

b) Substituting $x = 1$ into the expression found in part (a) we have:

$E(1) = \dfrac{2 \cdot (1)}{1+3} = \dfrac{2}{4} = \dfrac{1}{2}$

Since $E(1) < 1$, demand is inelastic.

c) Solve $E(x) = 1$

$\dfrac{2x}{x+3} = 1$

$2x = x+3$

$x = 3$

A price of $3 will maximize total revenue.

13. $q = D(x) = 967 - 25x$

a) $D'(x) = -25$

$E(x) = -\dfrac{x \cdot D'(x)}{D(x)} = -\dfrac{x \cdot (-25)}{967 - 25x}$

$= \dfrac{25x}{967 - 25x}$

b) We set $E(x) = 1$ and solve for x.

$\dfrac{25x}{967 - 25x} = 1$

$25x = 967 - 25x$

$50x = 967$

$x = \dfrac{967}{50}$

$x = 19.34$

Demand is unitary elastic when price is 19.34 cents.

c) Demand is elastic when $E(x) > 1$.

Testing a value on each side of 19.34 cents, we have:

$E(19) = \dfrac{25 \cdot 19}{967 - 25 \cdot 19} \approx 0.97 < 1$

$E(20) = \dfrac{25 \cdot 20}{967 - 25 \cdot 20} \approx 1.07 > 1$

Therefore, the demand for cookies is elastic for prices greater than 19.34 cents.

d) Demand is inelastic when $E(x) < 1$.

Using the calculations from part (c), we see that the demand for cookies is inelastic for prices less than 19.34 cents.

e) Total revenue is maximized when $E(x) = 1$.

In part (b) we showed that $E(x) = 1$ when price was 19.34 cents. Therefore, revenue will be maximized when price is 19.34 cents.

f) We have shown that the demand for cookies is elastic when the price of cookies is 20 cents. Therefore a small increase in price will cause total revenue to decrease.

15. $q = D(x) = \sqrt{200 - x^3}$

a) First, we rewrite the demand function to make it easier to find the derivative.

$$D(x) = (200 - x^3)^{\frac{1}{2}}.$$

Next, using the Chain Rule, we have:

$$D'(x) = \frac{1}{2}(200 - x^3)^{-\frac{1}{2}}(-3x^2)$$

$$= \frac{-3x^2}{2\sqrt{200 - x^3}}$$

Now, substituting in the elasticity function we get:

$$E(x) = -\frac{x \cdot D'(x)}{D(x)} = -\frac{x \cdot \left(\dfrac{-3x^2}{2\sqrt{200 - x^3}}\right)}{\sqrt{200 - x^3}}$$

$$= \frac{3x^3}{2\left(\sqrt{200 - x^3}\right)^2}$$

$$= \frac{3x^3}{2(200 - x^3)}$$

$$= \frac{3x^3}{400 - 2x^3}$$

b) $E(3) = \dfrac{3(3)^3}{400 - 2(3)^3}$

$$= \frac{81}{346}$$

$$\approx 0.2341$$

Since $E(3) < 1$, the demand for computer games is inelastic when price is \$3.

c) From part (b) we know that the demand for computer games is inelastic at a price of \$3. Therefore an increase in the price of computer games will lead to an increase in the total revenue.

17. $q = D(x) = \dfrac{k}{x^n}$

a) First, we rewrite the demand function to make it easier to find the derivative.

$$D(x) = k \cdot x^{-n}$$

Using the Power Rule, we have:

$$D'(x) = k \cdot (-n) x^{-n-1} = -nkx^{-n-1}$$

Substituting into the elasticity function we have:

$$E(x) = -\frac{x \cdot D'(x)}{D(x)} = -\frac{x \cdot \left(-nkx^{-n-1}\right)}{kx^{-n}}$$

$$= \frac{nk\left(x \cdot x^{-n-1}\right)}{kx^{-n}}$$

$$= \frac{nk\left(x^{-n}\right)}{kx^{-n}}$$

$$= n$$

b) No, the elasticity of demand is constant for all prices. $E(x) = n$.

c) Total revenue is maximized when $E(x) = 1$. Since $E(x) = n$, total revenue will be maximized when $n = 1$.

19. $L(x) = \ln D(x)$

$$L'(x) = \frac{1}{D(x)} \cdot D'(x) = \frac{D'(x)}{D(x)}$$

The formula for elasticity of demand is:

$$E(x) = -\frac{x \cdot D'(x)}{D(x)} = -x\left(\frac{D'(x)}{D(x)}\right)$$

Substituting $L'(x)$ for $\dfrac{D'(x)}{D(x)}$ we have:

$$E(x) = -x \cdot L'(x)$$

21. $\boxed{tw}$

Chapter 4

Integration

Exercise Set 4.1

1. $c(x) = -0.012x + 6.50,$ for $x \leq 300$

Note that the cost per pound of roasting coffee decreases as the number of pounds of coffee increases. We use the area under the graph to find the total cost of roasting 200 lb of coffee. Shading the area under $c(x)$ on the interval

$0 \leq x \leq 200$ we see:

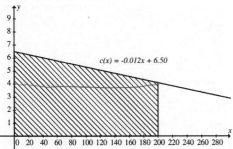

The area under the curve is a trapezoid; therefore, calculating the total cost of roasting 200 lb of coffee will require calculating the area of the trapezoid.

The formula for calculating the area of a

trapezoid is $A = \frac{1}{2}h(b_1 + b_2)$, where h is the

height of the trapezoid and b_1 and b_2 are the lengths of the respective bases. If we view the trapezoid sideways, we see that

$h = 200$

$b_1 = c(0) = -0.012(0) + 6.50 = 6.50$

$b_2 = c(200) = -0.012(200) + 6.50 = 4.1$

Substituting these values into the formula, we have:

Total Cost = Area of trapezoid

$$= \frac{1}{2}h(b_1 + b_2)$$

$$= \frac{1}{2}(200)(6.5 + 4.1)$$

$$= 100(10.6)$$

$$= 1060$$

The total cost of roasting 200 pounds of coffee is $1060.

3. $c(x) = -0.04x + 85,$ for $x \leq 1000$

Note that the cost per card decreases as the number of cards produced increases. We use the area under the graph to find the total cost of producing 650 cards.

Shading the area $c(x)$ on the interval

$0 \leq x \leq 650$ we see:

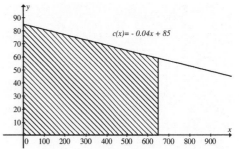

The area under the curve is a trapezoid; therefore, calculating the total cost of producing 650 note cards will require calculating the area of the trapezoid.

The formula for calculating the area of a

trapezoid is $A = \frac{1}{2}h(b_1 + b_2)$, where h is the

height of the trapezoid and b_1 and b_2 are the lengths of the respective bases. If we view the trapezoid sideways, we see that

$h = 650$

$b_1 = c(0) = -0.04(0) + 85 = 85$

$b_2 = c(650) = -0.04(650) + 85 = 59$

Substituting these values into the formula, we have:

Total Cost = Area of trapezoid

$$= \frac{1}{2}h(b_1 + b_2)$$

$$= \frac{1}{2}(650)(85 + 59)$$

$$= 325(144)$$

$$= 46,800$$

The total cost of producing 650 cards is 46,800 cents or $468.

5. $R'(x) = 2x - 1150,\ x \geq 0$

Note that the marginal revenue is rate of change of total revenue with respect to the number of tickets. We use the area under the graph to find the total revenue of selling 300 tickets.

Shading the area under $R'(x)$ on the interval $0 \leq x \leq 300$ we see:

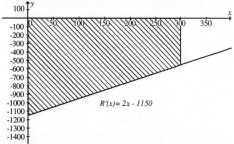

The area between the x-axis and the curve is a trapezoid; therefore, the total revenue will be equal to the area of the trapezoid.

The formula for calculating the area of a trapezoid is $A = \dfrac{1}{2}h(b_1 + b_2)$, where h is the height of the trapezoid and b_1 and b_2 are the lengths of the respective bases. If we view the trapezoid sideways, we see that
$h = 200$

$b_1 = R'(0) = 2(0) - 1150 = -1150$

$b_2 = R'(300) = 2(300) - 1150 = -550$

Because the marginal revenue function is below the x-axis, the values for b_1 and b_2 are negative.

Due to the nature of the application, we will keep these negative values. The result will be a negative area of the trapezoid. This simply means that the total revenue will be negative, which is a counterintuitive result, but we can calculate it just the same.

Substituting these values into the formula, we have:

Total Cost = Area of trapezoid

$$= \frac{1}{2}h(b_1 + b_2)$$

$$= \frac{1}{2}(300)\big(-1150 + (-550)\big)$$

$$= 150(-1700)$$

$$= -255{,}000$$

The total revenue from the sale of the first 300 tickets is $-\$255{,}000$.

7. $C'(x) = -\dfrac{2}{25}x + 50, \quad \text{for } x \leq 450$

We use the area under the graph to find the total cost of producing the first 200 dresses.

Shading the area under $C'(x)$ on the interval $0 \leq x \leq 200$ we see:

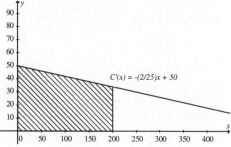

The area under the curve is a trapezoid; therefore, calculating the total cost of producing 200 dresses will require calculating the area of the trapezoid.

The formula for calculating the area of a trapezoid is $A = \dfrac{1}{2}h(b_1 + b_2)$, where h is the height of the trapezoid and b_1 and b_2 are the lengths of the respective bases. If we view the trapezoid sideways, we see that
$h = 200$

$b_1 = C'(0) = -\dfrac{2}{25}(0) + 50 = 50$

$b_2 = C'(200) = -\dfrac{2}{25}(200) + 50 = 34$

Substituting these values into the formula, we have:

Total Cost = Area of trapezoid

$$= \frac{1}{2}h(b_1 + b_2)$$

$$= \frac{1}{2}(200)(50 + 34)$$

$$= 100(84)$$

$$= 8400$$

The total cost of producing the first 200 dresses is $8400.

9. $C'(x) = -0.00002x^2 - 0.04x + 45$

We divide the interval $[0, 800]$ into five subintervals, each of width $\Delta x = \dfrac{800}{5} = 160$.

To determine the height of each rectangle, we use the left endpoint of each subinterval. We illustrate the rectangles with a graph.

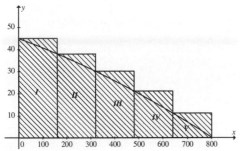

Then, we have

Total Cost ≈ Area I + Area II + Area III +

Area IV + Area V

The five left endpoints are 0, 160, 320, 480, and 640. Evaluating $C'(x)$ at each of these endpoints will determine the height of each rectangle and the width was determined to be 160 ft. Therefore, the area of each rectangle is:

Area I $= C'(0) \cdot 160 = 45 \cdot 160 = 7200$

Area II $= C'(160) \cdot 160 = 38.088 \cdot 160 = 6094.08$

Area III $= C'(320) \cdot 160 = 30.152 \cdot 160 = 4824.32$

Area IV $= C'(480) \cdot 160 = 21.192 \cdot 160 = 3390.72$

Area V $= C'(640) \cdot 160 = 11.208 \cdot 160 = 1793.28$

Summing the area of each of the five rectangles yields:

Total Cost ≈ 7200 + 6094.08 + 4824.32 +

3390.72 + 1793.28

≈ 23,302.4

Thus, the total cost of manufacturing 800 feet of molding is approximately 23,302.4 cents or about $233.02.

11. $C'(x) = 0.000008x^2 - 0.004x + 2$, for $x \le 350$

We divide the interval $[0, 270]$ into three

subintervals, each of width $\Delta x = \dfrac{270}{3} = 90$. To

determine the height of each rectangle, we use the left endpoint of each subinterval. We illustrate the rectangles with a graph.

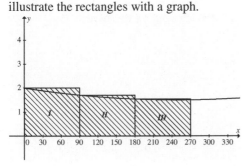

Then, we have

Total Cost ≈ Area I + Area II + Area III

The three left endpoints are 0, 90, and 180. Evaluating $C'(x)$ at each of these endpoints will determine the height of each rectangle and the width was determined to be 160 ft. Therefore, the area of each rectangle is:

Area I $= C'(0) \cdot 90 = 2 \cdot 90 = 180$

Area II $= C'(90) \cdot 90 = 1.7048 \cdot 90 = 153.432$

Area III $= C'(180) \cdot 90 = 1.5392 \cdot 90 = 138.528$

Summing the area of each of the three rectangles yields:

Total Cost ≈ 180 + 153.432 + 138.528

≈ 471.96

Thus, the total cost of producing 270 pints of fresh-squeezed orange juice is approximately $471.96.

13. Note that we are adding consecutive multiples of 3:

$$3 + 6 + 9 + 12 + 15 + 18 = \sum_{i=1}^{6} 3i.$$

15. $f(x_1) + f(x_2) + f(x_3) + f(x_4) = \displaystyle\sum_{i=1}^{4} f(x_i).$

17. $G(x_1) + G(x_2) + \cdots + G(x_{15}) = \displaystyle\sum_{i=1}^{15} G(x_i).$

19. $\displaystyle\sum_{i=1}^{4} 2^i = 2^1 + 2^2 + 2^3 + 2^4$

$= 2 + 4 + 8 + 16,$ or 30

21. $\displaystyle\sum_{i=1}^{5} f(x_i)$

$= f(x_1) + f(x_2) + f(x_3) + f(x_4) + f(x_5).$

23. $f(x) = \dfrac{1}{x^2}$

a) In the drawing in the text the interval $[1,7]$ has been divided into 6 subintervals, each having width 1 $\left[\Delta x = \dfrac{7-1}{6} = 1\right]$.

The heights of the rectangles shown are

$f(1) = \dfrac{1}{1^2} = 1$

$f(2) = \dfrac{1}{2^2} = \dfrac{1}{4} = 0.2500$

$f(3) = \dfrac{1}{3^2} = \dfrac{1}{9} \approx 0.1111$

$f(4) = \dfrac{1}{4^2} = \dfrac{1}{16} = 0.0625$

$f(5) = \dfrac{1}{5^2} = \dfrac{1}{25} = 0.0400$

$f(6) = \dfrac{1}{6^2} = \dfrac{1}{36} \approx 0.0278$

Therefore, the area of each rectangle is:

Rectangle I $= f(1) \cdot \Delta x = 1 \cdot 1 = 1$

Rectangle II $= f(2) \cdot \Delta x = 0.2500 \cdot 1 = 0.2500$

Rectangle III $= f(3) \cdot \Delta x = 0.1111 \cdot 1 = 0.1111$

Rectangle IV $= f(4) \cdot \Delta x = 0.0625 \cdot 1 = 0.0625$

Rectangle V $= f(5) \cdot \Delta x = 1 \cdot 0.0400 = 0.0400$

Rectangle VI $= f(6) \cdot \Delta x = 1 \cdot 0.0278 = 0.0278$

The area of the region under the curve over $[1,7]$ is approximately the sum of the areas of the 6 rectangles. Therefore, the total area is approximately:

$1 + 0.2500 + 0.1111 + 0.0625 +$

$0.0400 + 0.0278 \approx 1.4914.$

b) In the drawing in the text the interval $[1,7]$ has been divided into 12 subintervals, each having width 0.5 $\left[\Delta x = \dfrac{7-1}{12} = \dfrac{6}{12} = 0.5\right]$.

The heights of 6 of the rectangles were computed in part (a). The heights of the other 6 rectangles are computed as follows:

$f(1.5) = \dfrac{1}{1.5^2} = \dfrac{1}{2.25} \approx 0.4444$

$f(2.5) = \dfrac{1}{2.5^2} = \dfrac{1}{6.25} = 0.1600$

$f(3.5) = \dfrac{1}{3.5^2} = \dfrac{1}{12.25} \approx 0.00816$

$f(4.5) = \dfrac{1}{4.5^2} = \dfrac{1}{20.25} \approx 0.0494$

$f(5.5) = \dfrac{1}{5.5^2} = \dfrac{1}{30.25} \approx 0.0331$

$f(6.5) = \dfrac{1}{6.5^2} = \dfrac{1}{42.25} \approx 0.0237$

Therefore, the area of each rectangle is:

Rectangle I: $f(1) \cdot \Delta x = 1(0.5) = 0.5000$

Rectangle II: $f(1.5) \cdot \Delta x = 0.4444(0.5) \approx 0.2222$

Rectangle III: $f(2) \cdot \Delta x = 0.2500(0.5) = 0.1250$

Rectangle IV: $f(2.5) \cdot \Delta x = 0.1600(0.5) = 0.0800$

Rectangle V: $f(3) \cdot \Delta x = 0.1111(0.5) \approx 0.0556$

Rectangle VI: $f(3.5) \cdot \Delta x = 0.0816(0.5) \approx 0.0408$

Rectangle VII: $f(4) \cdot \Delta x = 0.0625(0.5) \approx 0.0313$

Rectangle VIII: $f(4.5) \cdot \Delta x = 0.0494(0.5) \approx 0.0247$

Rectangle IX: $f(5) \cdot \Delta x = 0.0400(0.5) = 0.0200$

Rectangle X: $f(5.5) \cdot \Delta x = 0.0331 \approx 0.0165$

Rectangle XI: $f(6) \cdot \Delta x = 0.0278(0.5) \approx 0.0139$

Rectangle XII: $f(6.5) \cdot \Delta x = 0.0237(0.5) \approx 0.0118$

The area of the region under the curve over $[1,7]$ is approximately the sum of the areas of the 12 rectangles. Therefore, the total area is approximately:

$0.5 + 0.2222 + 0.1250 + 0.0800 +$

$0.0556 + 0.0408 + 0.0313 +$

$0.0247 + 0.0200 + 0.0165 +$

$0.0139 + 0.0118 \approx 1.1418.$

Answers may vary slightly depending on when rounding was done.

25. $P'(x) = -0.0006x^3 + 0.28x^2 + 55.6x$

We divide $[0, 300]$ into 6 subintervals of width $\Delta x = 50$. Therefore, the values of x_i are:

$x_1 = 0$; $x_2 = 50$; $x_3 = 100$;

$x_4 = 150$; $x_5 = 200$; $x_6 = 250$.

Then, the area under the curve is approximately:

$\displaystyle\sum_{i=1}^{6} P'(x_i) \Delta x = P'(x_1) \cdot 50 + P'(x_2) \cdot 50 +$

$\qquad P'(x_3) \cdot 50 + P'(x_4) \cdot 50 +$

$\qquad P'(x_5) \cdot 50 + P'(x_6) \cdot 50$

$\qquad = P'(0) \cdot 50 + P'(50) \cdot 50 +$

$\qquad P'(100) \cdot 50 + P'(150) \cdot 50 +$

$\qquad P'(200) \cdot 50 + P'(250) \cdot 50$

$\qquad = 0 \cdot 50 + 3405 \cdot 50 + 7760 \cdot 50 +$

$\qquad 12{,}615 \cdot 50 + 17{,}520 \cdot 50 +$

$\qquad 22{,}025 \cdot 50$

$\qquad = 0 + 170{,}250 + 388{,}000 +$

$\qquad 630{,}750 + 876{,}000 + 1{,}101{,}250$

$\qquad = 3{,}166{,}250.$

The health club's total profit when 300 members are enrolled is approximately 3,166,250 cents or \$31,662.50.

27. $f(x) = 0.01x^4 - 1.44x^2 + 60$

Dividing the interval $[2, 10]$ into four subintervals, we calculate the width of each subinterval to be $\Delta x = \dfrac{10 - 2}{4} = \dfrac{8}{4} = 2$, with x_i ranging from $x_1 = 2$ to $x_4 = 8$. Although a drawing is not required, we can make one to help visualize the area.

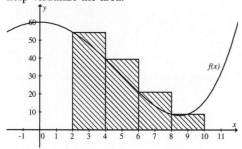

The area under the curve from 2 to 10 is approximately

$\displaystyle\sum_{i=1}^{4} f(x_i) \Delta x = f(2) \cdot 2 + f(4) \cdot 2 +$

$\qquad f(6) \cdot 2 + f(8) \cdot 2$

$\qquad = 54.4 \cdot 2 + 39.52 \cdot 2 +$

$\qquad 21.12 \cdot 2 + 8.8 \cdot 2$

$\qquad = 247.68.$

29. $F(x) = 0.2x^3 + 2x^2 - 0.2x - 2$

Dividing the interval $[-8, -3]$ into five subintervals, we calculate the width of each subinterval to be $\Delta x = \dfrac{-3 - (-8)}{5} = \dfrac{5}{5} = 1$, with x_i ranging from $x_1 = -8$ to $x_5 = -4$.

Although a drawing is not required, we can make one to help visualize the area.

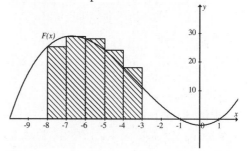

The area under the curve from -8 to -3 is approximately

$\displaystyle\sum_{i=1}^{5} F(x_i) \Delta x = F(-8) \cdot 1 + F(-7) \cdot 1 + F(-6) \cdot 1 +$

$\qquad F(-5) \cdot 1 + F(-4) \cdot 1$

$\qquad = 25.2 \cdot 1 + 28.8 \cdot 1 + 28 \cdot 1 +$

$\qquad 24 \cdot 1 + 18 \cdot 1$

$\qquad = 124.$

31. For the specific case, begin by expanding the sum:

$\displaystyle\sum_{i=1}^{4} k \cdot f(x_i) = k \cdot f(x_1) + k \cdot f(x_2) +$

$\qquad k \cdot f(x_3) + k \cdot f(x_4)$

Next, we factor out the common factor of k to get:

$\qquad = k \big[f(x_1) + f(x_2) + f(x_3) + f(x_4) \big]$

$\qquad = k \cdot \left[\displaystyle\sum_{i=1}^{4} f(x_i) \right]$

Thus:

$\displaystyle\sum_{i=1}^{4} k \cdot f(x_i) = k \sum_{i=1}^{4} f(x_i).$

Similarly for the general case, we have:

$$\sum_{i=1}^{n} k \cdot f(x_i)$$

$$= k \cdot f(x_1) + k \cdot f(x_2) + \cdots + k \cdot f(x_n)$$

$$= k[f(x_1) + f(x_2) + \cdots + f(x_n)]$$

$$= k\left[\sum_{i=1}^{n} f(x_i)\right]$$

Thus:

$$\sum_{i=1}^{n} k \cdot f(x_i) = k\sum_{i=1}^{n} f(x_i).$$

33. $f(x) = x^2 + 1$

$$\Delta x = \frac{5-0}{5} = 1$$

Using the Trapezoidal Rule, the area under the graph of $f(x)$ over the interval $[0,5]$ is approximately:

$$\text{Area} \approx \Delta x\left[\frac{f(0)}{2} + f(1) + \cdots + f(4) + \frac{f(5)}{2}\right]$$

The function values are:

$f(0) = 0^2 + 1 = 1$

$f(1) = 1^2 + 1 = 2$

$f(2) = 2^2 + 1 = 5$

$f(3) = 3^2 + 1 = 10$

$f(4) = 4^2 + 1 = 17$

$f(5) = 5^2 + 1 = 26$

Substituting these values into the Trapezoidal Rule, we have:

$$\text{Area} \approx 1 \cdot \left[\frac{1}{2} + 2 + 5 + 10 + 17 + \frac{26}{2}\right]$$

$$\approx \frac{1}{2} + 2 + 5 + 10 + 17 + 13$$

$$\approx 47.5$$

35. $g(x) = \sqrt{49 - x^2}$

Notice that this semi-circle has a radius of five. Therefore the exact area is given by:

$$A = \frac{1}{2}\pi(7)^2 = \frac{49}{2}\pi = 24.5\pi$$

To approximate the area under the graph of $g(x) = \sqrt{49 - x^2}$ using 10 rectangles, we first find the width of each rectangle.

$$\Delta x = \frac{7 - (-7)}{14} = \frac{14}{14} = 1.$$

The x_i will range from $x_1 = -7$ to $x_{10} = 6$. Although a drawing is not required, we can make one to help visualize the area.

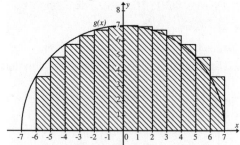

The area under the curve from -7 to 7 is approximately

$$\sum_{i=1}^{14} f(x_i)\Delta x = f(-7)\cdot 1 + f(-6)\cdot 1 +$$

$$f(-5)\cdot 1 + f(-4)\cdot 1 +$$

$$f(-3)\cdot 1 + f(-2)\cdot 1 +$$

$$f(-1)\cdot 1 + f(0)\cdot 1 + f(1)\cdot 1 +$$

$$f(2)\cdot 1 + f(3)\cdot 1 + f(4)\cdot 1 +$$

$$f(5)\cdot 1 + f(6)\cdot 1$$

$$\approx 75.4201$$

In order to compare, if we round to 4 decimal places we have:

$A = 24.5\pi = 76.9690$.

Exercise Set 4.2

1. $\displaystyle\int x^6\, dx$

$$= \frac{x^{6+1}}{6+1} + C \qquad \left[\int x^r dx = \frac{x^{r+1}}{r+1} + C\right]$$

$$= \frac{x^7}{7} + C \qquad \text{Don't forget the C.}$$

3. $\displaystyle\int 2\, dx$

$$= 2x + C \qquad \left[\int k\, dx = kx + C\right]$$

5. $\int x^{\frac{1}{4}} dx$

$= \dfrac{x^{\frac{1}{4}+1}}{\frac{1}{4}+1} + C \qquad \left[\int x^r dx = \dfrac{x^{r+1}}{r+1} + C \right]$

$= \dfrac{x^{\frac{5}{4}}}{\frac{5}{4}} + C$

$= \dfrac{4}{5} x^{\frac{5}{4}} + C$

7. $\int (x^2 + x - 1) dx$

$= \int x^2 dx + \int x dx - \int 1 dx \qquad$ The integral of a sum is the sum of the integrals.

$= \dfrac{x^{2+1}}{2+1} + \dfrac{x^{1+1}}{1+1} - x + C \leftarrow$ DON'T FORGET THE C!

$\left[\int x^r dx = \dfrac{x^{r+1}}{r+1} + C \right]$

$\left[\int k dx = kx + C \right]$

$= \dfrac{x^3}{3} + \dfrac{x^2}{2} - x + C$

9. $\int (2t^2 + 5t - 3) dt$

$= \int 2t^2 dt + \int 5t dt - \int 3 dt \qquad$ The integral of a sum is the sum of the integrals.

$= 2 \cdot \dfrac{t^{2+1}}{2+1} + 5 \cdot \dfrac{t^{1+1}}{1+1} - 3t + C$

$\left[\int x^r dx = \dfrac{x^{r+1}}{r+1} + C \right]$

$\left[\int k dx = kx + C \right]$

$= \dfrac{2}{3} t^3 + \dfrac{5}{2} t^2 - 3t + C$

11. $\int \dfrac{1}{x^3} dx = \int x^{-3} dx$

$= \dfrac{x^{-3+1}}{-3+1} + C \qquad \left[\int x^r dx = \dfrac{x^{r+1}}{r+1} + C \right]$

$= -\dfrac{x^{-2}}{2} + C$

13. $\int \sqrt[3]{x} dx = \int x^{\frac{1}{3}} dx$

$= \dfrac{x^{\frac{1}{3}+1}}{\frac{1}{3}+1} + C \qquad \left[\int x^r dx = \dfrac{x^{r+1}}{r+1} + C \right]$

$= \dfrac{x^{\frac{4}{3}}}{\frac{4}{3}} + C$

$= \dfrac{3}{4} x^{\frac{4}{3}} + C$

15. $\int \sqrt{x^5} dx = \int x^{\frac{5}{2}} dx$

$= \dfrac{x^{\frac{5}{2}+1}}{\frac{5}{2}+1} + C \qquad \left[\int x^r dx = \dfrac{x^{r+1}}{r+1} + C \right]$

$= \dfrac{x^{\frac{7}{2}}}{\frac{7}{2}} + C$

$= \dfrac{2}{7} x^{\frac{7}{2}} + C$

17. $\int \dfrac{dx}{x^4} = \int \dfrac{1}{x^4} dx = \int x^{-4} dx$

$= \dfrac{x^{-4+1}}{-4+1} + C \qquad \left[\int x^r dx = \dfrac{x^{r+1}}{r+1} + C \right]$

$= -\dfrac{x^{-3}}{3} + C$

19. $\int \dfrac{1}{x} dx$

$= \ln x + C, \ x > 0 \qquad \left[\int \dfrac{1}{x} dx = \ln x + C, x > 0 \right]$

21. $\int \left(\dfrac{3}{x} + \dfrac{5}{x^2} \right) dx$

$= \int \dfrac{3}{x} dx + \int \dfrac{5}{x^2} dx \qquad$ The integral of a sum is the sum of the integrals.

$= 3 \int x^{-1} dx + \int 5 x^{-2} dx$

$= 3 \cdot \ln x + 5 \cdot \dfrac{x^{-2+1}}{-2+1} + C, \quad x > 0$

$\left[\int x^{-1} dx = \ln x, x > 0 \right]$

$\left[\int x^r dx = \dfrac{x^{r+1}}{r+1} + C \right]$

$= 3 \ln x - 5 x^{-1} + C$

23. $\displaystyle\int \frac{-7}{\sqrt[3]{x^2}}\,dx = \int \frac{-7}{x^{2/3}} = \int -7x^{-2/3}\,dx$

$\displaystyle -7\int x^{-2/3}\,dx$

$\displaystyle = -7\cdot\frac{x^{-2/3+1}}{-\frac{2}{3}+1}+C \qquad \left[\int x^r\,dx = \frac{x^{r+1}}{r+1}+C\right]$

$\displaystyle = -7\cdot\frac{x^{1/3}}{\frac{1}{3}}+C$

$\displaystyle = -21x^{1/3}+C$

25. $\displaystyle\int 2e^{2x}\,dx$

$\displaystyle = \frac{2}{2}e^{2x}+C \qquad \left[\int be^{ax}\,dx = \frac{b}{a}e^{ax}+C\right]$

$\displaystyle = e^{2x}+C$

27. $\displaystyle\int e^{3x}\,dx$

$\displaystyle = \frac{1}{3}e^{3x}+C \qquad \left[\int be^{ax}\,dx = \frac{b}{a}e^{ax}+C\right]$

29. $\displaystyle\int e^{7x}\,dx$

$\displaystyle = \frac{1}{7}e^{7x}+C \qquad \left[\int be^{ax}\,dx = \frac{b}{a}e^{ax}+C\right]$

31. $\displaystyle\int 5e^{3x}\,dx$

$\displaystyle = \frac{5}{3}e^{3x}+C \qquad \left[\int be^{ax}\,dx = \frac{b}{a}e^{ax}+C\right]$

33. $\displaystyle\int 6e^{8x}\,dx$

$\displaystyle = \frac{6}{8}e^{8x}+C \qquad \left[\int be^{ax}\,dx = \frac{b}{a}e^{ax}+C\right]$

$\displaystyle = \frac{3}{4}e^{8x}+C$

35. $\displaystyle\int \frac{2}{3}e^{-9x}\,dx$

$\displaystyle = \frac{2}{3}\cdot\frac{1}{-9}e^{-9x}+C \qquad \left[\int be^{ax}\,dx = \frac{b}{a}e^{ax}+C\right]$

$\displaystyle = -\frac{2}{27}e^{-9x}+C$

37. $\displaystyle\int\left(5x^2 - 2e^{7x}\right)dx$

$\displaystyle = \int 5x^2\,dx - \int 2e^{7x}\,dx \qquad \text{The integral of a sum is the sum of the integrals.}$

$\displaystyle = 5\cdot\frac{x^{2+1}}{2+1} - \frac{2}{7}e^{7x}+C$

$\displaystyle \left[\int x^r\,dx = \frac{x^{r+1}}{r+1}+C\right]$

$\displaystyle \left[\int be^{ax}\,dx = \frac{b}{a}e^{ax}+C\right]$

$\displaystyle = \frac{5}{3}x^3 - \frac{2}{7}e^{7x}+C$

39. $\displaystyle\int\left(x^2 - \frac{3}{2}\sqrt{x} + x^{-4/3}\right)dx$

$\displaystyle = \int\left(x^2 - \frac{3}{2}x^{1/2} + x^{-4/3}\right)dx$

$\displaystyle = \int x^2\,dx - \int\frac{3}{2}x^{1/2}\,dx + \int x^{-4/3}\,dx$

$\displaystyle = \frac{x^{2+1}}{2+1} - \frac{3}{2}\cdot\frac{x^{1/2+1}}{\frac{1}{2}+1} + \frac{x^{-4/3+1}}{-\frac{4}{3}+1}+C$

$\displaystyle \left[\int x^r\,dx = \frac{x^{r+1}}{r+1}+C\right]$

$\displaystyle = \frac{x^3}{3} - \frac{3}{2}\cdot\frac{x^{3/2}}{\frac{3}{2}} + \frac{x^{-1/3}}{-\frac{1}{3}}+C$

$\displaystyle = \frac{x^3}{3} - x^{3/2} - 3x^{-1/3}+C$

41. $\displaystyle\int(3x+2)^2\,dx = \int\left(9x^2 + 12x + 4\right)dx$

$\displaystyle = \int 9x^2\,dx + \int 12x\,dx + \int 4\,dx$

$\displaystyle = 9\cdot\frac{x^{2+1}}{2+1} + 12\cdot\frac{x^{1+1}}{1+1} + 4x+C$

$\displaystyle = \frac{9}{3}x^3 + \frac{12}{2}x^2 + 4x+C$

$\displaystyle = 3x^3 + 6x^2 + 4x+C$

43. $\int\left(\dfrac{3}{x}-5e^{2x}+\sqrt{x^{7}}\right)dx$

$=\int\left(\dfrac{3}{x}-5e^{2x}+x^{7\!/\!2}\right)dx$

$=\int\dfrac{3}{x}dx-\int 5e^{2x}dx+\int x^{7\!/\!2}dx$

$=3\ln x-\dfrac{5}{2}\cdot e^{2x}+\dfrac{x^{7\!/\!2+1}}{\dfrac{7}{2}+1}+C$

$=3\ln x-\dfrac{5}{2}e^{2x}+\dfrac{2}{9}x^{9\!/\!2}+C$

45. $\int\left(\dfrac{7}{\sqrt{x}}-\dfrac{2}{3}e^{5x}-\dfrac{8}{x}\right)dx$

$=\int\left(7x^{-1\!/\!2}-\dfrac{2}{3}e^{5x}-\dfrac{8}{x}\right)dx$

$=\int 7x^{-1\!/\!2}dx-\int\dfrac{2}{3}e^{5x}dx-\int\dfrac{8}{x}dx$

$=7\cdot\dfrac{x^{-1\!/\!2+1}}{-\dfrac{1}{2}+1}-\dfrac{2}{3}\cdot\dfrac{1}{5}e^{5x}-8\ln x+C$

$=14x^{1\!/\!2}-\dfrac{2}{15}e^{5x}-8\ln x+C$

47. Find the function $f(x)$, such that

$f'(x)=x-3,\ f(2)=9$

We first find $f(x)$ by integrating:

$f(x)=\int(x-3)dx$

$=\int xdx-\int 3dx$

$=\dfrac{1}{2}x^{2}-3x+C.$

The condition $f(2)=9$ allows us to find C:

$f(2)=9$

$\dfrac{1}{2}(2)^{2}-3(2)+C=9$

$2-6+C=9$

$-4+C=9$

$C=13.$

Thus, $f(x)=\dfrac{1}{2}x^{2}-3x+13.$

49. Find the function $f(x)$, such that

$f'(x)=x^{2}-4,\ f(0)=7.$

We first find $f(x)$ by integrating:

$f(x)=\int\left(x^{2}-4\right)dx$

$=\int x^{2}dx-\int 4dx$

$=\dfrac{1}{3}x^{3}-4x+C.$

The condition $f(0)=7$ allows us to find C:

$f(0)=7$

$\dfrac{1}{3}(0)^{3}-4(0)+C=7$

$C=7.$

Thus, $f(x)=\dfrac{1}{3}x^{3}-4x+7.$

51. Find the function $f(x)$, such that

$f'(x)=5x^{2}+3x-7,\ f(0)=9.$

We first find $f(x)$ by integrating:

$f(x)=\int\left(5x^{2}+3x-7\right)dx$

$=\int 5x^{2}dx+\int 3xdx-\int 7dx$

$=\dfrac{5}{3}x^{3}+\dfrac{3}{2}x^{2}-7x+C.$

The condition $f(0)=9$ allows us to find C:

$f(0)=9$

$\dfrac{5}{3}(0)^{3}+\dfrac{3}{2}(0)^{2}-7(0)+C=9$

$C=9.$

Thus, $f(x)=\dfrac{5}{3}x^{3}+\dfrac{3}{2}x^{2}-7x+9.$

53. Find the function $f(x)$, such that

$f'(x)=3x^{2}-5x+1,\ f(1)=\dfrac{7}{2}.$

We first find $f(x)$ by integrating:

$f(x)=\int\left(3x^{2}-5x+1\right)dx$

$=\int 3x^{2}dx-\int 5xdx+\int dx$

$=x^{3}-\dfrac{5}{2}x^{2}+x+C.$

The condition $f(1)=\dfrac{7}{2}$ allows us to find C:

$$f(1) = \frac{7}{2}$$

$$(1)^3 - \frac{5}{2}(1)^2 + (1) + C = \frac{7}{2}$$

$$-\frac{1}{2} + C = \frac{7}{2}$$

$$C = 4.$$

Thus, $f(x) = x^3 - \frac{5}{2}x^2 + x + 4.$

55. Find the function $f(x)$, such that

$$f'(x) = 5e^{2x}, \quad f(0) = \frac{1}{2}.$$

We first find $f(x)$ by integrating:

$$f(x) = \int 5e^{2x} dx$$

$$= \frac{5}{2}e^{2x} + C.$$

The condition $f(0) = \frac{1}{2}$ allows us to find C:

$$f(0) = \frac{1}{2}$$

$$\frac{5}{2}e^{2(0)} + C = \frac{1}{2}$$

$$\frac{5}{2}e^0 + C = \frac{1}{2}$$

$$\frac{5}{2} \cdot 1 + C = \frac{1}{2}$$

$$C = -\frac{4}{2}$$

$$C = -2.$$

Thus, $f(x) = \frac{5}{2}e^{2x} - 2.$

57. Find the function $f(x)$, such that

$$f'(x) = \frac{4}{\sqrt{x}}, \quad f(1) = -5.$$

We first find $f(x)$ by integrating:

$$f(x) = \int \frac{4}{\sqrt{x}} dx$$

$$= \int 4x^{-1/2} dx$$

$$= 8x^{1/2} + C.$$

The condition $f(1) = -5$ allows us to find C:

$$f(1) = -5$$

$$8(1)^{1/2} + C = -5$$

$$8 + C = -5$$

$$C = -13.$$

Thus, $f(x) = 8x^{1/2} - 13.$

59. $D'(t) = 1975 - 1190t + 597t^2 - 71.3t^3$

We integrate to find $D(t)$.

$$D(t) = \int \left(1975 - 1190t + 597t^2 - 71.3t^3\right) dt$$

$$= 1975t - \frac{1190}{2}t^2 + \frac{597}{3}t^3 - \frac{71.3}{4}t^4 + C$$

$$= 1975t - 595t^2 + 199t^3 - 17.825t^4 + C$$

The condition $D(0) = 17{,}198$ allows us to find C. Substituting, we have

$$D(0) = 17{,}198$$

$$1975(0) - 595(0)^2 + 199(0)^3 - 17.825(0)^4 + C = 17{,}198$$

$$C = 17{,}198$$

Thus,

$$D(t) = 17{,}198 + 1975t - 595t^2 + 199t^3 - 17.825t^4$$

61. $C'(x) = x^3 - 2x$

We integrate to find $C(x)$, we use K for the constant of integration to avoid confusion with the cost function $C(x)$.

$$C(x) = \int C'(x) dx$$

$$= \int (x^3 - 2x) dx$$

$$= \frac{x^4}{4} - x^2 + K$$

Fixed costs are $7000. This means $C(0) = 7000$. This allows us to determine K.

$$C(0) = 7000$$

$$\frac{(0)^4}{4} - (0)^2 + K = 7000$$

$$K = 7000$$

Thus, the total cost function is

$$C(x) = \frac{x^4}{4} - x^2 + 7000.$$

63. $R'(x) = x^2 - 3$

a) We integrate to find $R(x)$.

$$R(x) = \int R'(x)\,dx$$

$$= \int (x^2 - 3)\,dx$$

$$= \frac{x^3}{3} - 3x + C$$

The condition $R(0) = 0$ allows us to find C.

$$R(0) = 0$$

$$\frac{(0)^3}{3} - 3(0) + C = 0$$

$$C = 0$$

Thus, the total revenue function is

$$R(x) = \frac{x^3}{3} - 3x.$$

b) $\boxed{tw}$

65. $D'(x) = -\dfrac{4000}{x^2} = -4000x^{-2}$

We integrate to find $D(x)$.

$$D(x) = \int D'(x)\,dx$$

$$= \int -4000x^{-2}\,dx$$

$$= -4000 \cdot \frac{x^{-1}}{-1} + C$$

$$= 4000x^{-1} + C$$

$$= \frac{4000}{x} + C$$

When the price is \$4 per unit, the demand is 1003 units. This means $D(4) = 1003$.

Substituting 4 for x and 1003 for $D(x)$ we can determine C as follows:

$$D(4) = 1003$$

$$\frac{4000}{4} + C = 1003$$

$$1000 + C = 1003$$

$$C = 3.$$

Thus, the demand function is $D(x) = \dfrac{4000}{x} + 3.$

67. $\dfrac{dE}{dt} = 30 - 10t$

a) We find $E(t)$ by integrating

$$E(t) = \int E'(t)\,dt$$

$$= \int (30 - 10t)\,dt$$

$$= 30t - 10 \cdot \frac{t^2}{2} + C$$

$$= 30t - 5t^2 + C$$

The condition $E(2) = 72$ allows us to find C as follows:

$$E(2) = 72$$

$$30(2) - 5(2)^2 + C = 72 \qquad \text{Substituting}$$

$$60 - 20 + C = 72$$

$$40 + C = 72$$

$$C = 32.$$

Thus, $E(t) = 30t - 5t^2 + 32.$

b) $E(t) = 32 + 30t - 5t^2$

Substituting 3 for t, we have:

$$E(3) = 32 + 30(3) - 5(3)^2$$

$$= 32 + 90 - 45$$

$$= 77.$$

After 3 hours, the operator's efficiency is 77%.

Substituting 5 for t, we have:

$$E(5) = 32 + 30(5) - 5(5)^2$$

$$= 32 + 150 - 125$$

$$= 57.$$

After 5 hours, the operator's efficiency is 57%.

69. $I'(t) = 3.389e^{0.1049t}$

a) We integrate to find $I(t)$.

$$I(t) = \int I'(t)\,dt$$

$$= \int 3.389e^{0.1049t}\,dt$$

$$= \frac{3.389}{0.1049}e^{0.1049t} + C$$

$$\approx 32.31e^{0.1049t} + C$$

The condition $I(0) = 0$ allows us to find C.

$$I(0) = 0$$

$$32.31e^{0.1049(0)} + C = 0 \qquad \text{Substituting}$$

$$32.31e^0 + C = 0$$

$$32.31 + C = 0$$

$$C = -32.31$$

The total number per 100,000 who have contracted influenza by time t is given by
$$I(t) = 32.31e^{0.1049t} - 32.31.$$

b) Using the function found in part (a), we substitute 27 for t.

$$I(27) = 32.31e^{0.1049(27)} - 32.31$$

$$= 32.31e^{2.8323} - 32.31$$

$$\approx 516.459$$

After 27 weeks, approximately 516 per 100,000 people have contracted influenza.

c) Using the function found in part (a), we substitute 34 for t.

$$I(34) = 32.31e^{0.1049(34)} - 32.31$$

$$= 32.31e^{3.5666} - 32.31$$

$$\approx 1111.34$$

After 34 weeks, approximately 1111 per 100,000 people have contracted influenza.

d) The number of people that contracted influenza during the last 7 weeks of the first 34 weeks is given by

$$I(34) - I(27) \approx 1111 - 516 \approx 595.$$

Approximately 595 per 100,000 people contracted influenza during the last 7 of the 34 weeks.

71. Find the function $f(t)$, such that

$$f'(t) = \sqrt{t} + \frac{1}{\sqrt{t}}, \quad f(4) = 0.$$

We first find $f(t)$ by integrating:

$$f(t) = \int \left(\sqrt{t} + \frac{1}{\sqrt{t}} \right) dt$$

$$= \int \left(t^{\frac{1}{2}} + t^{-\frac{1}{2}} \right) dt$$

$$= \frac{2}{3} t^{\frac{3}{2}} + 2t^{\frac{1}{2}} + C.$$

The condition $f(4) = 0$ allows us to find C:

$$f(4) = 0$$

$$\frac{2}{3}(4)^{\frac{3}{2}} + 2(4)^{\frac{1}{2}} + C = 0$$

$$\frac{2}{3}(8) + 4 + C = 0$$

$$\frac{16}{3} + 4 + C = 0$$

$$C = -\frac{28}{3}.$$

Thus, $f(t) = \frac{2}{3} t^{\frac{3}{2}} + 2t^{\frac{1}{2}} - \frac{28}{3}.$

73. $\int (5t+4)^2 t^4 \, dx$

First, we will expand the binomial.

$$= \int (5t+4)(5t+4)t^4 \, dt$$

$$= \int (25t^2 + 40t + 16)t^4 \, dt$$

Next, we distribute t^4.

$$\int (25t^6 + 40t^5 + 16t^4) \, dt$$

Now we can integrate as follows:

$$= \int 25t^6 \, dt + \int 40t^5 \, dt + \int 16t^4 \, dt \quad \begin{smallmatrix}\text{The integral of a sum is} \\ \text{the sum of the integrals.}\end{smallmatrix}$$

$$= 25 \cdot \frac{t^{6+1}}{6+1} + 40 \cdot \frac{t^{5+1}}{5+1} + 16 \cdot \frac{t^{4+1}}{4+1} + C$$

$$\left[\int x^r \, dx = \frac{x^{r+1}}{r+1} + C \right]$$

$$= \frac{25}{7} t^7 + \frac{40}{6} t^6 + \frac{16}{5} t^5 + C$$

$$= \frac{25}{7} t^7 + \frac{20}{3} t^6 + \frac{16}{5} t^5 + C$$

75. $\int (1-t)\sqrt{t} \, dt$

$$= \int \left(\sqrt{t} - t\sqrt{t} \right) dt \qquad \text{Distributing } \sqrt{t}$$

$$= \int \left(t^{\frac{1}{2}} - t^{\frac{3}{2}} \right) dt$$

$$= \frac{t^{\frac{3}{2}}}{\frac{3}{2}} - \frac{t^{\frac{5}{2}}}{\frac{5}{2}} + C$$

$$= \frac{2}{3} t^{\frac{3}{2}} - \frac{2}{5} t^{\frac{5}{2}} + C$$

77. $\int \dfrac{x^4 - 6x^2 - 7}{x^3} \, dx$

$= \int \left(x^4 - 6x^2 - 7\right) x^{-3} dx$

$= \int \left(x - 6x^{-1} - 7x^{-3}\right) dx$

$= \dfrac{x^2}{2} - 6 \ln x - 7 \cdot \dfrac{x^{-2}}{-2} + C$

$= \dfrac{x^2}{2} - 6 \ln x + \dfrac{7}{2} x^{-2} + C$

79. $\int \dfrac{1}{\ln 10} \cdot \dfrac{dx}{x}$

$= \dfrac{1}{\ln 10} \int \dfrac{1}{x} dx$

$= \dfrac{1}{\ln 10} \cdot \ln x + C$

$= \log x + C \qquad$ Properties of logarithms.

81. $\int (3x - 5)(2x + 1)^2 \, dx$

$= \int (3x - 5)\left(4x^2 + 4x + 1\right) dx$

$= \int \left(12x^3 - 8x^2 - 17x - 5\right) dx$

$= 12 \cdot \dfrac{x^4}{4} - 8 \cdot \dfrac{x^3}{3} - 17 \cdot \dfrac{x^2}{2} - 5x + C$

$= 3x^4 - \dfrac{8}{3} x^3 - \dfrac{17}{2} x^2 - 5x + C$

83. $\int \dfrac{x^2 - 1}{x + 1} \, dx$

$= \int \dfrac{(x - 1)(x + 1)}{x + 1} \, dx$

$= \int (x - 1) \, dx$

$= \dfrac{x^2}{2} - x + C$

85. $\boxed{tw}$

Exercise Set 4.3

1. Find any antiderivative $F(x)$ of $y = 4$. We choose the simplest one for which the constant of integration is 0:

$F(x) = \int 4 \, dx$

$\qquad = 4x + C$

$\qquad = 4x. \qquad [C = 0]$

The area under the curve over the interval $[1, 3]$ is given by $F(3) - F(1)$. Substitute 3 and 1, and find the difference:

$F(3) - F(1) = 4(3) - 4(1)$

$\qquad\qquad = 12 - 4$

$\qquad\qquad = 8.$

3. Find any antiderivative $F(x)$ of $y = 2x$. We choose the simplest one for which the constant of integration is 0:

$F(x) = \int 2x \, dx$

$\qquad = x^2 + C$

$\qquad = x^2. \qquad [C = 0]$

The area under the curve over the interval $[1, 3]$ is given by $F(3) - F(1)$. Substitute 3 and 1, and find the difference:

$F(3) - F(1) = (3)^2 - (1)^2$

$\qquad\qquad = 9 - 1$

$\qquad\qquad = 8.$

5. Find any antiderivative $F(x)$ of $y = x^2$. We choose the simplest one for which the constant of integration is 0:

$F(x) = \int x^2 \, dx$

$\qquad = \dfrac{x^3}{3} + C$

$\qquad = \dfrac{x^3}{3}. \qquad [C = 0]$

The area under the curve over the interval $[0, 5]$ is given by $F(5) - F(0)$. Substitute 5 and 0, and find the difference:

$$F(5) - F(0) = \frac{(5)^3}{3} - \frac{(0)^3}{3}$$

$$= \frac{125}{3}$$

$$41\tfrac{2}{3}.$$

7. Find any antiderivative $F(x)$ of $y = x^3$. We choose the simplest one for which the constant of integration is 0:

$$F(x) = \int x^3 dx$$

$$= \frac{x^4}{4} + C$$

$$= \frac{x^4}{4}. \qquad [C = 0]$$

The area under the curve over the interval $[0,1]$ is given by $F(1) - F(0)$. Substitute 1 and 0, and find the difference:

$$F(1) - F(0) = \frac{(1)^4}{4} - \frac{(0)^4}{4}$$

$$= \frac{1}{4}.$$

9. Find any antiderivative $F(x)$ of $y = 4 - x^2$. We choose the simplest one for which the constant of integration is 0:

$$F(x) = \int (4 - x^2) dx$$

$$= 4x - \frac{x^3}{3} + C$$

$$= 4x - \frac{x^3}{3}. \qquad [C = 0]$$

The area under the curve over the interval $[-2,2]$ is given by $F(2) - F(-2)$. Substitute 2 and -2, and find the difference:

$$F(2) - F(-2)$$

$$= \left(4(2) - \frac{(2)^3}{3}\right) - \left(4(-2) - \frac{(-2)^3}{3}\right)$$

$$= \left(8 - \frac{8}{3}\right) - \left(-8 + \frac{8}{3}\right)$$

$$= \frac{16}{3} - \left(-\frac{16}{3}\right)$$

$$= \frac{32}{3}$$

$$= 10\tfrac{2}{3}.$$

11. Find any antiderivative $F(x)$ of $y = e^x$. We choose the simplest one for which the constant of integration is 0:

$$F(x) = \int e^x dx$$

$$= e^x + C$$

$$= e^x. \qquad [C = 0]$$

The area under the curve over the interval $[0,3]$ is given by $F(3) - F(0)$. Substitute 3 and 0, and find the difference:

$$F(3) - F(0) = (e^3) - (e^0)$$

$$= e^3 - 1$$

$$\approx 19.086.$$

13. Find any antiderivative $F(x)$ of $y = \dfrac{3}{x}$. We choose the simplest one for which the constant of integration is 0:

$$F(x) = \int \frac{3}{x} dx$$

$$= 3\ln x + C$$

$$= 3\ln x. \qquad [C = 0]$$

The area under the curve over the interval $[1,6]$ is given by $F(6) - F(1)$. Substitute 6 and 1, and find the difference:

$$F(6) - F(1) = (3\ln 6) - (3\ln 1)$$

$$= 3\ln 6 - 0$$

$$\approx 5.375.$$

15. The height of the area under the curve is total cost per day and the width of the area under the curve is time in days. Thus, area under the curve represents total cost in dollars, for t days.

$$\frac{\text{Total Cost}}{\text{days}} \cdot \text{days} = \text{Total cost}.$$

17. The height of the area under the curve is total number of kilowatts used per hour and the width of the area under the curve is time in hours. Thus, area under the curve represents Total number of kilowatts used in t hours.

$$\frac{\#\text{KW}}{\text{Hour}} \cdot \text{Hours} = \#\text{KW}.$$

19. The height of the area under the curve is revenue in dollars per unit and the width of the area under the curve is number of units. Thus, area under the curve represents total revenue, in dollars, for x units produced. $\dfrac{\$}{\text{Unit}} \cdot \text{Units} = \$.$

21. The height of the area under the curve is milligrams per cubic centimeter and the width of the area under the curve is cubic centimeters. Thus, area under the curve represents total concentration of a drug, in milligrams, in v cubic centimeters of blood. $\dfrac{\text{mg}}{cm^3} \cdot cm^3 = \text{mg}.$

23. The height of the area under the curve is number of memorized words per minute and the width of the area under the curve is time in minutes. Thus, area under the curve represents the total number of words memorized in t minutes.

$\dfrac{\text{Words memorized}}{\text{Minute}} \cdot \text{Minutes} = \text{Words memorized}.$

25. Find any antiderivative $F(x)$ of $y = x^3$. We choose the simplest one for which the constant of integration is 0:

$F(x) = \int x^3 dx$

$= \dfrac{x^4}{4} + C$

$= \dfrac{x^4}{4}.$ $\qquad [C = 0]$

The area under the curve over the interval $[0,2]$ is given by $F(2) - F(0)$. Substitute 2 and 0, and find the difference:

$F(2) - F(0) = \dfrac{(2)^4}{4} - \dfrac{(0)^4}{4}$

$= \dfrac{16}{4}$

$= 4.$

27. Find any antiderivative $F(x)$ of $y = x^2 + x + 1$.

We choose the simplest one for which the constant of integration is 0:

$F(x) = \int x^2 + x + 1 dx$

$= \dfrac{x^3}{3} + \dfrac{x^2}{2} + x + C$

$= \dfrac{x^3}{3} + \dfrac{x^2}{2} + x.$ $\qquad [C = 0]$

The area under the curve over the interval $[2,3]$ is given by $F(3) - F(2)$. Substitute 3 and 2, and find the difference:

$F(3) - F(2)$

$= \dfrac{(3)^3}{3} + \dfrac{(3)^2}{2} + (3) - \left(\dfrac{(2)^3}{3} + \dfrac{(2)^2}{2} + (2) \right)$

$= \dfrac{27}{3} + \dfrac{9}{2} + 3 - \left(\dfrac{8}{3} + \dfrac{4}{2} + 2 \right)$

$= \dfrac{54}{6} + \dfrac{27}{6} + \dfrac{18}{6} - \left(\dfrac{16}{6} + \dfrac{12}{6} + \dfrac{12}{6} \right)$

$= \dfrac{99}{6} - \left(\dfrac{40}{6} \right)$

$= \dfrac{59}{6}$

$= 9\tfrac{5}{6}.$

29. Find any antiderivative $F(x)$ of $y = 5 - x^2$. We choose the simplest one for which the constant of integration is 0:

$F(x) = \int (5 - x^2) dx$

$= 5x - \dfrac{x^3}{3} + C$

$= 5x - \dfrac{x^3}{3}.$ $\qquad [C = 0]$

The area under the curve over the interval $[-1,2]$ is given by $F(2) - F(-1)$. Substitute 2 and -1, and find the difference:

$F(2) - F(-1)$

$= \left(5(2) - \dfrac{(2)^3}{3} \right) - \left(5(-1) - \dfrac{(-1)^3}{3} \right)$

$= \left(10 - \dfrac{8}{3} \right) - \left(-5 + \dfrac{1}{3} \right)$

$= \dfrac{22}{3} - \left(-\dfrac{14}{3} \right)$

$= \dfrac{36}{3}$

$= 12.$

31. Find any antiderivative $F(x)$ of $y = e^x$. We choose the simplest one for which the constant of integration is 0:

$F(x) = \int e^x \, dx$

$\quad = e^x + C$

$\quad = e^x. \qquad [C = 0]$

The area under the curve over the interval $[-1, 5]$ is given by $F(5) - F(-1)$. Substitute 5 and -1, and find the difference:

$F(5) - F(-1) = \left(e^5 \right) - \left(e^{-1} \right)$

$\qquad\qquad\quad = e^5 - e^{-1}$

$\qquad\qquad\quad \approx 148.045.$

33. $\boxed{tw}$

35. $\displaystyle\int_0^{1.5} \left(x - x^2 \right) dx$

$= \left[\dfrac{x^2}{2} - \dfrac{x^3}{3} \right]_0^{1.5}$

$= \left(\dfrac{(1.5)^2}{2} - \dfrac{(1.5)^3}{3} \right) - \left(\dfrac{(0)^2}{2} - \dfrac{(0)^3}{3} \right)$

$= \left(\dfrac{2.25}{2} - \dfrac{3.375}{3} \right) - 0$

$= 1.125 - 1.125$

$= 0$

The area above the x-axis is equal to the area below the x-axis.

37. $\displaystyle\int_{-1}^1 \left(x^4 - x^2 \right) dx$

$= \left[\dfrac{x^5}{5} - \dfrac{x^3}{3} \right]_{-1}^1$

$= \left(\dfrac{(1)^5}{5} - \dfrac{(1)^3}{3} \right) - \left(\dfrac{(-1)^5}{5} - \dfrac{(-1)^3}{3} \right)$

$= \left(\dfrac{1}{5} - \dfrac{1}{3} \right) - \left(\dfrac{-1}{5} - \dfrac{-1}{3} \right)$

$= \dfrac{-2}{15} - \left(\dfrac{2}{15} \right)$

$= -\dfrac{4}{15}$

The area below the x-axis is greater than the area above the x-axis.

39 – 41. Left to the student.

43. $\displaystyle\int_1^3 \left(3t^2 + 7 \right) dt$

$= \left[t^3 + 7t \right]_1^3$

$= \left(3^3 + 7(3) \right) - \left(1^3 + 7(1) \right)$

$= 48 - (8)$

$= 40$

45. $\displaystyle\int_1^4 \left(\sqrt{x} - 1 \right) dx = \int_1^4 \left(x^{1/2} - 1 \right) dx$

$= \left[\dfrac{2}{3} x^{3/2} - x \right]_1^4$

$= \left(\dfrac{2}{3}(4)^{3/2} - 4 \right) - \left(\dfrac{2}{3}(1)^{3/2} - 1 \right)$

$= \left(\dfrac{16}{3} - 4 \right) - \left(\dfrac{2}{3} - 1 \right)$

$= \left(\dfrac{4}{3} \right) - \left(-\dfrac{1}{3} \right)$

$= \dfrac{5}{3}$

47. $\int_{-2}^{5} \left(2x^2 - 3x + 7\right) dx$

$= \left[\frac{2}{3}x^3 - \frac{3}{2}x^2 + 7x\right]_{-2}^{5}$

$= \left(\frac{2}{3}(5)^3 - \frac{3}{2}(5)^2 + 7(5)\right) -$

$\qquad \left(\frac{2}{3}(-2)^3 - \frac{3}{2}(-2)^2 + 7(-2)\right)$

$= \frac{485}{6} - \left(-\frac{152}{6}\right)$

$= \frac{637}{6}$

49. $\int_{-5}^{2} e^t dt$

$= \left[e^t\right]_{-5}^{2}$

$= e^2 - e^{-5}$

≈ 7.382

51. $\int_{a}^{b} \frac{1}{2}x^2 dx$

$= \left[\frac{1}{6}x^3\right]_{a}^{b}$

$= \frac{1}{6}(b)^3 - \frac{1}{6}(a)^3$

$= \frac{b^3 - a^3}{6}$

53. $\int_{a}^{b} e^{2t} dt$

$= \left[\frac{1}{2}e^{2t}\right]_{a}^{b}$

$= = \frac{1}{2}e^{2b} - \frac{1}{2}e^{2a}$

$= \frac{e^{2b} - e^{2a}}{2}$

55. $\int_{1}^{e} \left(x + \frac{1}{x}\right) dx$

$= \left[\frac{x^2}{2} + \ln x\right]_{1}^{e}$

$= \left(\frac{e^2}{2} + \ln e\right) - \left(\frac{1^2}{2} + \ln 1\right)$

$= \frac{e^2}{2} + 1 - \left(\frac{1}{2} + 0\right)$

$= \frac{e^2}{2} + \frac{1}{2}$

$= \frac{e^2 + 1}{2}$

≈ 4.195

57. $\int_{0}^{2} \sqrt{2x}\, dx = \int_{0}^{2} \sqrt{2} \cdot x^{1/2} dx = \sqrt{2} \int_{0}^{2} x^{1/2} dx$

$= \sqrt{2} \left[\frac{2}{3}x^{3/2}\right]_{0}^{2}$

$= \sqrt{2} \left[\frac{2}{3}(2)^{3/2} - \frac{2}{3}(0)^{3/2}\right]$

$= \sqrt{2} \left[\frac{2}{3}\sqrt{2^3}\right]$

$= \sqrt{2} \left[\frac{2}{3}\sqrt{8}\right]$

$= \frac{2}{3}\sqrt{16}$

$= \frac{8}{3}$

59. We integrate to find:

$P(250) = \int_{0}^{250} P'(x)\, dx$

$= \int_{0}^{250} \sqrt[5]{x}\, dx$

$= \int_{0}^{250} x^{1/5} dx$

$= \left[\frac{5}{6}x^{6/5}\right]_{0}^{250}$

$= \frac{5}{6}(250)^{6/5} - \frac{5}{6}(0)^{6/5}$

$\approx 628.56.$ Using a calculator.

When a 250 foot well is drilled, Pure Water Enterprises profit is $628.56.

61. In order find the cost of producing an additional 14 feet of counter top after 50 feet have already been produced, we integrate $C'(x)$ over the interval $[50, 64]$.

$$C(64) - C(50) = \int_{50}^{64} C'(x)\, dx$$
$$= \int_{50}^{64} 8x^{-\frac{1}{3}}\, dx$$
$$= \left[12x^{\frac{2}{3}} \right]_{50}^{64}$$
$$= 12(64)^{\frac{2}{3}} - 12(50)^{\frac{2}{3}}$$
$$\approx 29.13. \qquad \text{Using a calculator.}$$

The cost of installing an extra 14 feet of counter top after 50 feet has already been ordered is $29.13.

63. $S'(t) = 20e^t$

a) We integrate $S'(t)$ over the interval $[0, 5]$ to find the accumulated sales.

$$S(5) = \int_0^5 S'(t)\, dt$$
$$= \int_0^5 20e^t\, dt$$
$$= \left[20e^t \right]_0^5$$
$$= 20e^5 - 20e^0$$
$$= 20e^5 - 20 \cdot 1$$
$$\approx 2948.26$$

The accumulated sales for the first 5 days are approximately $2948.26.

b) We integrate $S'(t)$ over the interval $[1, 5]$ to find the accumulated sales for the 2nd day through the 5th day.

$$S(5) = \int_1^5 S'(t)\, dt$$
$$= \int_1^5 20e^t\, dt$$
$$= \left[20e^t \right]_1^5$$
$$= 20e^5 - 20e^1$$
$$\approx 2913.90$$

The sales from the 2nd day through the 5th day are approximately $2913.90.

65. In 1996, $t = 1$ and in 2000, $t = 5$. Therefore, we integrate $D'(t)$ over the interval $[1, 5]$.

$$\int_1^5 D'(t)\, dt$$
$$= \int_1^5 \left(857.98 + 829.66t - 197.34t^2 + 15.36t^3 \right) dt$$
$$= \left[857.98t + 414.83t^2 - 65.78t^3 + 3.84t^4 \right]_1^5$$
$$\approx 7627.28$$

The credit market debt increase $7627.28 billion from 1996 to 2000.

67. We integrate $T(x)$ over the interval $[1, 10]$.

$$\int_1^{10} T(x)\, dx$$
$$= \int_1^{10} \left(2 + 0.3x^{-1} \right) dx$$
$$= \left[2x + 0.3 \ln x \right]_1^{10}$$
$$= 2(10) + 0.3 \ln(10) - \left(2(1) + 0.3 \ln(1) \right)$$
$$= 18 + 0.3 \ln 10$$
$$\approx 18.69$$

It takes 18.69 hours for a new worker to produce units 1 through 10.

To find the time it takes a new worker to produce units 20 through 30, we integrate $T(x)$ over the interval $[20, 30]$.

$$\int_{20}^{30} T(x)\, dx$$
$$= \int_{20}^{30} \left(2 + 0.3x^{-1} \right) dx$$
$$= \left[2x + 0.3 \ln x \right]_{20}^{30}$$
$$= 2(30) + 0.3 \ln(30) - \left(2(20) + 0.3 \ln(20) \right)$$
$$\approx 20.12$$

It takes 20.12 hours for a new worker to produce units 20 through 30.

69. We integrate $M'(t)$ over the interval $[0,10]$.

$$M(10) = \int_0^{10} M'(t)$$

$$= \int_0^{10} \left(-0.009t^2 + 0.2t\right) dt$$

$$= \left[-0.003t^3 + 0.1t^2\right]_0^{10}$$

$$= \left(-0.003(10)^3 + 0.1(10)^2\right)$$

$$\qquad - \left(-0.003(0)^3 + 0.1(0)^2\right)$$

$$= 7 - 0$$

$$= 7$$

In the first 10 minutes, 7 words are memorized.

71. We integrate $M'(t)$ over the interval $[10,15]$.

$$M(15) - M(10) = \int_{10}^{15} M'(t)$$

$$= \int_{10}^{15} \left(-0.009t^2 + 0.2t\right) dt$$

$$= \left[-0.003t^3 + 0.1t^2\right]_{10}^{15}$$

$$= \left(-0.003(15)^3 + 0.1(15)^2\right)$$

$$\qquad - \left(-0.003(10)^3 + 0.1(10)^2\right)$$

$$= 12.375 - 7$$

$$= 5.375$$

About 5 words are memorized during minutes 10 – 15.

73. We first find $s(t)$ by integrating:

$$s(t) = \int v(t)\,dt = \int 3t^2\,dt = t^3 + C.$$

Next we determine C by using the initial condition $s(0) = 4$, which is the starting position for s at time $t = 0$:

$$s(0) = 4$$

$$0^3 + C = 4$$

$$C = 4.$$

Thus, $s(t) = t^3 + 4$.

75. We first find $v(t)$ by integrating:

$$v(t) = \int a(t)\,dt = \int 4t\,dt = 2t^2 + C.$$

Next we determine C by using the initial condition $v(0) = 20$:

$$v(0) = 20$$

$$2 \cdot 0^2 + C = 20$$

$$C = 20.$$

Thus, $v(t) = 2t^2 + 20$.

77. We first find $v(t)$ by integrating:

$$v(t) = \int a(t)\,dt$$

$$= \int (-2t + 6)\,dt$$

$$= -t^2 + 6t + C_1.$$

Next we determine C_1 by using the initial condition $v(0) = 6$:

$$v(0) = 6$$

$$-(0)^2 + 6(0) + C_1 = 6$$

$$C_1 = 6.$$

Thus, $v(t) = -t^2 + 6t + 6$.

Next, we find $s(t)$ by integrating:

$$s(t) = \int v(t)\,dt$$

$$= \int \left(-t^2 + 6t + 6\right)\,dt$$

$$= -\frac{1}{3}t^3 + 3t^2 + 6t + C_2$$

Next we determine C_2 by using the initial condition $s(0) = 10$:

$$s(0) = 10$$

$$-\frac{1}{3}(0)^3 + 3(0)^2 + 6(0) + C_2 = 10$$

$$C_2 = 10.$$

Thus, $s(t) = -\frac{1}{3}t^3 + 3t^2 + 6t + 10$.

79. a) We integrate $v(t)$ over the interval $[0,5]$:

$$s(5) = \int_0^5 v(t)\,dt$$

$$= \int_0^5 \left(-0.5t^2 + 10t\right)\,dt$$

$$= \left[-\frac{1}{6}t^3 + 5t^2\right]_0^5$$

$$= \left(-\frac{1}{6}(5)^3 + 5(5)^2\right) - \left(-\frac{1}{6}(0)^3 + 5(0)^2\right)$$

$$= \frac{625}{6} - 0$$

$$\approx 104.17$$

The particle travels approximately 104.17 meters during the first 5 seconds.

b) We integrate $v(t)$ over the interval $[5, 10]$:

$$s(10) - s(5) = \int_5^{10} v(t)\, dt$$

$$= \left[-\frac{1}{6}t^3 + 5t^2 \right]_5^{10} \quad \text{From part (a)}.$$

$$\approx 229.17$$

The particle travels approximately 229.17 meters during the second 5 seconds.

81. a) Converting 15 seconds into hours, we have

$\dfrac{15}{3600} = \dfrac{1}{240}$. Thus, the motorcycle's acceleration function is

$$a(t) = \frac{60 - 0}{\dfrac{1}{240} - 0} = 14{,}400,$$ where $a(t)$ is in

mile per hour squared, and t is in hours. Thus, we can find the velocity function by integrating $a(t)$.

$$v(t) = \int a(t)\, dt = \int 14{,}400\, dt = 14{,}400t + C$$

We use the initial condition $v(0) = 0$ to find C.

$$v(0) = 0$$

$$14{,}400(0) + C = 0$$

$$C = 0$$

Thus, $v(t) = 14{,}400t$, where $v(t)$ is in miles per hour and t is in hours.

Now, substituting $\dfrac{1}{240}$ hour (15 seconds) for

t, we have $v\left(\dfrac{1}{240}\right) = 14{,}400\left(\dfrac{1}{240}\right) = 60.$

The motorcycle is traveling at a speed of 60 miles per hour after 15 seconds.
Note: the intuitive solution to this problem is if the motorcycle accelerates at a constant rate from 0 mph to 60 mph in 15 seconds, then the motorcycle is obviously traveling at 60 mph after 15 seconds. We derive the velocity function to find the distance in part (b)

b) Using the information in Part (a), we

integrate $v(t)$ over the interval $\left[0, \dfrac{1}{240}\right]$.

$$s\left(\frac{1}{240}\right) = \int_0^{1/240} 14{,}400t\, dt$$

$$= \left[7200t^2 \right]_0^{1/240}$$

$$= 7200\left(\frac{1}{240}\right)^2 - 7200(0)$$

$$= \frac{1}{8}.$$

The motorcycle has traveled $1/8^{\text{th}}$ of a mile after 15 seconds.

83. a) Converting 45 seconds into hours, we have

$\dfrac{45}{3600} = \dfrac{1}{80}$. Thus, the cyclist's acceleration

function is $a(t) = \dfrac{30 - 0}{\dfrac{1}{80} - 0} = 2400$, where

$a(t)$ is in kilometers per hour squared, and t is in hours. Thus, we can find the velocity function by integrating $a(t)$.

$$v(t) = \int a(t)\, dt = \int 2400\, dt = 2400t + C$$

We use the initial condition $v(0) = 0$ to find C.

$$v(0) = 0$$

$$2400(0) + C = 0$$

$$C = 0$$

Thus, $v(t) = 2400t$, where $v(t)$ is in kilometers per hour and t is in hours.

Now, substituting $\dfrac{1}{180}$ hour (20 seconds) for

t, we have $v\left(\dfrac{1}{180}\right) = 2400\left(\dfrac{1}{180}\right) = 13.33$

The cyclist is traveling at a speed of 13.33 kilometers per hour after 20 seconds.

b) Using the information in Part (a), we integrate $v(t)$ over the interval $\left[0, \dfrac{1}{80}\right]$.

$$s\left(\frac{1}{80}\right) = \int_0^{1/80} 2400t\,dt$$

$$= \left[1200t^2\right]_0^{1/80}$$

$$= 1200\left(\frac{1}{80}\right)^2 - 1200(0)^2$$

$$= \frac{3}{16} \approx 0.1875.$$

The motorcycle has traveled 0.1875 kilometers after 45 seconds.

85. We first find $v(t)$ by integrating:

$$v(t) = \int a(t)\,dt$$

$$= \int (-32)\,dt$$

$$= -32t + C_1.$$

Next we determine C_1 by using the initial condition $v(0) = v_0$:

$$v(0) = v_0$$

$$-32(0) + C_1 = v_0$$

$$C_1 = v_0.$$

Thus, $v(t) = -32t + v_0$.

Next, we find $s(t)$ by integrating:

$$s(t) = \int v(t)\,dt$$

$$= \int (-32t + v_0)\,dt$$

$$= -16t^2 + v_0 \cdot t + C_2.$$

Next we determine C_2 by using the initial condition $s(0) = s_0$:

$$s(0) = s_0$$

$$-16(0)^2 + v_0(0) + C_2 = s_0$$

$$C_2 = s_0.$$

Thus, $s(t) = -16t^2 + v_0 t + s_0$.

87. $a(t) = 7200$

$$v(t) = \int a(t)\,dt$$

$$= \int 7200\,dt$$

$$= 7200t + C$$

Since $v(0) = 0$, we have

$$v(0) = 0$$

$$7200(0) + C = 0$$

$$C = 0$$

Thus, $v(t) = 7200t$

Integrating $v(t)$ from $\left[0, \dfrac{1}{120}\right]$ we have

$$s\left(\frac{1}{120}\right) = \int_0^{1/120} 7200t\,dt$$

$$= \left[3600t^2\right]_0^{1/120}$$

$$= \frac{1}{4}.$$

The car travels $\dfrac{1}{4}$ mile in $\dfrac{1}{2}$ minute.

89. We integrate $v(t)$ over the interval $[1, 5]$:

$$s(5) - s(1) = \int_1^5 v(t)\,dt$$

$$= \int_1^5 (3t^2 + 2t)\,dt$$

$$= \left[t^3 + t^2\right]_1^5$$

$$= (5^3 + 5^2) - (1^3 + 1^2)$$

$$= 150 - 2$$

$$= 148$$

The particle travels approximately 148 miles from the 2nd hour to through the 5th hour.

91. $S(t) = \int S'(t)\,dt$

$$S(t) = \int 0.5e^t\,dt = 0.5e^t + C$$

Assuming $S(0) = 0$, we have:

$$S(0) = 0$$

$$0.5e^0 + C = 0$$

$$0.5 + C = 0$$

$$C = -0.5$$

Thus, $S(t) = 0.5e^t - 0.5$

When Bluetape reaches \$10,000 in sales, $S(t) = 10,000$. We solve the equation for t.

$$0.5e^t - 0.5 = 10,000$$

$$0.5e^t = 10,000.5$$

$$e^t = 20,001$$

$$\ln e^t = \ln 20,001$$

$$t = 9.9035$$

Therefore, they will reach \$10,000 in sales on the 10th day.

93. $\int_2^3 \frac{x^2-1}{x-1}\,dx = \int_2^3 \frac{(x-1)(x+1)}{(x-1)}\,dx$

$= \int_2^3 (x+1)\,dx$

$= \left[\frac{x^2}{2} + x\right]_2^3$

$= \left(\frac{3^2}{2} + 3\right) - \left(\frac{2^2}{2} + 2\right)$

$= \frac{15}{2} - 4$

$= \frac{7}{2} = 3.5$

95. $\int_4^{16} (x-1)\sqrt{x}\,dx = \int_4^{16}\left(x^{3/2} - x^{1/2}\right)dx$

$= \left[\frac{2}{5}x^{5/2} - \frac{2}{3}x^{3/2}\right]_4^{16}$

$= \left(\frac{2}{5}(16)^{5/2} - \frac{2}{3}(16)^{3/2}\right) - \left(\frac{2}{5}(4)^{5/2} - \frac{2}{3}(4)^{3/2}\right)$

$= \left(\frac{2048}{5} - \frac{128}{3}\right) - \left(\frac{64}{5} - \frac{16}{3}\right)$

$= \frac{5392}{15}$

$= 359\tfrac{7}{15}$

97. $\int_1^8 \frac{\sqrt[3]{x^2}-1}{\sqrt[3]{x}}\,dx = \int_1^8 \left(x^{2/3} - 1\right)x^{-1/3}\,dx$

$= \int_1^8 \left(x^{1/3} - x^{-1/3}\right)dx$

$= \left[\frac{3}{4}x^{4/3} - \frac{3}{2}x^{2/3}\right]_1^8$

$= \left(\frac{3}{4}(8)^{4/3} - \frac{3}{2}(8)^{2/3}\right) - \left(\frac{3}{4}(1)^{4/3} - \frac{3}{2}(1)^{2/3}\right)$

$= (12-6) - \left(\frac{3}{4} - \frac{3}{2}\right)$

$= 6 - \left(-\frac{3}{4}\right)$

$= \frac{27}{4}$

$= 6.75$

99. $\int_2^5 (t+\sqrt{3})(t-\sqrt{3})\,dt = \int_2^5 (t^2 - 3)\,dt$

$= \left[\frac{t^3}{3} - 3t\right]_2^5$

$= \left(\frac{(5)^3}{3} - 3(5)\right) - \left(\frac{(2)^3}{3} - 3(2)\right)$

$= \frac{80}{3} - \left(-\frac{10}{3}\right)$

$= \frac{90}{3}$

$= 30$

101. $\int_1^3 \left(x - \frac{1}{x}\right)^2 dx = \int_1^3 \left(x - x^{-1}\right)^2 dx$

$= \int_1^3 \left(x^2 - 2 + x^{-2}\right)dx \qquad \text{Expanding } \left(x - x^{-1}\right)^2$

$= \left[\frac{x^3}{3} - 2x - x^{-1}\right]_1^3$

$= \left(\frac{(3)^3}{3} - 2(3) - (3)^{-1}\right) - \left(\frac{(1)^3}{3} - 2(1) - (1)^{-1}\right)$

$= \frac{8}{3} - \left(-\frac{8}{3}\right)$

$= \frac{16}{3}$

$= 5\tfrac{1}{3}$

103. $\int_4^9 \frac{t+1}{\sqrt{t}}\,dt = \int_4^9 (t+1)t^{-1/2}\,dt$

$= \int_4^9 \left(t^{1/2} + t^{-1/2}\right)dt$

$= \left[\frac{2}{3}t^{3/2} + 2t^{1/2}\right]_4^9$

$= \left(\frac{2}{3}(9)^{3/2} + 2(9)^{1/2}\right) - \left(\frac{2}{3}(4)^{3/2} + 2(4)^{1/2}\right)$

$= 24 - \frac{28}{3}$

$= \frac{44}{3}$

$= 14\tfrac{2}{3}$

105. $\boxed{tw}$

107. Using the fnInt feature on a calculator, we get:

$$\int_{-8}^{1.4} \left(x^4 + 4x^3 - 36x^2 - 160x + 300\right) dx \approx 4068.789.$$

109. Using the fnInt feature on a calculator, we get:

$$\int_{-1}^{1} \left(3 + \sqrt{1-x^2}\right) dx \approx 7.571.$$

111. Using the fnInt feature on a calculator, we get:

$$\int_{-2}^{2} x^{2/3}\left(\frac{5}{2} - x\right) dx \approx 9.524.$$

113. Using the fnInt feature on a calculator, we get:

$$\int_{-10}^{10} \frac{8}{x^2 + 4}\, dx \approx 10.987.$$

Exercise Set 4.4

1. $\displaystyle\int_{1}^{5} f(x)\,dx = \int_{1}^{3} f(x)\,dx + \int_{3}^{5} f(x)\,dx$

$= \displaystyle\int_{1}^{3} (2x+1)\,dx + \int_{3}^{5}(10-x)\,dx$

$= \left[x^2 + x\right]_{1}^{3} + \left[10x - \dfrac{1}{2}x^2\right]_{3}^{5}$

$= \left[\left((3)^2 + (3)\right) - \left((1)^2 + (1)\right)\right] +$

$\quad \left[\left(10(5) - \dfrac{1}{2}(5)^2\right) - \left(10(3) - \dfrac{1}{2}(3)^2\right)\right]$

$= (12 - 2) + \left(\dfrac{75}{2} - \dfrac{51}{2}\right)$

$= 10 + 12$

$= 22$

3. $\displaystyle\int_{-2}^{3} g(x)\,dx = \int_{-2}^{0} g(x)\,dx + \int_{0}^{3} g(x)\,dx$

$= \displaystyle\int_{-2}^{0} (x^2 + 4)\,dx + \int_{0}^{3}(4 - x)\,dx$

$= \left[\dfrac{x^3}{3} + 4x\right]_{-2}^{0} + \left[4x - \dfrac{1}{2}x^2\right]_{0}^{3}$

$= \left[\left(\dfrac{(0)^3}{3} + 4(0)\right) - \left(\dfrac{(-2)^3}{3} + 4(-2)\right)\right] +$

$\quad \left[\left(4(3) - \dfrac{1}{2}(3)^2\right) - \left(4(0) - \dfrac{1}{2}(0)^2\right)\right]$

$= \left(0 - \left(-\dfrac{32}{3}\right)\right) + \left(\dfrac{15}{2} - 0\right)$

$= \dfrac{109}{6}$

$= 18\tfrac{1}{6}$

5. $\displaystyle\int_{-6}^{4} f(x)\,dx = \int_{-6}^{1} f(x)\,dx + \int_{1}^{4} f(x)\,dx$

$= \displaystyle\int_{-6}^{1} (-x^2 - 6x + 7)\,dx + \int_{1}^{4}\left(\dfrac{3}{2}x - 1\right)dx$

$= \left[-\dfrac{x^3}{3} - 3x^2 + 7x\right]_{-6}^{1} + \left[\dfrac{3}{4}x^2 - x\right]_{1}^{4}$

$= \left[\left(-\dfrac{(1)^3}{3} - 3(1)^2 + 7(1)\right) -\right.$

$\quad \left. \left(-\dfrac{(-6)^3}{3} - 3(-6)^2 + 7(-6)\right)\right] +$

$\quad \left[\left(\dfrac{3}{4}(4)^2 - (4)\right) - \left(\dfrac{3}{4}(1)^2 - (1)\right)\right]$

$= \left(\dfrac{11}{3} - (-78)\right) + \left(8 - \left(-\dfrac{1}{4}\right)\right)$

$= \dfrac{245}{3} + \dfrac{33}{4}$

$= \dfrac{1079}{12}$

$= 89\tfrac{11}{12}$

7. To calculate the points of intersection, we set $f(x)$ equal to $g(x)$ and solve.

$$f(x) = g(x)$$
$$9 = x^2$$
$$\sqrt{9} = \sqrt{x^2}$$
$$\pm 3 = x$$

The graphs intersect at $x = -3$ and $x = 3$.

9. To calculate the points of intersection, we set $f(x)$ equal to $g(x)$ and solve.

$$f(x) = g(x)$$
$$7 = x^2 - 3x + 2$$
$$0 = x^2 - 3x - 5$$

Applying the quadratic formula we have:

$$x = \frac{-(-3) \pm \sqrt{(-3)^2 - 4(1)(-5)}}{2(1)}$$
$$= \frac{3 \pm \sqrt{29}}{2}$$

The graphs intersect at

$$x = \frac{3 + \sqrt{29}}{2} \text{ and } x = \frac{3 - \sqrt{29}}{2}.$$

11. To calculate the points of intersection, we set $f(x)$ equal to $g(x)$ and solve.

$$f(x) = g(x)$$
$$x^2 - x - 5 = x + 10$$
$$x^2 - 2x - 15 = 0$$
$$(x - 5)(x + 3) = 0$$
$$x - 5 = 0 \quad \text{or} \quad x + 3 = 0$$
$$x = 5 \quad \text{or} \quad x = -3$$

The graphs intersect at $x = -3$ and $x = 5$.

13. $g(x) \geq f(x)$ on $[-1, 0]$ and $f(x) \geq g(x)$ on $[0, 1]$. We use two integrals to find the total area.

$$\int_{-1}^{0} \left[0 - \left(2x + x^2 - x^3 \right) \right] dx +$$
$$\int_{0}^{1} \left[\left(2x + x^2 - x^3 \right) - 0 \right] dx$$
$$\int_{-1}^{0} \left[0 - \left(2x + x^2 - x^3 \right) \right] dx +$$
$$\int_{0}^{1} \left[\left(2x + x^2 - x^3 \right) - 0 \right] dx$$
$$= \int_{-1}^{0} \left(-2x - x^2 + x^3 \right) dx +$$
$$\int_{0}^{1} \left(2x + x^2 - x^3 \right) dx$$
$$= \left[-x^2 - \frac{x^3}{3} + \frac{x^4}{4} \right]_{-1}^{0} + \left[x^2 + \frac{x^3}{3} - \frac{x^4}{4} \right]_{0}^{1}$$
$$= \left[-(0)^2 - \frac{(0)^3}{3} + \frac{(0)^4}{4} \right] -$$
$$\left[-(-1)^2 - \frac{(-1)^3}{3} + \frac{(-1)^4}{4} \right] +$$
$$\left[(1)^2 + \frac{(1)^3}{3} - \frac{(1)^4}{4} \right] - \left[(0)^2 + \frac{(0)^3}{3} - \frac{(0)^4}{4} \right]$$
$$= \left[0 - \left(-\frac{5}{12} \right) \right] + \left[\frac{13}{12} - 0 \right]$$
$$= \frac{5}{12} + \frac{13}{12}$$
$$= \frac{18}{12}$$
$$= \frac{3}{2}$$

15. $g(x) \geq f(x)$ over the entire region. We find the area.

$$\int_{-1}^{4}\left[(x+28)-\left(x^4-8x^3+18x^2\right)\right]dx$$

$$=\int_{-1}^{4}\left(-x^4+8x^3-18x^2+x+28\right)dx$$

$$=\left[-\frac{x^5}{5}+2x^4-6x^3+\frac{x^2}{2}+28x\right]_{-1}^{4}$$

$$=\left[-\frac{(4)^5}{5}+2(4)^4-6(4)^3+\frac{(4)^2}{2}+28(4)\right]-$$

$$\left[-\frac{(-1)^5}{5}+2(-1)^4-6(-1)^3+\frac{(-1)^2}{2}+28(-1)\right]$$

$$=\frac{216}{5}+\frac{193}{10}$$

$$=\frac{625}{10}$$

$$=62\tfrac{1}{2}$$

17. First graph the system of equations and shade the region bounded by the graphs.

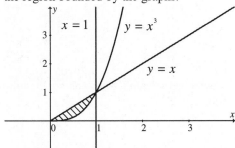

Here the boundaries are easily determined by looking at the graph, or by solving the following:

$x = x^3$

$0 = x^3 - x$

$0 = x\left(x^2-1\right)$

$x = 0 \quad$ or $\quad x^2 - 1 = 0$

$x = 0 \quad$ or $\quad x = \pm 1$

$x = 0 \quad$ or $\quad x = 1$

Note $x \geq x^3$ over the interval $[0,1]$. We compute the area as follows:

$$\int_{0}^{1}\left(x-x^3\right)dx$$

$$=\left[\frac{x^2}{2}-\frac{x^4}{4}\right]_{0}^{1}$$

$$=\left(\frac{(1)^2}{2}-\frac{(1)^4}{4}\right)-\left(\frac{(0)^2}{2}-\frac{(0)^4}{4}\right)$$

$$=\frac{1}{2}-\frac{1}{4}$$

$$=\frac{1}{4}.$$

19. First graph the system of equations and shade the region bounded by the graphs.

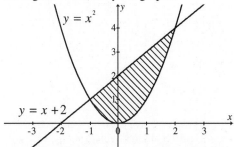

Here the boundaries are easily determined by looking at the graph, or solving the following equation:

$$x^2 = x+2$$

$$x^2 - x - 2 = 0$$

$$(x-2)(x+1) = 0$$

$$x = -1 \quad \text{or} \quad x = 2$$

Note $(x+2) \geq x^2$ over the interval $[-1,2]$. We compute the area as follows:

$\int_{-1}^{2}\left((x+2)-x^2\right)dx$

$=\int_{-1}^{2}\left(-x^2+x+2\right)dx$

$=\left[-\dfrac{x^3}{3}+\dfrac{x^2}{2}+2x\right]_{-1}^{2}$

$=\left(-\dfrac{(2)^3}{3}+\dfrac{(2)^2}{2}+2(2)\right)-$

$\qquad\left(-\dfrac{(-1)^3}{3}+\dfrac{(-1)^2}{2}+2(-1)\right)$

$=\left(-\dfrac{8}{3}+2+4\right)-\left(\dfrac{1}{3}+\dfrac{1}{2}-2\right)$

$=\dfrac{10}{3}-\left(-\dfrac{7}{6}\right)$

$=\dfrac{27}{6}$

$=4\tfrac{1}{2}.$

21. First graph the system of equations and shade the region bounded by the graphs.

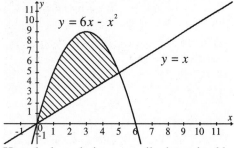

Here the boundaries are easily determined by looking at the graph, or by solving the following equation:

$\qquad x=6x-x^2$

$x^2-5x=0$

$x(x-5)=0$

$x=0\ \ \text{or}\ \ x=5$

Note $\left(6x-x^2\right)\ge x$ over the interval $[0,5]$. We compute the area as follows:

$\int_{0}^{5}\left((6x-x^2)-x\right)dx$

$=\int_{0}^{5}\left(-x^2+5x\right)dx$

$=\left[-\dfrac{x^3}{3}+\dfrac{5}{2}x^2\right]_{0}^{5}$

$=\left(-\dfrac{(5)^3}{3}+\dfrac{5}{2}(5)^2\right)-\left(-\dfrac{(0)^3}{3}+\dfrac{5}{2}(0)^2\right)$

$=\left(-\dfrac{125}{3}+\dfrac{125}{2}\right)-0$

$=\dfrac{125}{6}$

$=20\tfrac{5}{6}.$

23. First graph the system of equations and shade the region bounded by the graphs.

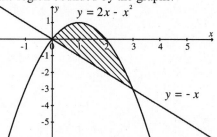

Here the boundaries are easily determined by looking at the graph, or by solving the following equation:

$\qquad -x=2x-x^2$

$x^2-3x=0$

$x(x-3)=0$

$x=0\ \ \text{or}\ \ x=3$

Note $\left(2x-x^2\right)\ge-x$ over the interval $[0,3]$.

We compute the area as follows:

$\int_{0}^{3}\left((2x-x^2)-(-x)\right)dx$

$=\int_{0}^{3}\left(-x^2+3x\right)dx$

$=\left[-\dfrac{x^3}{3}+\dfrac{3}{2}x^2\right]_{0}^{3}$

$=\left(-\dfrac{(3)^3}{3}+\dfrac{3}{2}(3)^2\right)-\left(-\dfrac{(0)^3}{3}+\dfrac{3}{2}(0)^2\right)$

$=\left(-\dfrac{27}{3}+\dfrac{27}{2}\right)-0$

$=\dfrac{27}{6}$

$=4\tfrac{1}{2}.$

25. First graph the system of equations and shade the region bounded by the graphs.

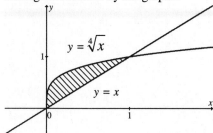

Here the boundaries are easily determined by looking at the graph, or by solving the following equation:

$$x = \sqrt[4]{x}$$

$$x^4 = x$$

$$x^4 - x = 0$$

$$x(x^3 - 1) = 0$$

$$x = 0 \text{ or } x = 1$$

Note $\sqrt[4]{x} \geq x$ over the interval $[0,1]$. We compute the area as follows:

$$\int_0^1 \left(\sqrt[4]{x} - x\right) dx$$

$$= \int_0^1 \left(x^{\frac{1}{4}} - x\right) dx$$

$$= \left[\frac{4}{5} x^{\frac{5}{4}} - \frac{x^2}{2}\right]_0^1$$

$$= \left(\frac{4}{5}(1)^{\frac{5}{4}} - \frac{(1)^2}{2}\right) - \left(\frac{4}{5}(0)^{\frac{5}{4}} - \frac{(0)^2}{2}\right)$$

$$= \left(\frac{4}{5} - \frac{1}{2}\right) - 0$$

$$= \frac{3}{10}.$$

27. First graph the system of equations and shade the region bounded by the graphs.

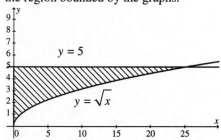

Here the boundaries are easily determined by looking at the graph. Note $5 \geq \sqrt{x}$ over the interval $[0,25]$. We compute the area as follows:

$$\int_0^{25} \left(5 - \sqrt{x}\right) dx$$

$$= \int_0^{25} \left(5 - x^{\frac{1}{2}}\right) dx$$

$$= \left[5x - \frac{2}{3} x^{\frac{3}{2}}\right]_0^{25}$$

$$= \left(5(25) - \frac{2}{3}(25)^{\frac{3}{2}}\right) - \left(5(0) - \frac{2}{3}(0)^{\frac{3}{2}}\right)$$

$$= \left(125 - \frac{250}{3}\right) - 0$$

$$= \frac{125}{3}$$

$$= 41\frac{2}{3}.$$

29. First graph the system of equations and shade the region bounded by the graphs.

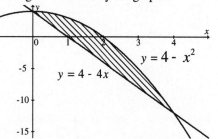

Here the boundaries are easily determined by looking at the graph, or by solving the following:

$$4 - 4x = 4 - x^2$$

$$x^2 - 4x = 0$$

$$x(x - 4) = 0$$

$$x = 0 \text{ or } x = 4.$$

Note $(4 - x^2) \geq (4 - 4x)$ over the interval $[0,4]$. We compute the area as follows:

$$\int_0^4 \left[(4 - x^2) - (4 - 4x)\right] dx$$

$$= \int_0^4 \left(-x^2 + 4x\right) dx$$

$$= \left[-\frac{x^3}{3} + 2x^2\right]_0^4$$

$$= \left(-\frac{(4)^3}{3} + 2(4)^2\right) - \left(-\frac{(0)^3}{3} + 2(0)^2\right)$$

$$= \left(-\frac{64}{3} + 32\right) - 0$$

$$= \frac{32}{3}$$

$$= 10\frac{2}{3}.$$

31. First graph the system of equations and shade the region bounded by the graphs.

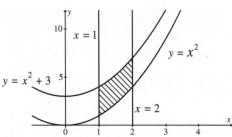

Here the boundaries are easily determined by looking at the graph. Note $(x^2+3) \geq (x^2)$ over the interval $[1,2]$. We compute the area as follows:

$$\int_1^2 \left[(x^2+3) - (x^2) \right] dx$$

$$= \int_1^2 3\,dx$$

$$= [3x]_1^2$$

$$= (3(2) - 3(1))$$

$$= 3.$$

33. First graph the system of equations and shade the region bounded by the graphs.

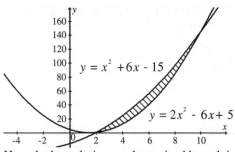

Here the boundaries are determined by solving the following equation:

$$2x^2 - 6x + 5 = x^2 + 6x - 15$$

$$x^2 - 12x + 20 = 0$$

$$(x-2)(x-10) = 0$$

$$x = 2 \quad \text{or} \quad x = 10$$

Note $(x^2 + 6x - 15) \geq (2x^2 - 6x + 5)$ over the interval $[2,10]$. We compute the area as follows:

$$\int_2^{10} \left[(x^2 + 6x - 15) - (2x^2 - 6x + 5) \right] dx$$

$$= \int_2^{10} \left(-x^2 + 12x - 20 \right) dx$$

$$= \left[-\frac{x^3}{3} + 6x^2 - 20x \right]_2^{10}$$

$$= \left(-\frac{(10)^3}{3} + 6(10)^2 - 20(10) \right) -$$

$$\left(-\frac{(2)^3}{3} + 6(2)^2 - 20(2) \right)$$

$$= \frac{200}{3} - \left(-\frac{56}{3} \right)$$

$$= \frac{256}{3}$$

$$= 85\tfrac{1}{3}.$$

35. The average value is:

$$y_{av} = \frac{1}{b-a} \int_a^b f(x)\,dx$$

$$= \frac{1}{2-(-2)} \int_{-2}^2 \left(4 - x^2 \right) dx$$

$$= \frac{1}{4} \left[4x - \frac{x^3}{3} \right]_{-2}^2$$

$$= \frac{1}{4} \left[\left(4(2) - \frac{(2)^3}{3} \right) - \left(4(-2) - \frac{(-2)^3}{3} \right) \right]$$

$$= \frac{1}{4} \left[\frac{16}{3} - \left(-\frac{16}{3} \right) \right]$$

$$= \frac{8}{3}.$$

37. The average value is:

$$y_{av} = \frac{1}{b-a} \int_a^b f(x)\,dx$$

$$= \frac{1}{1-(0)} \int_0^1 e^{-x}\,dx$$

$$= \frac{1}{1} \left[-e^{-x} \right]_0^1$$

$$= \left[-e^{-1} - \left(-e^0 \right) \right]$$

$$= -e^{-1} + 1, \text{ or approximately } 0.632.$$

39. The average value is:

$$y_{av} = \frac{1}{b-a}\int_a^b f(x)\,dx$$

$$= \frac{1}{4-(0)}\int_0^4 \left(x^2 + x - 2\right)dx$$

$$= \frac{1}{4}\left[\frac{x^3}{3} + \frac{x^2}{2} - 2x\right]_0^4$$

$$= \frac{1}{4}\left[\left(\frac{(4)^3}{3} + \frac{(4)^2}{2} - 2(4)\right) - \right.$$

$$\left.\left(\frac{(0)^3}{3} + \frac{(0)^2}{2} - 2(0)\right)\right]$$

$$= \frac{1}{4}\left[\frac{64}{3} - 0\right]$$

$$= \frac{16}{3}.$$

41. The average value is:

$$y_{av} = \frac{1}{b-a}\int_a^b f(x)\,dx$$

$$= \frac{1}{a-(0)}\int_0^a \left(4x + 5\right)dx$$

$$= \frac{1}{a}\left[2x^2 + 5x\right]_0^a$$

$$= \frac{1}{a}\left[\left(2(a)^2 + 5(a)\right) - \left(2(0)^2 + 5(0)\right)\right]$$

$$= \frac{1}{a}\left[\left(2a^2 + 5a\right) - 0\right]$$

$$= 2a + 5.$$

43. The average value is:

$$y_{av} = \frac{1}{b-a}\int_a^b f(x)\,dx$$

$$= \frac{1}{2-(1)}\int_1^2 x^n\,dx, \qquad n \neq 0$$

$$= \frac{1}{1}\left[\frac{1}{n+1}x^{n+1}\right]_1^2$$

$$= \left[\frac{1}{n+1}(2)^{n+1} - \frac{1}{n+1}(1)^{n+1}\right]$$

$$= \frac{2^{n+1}}{n+1} - \frac{1}{n+1}$$

$$= \frac{2^{n+1}-1}{n+1}.$$

45. a) We find total profit by integrating:

$$P(10) = R(10) - C(10)$$

$$= \int_0^{10}\left[R'(t) - C'(t)\right]dt$$

$$= \int_0^{10}\left[\left(100e^t\right) - \left(100 - 0.2t\right)\right]dt$$

$$= \int_0^{10}\left[100e^t - 100 + 0.2t\right]dt$$

$$= \left[100e^t - 100t + 0.1t^2\right]_0^{10}$$

$$= \left(100e^{10} - 100(10) + 0.1(10)^2\right) -$$

$$\left(100e^0 - 100(0) + 0.1(0)^2\right)$$

$$= \left(100e^{10} - 990\right) - \left(100\right)$$

$$= 100e^{10} - 1090$$

$$\approx 2,201,556.58$$

The total profit for the first 10 days is approximately \$2,201,556.58.

b) The average daily profit for the first ten days is given by

$$P_{av} = \frac{1}{10-0}\int_0^{10}\left[R'(t) - C'(t)\right]dt$$

$$\approx \frac{1}{10}\left[2,201,556.58\right] \qquad \text{From part (a).}$$

$$\approx 220,155.66$$

The average daily profit over the first 10 days is approximately \$220,155.66.

47. We find the average weekly sales for the first 5 weeks as follows:

$$S_{av} = \frac{1}{5-0}\int_0^5 S(t)\,dt$$

$$= \frac{1}{5}\int_0^5 9e^t\,dt$$

$$= \frac{1}{5}\left[9e^t\right]_0^5$$

$$= \frac{1}{5}\left[9e^5 - 9e^0\right]$$

$$= \frac{1}{5}\left[9e^5 - 9\right]$$

$$= \frac{9}{5}\left[e^5 - 1\right]$$

$$\approx 265.3437.$$

Therefore, average weekly sales for the first 5 weeks are \$265.3437 hundred, or \$26,534.37.

49. We find the average weekly sales for weeks 2 through 5 by integrating over the interval $[1,5]$ as follows:

$$S_{av} = \frac{1}{5-1}\int_1^5 S(t)\,dt$$

$$= \frac{1}{4}\int_1^5 9e^t\,dt$$

$$= \frac{1}{4}\left[9e^t\right]_1^5$$

$$= \frac{1}{4}\left[9e^5 - 9e^1\right]$$

$$= \frac{9}{4}\left[e^5 - e\right]$$

$$\approx 327.8135.$$

Therefore, average weekly sales for weeks 2 though 5 week are \$327.8135 hundred, or \$32,781.35.

51. a) We notice $-0.003t^2 \geq -0.009t^2$, which means $M'(t) \geq m'(t)$ so Ben has the higher rate of memorization.

b) $M(10) - m(10) = \int_0^{10}\left[M'(t) - m'(t)\right]dt$

$$\int_0^{10}\left[\left(-0.003t^2 + 0.2t\right) - \left(-0.009t^2 + 0.2t\right)\right]dt$$

$$= \int_0^{10} 0.006t^2\,dt$$

$$= \left[0.002t^3\right]_0^{10}$$

$$= 0.002(10)^3 - 0.002(0)^3$$

$$= 2$$

c) $m_{av} = \frac{1}{10-0}\int_0^{10} m'(t)\,dt$

$$m_{av} = \frac{1}{10-0}\int_0^{10}\left(-0.009t^2 + 0.2t\right)dt$$

$$= \frac{1}{10}\left[-0.003t^3 + 0.1t^2\right]_0^{10}$$

$$= \frac{1}{10}\left[\left(-0.003(10)^3 + 0.1(10)^2\right) - \left(-0.003(0)^3 + 0.1(0)^2\right)\right]$$

$$= \frac{1}{10}\left[(-3+10) - (0+0)\right]$$

$$= \frac{1}{10}(7)$$

$$= \frac{7}{10} = 0.7$$

Alice averaged memorizing about 0.7 words per minute during the first 10 minutes.

d) $m_{av} = \frac{1}{10-0}\int_0^{10} M'(t)\,dt$

$$m_{av} = \frac{1}{10-0}\int_0^{10}\left(-0.003t^2 + 0.2t\right)dt$$

$$= \frac{1}{10}\left[-0.001t^3 + 0.1t^2\right]_0^{10}$$

$$= \frac{1}{10}\left[\left(-0.001(10)^3 + 0.1(10)^2\right) - \left(-0.001(0)^3 + 0.1(0)^2\right)\right]$$

$$= \frac{1}{10}\left[(-1+10) - (0+0)\right]$$

$$= \frac{1}{10}(9)$$

$$= \frac{9}{10} = 0.9$$

Ben averaged memorizing about 0.9 words per minute during the first 10 minutes.

53. a) $W(0) = -6(0)^2 + 12(0) + 90 = 90.$

At the beginning of the interval, the keyboarder's speed is 90 words per minute.

b) First, we find the derivative.

$W'(t) = -12t + 12$

Next, we set the derivative equal to zero to find the critical value.

$$W'(t) = 0$$

$$-12t + 12 = 0$$

$$t = 1$$

The only critical value occurs when $t = 0$. We also know that $W''(t) = -12 < 0$.

Therefore, by the Max-Min Principle 2, we know that an absolute maximum occurs at $t = 1$.

$W(1) = -6(1)^2 + 12(1) + 90 = 96$

Thus, the maximum speed is 96 words per minute, occurring 1 minute into the interval.

c) $W_{av} = \dfrac{1}{5-0}\displaystyle\int_0^5 W(t)\,dt$

$\quad W_{av} = \dfrac{1}{5-0}\displaystyle\int_0^5 \left(-6t^2 + 12t + 90\right)dt$

$\quad = \dfrac{1}{5}\left[-2t^3 + 6t^2 + 90t\right]_0^5$

$\quad = \dfrac{1}{5}\left[\left(-2(5)^3 + 6(5)^2 + 90(5)\right) - \right.$

$\qquad\qquad \left.\left(-2(0)^3 + 6(0)^2 + 90(0)\right)\right]$

$\quad = \dfrac{1}{5}[350 - 0]$

$\quad = 70$

The keyboarder's average speed over the 5 minute interval is 70 words per minute.

55. a) $C(0) = 42.03e^{-0.01050(0)} = 42.03$

The initial dosage is 42.03 micorgrams per milliliter.

b) $C_{av} = \dfrac{1}{120-10}\displaystyle\int_{10}^{120} C(t)\,dt$

$\quad C_{av} = \dfrac{1}{110}\displaystyle\int_{10}^{120}\left(42.03e^{-0.01050t}\right)dt$

$\quad = \dfrac{1}{110}\left[\dfrac{42.03}{-0.01050}e^{-0.01050t}\right]_{10}^{120}$

$\quad = \dfrac{1}{110}\left[\left(\dfrac{42.03}{-0.01050}e^{-0.01050(120)}\right) - \right.$

$\qquad\qquad \left.\dfrac{42.03}{-0.01050}e^{-0.01050(10)}\right]$

$\quad \approx \dfrac{1}{110}[2468.4439]$

$\quad \approx 22.44$

The average amount of phenylbutazone in the calf's body for the time between 10 and 120 hours is about 22.44 micrograms per milliliter.

57. a) We determine the average temperature as follows:

$\dfrac{1}{10-0}\displaystyle\int_0^{10}\left(-t^2 + 5t + 40\right)dt$

$= \dfrac{1}{10}\left[-\dfrac{t^3}{3} + \dfrac{5}{2}t^2 + 40t\right]_0^{10}$

$= \dfrac{1}{10}\left[\left(-\dfrac{(10)^3}{3} + \dfrac{5}{2}(10)^2 + 40(10)\right) - \right.$

$\qquad\qquad \left.\left(-\dfrac{(0)^3}{3} + \dfrac{5}{2}(0)^2 + 40(0)\right)\right]$

$= \dfrac{1}{10}\left[\dfrac{950}{3} - 0\right]$

$= \dfrac{95}{3} \approx 31.67$

The average temperature over the 10 hour period is 31.7 degrees.

b) First, we find the critical values.

$f'(t) = -2t + 5$

$\quad f'(t) = 0$

$-2t + 5 = 0$

$\qquad t = \dfrac{5}{2} = 2.5$

Since there is only one critical value and $f''(t) = -2 < 0$, we know that it represents a maximum over the closed interval. Therefore, the minimum temperature must be at an endpoint. We have:

$f(0) = -(0)^2 + 5(0) + 40 = 40$

$f(10) = -(10)^2 + 5(10) + 40 = -10$

The minimum temperature over the 10 hour period is minus 10 degrees.

c) From part (b), we know that the maximum temperature occurs when $t = 2.5$. Substituting we have:

$f(2.5) = -(2.5)^2 + 5(2.5) + 40 = 46.25$

Therefore, the maximum temperature over the 10 hour period is 46.25 degrees.

59. First graph the system of equations and shade the region bounded by the graphs.

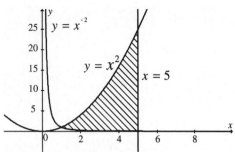

Here the boundaries are determined by looking at the graph. Note $x^2 \geq x^{-2}$ over the interval $[1,5]$. We compute the area as follows:

$$\int_1^5 \left(x^2 - x^{-2}\right) dx$$

$$= \left[\frac{x^3}{3} + x^{-1}\right]_1^5$$

$$= \left(\frac{(5)^3}{3} + (5)^{-1}\right) - \left(\frac{(1)^3}{3} + (1)^{-1}\right)$$

$$= \frac{608}{15}$$

$$= 40\tfrac{8}{15}.$$

61. First graph the system of equations and shade the region bounded by the graphs.

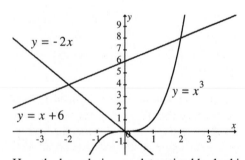

Here the boundaries are determined by looking at the graph. We will split the interval $[-2,2]$ up into two parts.
On the interval $[-2,0]$ we have $x+6 \geq -2x$.

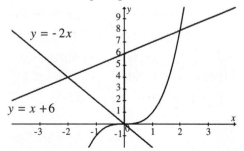

On the interval $[0,2]$ we have $x+6 \geq x^3$.

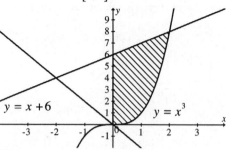

We compute the area as follows:

$$\int_{-2}^0 \left[(x+6)-(-2x)\right] dx + \int_0^2 \left[(x+6)-x^3\right] dx$$

$$= \int_{-2}^0 (3x+6)\, dx + \int_0^2 \left[-x^3 + x + 6\right] dx$$

$$= \left[\frac{3}{2}x^2 + 6x\right]_{-2}^0 + \left[-\frac{x^4}{4} + \frac{x^2}{2} + 6x\right]_0^2$$

$$= \left(0-(-6)\right) + \left(10-0\right)$$

$$= 16$$

63. First graph the system of equations and shade the region bounded by the graphs.

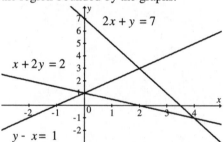

Here the boundaries are determined by looking at the graph. We will split the interval $[0,4]$ up into two parts. Solving each equation for y, we have

$$x+2y = 2 \rightarrow y = -\frac{x}{2}+1$$

$$y - x = 1 \rightarrow y = x+1$$

$$2x+y = 7 \rightarrow y = -2x+7$$

On the interval $[0,2]$ we have $x+1 \geq -\frac{x}{2}+1$.

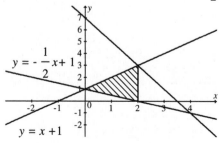

On the interval $[2,4]$ we have $x+1 \geq -2x+7$.

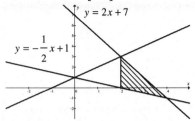

We compute the area as follows:

$$\int_0^2 \left[(x+1) - \left(-\frac{1}{2}x+1 \right) \right] dx +$$

$$\int_2^4 \left[(-2x+7) - \left(-\frac{1}{2}x+1 \right) \right] dx$$

$$= \int_0^2 \frac{3}{2}x\,dx + \int_2^4 \left[-\frac{3}{2}x+6 \right] dx$$

$$= \left[\frac{3}{4}x^2 \right]_0^2 + \left[-\frac{3}{4}x^2 + 6x \right]_2^4$$

$$= (3-0) + (12-9)$$

$$= 6$$

65. First we find the coordinates of the relative extrema.

$$y = x^3 - 3x + 2$$

$$y' = 3x^2 - 3$$

The derivative exists for all real numbers. We solve $y' = 0$.

$$3x^2 - 3 = 0$$

$$x^2 - 1 = 0$$

$$x = \pm 1$$

$x = -1$ or $x = 1$

We use the Second Derivative Test.

$$y'' = 6x$$

When $x = -1, y'' = 6(-1) - 6 < 0$

Therefore there is a relative maximum at $x = -1$.

When $x = 1, y'' = 6(1) = 6 > 0$

Therefore there is a relative minimum at $x = 1$. We graph the region noting that the equation of the x-axis is $y = 0$.

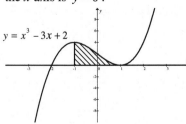

On the interval $[-1,1]$, $x^3 - 3x + 2 \geq 0$. We compute the area as follows:

$$\int_{-1}^1 (x^3 - 3x + 2)\,dx$$

$$= \left[\frac{x^4}{4} - \frac{3}{2}x^2 + 2x \right]_{-1}^1$$

$$= \left(\frac{(1)^4}{4} - \frac{3}{2}(1)^2 + 2(1) \right) -$$

$$\left(\frac{(-1)^4}{4} - \frac{3}{2}(-1)^2 + 2(-1) \right)$$

$$= \left(\frac{3}{4} \right) - \left(-\frac{13}{4} \right)$$

$$= 4.$$

67. $\int_1^2 \left[(3x^2 + 5x) - (3x + K) \right] dx$

$$= \int_1^2 (3x^2 + 2x - K)\,dx$$

$$= \left[x^3 + x^2 - Kx \right]_1^2$$

$$= \left[2^3 + 2^2 - K \cdot 2 \right] - \left[1^3 + 1^2 - K \right]$$

$$= \left[12 - 2K \right] - \left[2 - K \right]$$

$$= 10 - K$$

Since

$$\int_1^2 \left[(3x^2 + 5x) - (3x + K) \right] dx = 6$$

We solve for K by substituting.

$$10 - K = 6$$

$$-K = -4$$

$$K = 4$$

69. From the graph we find that the interval we are concerned with is $[0,2]$. Over this interval

$x\sqrt{4 - x^2} \geq \dfrac{-4x}{x^2 + 1}$. We use the fnInt function on the calculator to find

$$\int_0^2 \left[x\sqrt{4 - x^2} - \frac{-4x}{x^2 + 1} \right] dx \approx 5.8855.$$

71. From the graph we find that the interval we are concerned with is $[-1,1]$. Over this interval

$\sqrt{1 - x^2} \geq 1 - x^2$. We use the fnInt function on the calculator to find

$$\int_{-1}^1 \left[\sqrt{1 - x^2} - (1 - x^2) \right] dx \approx 0.2375.$$

Exercise Set 4.5

1. $\int \left(8+x^3\right)^5 3x^2 dx$

Let $u=8+x^3$, then $du=3x^2 dx$.

$= \int u^5 du$ Substitution: $\begin{smallmatrix}u=8+x^3\\du=3x^2dx\end{smallmatrix}$

$= \dfrac{1}{6}u^6 + C$ Formula A

$= \dfrac{1}{6}\left(8+x^3\right)^6 + C$ Reverse substitution

3. $\int \left(x^2-6\right)^7 x\,dx$

Let $u=x^2-6$, then $du=2x\,dx$. We do not have $2x\,dx$. We only have $x\,dx$ and need to supply a 2. We do this by multiplying by $\frac{1}{2}\cdot 2$ as follows.

$= \dfrac{1}{2}\cdot 2\int\left(x^2-6\right)^7 x\,dx$ Multiplying by 1

$= \dfrac{1}{2}\int\left(x^2-6\right)^7 2x\,dx$ $\left[a\int f(x)\,dx = \int af(x)\,dx\right]$

$= \dfrac{1}{2}\int u^7 du$ Substitution: $\begin{smallmatrix}u=x^2-6\\du=2xdx\end{smallmatrix}$

$= \dfrac{1}{16}u^8 + C$ Formula A

$= \dfrac{1}{16}\left(x^2-6\right)^8 + C$ Reverse substitution

5. $\int \left(3t^4+2\right)t^3 dt$

Let $u=3t^4+2$, then $du=12t^3 dt$. We do not have $12t^3 dt$. We only have $t^3 dt$ and need to supply a 12. We do this by multiplying by $\frac{1}{12}\cdot 12$ as follows.

$= \dfrac{1}{12}\cdot 12\int\left(3t^4+2\right)^7 t^3 dt$ Multiplying by 1

$= \dfrac{1}{12}\int\left(3t^4+2\right)12t^3 dt$

$= \dfrac{1}{12}\int u\,du$ Substitution: $\begin{smallmatrix}u=3t^4+2\\du=12t^3dt\end{smallmatrix}$

$= \dfrac{1}{24}u^2 + C$ Formula A

$= \dfrac{1}{24}\left(3t^4+2\right)^2 + C$ Reverse substitution

7. $\int \dfrac{2}{1+2x}dx$

Let $u=1+2x$, then $du=2dx$.

$= \int \dfrac{du}{u}$ Substitution: $\begin{smallmatrix}u=1+2x\\du=2dx\end{smallmatrix}$

$= \ln u + C$ Formula C, $u>0$.

$= \ln(1+2x) + C$ Reverse substitution

9. $\int \left(\ln x\right)^3 \dfrac{1}{x}dx$

Let $u=\ln x$, then $du=\dfrac{1}{x}dx$.

$= \int u^3 du$ Substitution: $\begin{smallmatrix}u=\ln x\\du=\frac{1}{x}dx\end{smallmatrix}$

$= \dfrac{1}{4}u^4 + C$ Formula A

$= \dfrac{1}{4}\left(\ln x\right)^4 + C$ Reverse substitution

11. $\int e^{3x} dx$

Let $u=3x$, then $du=3dx$. We do not have $3dx$. We only have dx and need to supply a 3. We do this by multiplying by $\frac{1}{3}\cdot 3$ as follows.

$= \dfrac{1}{3}\cdot 3\int e^{3x} dx$ Multiplying by 1

$= \dfrac{1}{3}\int e^{3x} \cdot 3dx$

$= \dfrac{1}{3}\int e^u du$ Substitution: $\begin{smallmatrix}u=3x\\du=3dx\end{smallmatrix}$

$= \dfrac{1}{3}e^u + C$ Formula B

$= \dfrac{1}{3}e^{3x} + C$ Reverse substitution

13. $\int e^{x/3} dx$

Let $u=\dfrac{x}{3}$, then $du=\dfrac{1}{3}dx$. We do not have $\dfrac{1}{3}dx$. We only have dx and need to supply a $\dfrac{1}{3}$. We do this by multiplying by $\frac{1}{3}\cdot 3$ as follows.

$= \dfrac{1}{3} \cdot 3 \displaystyle\int e^{x/3} \, dx$ Multiplying by 1

$= 3 \displaystyle\int e^{x/3} \cdot \dfrac{1}{3} \, dx$

$= 3 \displaystyle\int e^u \, du$ Substitution: $\substack{u = x/3 \\ du = 1/3 \, dx}$

$= 3 e^u + C$ Formula B

$= 3 e^{x/3} + C$ Reverse substitution

15. $\displaystyle\int x^4 e^{x^5} \, dx$

Let $u = x^5$, then $du = 5x^4 \, dx$. We do not have $5x^4 \, dx$. We only have $x^4 \, dx$ and need to supply a 5. We do this by multiplying by $\tfrac{1}{5} \cdot 5$ as follows.

$= \dfrac{1}{5} \cdot 5 \displaystyle\int x^4 e^{x^5} \, dx$ Multiplying by 1

$= \dfrac{1}{5} \displaystyle\int 5 x^4 e^{x^5} \, dx$

$= \dfrac{1}{5} \displaystyle\int e^u \, du$ Substitution: $\substack{u = x^5 \\ du = 5x^4 \, dx}$

$= \dfrac{1}{5} e^u + C$ Formula B

$= \dfrac{1}{5} e^{x^5} + C$ Reverse substitution

17. $\displaystyle\int t e^{-t^2} \, dt$

Let $u = -t^2$, then $du = -2t \, dt$. We do not have $-2t \, dt$. We only have $t \, dt$ and need to supply a -2. We do this by multiplying by $\left(-\tfrac{1}{2}\right) \cdot (-2)$ as follows.

$= \left(-\dfrac{1}{2}\right) \cdot (-2) \displaystyle\int t e^{-t^2} \, dt$ Multiplying by 1

$= -\dfrac{1}{2} \displaystyle\int (-2t) e^{-t^2} \, dt$

$= -\dfrac{1}{2} \displaystyle\int e^u \, du$ Substitution: $\substack{u = -t^2 \\ du = -2t \, dt}$

$= -\dfrac{1}{2} e^u + C$ Formula B

$= -\dfrac{1}{2} e^{-t^2} + C$ Reverse substitution

19. $\displaystyle\int \dfrac{1}{5 + 2x} \, dx$

Let $u = 5 + 2x$, then $du = 2 \, dx$. We do not have $2 \, dx$. We only have dx and need to supply a 2. We do this by multiplying by $\tfrac{1}{2} \cdot 2$ as follows.

$= \dfrac{1}{2} \cdot 2 \displaystyle\int \dfrac{1}{5 + 2x} \, dx$ Multiplying by 1

$= \dfrac{1}{2} \displaystyle\int \dfrac{2 \, dx}{5 + 2x}$

$= \dfrac{1}{2} \displaystyle\int \dfrac{du}{u}$ Substitution: $\substack{u = 5 + 2x \\ du = 2 \, dx}$

$= \dfrac{1}{2} \ln u + C$ Formula C, $u > 0$.

$= \dfrac{1}{2} \ln(5 + 2x) + C$ Reverse substitution

21. $\displaystyle\int \dfrac{dx}{12 + 3x}$

Let $u = 12 + 3x$, then $du = 3 \, dx$. We do not have $3 \, dx$. We only have dx and need to supply a 3. We do this by multiplying by $\tfrac{1}{3} \cdot 3$ as follows.

$= \dfrac{1}{3} \cdot 3 \displaystyle\int \dfrac{dx}{12 + 3x}$ Multiplying by 1

$= \dfrac{1}{3} \displaystyle\int \dfrac{3 \, dx}{12 + 3x}$

$= \dfrac{1}{3} \displaystyle\int \dfrac{du}{u}$ Substitution: $\substack{u = 12 + 3x \\ du = 3 \, dx}$

$= \dfrac{1}{3} \ln u + C$ Formula C, $u > 0$.

$= \dfrac{1}{3} \ln(12 + 3x) + C$ Reverse substitution

23. $\displaystyle\int \dfrac{dx}{1 - x}$

Let $u = 1 - x$, then $du = -dx$. We do not have $-dx$. We only have dx and need to supply a -1. We do this by multiplying by $(-1)(-1)$ as follows.

$= (-1) \cdot (-1) \displaystyle\int \dfrac{dx}{1 - x}$ Multiplying by 1

$= -1 \cdot \displaystyle\int \dfrac{-1 \, dx}{1 - x}$

$= -\displaystyle\int \dfrac{du}{u}$ Substitution: $\substack{u = 1 - x \\ du = -dx}$

$= -\ln u + C$ Formula C, $u > 0$.

$= -\ln(1 - x) + C$ Reverse substitution

25. $\displaystyle\int t \left(t^2 - 1\right)^5 \, dt$

Let $u = t^2 - 1$, then $du = 2t \, dt$. We do not have $2t \, dt$. We only have $t \, dt$ and need to supply a 2. We do this by multiplying by $\tfrac{1}{2} \cdot 2$ as follows.

$$= \frac{1}{2} \cdot 2 \int t \left(t^2 - 1\right)^5 dt \quad \text{Multiplying by 1}$$

$$= \frac{1}{2} \int 2t \left(t^2 - 1\right)^5 dt$$

$$= \frac{1}{2} \int u^5 du \qquad \text{Substitution: } {}^{u=t^2-1}_{du=2tdt}$$

$$= \frac{1}{12} u^6 + C \qquad \text{Formula A}$$

$$= \frac{1}{12} \left(t^2 - 1\right)^6 + C \qquad \text{Reverse substitution}$$

27. $\int \left(x^4 + x^3 + x^2\right)^7 \left(4x^3 + 3x^2 + 2x\right) dx$

Let $u = x^4 + x^3 + x^2$, then

$du = \left(4x^3 + 3x^2 + 2x\right) dx$.

$$= \int u^7 du \qquad \text{Substitution: } {}^{u=x^4+x^3+x^2}_{du=\left(4x^3+3x^2+2x\right)dx}$$

$$= \frac{1}{8} u^8 + C \qquad \text{Formula A}$$

$$= \frac{1}{8} \left(x^4 + x^3 + x^2\right)^8 + C \qquad \text{Reverse substitution}$$

29. $\int \frac{e^x dx}{4 + e^x}$

Let $u = 4 + e^x$, then $du = e^x dx$.

$$= \int \frac{du}{u} \qquad \text{Substitution: } {}^{u=4+e^x}_{du=e^x dx}$$

$$= \ln u + C \qquad \text{Formula C, } u > 0.$$

$$= \ln \left(4 + e^x\right) + C \qquad \text{Reverse substitution}$$

31. $\int \frac{\ln x^2}{x} dx$

Using properties of logarithms, we have

$$\int \frac{2 \cdot \ln x}{x} dx = 2 \int \frac{\ln x}{x} dx.$$

Let $u = \ln x$, then $du = \frac{1}{x} dx$.

$$= 2 \int u \, du \qquad \text{Substitution: } {}^{u=\ln x}_{du=\frac{1}{x}dx}$$

$$= u^2 + C \qquad \text{Formula A}$$

$$= \left(\ln x\right)^2 + C \qquad \text{Reverse substitution}$$

33. $\int \frac{dx}{x \ln x}$

Let $u = \ln x$, then $du = \frac{1}{x} dx$.

$$= \int \frac{1}{\ln x} \cdot \left(\frac{1}{x} dx\right)$$

$$= \int \frac{1}{u} du \qquad \text{Substitution: } {}^{u=\ln x}_{du=\frac{1}{x}dx}$$

$$= \ln u + C \qquad \text{Formula C, } u > 0.$$

$$= \ln \left(\ln x\right) + C \qquad \text{Reverse substitution}$$

35. $\int x\sqrt{ax^2 + b} \, dx$

Let $u = ax^2 + b$, then $du = 2ax dx$. We do not have $2ax dx$. We only have $x dx$ and need to supply a $2a$. We do this by multiplying by $\frac{1}{2a} \cdot 2a$ as follows.

$$= \frac{1}{2a} \cdot 2a \int x\sqrt{ax^2 + b} \, dx \quad \text{Multiplying by 1}$$

$$= \frac{1}{2a} \int 2ax\sqrt{ax^2 + b} \, dx$$

$$= \frac{1}{2a} \int \sqrt{u} \, du \qquad \text{Substitution: } {}^{u=ax^2+b}_{du=2axdx}$$

$$= \frac{1}{2a} \int u^{\frac{1}{2}} du$$

$$= \frac{1}{2a} \cdot \frac{2}{3} u^{\frac{3}{2}} + C \qquad \text{Formula A}$$

$$= \frac{1}{3a} \left(ax^2 + b\right)^{\frac{3}{2}} + C \qquad \text{Reverse substitution}$$

37. $\int P_0 e^{kt} dt = P_0 \cdot \int e^{kt} dt$

Let $u = kt$, then $du = kdt$. We do not have kdt. We only have dt and need to supply a k. We do this by multiplying by $\frac{1}{k} \cdot k$ as follows.

$$= \frac{1}{k} \cdot k \cdot P_0 \int e^{kt} dt \quad \text{Multiplying by 1}$$

$$= \frac{1}{k} \cdot P_0 \int e^{kt} \cdot kdt$$

$$= \frac{P_0}{k} \int e^u du \qquad \text{Substitution: } {}^{u=kt}_{du=kdt}$$

$$= \frac{P_0}{k} e^u + C \qquad \text{Formula B}$$

$$= \frac{P_0}{k} e^{kt} + C \qquad \text{Reverse substitution}$$

39. $\displaystyle\int \frac{x^3\,dx}{\left(2-x^4\right)^7}$

Let $u = 2 - x^4$, then $du = -4x^3\,dx$. We do not have $-4x^3\,dx$. We only have $x^3\,dx$ and need to supply a -4. We do this by multiplying by $\frac{1}{-4}\cdot -4$ as follows.

$\displaystyle =\left(-\frac{1}{4}\right)\cdot(-4)\int \frac{x^3\,dx}{\left(2-x^4\right)^7}$ Multiplying by 1

$\displaystyle =-\frac{1}{4}\int \frac{-4x^3\,dx}{\left(2-x^4\right)^7}$

$\displaystyle =-\frac{1}{4}\int \frac{du}{u^7}$ Substitution: $\begin{matrix}u=2-x^4\\ du=-4x^3dx\end{matrix}$

$\displaystyle =-\frac{1}{4}\int u^{-7}\,du$

$\displaystyle =-\frac{1}{4}\cdot\frac{1}{-6}u^{-6}+C$ Formula A

$\displaystyle =\frac{1}{24}\left(2-x^4\right)^{-6}+C$ Reverse substitution

$\displaystyle =\frac{1}{24\left(2-x^4\right)^6}+C$

41. $\displaystyle\int 12x\sqrt[5]{1+6x^2}\,dx$

Let $u = 1 + 6x^2$, then $du = 12x\,dx$.

$\displaystyle =\int \sqrt[5]{u}\,du$ Substitution: $\begin{matrix}u=1+6x^2\\ du=12xdx\end{matrix}$

$\displaystyle =\int u^{1/5}\,du$

$\displaystyle =\frac{5}{6}u^{6/5}+C$ Formula A

$\displaystyle =\frac{5}{6}\left(1+6x^2\right)^{6/5}+C$ Reverse substitution

43. $\displaystyle\int_0^1 2xe^{x^2}\,dx$

We first find the indefinite integral

$\displaystyle\int 2xe^{x^2}\,dx$

$\displaystyle =\int e^u\,du$ Substitution: $\begin{matrix}u=x^2\\ du=2xdx\end{matrix}$

$\displaystyle =e^u+C$ Formula B

$\displaystyle =e^{x^2}+C$ Reverse Substitution

Next, we evaluate the definite integral on $[0,1]$.

$\displaystyle\int_0^1 2xe^{x^2}\,dx=\left[e^{x^2}\right]_0^1$ Let $C = 0$.

$\displaystyle =e^{(1)^2}-e^{(0)^2}$

$\displaystyle =e^1-e^0$

$\displaystyle =e-1$

45. $\displaystyle\int_0^1 x\left(x^2+1\right)^5\,dx$

We first find the indefinite integral

$\displaystyle\int x\left(x^2+1\right)^5\,dx$

$\displaystyle =\frac{1}{2}\cdot 2\int x\left(x^2+1\right)^5\,dx$ Multiplying by 1

$\displaystyle =\frac{1}{2}\int 2x\left(x^2+1\right)^5\,dx$

$\displaystyle =\frac{1}{2}\int u^5\,du$ Substitution: $\begin{matrix}u=x^2+1\\ du=2xdx\end{matrix}$

$\displaystyle =\frac{1}{12}u^6+C$ Formula A

$\displaystyle =\frac{1}{12}\left(x^2+1\right)^6+C$ Reverse Substitution

Next, we evaluate the definite integral on $[0,1]$.

$\displaystyle\int_0^1 x\left(x^2+1\right)^5\,dx$

$\displaystyle =\left[\frac{1}{12}\left(x^2+1\right)^6\right]_0^1$ Let $C = 0$.

$\displaystyle =\left[\frac{1}{12}\left(1^2+1\right)^6-\frac{1}{12}\left(0^2+1\right)^6\right]$

$\displaystyle =\frac{1}{12}(2)^6-\frac{1}{12}(1)^6$

$\displaystyle =\frac{16}{3}-\frac{1}{12}$

$\displaystyle =\frac{21}{4}$

47. $\displaystyle\int_0^4 \frac{dt}{1+t}$

We first find the indefinite integral

$\displaystyle\int \frac{dt}{1+t}$

$\displaystyle =\int \frac{du}{u}$ Substitution: $\begin{matrix}u=1+t\\ du=dt\end{matrix}$

$\displaystyle =\ln u+C$ Formula C, $u > 0$

$\displaystyle =\ln(1+t)+C$ Reverse Substitution

Next, we evaluate the definite integral on $[0,4]$.

$\int_0^4 \dfrac{dt}{1+t}$

$= \left[\ln\left(1+t\right) \right]_0^4 \qquad$ Let $C = 0$.

$= \left[\ln\left(1+4\right) - \ln\left(1+0\right) \right]$

$= \ln\left(5\right) - \ln\left(1\right)$

$= \ln 5 - 0$

$= \ln 5$

49. $\int_1^4 \dfrac{2x+1}{x^2+x-1}\,dx$

We first find the indefinite integral

$\int \dfrac{2x+1}{x^2+x-1}\,dx$

$= \int \dfrac{du}{u} \qquad$ Substitution: $\begin{smallmatrix}u=x^2+x-1\\du=2x+1dx\end{smallmatrix}$

$= \ln u + C \qquad$ Formula C, $u > 0$

$= \ln\left(x^2+x-1\right)+C \qquad$ Reverse Substitution

Next, we evaluate the definite integral on $\left[1,4\right]$.

$\int_1^4 \dfrac{2x+1}{x^2+x-1}\,dx$

$= \left[\ln\left(x^2+x-1\right) \right]_1^4 \qquad$ Let $C = 0$.

$= \left[\ln\left(\left(4\right)^2+\left(4\right)-1\right) - \ln\left(\left(1\right)^2+\left(1\right)-1\right) \right]$

$= \ln\left(19\right) - \ln\left(1\right)$

$= \ln 19 - 0$

$= \ln 19$

51. $\int_0^b e^{-x}\,dx$

We first find the indefinite integral

$\int e^{-x}\,dx$

$= -\int -e^{-x}\,dx \qquad$ Multipling by 1

$= -\int e^u\,du \qquad$ Substitution: $\begin{smallmatrix}u=-x\\du=-dx\end{smallmatrix}$

$= -e^u + C \qquad$ Formula B

$= -e^{-x} + C \qquad$ Reverse Substitution

Next, we evaluate the definite integral on $\left[0,b\right]$.

$\int_0^b e^{-x}\,dx$

$= \left[-e^{-x} \right]_0^b \qquad$ Let $C = 0$.

$= -e^{-(b)} - \left(-e^{-(0)} \right)$

$= -e^{-b} + e^0$

$= 1 - e^{-b}$

53. $\int_0^b me^{-mx}\,dx$

We first find the indefinite integral

$\int me^{-mx}\,dx$

$= -\int -me^{-mx}\,dx \qquad$ Multipling by 1

$= -\int e^u\,du \qquad$ Substitution: $\begin{smallmatrix}u=-mx\\du=-mdx\end{smallmatrix}$

$= -e^u + C \qquad$ Formula B

$= -e^{-mx} + C \qquad$ Reverse Substitution

Next, we evaluate the definite integral on $\left[0,b\right]$.

$\int_0^b me^{-mx}\,dx$

$= \left[-e^{-mx} \right]_0^b \qquad$ Let $C = 0$.

$= -e^{-m(b)} - \left(-e^{-m(0)} \right)$

$= -e^{-mb} + e^0$

$= 1 - e^{-mb}$

55. $\int_0^4 \left(x-6\right)^2\,dx$

We first find the indefinite integral

$\int \left(x-6\right)^2\,dx$

$= \int u^2\,du \qquad$ Substitution: $\begin{smallmatrix}u=x-6\\du=dx\end{smallmatrix}$

$= \dfrac{1}{3}u^3 + C \qquad$ Formula A

$= \dfrac{1}{3}\left(x-6\right)^3 + C \qquad$ Reverse Substitution

Next, we evaluate the definite integral on $\left[0,4\right]$.

$\int_0^4 \left(x-6\right)^2\,dx$

$= \left[\dfrac{1}{3}\left(x-6\right)^3 \right]_0^4 \qquad$ Let $C = 0$.

$= \left[\dfrac{1}{3}\left(\left(4\right)-6\right)^3 - \dfrac{1}{3}\left(\left(0\right)-6\right)^3 \right]$

$= \dfrac{1}{3}\left(-2\right)^3 - \dfrac{1}{3}\left(-6\right)^3$

$= -\dfrac{8}{3} - \left(-\dfrac{216}{3} \right)$

$= \dfrac{208}{3}$

57. $\int_0^2 \dfrac{3x^2\,dx}{\left(1+x^3\right)^5}$

We first find the indefinite integral

$\displaystyle\int \dfrac{3x^2\,dx}{\left(1+x^3\right)^5}$

$= \displaystyle\int \dfrac{du}{u^5}$ Substitution: $\begin{array}{l} u=1+x^3 \\ du=3x^2\,dx \end{array}$

$= \displaystyle\int u^{-5}\,du$

$= -\dfrac{1}{4}u^{-4} + C$ Formula A

$= -\dfrac{1}{4}\left(1+x^3\right)^{-4} + C$ Reverse Substitution

$= -\dfrac{1}{4\left(1+x^3\right)^4} + C$

Next, we evaluate the definite integral on $[0,2]$.

$\displaystyle\int_0^2 \dfrac{3x^2\,dx}{\left(1+x^3\right)^5}$

$= \left[-\dfrac{1}{4\left(1+x^3\right)^4} \right]_0^2$ Let $C = 0$.

$= \left[\left(-\dfrac{1}{4\left(1+(2)^3\right)^4} \right) - \left(-\dfrac{1}{4\left(1+(0)^3\right)^4} \right) \right]$

$= \left(-\dfrac{1}{4(9)^4} \right) - \left(-\dfrac{1}{4(1)^4} \right)$

$= \left(-\dfrac{1}{26{,}244} \right) - \left(-\dfrac{1}{4} \right)$

$= -\dfrac{1}{26{,}244} + \dfrac{6561}{26{,}244}$

$= \dfrac{6560}{26{,}244}$

$= \dfrac{1640}{6561}$

59. $\int_0^{\sqrt{7}} 7x \cdot \sqrt[3]{1+x^2}\,dx$

We first find the indefinite integral

$\displaystyle\int 7x \cdot \sqrt[3]{1+x^2}\,dx$

$= 7\displaystyle\int x \cdot \sqrt[3]{1+x^2}\,dx$

$= \dfrac{7}{2}\displaystyle\int 2x \cdot \sqrt[3]{1+x^2}\,dx$

$= \dfrac{7}{2}\displaystyle\int \sqrt[3]{1+x^2}\,(2x\,dx)$

$= \dfrac{7}{2}\displaystyle\int \sqrt[3]{u}\,du$ Substitution: $\begin{array}{l} u=1+x^2 \\ du=2x\,dx \end{array}$

$= \dfrac{7}{2}\displaystyle\int u^{1/3}\,du$

$= \dfrac{21}{8}u^{4/3} + C$ Formula A

$= \dfrac{21}{8}\left(1+x^2\right)^{4/3} + C$ Reverse Substitution

We evaluate the definite integral on $\left[0,\sqrt{7}\right]$.

$\displaystyle\int_0^{\sqrt{7}} 7x\sqrt[3]{1+x^2}\,dx$

$= \left[\dfrac{21}{8}\left(1+x^2\right)^{4/3} \right]_0^{\sqrt{7}}$ Let $C = 0$.

$= \left[\left(\dfrac{21}{8}\left(1+\left(\sqrt{7}\right)^2\right)^{4/3} \right) - \left(\dfrac{21}{8}\left(1+(0)^2\right)^{4/3} \right) \right]$

$= \left(\dfrac{21}{8}(8)^{4/3} \right) - \left(\dfrac{21}{8}(1)^{4/3} \right)$

$= (42) - \left(\dfrac{21}{8} \right)$

$= \dfrac{315}{8}$

61. Left to the student.

63. a) We will evaluate each integral separately and then do the operations to find $P(t)$.

First, we will find revenue from the sale of products.

$R(t) = \displaystyle\int R'(t)\,dt$

$\quad\quad = \displaystyle\int 4000t\,dt$

$\quad\quad = 2000t^2 + C$

We use the initial condition to find C.

$R(0) = 0$

$2000(0)^2 + C = 0$

$\quad\quad\quad\quad C = 0$

Thus, the revenue from the sale of product is
$R(t) = 2000t^2$.

The revenue from the sale of the machine is the salvage value of the machine, which is given by $V(t) = 200,000 - 25,000e^{0.1t}$.

The cost of the machine was \$250,000. Putting these three functions together, the total profit from the machine, in dollars, is

$P(t) = R(t) + V(t) - C(t)$

$= (2000t^2) + (200,000 - 25,000e^{0.1t}) - $
$$(250,000)$$

$= 2000t^2 - 50,000 - 25,000e^{0.1t}$.

b) Substituting 10 for t, we have:

$P(10) = 2000t^2 - 25,000e^{0.1t} - 50,000$

$= 2000(10)^2 - 50,000 - 25,000e^{0.1(10)}$

$= 150,000 - 25,000e$

$\approx 82,042.95$

After 10 years the total profit from the machine is \$82,042.95.

65. a) $\int_0^{105} D(t)\,dt = \int_0^{105} 100,000e^{0.025t}\,dt$

Find the definite integral first.

$\int 100,000e^{0.025t}\,dt$

$= \dfrac{1}{0.025} \cdot (0.025) \int 100,000e^{0.025t}\,dt$

$= 4,000,000 \int e^{0.025t}(0.025)\,dt$

$= 4,000,000 \int e^u\,du$ Substitution: $\begin{smallmatrix}u=0.025t\\du=0.025dt\end{smallmatrix}$

$= 4,000,000e^u + C$

$= 4,000,000e^{0.025t} + C$

Next, we evaluate the definite integral on $[0,105]$.

$\int_0^{105} 100,000e^{0.025t}\,dt$

$= \left[4,000,000e^{0.025t}\right]_0^{105}$ Let $C = 0$.

$= 4,000,000\left[e^{0.025(105)} - e^{0.025(0)}\right]$

$= 4,000,000\left[e^{2.625} - e^0\right]$

$= 4,000,000\left[e^{2.625} - 1\right]$

$\approx 51,218,297$

There were approximately 51,218,297 divorces from 1900 to 2005.

b) We evaluated the indefinite integral in part (a), we simply need to evaluate the definite integral over the interval $[80,106]$

$\int_{80}^{106} 100,000e^{0.025t}\,dt$

$= \left[4,000,000e^{0.025t}\right]_{80}^{106}$ From part (a).

$= 4,000,000\left[e^{0.025(106)} - e^{0.025(80)}\right]$

$= 4,000,000\left[e^{2.65} - e^2\right]$

$\approx 4,000,000[6.764982546]$

$\approx 27,059,930$

There were approximately 27,059,930 divorces from 1980 to 2006.

67. $\displaystyle\int_{-4}^{4} x\sqrt{16 - x^2}\,dx$

We notice that $0 \geq x\sqrt{16 - x^2}$, on $[-4,0]$ and $x\sqrt{4 - x^2} \geq 0$, on $[0,4]$. We will divide the interval into two parts.

First on the interval $[-4,0]$, we find the indefinite integral.

$\int\left(0 - x\sqrt{16 - x^2}\right)dx$

$= \int\left(-x\sqrt{16 - x^2}\right)dx$

$= \dfrac{1}{2}\int\sqrt{16 - x^2}\,(-2x)\,dx$

$= \dfrac{1}{2}\int\sqrt{u}\,du$ Substitution: $\begin{smallmatrix}u=16-x^2\\du=-2xdx\end{smallmatrix}$

$= \dfrac{1}{2}\int u^{1/2}\,du$

$= \dfrac{1}{3}u^{3/2} + C$

$= \dfrac{1}{3}(16 - x^2)^{3/2} + C$

Next, we will evaluate the integral on $[-4,0]$.

$\int_{-4}^{0} -x\sqrt{16 - x^2}\,dx$

$= \left[\dfrac{1}{3}(16 - x^2)^{3/2}\right]_{-4}^{0}$

$= \left[\left(\dfrac{1}{3}(16 - (0)^2)^{3/2}\right) - \left(\dfrac{1}{3}(16 - (-4)^2)^{3/2}\right)\right]$

$= \dfrac{1}{3}(16)^{3/2} - \dfrac{1}{3}(0)^{3/2}$

$= \dfrac{64}{3}$

Next, on the interval $[0,4]$, we find the indefinite integral.

$$\int \left(x\sqrt{16-x^2} - 0 \right) dx$$

$$= \int \left(x\sqrt{16-x^2} \right) dx$$

$$= -\frac{1}{2}\int \sqrt{16-x^2}\,(-2x)\,dx$$

$$= -\frac{1}{2}\int \sqrt{u}\,du \qquad \text{Substitution: } \substack{u=16-x^2 \\ du=-2x\,dx}$$

$$= -\frac{1}{2}\int u^{1/2}\,du$$

$$= -\frac{1}{3}u^{3/2} + C$$

$$= -\frac{1}{3}\left(16-x^2\right)^{3/2} + C$$

Next, we will evaluate the integral on $[0,4]$.

$$\int_0^4 x\sqrt{16-x^2}\,dx$$

$$= \left[-\frac{1}{3}\left(16-x^2\right)^{3/2} \right]_0^4$$

$$= \left[\left(-\frac{1}{3}\left(16-(4)^2\right)^{3/2} \right) - \left(-\frac{1}{3}\left(16-(0)^2\right)^{3/2} \right) \right]$$

$$= -\frac{1}{3}(0)^{3/2} + \frac{1}{3}(16)^{3/2}$$

$$= \frac{64}{3}$$

Therefore, the total area is:

$$\int_{-4}^0 -x\sqrt{4-x^2}\,dx + \int_0^4 x\sqrt{4-x^2}\,dx$$

$$= \frac{64}{3} + \frac{64}{3} = \frac{128}{3}.$$

69. $\displaystyle \int 5x\sqrt{1-4x^2}\,dx = 5\int x\sqrt{1-4x^2}\,dx$

Let $u = 1-4x^2$, then $du = -8x\,dx$. We do not have $-8x\,dx$. We only have $x\,dx$ and need to supply a -8. We do this by multiplying by $\frac{1}{-8}\cdot -8$ as follows.

$$= \left(-\frac{1}{8} \right)(-8)5\int x\sqrt{1-4x^2}\,dx$$

$$= -\frac{5}{8}\int \sqrt{1-4x^2}\,(-8x)\,dx$$

$$= -\frac{5}{8}\int \sqrt{u}\,du \qquad \text{Substitution: } \substack{u=1-4x^2 \\ du=-8x\,dx}$$

$$= -\frac{5}{8}\int u^{1/2}\,du$$

$$= -\frac{5}{12}u^{3/2} + C \qquad \text{Formula A}$$

$$= -\frac{5}{12}\left(1-4x^2\right)^{3/2} + C \qquad \text{Reverse substitution}$$

71. $\displaystyle \int \frac{x^2}{e^{x^3}}\,dx = \int x^2 e^{-x^3}\,dx$

Let $u = -x^3$, then $du = -3x^2\,dx$. We do not have $-3x^2\,dx$. We only have $x^2\,dx$ and need to supply a -3. We do this by multiplying by $\frac{1}{-3}\cdot -3$ as follows.

$$= -\frac{1}{3}\cdot(-3)\int x^2 e^{-x^3}\,dx$$

$$= -\frac{1}{3}\int e^{-x^3}\left(-3x^2\right)dx$$

$$= -\frac{1}{3}\int e^u\,du \qquad \text{Substitution: } \substack{u=-x^3 \\ du=-3x^2\,dx}$$

$$= -\frac{1}{3}e^u + C \qquad \text{Formula B}$$

$$= -\frac{1}{3}e^{-x^3} + C \qquad \text{Reverse substitution}$$

73. $\displaystyle \int \frac{e^{1/t}}{t^2}\,dt$

Let $u = \frac{1}{t} = t^{-1}$, then $du = -t^{-2}dt = -\frac{1}{t^2}dt$. We do not have $-\frac{1}{t^2}dt$. We only have $\frac{1}{t^2}dt$. and need to supply a -1. We do this by multiplying by $(-1)\cdot(-1)$ as follows.

$$= (-1)\int (-1)\frac{e^{1/t}}{t^2}\,dt$$

$$= -\int e^u\,du \qquad \text{Substitution: } \substack{u=\frac{1}{t} \\ du=-\frac{1}{t^2}dt}$$

$$= -e^u + C \qquad \text{Formula B}$$

$$= -e^{1/t} + C \qquad \text{Reverse substitution}$$

75. $\int \dfrac{dx}{x(\ln x)^4}$

Let $u = \ln x$, then $du = \dfrac{1}{x}dx$.

$= \int \dfrac{du}{u^4}$ Substitution: $\begin{array}{l} u=\ln x \\ du=\frac{1}{x}dx \end{array}$

$= \int u^{-4}\,du$

$= \dfrac{1}{-3}u^{-3} + C$

$= -\dfrac{1}{3}(\ln x)^{-3} + C$

77. $\int x^2\sqrt{x^3+1}\,dx$

Let $u = x^3+1$, then $du = 3x^2\,dx$. We do not have $3x^2\,dx$ We only have $x^2\,dx$. and need to supply a 3. We do this by multiplying by $\frac{1}{3}\cdot 3$ as follows.

$= \dfrac{1}{3}\cdot 3\int x^2\sqrt{x^3+1}\,dx$

$= \dfrac{1}{3}\int \sqrt{x^3+1}\left(3x^2\right)dx$

$= \dfrac{1}{3}\int \sqrt{u}\,du$ Substitution: $\begin{array}{l} u=x^3+1 \\ du=3x^2dx \end{array}$

$= \dfrac{1}{3}\int u^{1/2}\,du$ Formula A

$= \dfrac{2}{9}u^{3/2} + C$

$= \dfrac{2}{9}\left(x^3+1\right)^{3/2} + C$

79. $\int \dfrac{x-3}{\left(x^2-6x\right)^{1/3}}\,dx$

Let $u = x^2-6x$, then $du = (2x-6)\,dx$. We do not have $(2x-6)\,dx$ We only have $(x-3)\,dx$. and need to supply a 2. We do this by multiplying by $\frac{1}{2}\cdot 2$ as follows.

$= \dfrac{1}{2}\cdot 2\int \dfrac{x-3}{\left(x^2-6x\right)^{1/3}}\,dx$

$= \dfrac{1}{2}\int \dfrac{2x-6}{\left(x^2-6x\right)^{1/3}}\,dx$

$= \dfrac{1}{2}\int \dfrac{1}{u^{1/3}}\,du$ Substitution: $\begin{array}{l} u=x^2-6x \\ du=(2x-6)dx \end{array}$

$= \dfrac{1}{2}\int u^{-1/3}\,du$

$= \dfrac{1}{2}\cdot \dfrac{u^{2/3}}{2/3} + C$

$= \dfrac{3}{4}\left(x^2-6x\right)^{2/3} + C$

81. $\int \dfrac{t^2+2t}{(t+1)^2}\,dt = \int\left[1 - \dfrac{1}{(t+1)^2}\right]dt$ See the hint.

$= \int dt - \int \dfrac{1}{(t+1)^2}$

Let $u = t+1$, then $du = dt$.

$= \int du - \int \dfrac{1}{u^2}\,du$ Substitution: $\begin{array}{l} u=t+1 \\ du=dt \end{array}$

$= \int du - \int u^{-2}\,du$

$= u - \left[-u^{-1}\right] + K$

$= t+1 + (t+1)^{-1} + K$

$= t + \dfrac{1}{t+1} + C$ $[C = K+1]$

83. $\int \dfrac{x+3}{x+1}\,dx = \int\left[1 + \dfrac{2}{x+1}\right]dx$ See the hint.

$= \int dx + \int \dfrac{2}{x+1}\,dx$

Let $u = x+1$, then $du = dx$.

$= \int du + 2\int \dfrac{du}{u}$ Substitution: $\begin{array}{l} u=x+1 \\ du=dx \end{array}$

$= u + 2\ln u + K$

$= x+1 + 2\ln(x+1) + K$

$= x + 2\ln(x+1) + C$ $[C = K+1]$

85. $\int \dfrac{dx}{x(\ln x)^n}, \qquad n \neq -1$

Let $u = \ln x$, then $du = \dfrac{1}{x}\,dx$.

$= \int \dfrac{du}{u^n}$ Substitution: $\begin{smallmatrix} u=\ln x \\ du=\frac{1}{x}dx \end{smallmatrix}$

$= \int u^{-n}\,du$

$= \dfrac{u^{-n+1}}{-n+1} + C$

$= \dfrac{(\ln x)^{-n+1}}{-n+1} + C$

87. $\int \dfrac{e^x - e^{-x}}{e^x + e^{-x}}\,dx$

Let $u = e^x + e^{-x}$, then $du = e^x - e^{-x}\,dx$.

$= \int \dfrac{du}{u}$ Substitution: $\begin{smallmatrix} u=e^x+e^{-x} \\ du=e^x-e^{-x}dx \end{smallmatrix}$

$= \ln u + C$

$= \ln(e^x + e^{-x}) + C$

89. $\int \dfrac{dx}{x \ln x [\ln(\ln x)]}$

Let $u = \ln(\ln x)$, then $du = \dfrac{1}{x \ln x}\,dx$.

$= \int \dfrac{du}{u}$ Substitution: $\begin{smallmatrix} u=\ln(\ln x) \\ du=\frac{1}{x\ln x}dx \end{smallmatrix}$

$= \ln u + C$

$= \ln[\ln(\ln x)] + C$

91. $\int 9x(7x^2 + 9)^n\,dx, \qquad n \neq -1$

$= 9\int x(7x^2 + 9)^n\,dx$

Let $u = 7x^2 + 9$, then $du = 14x\,dx$. We do not have $14x\,dx$. We only have $x\,dx$ and need to supply a 14. We do this by multiplying by $\frac{1}{14} \cdot 14$ as follows.

$= \dfrac{1}{14} \cdot 14 \cdot 9 \int x(7x^2 + 9)^n\,dx$

$= \dfrac{9}{14} \int 14x(7x^2 + 9)^n\,dx$

$= \dfrac{9}{14} \int u^n\,du$ Substitution: $\begin{smallmatrix} u=7x^2+9 \\ du=14xdx \end{smallmatrix}$

$= \dfrac{9}{14} \cdot \dfrac{u^{n+1}}{n+1} + C$

$= \dfrac{9}{14(n+1)}(7x^2 + 9)^{n+1} + C$

93. $\boxed{tw}$

Exercise Set 4.6

---

1. $\int 4xe^{4x}\,dx = \int x(4e^{4x}\,dx)$

Let

$\quad u = x \quad$ and $\quad dv = 4e^{4x}\,dx$

Then

$\quad du = dx \quad$ and $\quad v = e^{4x}$

Using the Integration-by-Parts Formula gives

$\quad u \qquad dv \qquad\quad u \cdot v \quad\quad v\ du$

$\int x(4e^{4x}\,dx) = x \cdot e^{4x} - \int e^{4x}\,dx$

$\qquad\qquad = xe^{4x} - \dfrac{1}{4}e^{4x} + C \,.$

3. $\int x^3(3x^2)\,dx$

Let

$\quad u = x^3 \quad$ and $\quad dv = 3x^2\,dx$

Then

$\quad du = 3x^2\,dx \quad$ and $\quad v = x^3$

Using the Integration-by-Parts Formula gives

$\quad u \qquad dv \qquad u \quad v \qquad v\ du$

$\int x^3(3x^2\,dx) = x^3 \cdot x^3 - \int x^3 3x^2\,dx$

$\qquad\qquad = x^6 - \int 3x^5\,dx$

$\qquad\qquad = x^6 - 3 \cdot \dfrac{x^6}{6} + C$

$\qquad\qquad = x^6 - \dfrac{x^6}{2} + C$

$\qquad\qquad = \dfrac{x^6}{2} + C\,.$

It should be noted that this problem can also be worked with substitution or formula A.

Integrate using Substitution

$$\int x^3\left(3x^2\right)dx$$

$$=\int u\,du \qquad \text{Substitution: } \begin{smallmatrix} u=x^3 \\ du=3x^2dx \end{smallmatrix}$$

$$=\frac{1}{2}u^2+C$$

$$=\frac{1}{2}\left(x^3\right)^2+C$$

$$=\frac{x^6}{2}+C$$

Integrate using formula A.

$$\int x^3\left(3x^2\right)dx$$

$$=\int 3x^5\,dx$$

$$=3\cdot\frac{x^6}{6}+C \qquad \text{Using Formula A}$$

$$=\frac{x^6}{2}+C$$

5. $\displaystyle\int xe^{5x}dx$

Let

$$u=x \qquad \text{and} \qquad dv=e^{5x}dx$$

Then

$$du=dx \qquad \text{and} \qquad v=\frac{1}{5}e^{5x}$$

We integrated dv using the formula

$$\int be^{ax}dx=\frac{b}{a}e^{ax}+C.$$

Using the Integration-by-Parts Formula gives

$$\overset{u}{}\quad\overset{dv}{}\quad\overset{u}{}\quad\overset{v}{}\quad\overset{v}{}\ \overset{du}{}$$

$$\int x\left(e^{5x}dx\right)=x\cdot\frac{1}{5}e^{5x}-\int\frac{1}{5}e^{5x}dx$$

$$=x\cdot\frac{1}{5}e^{5x}-\frac{1}{5}\cdot\frac{1}{5}e^{5x}+C$$

$$\left(\int be^{ax}dx=\frac{b}{a}e^{ax}+C\right)$$

$$=\frac{1}{5}xe^{5x}-\frac{1}{25}e^{5x}+C.$$

7. $\displaystyle\int xe^{-2x}dx$

Let

$$u=x \qquad \text{and} \qquad dv=e^{-2x}dx$$

Then

$$du=dx \qquad \text{and} \qquad v=-\frac{1}{2}e^{-2x}$$

We integrated dv using the formula

$$\int be^{ax}dx=\frac{b}{a}e^{ax}+C.$$

Using the Integration-by-Parts Formula gives

$$\overset{u}{}\quad\overset{dv}{}\quad\overset{u}{}\quad\overset{v}{}\qquad\overset{v}{}\qquad\overset{du}{}$$

$$\int x\left(e^{-2x}dx\right)=x\cdot\left(-\frac{1}{2}e^{-2x}\right)-\int\left(-\frac{1}{2}e^{-2x}\right)dx$$

$$=-\frac{1}{2}xe^{-2x}-\left(-\frac{1}{2}\right)\cdot\left(-\frac{1}{2}e^{-2x}\right)+C$$

$$\left(\int be^{ax}dx=\frac{b}{a}e^{ax}+C\right)$$

$$=-\frac{1}{2}xe^{-2x}-\frac{1}{4}e^{-2x}+C.$$

9. $\displaystyle\int x^2\ln x\,dx$

Let

$$u=\ln x \qquad \text{and} \qquad dv=x^2dx$$

Then

$$du=\frac{1}{x}dx \qquad \text{and} \qquad v=\frac{1}{3}x^3$$

Using the Integration-by-Parts Formula gives

$$\overset{u}{}\quad\overset{dv}{}\quad\overset{u}{}\quad\overset{v}{}\qquad\overset{v}{}\quad\overset{du}{}$$

$$\int \ln x\left(x^2\,dx\right)=\ln x\cdot\left(\frac{1}{3}x^3\right)-\int\left(\frac{1}{3}x^3\right)\left(\frac{1}{x}dx\right)$$

$$=\frac{1}{3}x^3\ln x-\int\frac{1}{3}x^2dx$$

$$=\frac{1}{3}x^3\ln x-\frac{1}{9}x^3+C$$

$$=\frac{x^3\ln x}{3}-\frac{x^3}{9}+C.$$

11. $\displaystyle\int x\ln\sqrt{x}\,dx=\int x\ln x^{1/2}\,dx$

Let

$$u=\ln x^{1/2} \qquad \text{and} \qquad dv=xdx$$

Then

$$du=\frac{1}{x^{1/2}}\cdot\frac{1}{2}x^{-1/2}dx=\frac{1}{2x}dx \qquad \text{and} \qquad v=\frac{1}{2}x^2$$

Using the Integration-by-Parts Formula gives

$$\overset{u}{}\qquad\overset{dv}{}\quad\overset{u}{}\quad\overset{v}{}\qquad\overset{v}{}\quad\overset{du}{}$$

$$\int \ln x^{1/2}\left(xdx\right)=\ln x^{1/2}\cdot\left(\frac{x^2}{2}\right)-\int\left(\frac{x^2}{2}\right)\left(\frac{1}{2x}dx\right)$$

$$=\frac{x^2\ln x^{1/2}}{2}-\int\frac{1}{4}xdx$$

$$=\frac{x^2\ln x^{1/2}}{2}-\frac{1}{8}x^2+C$$

$$=\frac{x^2\ln x}{4}-\frac{1}{8}x^2+C \qquad \left(\ln x^{1/2}=\frac{1}{2}\ln x\right)$$

13. $\int \ln(x+5)\,dx$

Let

$$u = \ln(x+5) \quad \text{and} \quad dv = dx$$

Then

$$du = \frac{1}{x+5}\,dx \quad \text{and} \quad v = x+5$$

<small>Choosing $x+5$ as an antiderivative of dv</small>

Using the Integration-by-Parts Formula gives

$$\overset{u}{} \quad \overset{dv}{}$$

$$\int \ln(x+5)\,dx =$$

$$\overset{u}{} \quad \overset{v}{} \quad \overset{v}{} \quad \overset{du}{}$$

$$\ln(x+5) \cdot (x+5) - \int (x+5)\left(\frac{1}{x+5}\,dx\right)$$

$$= (x+5)\ln(x+5) - \int dx$$

$$= (x+5)\ln(x+5) - x + C.$$

15. $\int (x+2)\ln x\,dx$

Let

$$u = \ln x \quad \text{and} \quad dv = (x+2)\,dx$$

Then

$$du = \frac{1}{x}\,dx \quad \text{and} \quad v = \frac{x^2}{2} + 2x$$

Using the Integration-by-Parts Formula gives

$$\overset{u}{} \quad \overset{dv}{}$$

$$\int \ln x \cdot (x+2)\,dx =$$

$$\overset{u}{} \quad \overset{v}{} \quad \overset{v}{} \quad \overset{du}{}$$

$$\ln(x) \cdot \left(\frac{x^2}{2} + 2x\right) - \int \left(\frac{x^2}{2} + 2x\right)\left(\frac{1}{x}\,dx\right)$$

$$= \left(\frac{x^2}{2} + 2x\right)\ln x - \int \left(\frac{x}{2} + 2\right)dx$$

$$= \left(\frac{x^2}{2} + 2x\right)\ln x - \int \frac{x}{2}\,dx - \int 2\,dx$$

$$= \left(\frac{x^2}{2} + 2x\right)\ln x - \frac{x^2}{4} - 2x + C.$$

17. $\int (x-1)\ln x\,dx$

Let

$$u = \ln x \quad \text{and} \quad dv = (x-1)\,dx$$

Then

$$du = \frac{1}{x}\,dx \quad \text{and} \quad v = \frac{x^2}{2} - x$$

Using the Integration-by-Parts Formula gives

$$\overset{u}{} \quad \overset{dv}{}$$

$$\int \ln x \cdot (x-1)\,dx =$$

$$\overset{u}{} \quad \overset{v}{} \quad \overset{v}{} \quad \overset{du}{}$$

$$\ln(x) \cdot \left(\frac{x^2}{2} - x\right) - \int \left(\frac{x^2}{2} - x\right)\left(\frac{1}{x}\,dx\right)$$

$$= \left(\frac{x^2}{2} - x\right)\ln x - \int \left(\frac{x}{2} - 1\right)dx$$

$$= \left(\frac{x^2}{2} - x\right)\ln x - \int \frac{x}{2}\,dx + \int 1\,dx$$

$$= \left(\frac{x^2}{2} - x\right)\ln x - \frac{x^2}{4} + x + C.$$

19. $\int x\sqrt{x+2}\,dx$

Let

$$u = x \quad \text{and} \quad dv = \sqrt{x+2}\,dx = (x+2)^{1/2}\,dx$$

Then

$$du = dx \quad \text{and} \quad v = \frac{(x+2)^{3/2}}{3/2} = \frac{2}{3}(x+2)^{3/2}$$

Using the Integration-by-Parts Formula gives

$$\overset{u}{} \quad \overset{dv}{}$$

$$\int x\left(\sqrt{x+2}\,dx\right) =$$

$$\overset{u}{} \quad \overset{v}{} \quad \overset{v}{} \quad \overset{du}{}$$

$$x \cdot \frac{2}{3}(x+2)^{3/2} - \int \frac{2}{3}(x+2)^{3/2}\,dx$$

$$= \frac{2}{3}x(x+2)^{3/2} - \frac{2}{3}\int (x+2)^{3/2}\,dx$$

$$= \frac{2}{3}x(x+2)^{3/2} - \frac{2}{3}\frac{(x+2)^{5/2}}{5/2} + C$$

$$= \frac{2}{3}x(x+2)^{3/2} - \frac{4}{15}(x+2)^{5/2} + C.$$

21. $\int x^3 \ln(2x)\,dx$

Let

$$u = \ln(2x) \quad \text{and} \quad dv = x^3\,dx$$

Then

$$du = \frac{1}{2x} \cdot 2\,dx = \frac{1}{x}\,dx \quad \text{and} \quad v = \frac{1}{4}x^4$$

Using the Integration-by-Parts Formula gives

$$\int \ln(2x)\cdot(x^3 dx) = \ln(2x)\cdot\left(\frac{x^4}{4}\right) - \int\left(\frac{x^4}{4}\right)\left(\frac{1}{x}dx\right)$$

$$= \frac{x^4}{4}\ln(2x) - \int \frac{x^3}{4}dx$$

$$= \frac{x^4}{4}\ln(2x) - \frac{x^4}{16} + C.$$

23. $\int x^2 e^x dx$

Let

$\quad u = x^2 \quad$ and $\quad dv = e^x dx$

Then

$\quad du = 2x dx \quad$ and $\quad v = e^x$

Using the Integration-by-Parts Formula gives

$$\int x^2 \left(e^x dx\right) = x^2 e^x - \int 2x e^x dx$$

We will use integration by parts again to evaluate $\int 2x e^x dx$

Let

$\quad u = 2x \quad$ and $\quad dv = e^x dx$

Then

$\quad du = 2dx \quad$ and $\quad v = e^x$

Using the Integration-by-Parts Formula gives

$$\int 2x\left(e^x dx\right) = 2x e^x - \int 2e^x dx$$

$$= 2x e^x - 2e^x + C_1$$

Thus,

$$\int x^2 e^x dx = x^2 e^x - \int 2x e^x dx$$

$$= x^2 e^x - \left(2x e^x - 2e^x + C_1\right)$$

$$= x^2 e^x - 2x e^x + 2e^x - C_1$$

$$= x^2 e^x - 2x e^x + 2e^x + C \quad [C = -C_1]$$

Note, Since we have an integral $\int f(x)g(x)dx$, where $f(x)$ can be differentiated repeatedly to a derivative that is eventually 0 and $g(x)$ can be integrated repeatedly easily, we can use tabular integration as shown below:

$f(x)$ and Repeated Derivatives	Sign of Product	$g(x)$ and Repeated Integrals
x^2	+	e^x
$2x$	$-$	e^x
2	+	e^x
0		e^x

We add the products along the arrows, making the alternate sign changes.

$$\int x^2 e^x dx = x^2 e^x - 2x e^x + 2e^x + C$$

25. $\int x^2 e^{2x} dx$

Let

$\quad u = x^2 \quad$ and $\quad dv = e^{2x} dx$

Then

$$du = 2x dx \text{ and } v = \frac{1}{2}e^{2x} \quad \left(\int b e^{ax} = \frac{b}{a}e^{ax} + C\right)$$

Using the Integration-by-Parts Formula gives

$$\int x^2 \left(e^{2x} dx\right) = x^2 \cdot \frac{1}{2}e^x - \int\left(\frac{1}{2}e^{2x}\right)\cdot 2x dx$$

$$\frac{1}{2}x^2 e^x - \int x e^{2x} dx$$

We will use integration by parts again to evaluate $\int x e^{2x} dx$

Let

$\quad u = x \quad$ and $\quad dv = e^{2x} dx$

Then

$$du = dx \text{ and } v = \frac{1}{2}e^{2x} \quad \left(\int b e^{ax} = \frac{b}{a}e^{ax} + C\right)$$

Using the Integration-by-Parts Formula gives

$$\int x\left(e^{2x} dx\right) = x\left(\frac{1}{2}e^{2x}\right) - \int\left(\frac{1}{2}e^{2x}\right)dx$$

$$= \frac{1}{2}x e^{2x} - \frac{1}{2}\cdot\frac{1}{2}e^{2x} + C_1$$

$$= \frac{1}{2}x e^{2x} - \frac{1}{4}e^{2x} + C_1$$

Thus,

$$\int x^2 \left(e^{2x} dx\right) = x^2 \cdot \frac{1}{2}e^x - \int \left(\frac{1}{2}e^{2x}\right) \cdot 2x\, dx$$

$$= x^2 \cdot \frac{1}{2}e^x - \int \left(xe^{2x}\right) dx$$

$$= \frac{1}{2}x^2 e^x - \left(\frac{1}{2}xe^{2x} - \frac{1}{4}e^{2x} + C_1\right)$$

$$= \frac{1}{2}x^2 e^x - \frac{1}{2}xe^{2x} + \frac{1}{4}e^{2x} - C_1$$

$$= \frac{1}{2}x^2 e^x - \frac{1}{2}xe^{2x} + \frac{1}{4}e^{2x} + C$$

$$\left[C = -C_1\right]$$

Alternatively, we can also use tabular integration as shown below:

$f(x)$ and Repeated Derivatives	Sign of Product	$g(x)$ and Repeated Integrals
x^2		e^{2x}
$2x$	$+$	$\frac{1}{2}e^{2x}$
2	$-$	$\frac{1}{4}e^{2x}$
0	$+$	$\frac{1}{8}e^{2x}$

We add the products along the arrows, making the alternate sign changes.

$$\int x^2 e^{2x} dx$$

$$= x^2 \left(\frac{1}{2}e^{2x}\right) - 2x\left(\frac{1}{4}e^{2x}\right) + 2\left(\frac{1}{8}e^{2x}\right) + C$$

$$= \frac{1}{2}x^2 e^{2x} - \frac{1}{2}xe^{2x} + \frac{1}{4}e^{2x} + C$$

27. $\int x^3 e^{-2x} dx$

This integral requires multiple applications of the Integration-by-Parts formula. However, since $f(x) = x^3$ is easily differentiable and to a derivative that is eventually 0 and $g(x) = e^{-2x}$ can be integrated easily using

$$\int be^{ax} = \frac{b}{a}e^{ax} + C$$ we will use tabular integration as shown to simplify our work.

$f(x)$ and Repeated Derivatives	Sign of Product	$g(x)$ and Repeated Integrals
x^3		e^{-2x}
$3x^2$	$+$	$-\frac{1}{2}e^{-2x}$
$6x$	$-$	$\frac{1}{4}e^{-2x}$
6	$+$	$-\frac{1}{8}e^{-2x}$
0	$-$	$\frac{1}{16}e^{-2x}$

We add the products along the arrows, making the alternate sign changes to obtain

$$\int x^3 e^{-2x} dx$$

$$= x^3\left(-\frac{1}{2}e^{-2x}\right) - 3x^2\left(\frac{1}{4}e^{-2x}\right) +$$

$$6x\left(-\frac{1}{8}e^{-2x}\right) - 6\left(\frac{1}{16}e^{-2x}\right) + C$$

$$= -\frac{1}{2}x^3 e^{-2x} - \frac{3}{4}x^2 e^{-2x} - \frac{3}{4}e^{-2x} - \frac{3}{8}e^{-2x} + C.$$

29. $\int \left(x^4 + 4\right)e^{3x} dx$

This integral requires multiple applications of the Integration-by-Parts formula. However, since $f(x) = x^4 + 4$ is easily differentiable and to a derivative that is eventually 0 and $g(x) = e^{3x}$ can be integrated easily using

$$\int be^{ax} = \frac{b}{a}e^{ax} + C$$ we will use tabular integration as shown to simplify our work.

$f(x)$ and Repeated Derivatives	Sign of Product	$g(x)$ and Repeated Integrals
$x^4 + 4$		e^{3x}
$4x^3$	$+$	$\frac{1}{3}e^{3x}$
$12x^2$	$-$	$\frac{1}{9}e^{3x}$
$24x$	$+$	$\frac{1}{27}e^{3x}$
24	$-$	$\frac{1}{81}e^{3x}$
0	$+$	$\frac{1}{243}e^{3x}$

We add the products along the arrows, making the alternate sign changes to obtain

$$\int (x^4 + 4)e^{3x}dx$$

$$= (x^4 + 4)\left(\frac{1}{3}e^{3x}\right) - 4x^3\left(\frac{1}{9}e^{3x}\right) +$$

$$12x^2\left(\frac{1}{27}e^{3x}\right) - 24x\left(\frac{1}{81}e^{3x}\right) +$$

$$24\left(\frac{1}{243}e^{3x}\right) + C$$

$$= \frac{1}{3}(x^4 + 4)e^{3x} - \frac{4}{9}x^3e^{3x} + \frac{4}{9}x^2e^{3x} -$$

$$\frac{8}{27}xe^{3x} + \frac{8}{81}e^{3x} + C.$$

31. $\int_1^2 x^2 \ln x\, dx$

In Exercise 9 we found the indefinite integral

$$\int x^2 \ln x\, dx = \frac{x^3 \ln x}{3} - \frac{x^3}{9} + C\ .$$

We use the indefinite integral to evaluate the definite integral as follows:

$$\int_1^2 x^2 \ln x\, dx$$

$$= \left[\frac{x^3 \ln x}{3} - \frac{x^3}{9}\right]_1^2 \qquad \text{Use } C = 0.$$

$$= \left[\frac{(2)^3 \ln(2)}{3} - \frac{(2)^3}{9}\right] - \left[\frac{(1)^3 \ln(1)}{3} - \frac{(1)^3}{9}\right]$$

$$= \left[\frac{8}{3}(\ln 2) - \frac{8}{9}\right] - \left[0 - \frac{1}{9}\right]$$

$$= \frac{8}{3}\ln 2 - \frac{8}{9} + \frac{1}{9}$$

$$= \frac{8}{3}\ln 2 - \frac{7}{9}.$$

33. $\int_2^6 \ln(x + 8)\, dx$

First, we find the indefinite integral.

$$\int \ln(x + 8)\, dx$$

Let

$$u = \ln(x + 8) \qquad \text{and} \qquad dv = dx$$

Then

$$du = \frac{1}{x + 8}dx \qquad \text{and} \qquad v = x + 8$$

Choosing $x+8$ as an antiderivative of dv

Using the Integration-by-Parts Formula gives

$$\begin{array}{cc} u & dv \\ \int \ln(x + 8)\, dx = \end{array}$$

$$\begin{array}{cccc} u & v & v & du \\ \ln(x + 8) \cdot (x + 8) - \int (x + 8)\left(\frac{1}{x + 8}dx\right) \end{array}$$

$$= (x + 8)\ln(x + 8) - \int dx$$

$$= (x + 8)\ln(x + 8) - x + C.$$

We now use the indefinite integral to evaluate the definite integral. To simplify our work, we choose the constant of integration C to be 0.

$$\int_2^6 \ln(x + 8)\, dx$$

$$= \left[(x + 8)\ln(x + 8) - x\right]_2^6$$

$$= \left[((6) + 8)\ln((6) + 8) - (6)\right] -$$

$$\left[((2) + 8)\ln((2) + 8) - (2)\right]$$

$$= \left[14\ln 14 - 6\right] - \left[10\ln 10 - 2\right]$$

$$= 14\ln 14 - 10\ln 10 - 4.$$

35. $\int_0^1 xe^x dx$

First, we find the indefinite integral $\int xe^x dx$.

Let

$$u = x \qquad \text{and} \qquad dv = e^x dx$$

Then

$$du = dx \qquad \text{and} \qquad v = e^x$$

Using the Integration-by-Parts Formula gives

$$\begin{array}{cccc} u & dv & u\ v & v\ du \\ \int x(e^x dx) = x \cdot e^x - \int e^x dx \end{array}$$

$$= xe^x - e^x + C.$$

We now use the indefinite integral to evaluate the definite integral. To simplify our work, we choose the constant of integration C to be 0.

$$\int_0^1 xe^x dx$$

$$= \left[xe^x - e^x\right]_0^1$$

$$= \left[(1)e^{(1)} - e^{(1)}\right] - \left[(0)e^{(0)} - e^{(0)}\right]$$

$$= [e - e] - [0 - 1]$$

$$= 1.$$

37. $C(x) = \int C'(x)\,dx = \int 4x\sqrt{x+3}\,dx$

Let

$$u = 4x \quad \text{and} \quad dv = \sqrt{x+3}\,dx = (x+3)^{\frac{1}{2}}\,dx$$

Then

$$du = 4dx \quad \text{and} \quad v = \frac{(x+3)^{\frac{3}{2}}}{\frac{3}{2}} = \frac{2}{3}(x+3)^{\frac{3}{2}}$$

Using the Integration-by-Parts Formula gives

$$\int 4x\left(\sqrt{x+3}\,dx\right)$$

$$= 4x \cdot \frac{2}{3}(x+3)^{\frac{3}{2}} - \int \frac{2}{3}(x+3)^{\frac{3}{2}} \cdot 4dx$$

$$= \frac{8}{3}x(x+3)^{\frac{3}{2}} - \frac{8}{3}\int (x+3)^{\frac{3}{2}}\,dx$$

$$= \frac{8}{3}x(x+3)^{\frac{3}{2}} - \frac{8}{3}\frac{(x+3)^{\frac{5}{2}}}{\frac{5}{2}} + K \quad \begin{array}{l}\text{Use } K \text{ to avoid}\\ \text{confusion with cost.}\end{array}$$

$$= \frac{8}{3}x(x+3)^{\frac{3}{2}} - \frac{16}{15}(x+3)^{\frac{5}{2}} + K.$$

Next, we use the condition $C(13) = 1126.40$ to find the constant of integration K.

$$C(13) = 1126.40$$

$$\frac{8}{3}(13)\left((13)+3\right)^{\frac{3}{2}} - \frac{16}{15}\left((13)+3\right)^{\frac{5}{2}} + K = 1126.40$$

$$\frac{104}{3} \cdot (16)^{\frac{3}{2}} - \frac{16}{15}(16)^{\frac{5}{2}} + K = 1126.40$$

$$\frac{104}{3}(64) - \frac{16}{15}(1024) + K = 1126.40$$

$$\frac{6656}{3} - \frac{16,384}{15} + K = 1126.40$$

$$\frac{16,896}{15} + K = 1126.40$$

$$1126.40 + K = 1126.40$$

$$K = 0$$

Therefore, the total cost function is

$$C(x) = \frac{8}{3}x(x+3)^{\frac{3}{2}} - \frac{16}{15}(x+3)^{\frac{5}{2}}$$

39. a) $K(t) = 10te^{-t}$

The number of kilowatt hours the family uses in the first T hours of the day is given by $\int_0^T K(t)\,dt$. We find the indefinite integral first.

$$\int K(t)\,dt = \int 10te^{-t}\,dt$$

Let

$$u = 10t \quad \text{and} \quad dv = e^{-t}\,dt$$

Then

$$du = 10dt \quad \text{and} \quad v = -e^{-t}$$

Using the Integration-by-Parts Formula gives

$$\int 10te^{-t}\,dt = 10t \cdot \left(-e^{-t}\right) - \int -e^{-t} \cdot 10dt$$

$$= -10te^{-t} + 10\int e^{-t}\,dt$$

$$= -10te^{-t} - 10e^{-t} + C$$

Next, we evaluate the definite integral.

$$\int_0^T K(t)\,dt$$

$$= \left[-10te^{-t} - 10e^{-t}\right]_0^T$$

$$= \left[-10Te^{-T} - 10e^{-T}\right] - \left[-10(0)e^{-0} - 10e^{-0}\right]$$

$$= -10Te^{-T} - 10e^{-T} + 10.$$

b) Substitute 4 for T.

$$\int_0^4 K(t)\,dt = -10(4)e^{-(4)} - 10e^{-(4)} + 10$$

$$= -40e^{-4} - 10e^{-4} + 10$$

$$\approx 9.084$$

The family uses approximately 9.084 kW-h during the first 4 hours of the day.

41. $\int \sqrt{x}\ln x\,dx = \int x^{\frac{1}{2}}\ln x\,dx$

Let

$$u = \ln x \quad \text{and} \quad dv = x^{\frac{1}{2}}dx$$

Then

$$du = \frac{1}{x}dx \quad \text{and} \quad v = \frac{2}{3}x^{\frac{3}{2}}$$

Using the Integration-by-Parts Formula gives

$$\int \ln x\left(x^{\frac{1}{2}}\,dx\right)$$

$$= \ln x \cdot \left(\frac{2}{3}x^{\frac{3}{2}}\right) - \int \left(\frac{2}{3}x^{\frac{3}{2}}\right)\left(\frac{1}{x}dx\right)$$

$$= \frac{2}{3}x^{\frac{3}{2}}\ln x - \frac{2}{3}\int x^{\frac{1}{2}}dx$$

$$= \frac{2}{3}x^{\frac{3}{2}}\ln x - \frac{2}{3}\left(\frac{2}{3}x^{\frac{3}{2}}\right) + C$$

$$= \frac{2}{3}x^{\frac{3}{2}}\ln x - \frac{4}{9}x^{\frac{3}{2}} + C.$$

43. $\int \frac{\ln x}{\sqrt{x}}\,dx = \int x^{-\frac{1}{2}}\ln x\,dx$

Let

$$u = \ln x \quad \text{and} \quad dv = x^{-\frac{1}{2}}dx$$

Then

$$du = \frac{1}{x}dx \quad \text{and} \quad v = 2x^{\frac{1}{2}}$$

Using the Integration-by-Parts Formula gives

$$\int \ln x \left(x^{-\frac{1}{2}} \, dx \right)$$

$$= \ln x \cdot \left(2x^{\frac{1}{2}} \right) - \int \left(2x^{\frac{1}{2}} \right) \left(\frac{1}{x} \, dx \right)$$

$$= 2x^{\frac{1}{2}} \ln x - 2 \int x^{-\frac{1}{2}} \, dx$$

$$= 2x^{\frac{1}{2}} \ln x - 2 \left(2x^{\frac{1}{2}} \right) + C$$

$$= 2x^{\frac{1}{2}} \ln x - 4x^{\frac{1}{2}} + C$$

$$= 2\sqrt{x} \left(\ln x \right) - 4\sqrt{x} + C.$$

45. $\int \left(27x^3 + 83x - 2 \right) \sqrt[6]{3x + 8} \, dx$

$$= \left(27x^3 + 83x - 2 \right) \left(3x + 8 \right)^{\frac{1}{6}} \, dx$$

We will use tabular integration as shown to simplify our work.

$f(x)$ and Repeated Derivatives	Sign of Product	$g(x)$ and Repeated Integrals
$27x^3 + 83x - 2$	+	$(3x+8)^{\frac{1}{6}}$
$81x^2 + 83$	−	$\frac{2}{7}(3x+8)^{\frac{7}{6}}$
$162x$	+	$\frac{4}{91}(3x+8)^{\frac{13}{6}}$
162	−	$\frac{8}{1729}(3x+8)^{\frac{19}{6}}$
0		$\frac{16}{43,225}(3x+8)^{\frac{25}{6}}$

$$\int \left(27x^3 + 83x - 2 \right) \sqrt[6]{3x+8} \, dx$$

$$= \frac{2}{7} \left(27x^3 + 83x - 2 \right) \left(3x+8 \right)^{\frac{7}{6}} -$$

$$\frac{4}{91} \left(81x^2 + 83 \right) \left(3x+8 \right)^{\frac{13}{6}} +$$

$$\frac{1296}{1729} x \left(3x+8 \right)^{\frac{19}{6}} - \frac{2592}{43,225} \left(3x+8 \right)^{\frac{25}{6}} + C$$

47. $\int x^n \left(\ln x \right)^2 \, dx, \qquad n \neq -1$

Let

$$u = \left(\ln x \right)^2 \qquad \text{and} \qquad dv = x^n dx$$

Then

$$du = \frac{2 \ln x}{x} \, dx \quad \text{and} \quad v = \frac{x^{n+1}}{n+1}$$

$$\int x^n \left(\ln x \right)^2 \, dx = \frac{x^{n+1} \left(\ln x \right)^2}{n+1} - \int \frac{2x^n \ln x}{n+1} \, dx$$

$$= \frac{x^{n+1} \left(\ln x \right)^2}{n+1} - \frac{2}{n+1} \int x^n \ln x \, dx$$

Use integration by parts again to evaluate $\int x^n \ln x \, dx$.

Let

$$u = \ln x \qquad \text{and} \qquad dv = x^n dx$$

Then

$$du = \frac{1}{x} \, dx \quad \text{and} \quad v = \frac{x^{n+1}}{n+1}$$

$$\int x^n \ln x \, dx = \frac{x^{n+1} \ln x}{n+1} - \int \frac{x^n}{n+1} \, dx$$

$$= \frac{x^{n+1} \ln x}{n+1} - \frac{x^{n+1}}{\left(n+1 \right)^2} + K$$

Therefore,

$$\int x^n \left(\ln x \right)^2 \, dx =$$

$$= \frac{x^{n+1} \left(\ln x \right)^2}{n+1} - \frac{2}{n+1} \left(\frac{x^{n+1} \ln x}{n+1} - \frac{x^{n+1}}{\left(n+1 \right)^2} + K \right)$$

$$= \frac{x^{n+1}}{n+1} \left(\ln x \right)^2 - \frac{2x^{n+1}}{\left(n+1 \right)^2} \left(\ln x \right) + \frac{2x^{n+1}}{\left(n+1 \right)^3} + C$$

$$\left(C = \frac{-2K}{n+1} \right)$$

49. $\int x^n e^x \, dx$

Let

$$u = x^n \qquad \text{and} \qquad dv = e^x dx$$

Then

$$du = nx^{n-1} dx \quad \text{and} \quad v = e^x$$

$$\int x^n e^x \, dx = x^n e^x - \int e^x \left(nx^{n-1} \right) dx$$

$$= x^n e^x - n \int x^{n-1} e^x \, dx.$$

51. $\boxed{tw}$

53. Using the fnInt feature on a calculator, we find that $\int_1^{10} x^5 \ln x \, dx \approx 355,986.$

Exercise Set 4.7

1. $\int xe^{-3x}dx$

This integral fits Formula 6 in Table 1.

$\int xe^{ax}dx = \dfrac{1}{a^2}\cdot e^{ax}(ax-1)+C$

In the given integral, $a=-3$, so we have, by the formula,

$\int xe^{-3x}dx = \dfrac{1}{(-3)^2}\cdot e^{(-3)x}\left((-3)x-1\right)+C$

$\qquad = \dfrac{1}{9}e^{-3x}(-3x-1)+C$

$\qquad = -\dfrac{1}{9}e^{-3x}(3x+1)+C.$

3. $\int 6^x dx$

This integral fits Formula 11 in Table 1.

$\int a^x dx = \dfrac{a^x}{\ln a}+C, \ \ a>0, a\neq 1$

In the given integral, $a=6$, so we have, by the formula,

$\int 6^x dx = \dfrac{6^x}{\ln 6}+C.$

5. $\int \dfrac{1}{25-x^2}dx$

This integral fits Formula 15 in Table 1.

$\int \dfrac{1}{a^2-x^2}dx = \dfrac{1}{2a}\ln\left|\dfrac{a+x}{a-x}\right|+C$

In the given integral, $a^2=25$, or $a=5$, so we have, by the formula,

$\int \dfrac{1}{25-x^2}dx = \int \dfrac{1}{5^2-x^2}dx$

$\qquad = \dfrac{1}{2\cdot 5}\ln\left|\dfrac{5+x}{5-x}\right|+C$

$\qquad = \dfrac{1}{10}\ln\left|\dfrac{5+x}{5-x}\right|+C.$

7. $\int \dfrac{x}{3-x}dx$

This integral fits Formula 18 in Table 1.

$\int \dfrac{x}{a+bx}dx = \dfrac{a}{b^2}+\dfrac{x}{b}-\dfrac{a}{b^2}\ln|a+bx|+C$

In the given integral, $a=3$ and $b=-1$, so we have, by the formula,

$\int \dfrac{x}{3+(-1)x}dx$

$= \dfrac{3}{(-1)^2}+\dfrac{x}{(-1)}-\dfrac{3}{(-1)^2}\ln|3-x|+C$

$= 3-x-3\ln|3-x|+C.$

9. $\int \dfrac{1}{x(8-x)^2}dx$

This integral fits Formula 21 in Table 1.

$\int \dfrac{1}{x(a+bx)^2}dx = \dfrac{1}{a(a+bx)}+\dfrac{1}{a^2}\ln\left|\dfrac{x}{a+bx}\right|+C$

In the given integral, $a=8$ and $b=-1$, so we have, by the formula,

$\int \dfrac{1}{x(8-x)^2}dx$

$= \dfrac{1}{8(8-x)}+\dfrac{1}{(8)^2}\ln\left|\dfrac{x}{8-x}\right|+C$

$= \dfrac{1}{8(8-x)}+\dfrac{1}{64}\ln\left|\dfrac{x}{8-x}\right|+C$

11. $\int \ln(3x)dx$

$= \int (\ln 3 + \ln x)dx$

$= \int \ln 3\,dx + \int \ln x\,dx$

$= \ln 3\int dx + \int \ln x\,dx$

The integral in the first term is the integral of a constant and we will integrate accordingly. The integral in the second term fits Formula 8 in Table 1.

$\int \ln x\,dx = x\ln x - x + C$

so we have, by the formula,

$\int \ln(3x)dx$

$= \ln 3\int dx + \int \ln x\,dx$

$= \ln 3\cdot x + C_1 + x\ln x - x + C_2$

$= (\ln 3)x + x\ln x - x + C \qquad (C = C_1 + C_2)$

13. $\int x^4\ln x\,dx$

This integral fits Formula 10 in Table 1.

$\int x^n\ln x\,dx = x^{n+1}\left[\dfrac{\ln x}{n+1}-\dfrac{1}{(n+1)^2}\right]+C$

In the given integral, $n=4$, so we have, by the formula,

$$\int x^4 \ln x \, dx = x^{4+1} \left[\frac{\ln x}{4+1} - \frac{1}{(4+1)^2} \right] + C$$

$$= x^5 \left[\frac{\ln x}{5} - \frac{1}{(5)^2} \right] + C$$

$$= \frac{x^5}{5}(\ln x) - \frac{x^5}{25} + C.$$

15. $\int x^3 \ln x \, dx$

This integral fits Formula 10 in Table 1.

$$\int x^n \ln x \, dx = x^{n+1} \left[\frac{\ln x}{n+1} - \frac{1}{(n+1)^2} \right] + C$$

In the given integral, $n = 3$, so we have, by the formula,

$$\int x^3 \ln x \, dx = x^{3+1} \left[\frac{\ln x}{3+1} - \frac{1}{(3+1)^2} \right] + C$$

$$= x^4 \left[\frac{\ln x}{4} - \frac{1}{(4)^2} \right] + C$$

$$= \frac{x^4}{4}(\ln x) - \frac{x^4}{16} + C.$$

17. $\int \dfrac{dx}{\sqrt{x^2 + 7}}$

This integral fits Formula 12 in Table 1.

$$\int \frac{1}{\sqrt{x^2 + a^2}} \, dx = \ln \left| x + \sqrt{x^2 + a^2} \right| + C$$

In the given integral, $a^2 = 7$, so we have, by the formula,

$$\int \frac{1}{\sqrt{x^2 + 7}} \, dx = \ln \left| x + \sqrt{x^2 + 7} \right| + C.$$

19. $\int \dfrac{10 \, dx}{x(5 - 7x)^2} = 10 \int \dfrac{dx}{x(5 - 7x)^2}$

This integral fits Formula 21 in Table 1.

$$\int \frac{1}{x(a + bx)^2} \, dx = \frac{1}{a(a + bx)} + \frac{1}{a^2} \ln \left| \frac{x}{a + bx} \right| + C$$

In the given integral, $a = 5$ and $b = -7$, so we have, by the formula,

$$10 \int \frac{1}{x(5 - 7x)^2} \, dx$$

$$= 10 \left[\frac{1}{5(5 - 7x)} + \frac{1}{(5)^2} \ln \left| \frac{x}{5 - 7x} \right| \right] + C$$

$$= \frac{2}{5 - 7x} + \frac{2}{5} \ln \left| \frac{x}{5 - 7x} \right| + C.$$

21. $\int \dfrac{-5}{4x^2 - 1} \, dx = -5 \int \dfrac{1}{4x^2 - 1} \, dx$

This integral almost fits Formula 14 in table 1.

$$\int \frac{1}{x^2 - a^2} \, dx = \frac{1}{2a} \ln \left| \frac{x - a}{x + a} \right| + C$$

However, the x^2 coefficient needs to be 1. We factor out 4 as follows and then we apply formula 14.

$$-5 \int \frac{1}{4x^2 - 1} \, dx = -5 \int \frac{1}{4 \left(x^2 - \frac{1}{4} \right)} \, dx$$

$$= -\frac{5}{4} \int \frac{1}{x^2 - \frac{1}{4}} \, dx$$

In the given integral, $a^2 = \frac{1}{4}$ and $a = \frac{1}{2}$, so we have, by the formula,

$$-\frac{5}{4} \int \frac{1}{x^2 - \frac{1}{4}} \, dx = -\frac{5}{4} \left[\frac{1}{2 \left(\frac{1}{2} \right)} \ln \left| \frac{x - \frac{1}{2}}{x + \frac{1}{2}} \right| \right] + C$$

$$= -\frac{5}{4} \ln \left| \frac{x - \frac{1}{2}}{x + \frac{1}{2}} \right| + C$$

23. $\int \sqrt{4m^2 + 16} \, dm$

This integral almost fits Formula 22 in table 1.

$$\int \sqrt{x^2 + a^2} \, dx$$

$$= \frac{1}{2} \left[x\sqrt{x^2 + a^2} + a^2 \ln \left| x + \sqrt{x^2 + a^2} \right| \right] + C$$

However, the m^2 coefficient needs to be 1. We factor out 4 as follows and then we apply formula 22.

$$\int \sqrt{4(m^2 + 4)} \, dm = \int 2\sqrt{m^2 + 4} \, dm$$

$$= 2 \int \sqrt{m^2 + 4} \, dm$$

In the given integral, $a^2 = 4$ and $a = 2$, so we have, by the formula,

$$2 \int \sqrt{m^2 + 4} \, dm$$

$$= 2 \cdot \frac{1}{2} \left[m\sqrt{m^2 + 4} + 4 \ln \left| m + \sqrt{m^2 + 4} \right| \right] + C$$

$$= m\sqrt{m^2 + 4} + 4 \ln \left| m + \sqrt{m^2 + 4} \right| + C.$$

25. $\int \dfrac{-5\ln x}{x^3}\,dx = -5\int x^{-3}\ln x\,dx$

This integral fits Formula 10 in Table 1.

$$\int x^n \ln x\,dx = x^{n+1}\left[\dfrac{\ln x}{n+1} - \dfrac{1}{(n+1)^2}\right] + C$$

In the given integral, $n = -3$, so we have, by the formula,

$-5\int x^{-3}\ln x\,dx$

$= -5 \cdot x^{-3+1}\left[\dfrac{\ln x}{-3+1} - \dfrac{1}{(-3+1)^2}\right] + C$

$= -5 \cdot x^{-2}\left[\dfrac{\ln x}{-2} - \dfrac{1}{(-2)^2}\right] + C$

$= \dfrac{5}{2x^2}(\ln x) + \dfrac{5}{4x^2} + C$

27. $\int \dfrac{e^x}{x^{-3}}\,dx = \int x^3 e^x\,dx$

This integral fits Formula 7 in Table 1.

$$\int x^n e^{ax}\,dx = \dfrac{x^n e^{ax}}{a} - \dfrac{n}{a}\int x^{n-1}e^{ax}\,dx + C$$

In the given integral, $n = 3$ and $a = 1$, so we have, by the formula,

$\int x^3 e^x\,dx = \dfrac{x^3 e^x}{1} - \dfrac{3}{1}\int x^{3-1}e^x\,dx + C$

$\qquad = x^3 e^x - 3\int x^2 e^x\,dx + C$

We will apply Formula 7 again, this time with $n = 2$ and $a = 1$

$x^3 e^x - 3\left[\dfrac{x^2 e^x}{1} - \dfrac{2}{1}\int x^{2-1}e^x\,dx\right] + C$

$= x^3 e^x - 3x^2 e^x + 6\int xe^x\,dx + C$

Now we apply Formula 6

$\int xe^{ax}\,dx = \dfrac{1}{a^2}\cdot e^{ax}(ax - 1) + C$

with $a = 1$

$= x^3 e^x - 3x^2 e^x + 6\left[\dfrac{1}{(1)^2}\cdot e^x(x - 1)\right] + C$

$= x^3 e^x - 3x^2 e^x + 6xe^x - 6e^x + C.$

29. $\int x\sqrt{1 + 2x}\,dx$

This integral fits Formula 23 in Table 1.

$$\int x\sqrt{a + bx}\,dx = \dfrac{2}{15b^2}(3bx - 2a)(a + bx)^{3/2} + C$$

In the given integral, $a = 1$ and $b = 2$, so we have, by the formula,

$\int x\sqrt{1 + 2x}\,dx$

$= \dfrac{2}{15(2)^2}\big(3(2)x - 2(1)\big)(1 + 2x)^{3/2} + C$

$= \dfrac{2}{60}(6x - 2)(1 + 2x)^{3/2} + C$

$= \dfrac{1}{30}\cdot 2(3x - 1)(1 + 2x)^{3/2} + C$

$= \dfrac{1}{15}(3x - 1)(1 + 2x)^{3/2} + C.$

31. $S(x) = \int S'(x)\,dx$

$= \int \dfrac{100x}{(20 - x)^2}\,dx = 100\int \dfrac{x}{(20 - x)^2}\,dx$

This integral fits Formula 19 in Table 1.

$$\int \dfrac{x}{(a + bx)^2}\,dx = \dfrac{a}{b^2(a + bx)} + \dfrac{1}{b^2}\ln|a + bx| + C$$

In the given integral, $a = 20$ and $b = -1$, so we have, by the formula,

$100\int \dfrac{x}{(20 - x)^2}\,dx$

$= 100\left[\dfrac{20}{(-1)^2(20 - x)} + \dfrac{1}{(-1)^2}\ln|20 - x|\right] + C$

$= \dfrac{2000}{(20 - x)} + 100\ln|20 - x| + C.$

Next, we use the condition $S(19) = 2000$ to determine C.

$$S(19) = 2000$$

$\dfrac{2000}{(20 - (19))} + 100\ln|20 - (19)| + C = 2000$

$\dfrac{2000}{1} + 100\ln|1| + C = 2000$

$2000 + C = 2000$

$C = 0$

Thus, the supply function is

$S(x) = \dfrac{2000}{(20 - x)} + 100\ln|20 - x|$

$= 100\left[\dfrac{20}{20 - x} + \ln|20 - x|\right].$

33. $\int \dfrac{8}{3x^2 - 2x}\,dx = 8\int \dfrac{1}{x(-2+3x)}\,dx$

This integral fits Formula 20 in Table 1.

$$\int \dfrac{1}{x(a+bx)}\,dx = \dfrac{1}{a}\ln\left|\dfrac{x}{a+bx}\right| + C$$

In the given integral, $a = -2$ and $b = 3$, so we have, by the formula,

$$8\int \dfrac{1}{x(-2+3x)}\,dx = 8\cdot\dfrac{1}{(-2)}\ln\left|\dfrac{x}{-2+3x}\right| + C$$

$$= -4\ln\left|\dfrac{x}{-2+3x}\right| + C.$$

35. $\int \dfrac{dx}{x^3 - 4x^2 + 4x}$

$$= \int \dfrac{dx}{x(x^2 - 4x + 4)}$$

$$= \int \dfrac{1}{x(x-2)^2}\,dx$$

$$= \int \dfrac{1}{x(-2+x)^2}\,dx$$

This integral fits Formula 21 in Table 1.

$$\int \dfrac{1}{x(a+bx)^2}\,dx = \dfrac{1}{a(a+bx)} + \dfrac{1}{a^2}\ln\left|\dfrac{x}{a+bx}\right| + C$$

In the given integral, $a = -2$ and $b = 1$, so we have, by the formula,

$$\int \dfrac{1}{x(-2+x)^2}\,dx$$

$$= \dfrac{1}{-2(-2+x)} + \dfrac{1}{(-2)^2}\ln\left|\dfrac{x}{-2+x}\right| + C$$

$$= \dfrac{1}{-2(x-2)} + \dfrac{1}{4}\ln\left|\dfrac{x}{x-2}\right| + C$$

$$= \dfrac{-1}{2(x-2)} + \dfrac{1}{4}\ln\left|\dfrac{x}{x-2}\right| + C.$$

37. $\int \dfrac{-e^{-2x}\,dx}{9 - 6e^{-x} + e^{-2x}}\,dx = \int \dfrac{e^{-x}\left(-e^{-x}\right)dx}{\left(-3+e^{-x}\right)^2}$

We substitute $u = e^{-x}$ and $du = -e^{-x}dx$, and get

$$\dfrac{u}{\left(-3+u\right)^2}\,du$$

Which fits Formula 19 in Table 1.

$$\int \dfrac{x}{(a+bx)^2}\,dx = \dfrac{a}{b^2(a+bx)} + \dfrac{1}{b^2}\ln|a+bx| + C$$

In the given integral, we have $u = x$, $a = -3$ and $b = 1$, so we have, by the formula,

$$\int \dfrac{e^{-x}\left(-e^{-x}\right)dx}{\left(-3+e^{-x}\right)^2} = \int \dfrac{u}{\left(-3+u\right)^2}\,du$$

$$= \dfrac{-3}{1^2\left(-3+u\right)} + \dfrac{1}{1^2}\ln\left|-3+u\right| + C$$

$$= \dfrac{-3}{1^2\left(-3+e^{-x}\right)} + \dfrac{1}{1^2}\ln\left|-3+e^{-x}\right| + C \quad \text{\small Substituting for } u$$

$$= \dfrac{-3}{e^{-x}-3} + \ln\left|e^{-x}-3\right| + C.$$

Chapter 5

Applications of Integration

1. $D(x) = -\dfrac{5}{6}x + 9,\ S(x) = \dfrac{1}{2}x + 1$

    a) To find the equilibrium point we set
        $D(x) = S(x)$ and solve.

        $$-\frac{5}{6}x + 9 = \frac{1}{2}x + 1$$

        $$9 - 1 = \frac{1}{2}x + \frac{5}{6}x$$

        $$8 = \frac{4}{3}x \qquad \left(\frac{1}{2} + \frac{5}{6} = \frac{8}{6} = \frac{4}{3}\right)$$

        $$\frac{3}{4} \cdot 8 = x \qquad \text{Multiplying by } \frac{3}{4}$$

        $$6 = x$$

        Thus $x_E = 6$ units. To find p_E we substitute
        x_E into $D(x)$ or $S(x)$. Here we use $D(x)$.

        $$p_E = D(x_E)$$
        $$= D(6)$$
        $$= -\frac{5}{6}(6) + 9$$
        $$= -5 + 9$$
        $$= 4$$

        When 6 units are sold the equilibrium price
        is \$4; therefore, the equilibrium point is
        $(6, \$4)$.

        Notice, we could have used $S(x)$ to find the
        equilibrium point as well. Substituting, we
        have:

        $$p_E = S(x_E)$$
        $$= S(6)$$
        $$= \frac{1}{2}(6) + 1$$
        $$= 3 + 1$$
        $$= 4$$

        This results in the same equilibrium point
        $(6, \$4)$.

    b) The consumer surplus is

        $$\int_0^{x_E} D(x)\, dx - x_E p_E.$$

        Substituting $-\dfrac{5}{6}x + 9$ for $D(x)$, 6 for x_E,
        and 4 for p_E we have:

        $$\int_0^6 \left(-\frac{5}{6}x + 9\right) dx - 6 \cdot 4$$

        $$= \left[-\frac{5x^2}{12} + 9x\right]_0^6 - 24$$

        $$= \left[\left(-\frac{5(6)^2}{12} + 9 \cdot 6\right) - \left(-\frac{5(0)^2}{12} + 9 \cdot 0\right)\right] - 24$$

        $$= \left[(-15 + 54) - (0)\right] - 24$$

        $$= 39 - 24$$

        $$= 15$$

        The consumer surplus at the equilibrium
        point is \$15.

    c) The producer surplus is

        $$x_E p_E - \int_0^{x_E} S(x)\, dx.$$

        Substituting $\dfrac{1}{2}x + 1$ for $S(x)$, 6 for x_E, and
        4 for p_E we have:

        $$6 \cdot 4 - \int_0^6 \left(\frac{1}{2}x + 1\right) dx$$

        $$= 24 - \left[\frac{x^2}{4} + x\right]_0^6$$

        $$= 24 - \left[\left(\frac{6^2}{4} + 6\right) - (0^2 + 0)\right]$$

        $$= 24 - \left[\frac{36}{4} + 6\right]$$

        $$= 24 - [9 + 6]$$

        $$= 24 - 15$$

        $$= 9$$

        The producer surplus at the equilibrium
        point is \$9.

3. $D(x) = (x-4)^2, \quad S(x) = x^2 + 2x + 6$

a) To find the equilibrium point we set
$D(x) = S(x)$ and solve.

$$(x-4)^2 = x^2 + 2x + 6$$
$$x^2 - 8x + 16 = x^2 + 2x + 6$$
$$-8x + 16 = 2x + 6$$
$$10 = 10x$$
$$1 = x$$

Thus $x_E = 1$ unit. To find p_E we substitute
x_E into $D(x)$ or $S(x)$. Here we use $D(x)$.

$$p_E = D(x_E)$$
$$= D(1)$$
$$= ((1) - 4)^2$$
$$= (-3)^2$$
$$= 9$$

When 1 unit is sold the equilibrium price is
$9; therefore, the equilibrium point is
$(1, \$9)$.

b) The consumer surplus is
$$\int_0^{x_E} D(x)\,dx - x_E p_E \,.$$

Substituting $(x-4)^2$ for $D(x)$, 1 for x_E,
and 9 for p_E we have:

$$\int_0^1 (x-4)^2 \, dx - 1 \cdot 9$$
$$= \int_0^1 (x^2 - 8x + 16)\, dx - 9$$
$$= \left[\frac{x^3}{3} - 4x^2 + 16x \right]_0^1 - 9$$
$$= \left[\left(\frac{1}{3} - 4 + 16 \right) - (0 - 0 + 0) \right] - 9$$
$$= \frac{37}{3} - 9$$
$$= \frac{10}{3} \approx 3.33$$

The consumer surplus at the equilibrium
point is $3.33.

c) The producer surplus is
$$x_E p_E - \int_0^{x_E} S(x)\,dx \,.$$

Substituting $x^2 + 2x + 6$ for $S(x)$, 1 for x_E,
and 9 for p_E we have:

$$1 \cdot 9 - \int_0^1 (x^2 + 2x + 6)\,dx$$
$$= 9 - \left[\frac{x^3}{3} + x^2 + 6x \right]_0^1$$
$$= 9 - \left[\left(\frac{1}{3} + 1 + 6 \right) - (0 + 0 + 0) \right]$$
$$= 9 - \left[\frac{22}{3} \right]$$
$$= \frac{5}{3} \approx 1.67$$

The producer surplus at the equilibrium
point is $1.67.

5. $D(x) = (x-6)^2, \quad S(x) = x^2$

a) To find the equilibrium point we set
$D(x) = S(x)$ and solve.

$$(x-6)^2 = x^2$$
$$x^2 - 12x + 36 = x^2$$
$$-12x + 36 = 0$$
$$36 = 12x$$
$$3 = x$$

Thus $x_E = 3$ units. To find p_E we substitute
x_E into $D(x)$ or $S(x)$. Here we use $S(x)$.

$$p_E = S(x_E) = S(3) = (3)^2 = 9$$

When 3 units are sold the equilibrium price
is $9; therefore, the equilibrium point is
$(3, \$9)$.

b) The consumer surplus is
$$\int_0^{x_E} D(x)\,dx - x_E p_E \,.$$

Substituting $(x-6)^2$ for $D(x)$, 3 for x_E,
and 9 for p_E we have:

$$\int_0^3 (x-6)^2 \, dx - 3 \cdot 9$$
$$= \int_0^3 (x^2 - 12x + 36)\, dx - 27$$
$$= \left[\frac{x^3}{3} - 6x^2 + 36x \right]_0^3 - 27$$
$$= \left[\left(\frac{3^3}{3} - 6 \cdot 3^2 + 36 \cdot 3 \right) - (0 - 0 + 0) \right] - 27$$
$$= \left[(9 - 54 + 108) - (0) \right] - 27$$
$$= 63 - 27$$
$$= 36$$

The consumer surplus at the equilibrium
point is $36.

c) The producer surplus is

$$x_E p_E - \int_0^{x_E} S(x)\,dx.$$

Substituting x^2 for $S(x)$, 3 for x_E, and 9 for p_E we have:

$$3 \cdot 9 - \int_0^3 x^2\,dx$$

$$= 27 - \left[\frac{x^3}{3} \right]_0^3$$

$$= 27 - \left[\left(\frac{3^3}{3} \right) - \left(\frac{0^3}{3} \right) \right]$$

$$= 27 - [9]$$

$$= 18$$

The producer surplus at the equilibrium point is $18.

7. $D(x) = 1000 - 10x,\ \ S(x) = 250 + 5x$

a) To find the equilibrium point we set $D(x) = S(x)$ and solve.

$$1000 - 10x = 250 + 5x$$

$$750 = 15x$$

$$50 = x$$

Thus $x_E = 50$ units. To find p_E we substitute x_E into $D(x)$ or $S(x)$. Here we use $D(x)$.

$$p_E = D(x_E)$$

$$= D(50)$$

$$= 1000 - 10 \cdot 50$$

$$= 500$$

When 50 units are sold the equilibrium price is $500; therefore, the equilibrium point is $(50, \$500)$.

b) The consumer surplus is

$$\int_0^{x_E} D(x)\,dx - x_E p_E.$$

Substituting $1000 - 10x$ for $D(x)$, 50 for x_E, and 500 for p_E we have:

$$\int_0^{50} (1000 - 10x)\,dx - 50 \cdot 500$$

$$= \left[1000x - 5x^2 \right]_0^{50} - 25,000$$

$$= \left[\left(1000 \cdot 50 - 5(50)^2 \right) - \left(1000 \cdot 0 - 5 \cdot 0^2 \right) \right]$$
$$\quad - 25,000$$

$$= [50,000 - 12,500] - 25,000$$

$$= 12,500$$

The consumer surplus at the equilibrium point is $12,500.

c) The producer surplus is

$$x_E p_E - \int_0^{x_E} S(x)\,dx.$$

Substituting $250 + 5x$ for $S(x)$, 50 for x_E, and 500 for p_E we have:

$$50 \cdot 500 - \int_0^{50} (250 + 5x)\,dx$$

$$= 25,000 - \left[250x + \frac{5x^2}{2} \right]_0^{50}$$

$$= 25,000 -$$

$$\left[\left(250 \cdot 50 + \frac{5(50)^2}{2} \right) - \left(250 \cdot 0 - \frac{5(0)^2}{2} \right) \right]$$

$$= 25,000 - [12,500 + 6250]$$

$$= 6250$$

The producer surplus at the equilibrium point is $6250.

9. $D(x) = 5 - x$, for $0 \le x \le 5$;

$$S(x) = \sqrt{x + 7}$$

a) To find the equilibrium point we set $D(x) = S(x)$ and solve.

$$5 - x = \sqrt{x + 7}$$

$$(5 - x)^2 = \left(\sqrt{x + 7} \right)^2$$

$$25 - 10x + x^2 = x + 7$$

$$x^2 - 11x + 18 = 0$$

$$(x - 2)(x - 9) = 0$$

$$x - 2 = 0 \quad \text{or} \quad x - 9 = 0$$

$$x = 2 \quad \text{or} \quad x = 9$$

Note, $x = 9$ is not a solution to the equation. Only $x = 2$ is in the domain of $D(x)$, thus $x_E = 2$ units. To find p_E we substitute x_E into $D(x)$ or $S(x)$. Here we use $D(x)$.

$$p_E = D(x_E) = D(2) = 5 - 2 = 3$$

When 2 units are sold the equilibrium price is \$3; therefore, the equilibrium point is $(2, \$3)$.

b) The consumer surplus is

$$\int_0^{x_E} D(x)\,dx - x_E p_E.$$

Substituting $5 - x$ for $D(x)$, 2 for x_E, and 3 for p_E we have:

$$\int_0^2 (5 - x)\,dx - 2 \cdot 3$$

$$= \left[5x - \frac{x^2}{2} \right]_0^2 - 6$$

$$= \left[\left(5 \cdot 2 - \frac{(2)^2}{2} \right) - \left(5 \cdot 0 - \frac{(0)^2}{0} \right) \right] - 6$$

$$= [10 - 2] - 6$$

$$= 2$$

The consumer surplus at the equilibrium point is \$2.

c) The producer surplus is

$$x_E p_E - \int_0^{x_E} S(x)\,dx.$$

Substituting $\sqrt{7+x}$ for $S(x)$, 2 for x_E, and 3 for p_E we have:

$$2 \cdot 3 - \int_0^2 \left(\sqrt{x+7} \right) dx$$

$$= 6 - \int_0^2 (x+7)^{\frac{1}{2}}\,dx$$

$$= 6 - \left[\frac{2}{3}(x+7)^{\frac{3}{2}} \right]_0^2$$

$$= 6 - \left[\left(\frac{2}{3}(2+7)^{\frac{3}{2}} \right) - \left(\frac{2}{3}(0+7)^{\frac{3}{2}} \right) \right]$$

$$= 6 - \left[\frac{2}{3}(9)^{\frac{3}{2}} - \frac{2}{3}(7)^{\frac{3}{2}} \right]$$

$$= 6 - \left[\frac{2}{3} \cdot (27) - \frac{2}{3}(7)^{\frac{3}{2}} \right]$$

$$\approx 0.35 \qquad\qquad \text{Using a calculator}$$

The producer surplus at the equilibrium point is \$0.35.

11. $D(x) = \dfrac{100}{\sqrt{x}}, \qquad S(x) = \sqrt{x}$

a) To find the equilibrium point we set $D(x) = S(x)$ and solve.

$$\frac{100}{\sqrt{x}} = \sqrt{x}$$

$$100 = \sqrt{x} \cdot \sqrt{x}$$

$$100 = x$$

Thus $x_E = 100$ units. To find p_E we substitute x_E into $D(x)$ or $S(x)$. Here we use $S(x)$.

$$p_E = S(x_E) = S(100) = \sqrt{100} = 10$$

When 100 units are sold the equilibrium price is \$10; therefore, the equilibrium point is $(100, \$10)$.

b) The consumer surplus is

$$\int_0^{x_E} D(x)\,dx - x_E p_E.$$

Substituting $\dfrac{100}{\sqrt{x}} = 100 x^{-\frac{1}{2}}$ for $D(x)$, 100 for x_E, and 10 for p_E we have:

$$\int_0^{100} \left(100 x^{-\frac{1}{2}} \right) dx - 100 \cdot 10$$

$$= \left[\frac{100 x^{\frac{1}{2}}}{\frac{1}{2}} \right]_0^{100} - 1000$$

$$= \left[\left(200(100)^{\frac{1}{2}} \right) - \left(200(0)^{\frac{1}{2}} \right) \right] - 1000$$

$$= [2000] - 1000$$

$$= 1000$$

The consumer surplus at the equilibrium point is \$1000.

c) The producer surplus is

$$x_E p_E - \int_0^{x_E} S(x)\,dx.$$

Substituting $\sqrt{x}$ for $S(x)$, 100 for x_E, and 10 for p_E we have:

$$100 \cdot 10 - \int_0^{100} \left(x^{1/2}\right) dx$$

$$= 1000 - \left[\frac{2}{3}(x)^{3/2}\right]_0^{100}$$

$$= 1000 - \left[\left(\frac{2}{3}(100)^{3/2}\right) - \left(\frac{2}{3}(0)^{3/2}\right)\right]$$

$$= 1000 - \left[\frac{2000}{3} - 0\right]$$

$$= 1000 - \frac{2000}{3}$$

$$= \frac{1000}{3} \approx 333.33$$

The producer surplus at the equilibrium point is $333.33.

13. $D(x) = (x-4)^2, \quad S(x) = x^2 + 2x + 8$

a) To find the equilibrium point we set $D(x) = S(x)$ and solve.

$$(x-4)^2 = x^2 + 2x + 8$$

$$x^2 - 8x + 16 = x^2 + 2x + 8$$

$$-8x + 16 = 2x + 8$$

$$8 = 10x$$

$$\frac{8}{10} = x$$

$$\frac{4}{5} = x$$

Thus $x_E = \frac{4}{5} = 0.8$ units. Assuming partial units are acceptable, to find p_E we substitute x_E into $D(x)$ or $S(x)$. Here we use $D(x)$.

$$p_E = D(x_E)$$

$$= D\left(\frac{4}{5}\right)$$

$$= \left(\frac{4}{5} - 4\right)^2$$

$$= \left(-\frac{16}{5}\right)^2$$

$$= \frac{256}{25}$$

$$= 10.24$$

When 0.8 units are sold the equilibrium price is $10.24; therefore, the equilibrium point is $(0.8, \$10.24)$.

b) The consumer surplus is

$$\int_0^{x_E} D(x)\,dx - x_E p_E.$$

Substituting $(x-4)^2$ for $D(x)$, 0.8 for x_E, and 10.24 for p_E we have:

$$\int_0^{0.8} (x-4)^2\,dx - 0.8 \cdot 10.24$$

$$= \left[\frac{(x-4)^3}{3}\right]_0^{0.8} - 8.192$$

$$= \left[\left(\frac{(0.8-4)^3}{3}\right) - \left(\frac{(0-4)^3}{3}\right)\right] - 8.192$$

$$= \left[\left(\frac{-32.768}{3}\right) - \left(\frac{-64}{3}\right)\right] - 8.192$$

$$\approx 2.22 \qquad \text{Using a calculator}$$

The consumer surplus at the equilibrium point is $2.22.

c) The producer surplus is

$$x_E p_E - \int_0^{x_E} S(x)\,dx.$$

Substituting $x^2 + 2x + 8$ for $S(x)$, 0.8 for x_E, and 10.24 for p_E we have:

$$0.8 \cdot 10.24 - \int_0^{0.8} (x^2 + 2x + 8)\,dx$$

$$= 8.192 - \left[\frac{x^3}{3} + x^2 + 8x\right]_0^{0.8}$$

$$= 8.192 - \left[\left(\frac{(0.8)^3}{3} + (0.8)^2 + 8(0.8)\right) - (0)\right]$$

$$\approx 8.192 - 7.21066667$$

$$\approx 0.98$$

The producer surplus at the equilibrium point is $0.98.

15. $D(x) = e^{-x+4.5}$, $S(x) = e^{x-5.5}$

   a) To find the equilibrium point we set
   $D(x) = S(x)$ and solve.

   $$e^{-x+4.5} = e^{x-5.5}$$
   $$\ln\left(e^{-x+4.5}\right) = \ln\left(e^{x-5.5}\right)$$
   $$-x + 4.5 = x - 5.5$$
   $$10 = 2x$$
   $$5 = x$$

   Thus $x_E = 5$ units. To find p_E we substitute
   x_E into $D(x)$ or $S(x)$. Here we use $D(x)$.

   $$p_E = D(x_E)$$
   $$= D(5) = e^{-5+4.5} = e^{-0.5} \approx 0.61$$

   When 5 units are sold the equilibrium price
   is \$0.61; therefore, the equilibrium point is
   $(5, \$0.61)$.

   b) The consumer surplus is

   $$\int_0^{x_E} D(x)\,dx - x_E p_E.$$

   Substituting $e^{-x+4.5}$ for $D(x)$, 5 for x_E, and
   0.61 for p_E we have:

   $$\int_0^5 e^{-x+4.5}\,dx - 5 \cdot 0.61$$
   $$= \left[-e^{-x+4.5}\right]_0^5 - 3.05$$
   $$= \left[\left(-e^{-5+4.5}\right) - \left(-e^{-0+4.5}\right)\right] - 3.05$$
   $$= \left[-e^{-0.5} + e^{4.5}\right] - 3.05$$
   $$\approx 86.36 \qquad \text{Using a calculator}$$

   The consumer surplus at the equilibrium
   point is \$86.36.

   c) The producer surplus is

   $$x_E p_E - \int_0^{x_E} S(x)\,dx.$$

   Substituting $e^{x-5.5}$ for $S(x)$, 5 for x_E, and
   0.61 for p_E we have:

   $$5 \cdot 0.61 - \int_0^5 \left(e^{x-5.5}\right)dx$$
   $$= 3.05 - \left[e^{x-5.5}\right]_0^5$$
   $$= 3.05 - \left[\left(e^{5-5.5}\right) - \left(e^{0-5.5}\right)\right]$$
   $$= 3.05 - \left[e^{-0.5} - e^{-5.5}\right]$$
   $$\approx 2.45$$

   The producer surplus at the equilibrium
   point is \$2.45.

17. $\boxed{tw}$

19. $D(x) = \dfrac{x+8}{x+1}$, $S(x) = \dfrac{x^2+4}{20}$

   a) Graphing the equations and using the
   INTERSECT feature we have:

   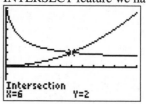

   The equilibrium point is $(6, \$2)$.

   b)

   | Consumer surplus | | | |
   | Producer surplus | | | |

   c) The consumer surplus is

   $$\int_0^{x_E} D(x)\,dx - x_E p_E.$$

   Substituting $\dfrac{x+8}{x+1}$ for $D(x)$, 6 for x_E, and
   2 for p_E we have:

   $$\int_0^6 \left(\frac{x+8}{x+1}\right)dx - 6 \cdot 2$$
   $$\approx 19.62137104 - 12 \qquad \text{Using a calculator}$$
   $$\approx 7.62$$

   The consumer surplus at the equilibrium
   point is \$7.62.

   d) The producer surplus is

   $$x_E p_E - \int_0^{x_E} S(x)\,dx.$$

   Substituting $\dfrac{x^2+4}{20}$ for $S(x)$, 6 for x_E, and
   2 for p_E we have:

   $$6 \cdot 2 - \int_0^6 \left(\frac{x^2+4}{20}\right)dx$$
   $$= 12 - \left[\frac{x^3}{60} + \frac{x}{5}\right]_0^6$$
   $$= 12 - \frac{24}{5}$$
   $$= \frac{36}{5} = 7.20$$

   The producer surplus at the equilibrium
   point is \$7.20.

21. a) Entering the data into the STAT editor on the calculator and plotting the points, we get the following scatter plot:

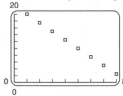

A linear function appears to be the best fit for the data.

b) Using the linear regression feature on the calculator, we get the equation:

$y = -2.5x + 22.5$.

c) The consumer surplus is

$$\int_0^x D(x)\,dx - x \cdot p\,.$$

Substituting $-2.5x + 22.5$ for $D(x)$, 6 for x, and 7.50 for p we have:

$$\int_0^6 (-2.5x + 22.5)\,dx - 6(7.50) = 45$$

The consumer surplus is $45.

d) First find the value for x when $y = 11.50$.

$11.50 = -2.5x + 22.5$

$-11 = -2.5x$

$4.4 = x$

Next, find the consumer surplus.

$$\int_0^{4.4} (-2.5x + 22.5)\,dx - 4.4(11.50) = 24.20$$

When the price of bungee jumping is $11.50 per half hour, Reggie's consumer surplus is $24.20.

Exercise Set 5.2

---

1. $P(t) = P_0 e^{kt}$

Substituting 3 for t, 700 for P_0 and 0.09 for k we have:

$P(3) = 700e^{0.09(3)}$

$= 700e^{0.27}$

$\approx 700(1.309964)$

≈ 916.98

The amount in the account is $916.98 after 3 years.

3. $\int_0^T R(t)e^{kt}\,dt$

Substituting 2000 for $R(t)$, 0.06 for k, and 8 for T we have

$$\int_0^8 (2000)e^{0.06t}\,dt = \left[\frac{2000}{0.06}e^{0.06t}\right]_0^8$$

$$= \frac{2000}{0.06}\left(e^{0.48} - 1\right)$$

$$\approx 20{,}535.81$$

The future value of the continuous money flow is $20,535.81.

5. $\int_0^T R(t)e^{kt}\,dt$

Substituting 1500 for $R(t)$, 0.05 for k, and 20 for T we have

$$\int_0^{20} (1500)e^{0.05t}\,dt = \left[\frac{1500}{0.05}e^{0.05t}\right]_0^{20}$$

$$= \frac{1500}{0.05}\left(e^1 - 1\right)$$

$$\approx 51{,}548.45$$

The future value of the continuous money flow is $51,548.45.

7. We find P_0 such that

$$50{,}000 = \int_0^{20} P_0 e^{0.085t}\,dt$$

$$50{,}000 = P_0\left[\frac{1}{0.085}e^{0.085t}\right]_0^{20}$$

$$50{,}000 = P_0\left[\frac{1}{0.085}e^{1.7} - \frac{1}{0.085}e^0\right]$$

$$50{,}000 = P_0\left[\frac{1}{0.085}\left(e^{1.7} - 1\right)\right]$$

$$50{,}000 \approx P_0\left[52.63467520\right]$$

$$\frac{50{,}000}{52.63467520} \approx P_0$$

$$949.94 \approx P_0$$

A continuous money flow of $949.94 per year, invested at 8.5%, compounded continuously for 20 years, will yield $50,000.

9. We find P_0 such that

$$40,000 = \int_0^{30} P_0 e^{0.09t}\, dt$$

$$40,000 = P_0 \left[\frac{1}{0.09} e^{0.09t} \right]_0^{30}$$

$$40,000 = P_0 \left[\frac{1}{0.09} e^{2.7} - \frac{1}{0.09} e^0 \right]$$

$$40,000 = P_0 \left[\frac{1}{0.09} \left(e^{2.7} - 1 \right) \right]$$

$$40,000 \approx P_0 \left[154.21924139 \right]$$

$$\frac{40,000}{154.21924139} \approx P_0$$

$$259.37 \approx P_0$$

A continuous money flow of $259.37 per year, invested at 9%, compounded continuously for 30 years, will yield $40,000.

11. $P = P_0 e^{kt}$

Therefore,

$$P_0 = \frac{P}{e^{kt}} = Pe^{-kt}$$

Substituting 50,000 for P, 0.07 for k and 20 for t, we have:

$$P_0 = 50,000 e^{-0.07(20)}$$

$$P_0 = 50,000 e^{-1.4}$$

$$P_0 \approx 50,000(0.24659696)$$

$$P_0 \approx 12,329.85$$

Dori and Dwayne should make an initial investment $12,329.85.

13. $P = P_0 e^{kt}$

Therefore,

$$P_0 = \frac{P}{e^{kt}} = Pe^{-kt}$$

Substituting 100,000 for P, 0.05 for k and 10 for t, we have:

$$P_0 = 100,000 e^{-0.05(10)}$$

$$P_0 = 100,000 e^{-0.5}$$

$$P_0 \approx 100,000(0.60653066)$$

$$P_0 \approx 60,653.07$$

The present value of $100,000 due in 10 years, at 5%, compounded continuously is $60,653.07.

15. The accumulated present value is

$$A = \int_0^T R(t) e^{-kt}\, dt .$$

Substituting 2700 for $R(t)$, 0.09 for k, and 10 for T, we have:

$$A = \int_0^{10} 2700 e^{-0.09t}\, dt$$

$$= \left[\frac{2700}{-0.09} e^{-0.09t} \right]_0^{10}$$

$$= \left[-30,000 \left(e^{-0.9} - e^0 \right) \right]$$

$$\approx -30,000(-0.59343034)$$

$$\approx 17,802.91$$

The accumulated present value is $17,802.91.

17. $A = \int_0^T R(t) e^{-kt}\, dt .$

Substituting 95,000 for $R(t)$, 0.06 for k, and 30 [65-35=30] for T, we have:

$$A = \int_0^{30} 95,000 e^{-0.06t}\, dt$$

$$= \left[\frac{95,000}{-0.06} e^{-0.06t} \right]_0^{30}$$

$$= \left[\frac{95,000}{-0.06} \left(e^{-1.8} - e^0 \right) \right]$$

$$\approx \frac{95,000}{-0.06}(-0.83470111)$$

$$\approx 1,321,610.09$$

The accumulated present value is $1,321.610.09.

19. $c = c_0 + \int_0^L m(t) e^{-rt}\, dt$

Substituting 500,000 for c_0, 20,000 for $m(t)$, 0.05 for r, and 20 for L, we have:

$$c = 500,000 + \int_0^{20} 20,000 e^{-0.05t}\, dt$$

$$= 500,000 + \left[\frac{20,000}{-0.05} e^{-0.05t} \right]_0^{20}$$

$$= 500,000 - 400,000 \left[e^{-0.05(20)} - e^{-0.05(0)} \right]$$

$$= 500,000 - 400,000 \left[e^{-1} - e^0 \right]$$

$$\approx 752,848$$

The capitalized cost under the given assumptions is $752,848.

21. $c = c_0 + \int_0^L m(t)e^{-rt}\,dt$

Substituting 600,000 for c_0, $40,000 + 1000e^{0.01t}$

for $m(t)$, 0.04 for r, and 40 for L, we have:

$c = 600,000 + \int_0^{40}\left(40,000 + 1000e^{0.01t}\right)e^{-0.04t}\,dt$

$= 600,000 + \int_0^{40}\left(40,000e^{-0.04t} + 1000e^{-0.03t}\right)dt$

$= 600,000 + \left[\dfrac{40,000}{-0.04}e^{-0.04t} + \dfrac{1000}{-0.03}e^{-0.03t}\right]_0^{40}$

$= 600,000 + \left[-1,000,000e^{-0.04t} - \right.$

$\left. \dfrac{100,000}{3}e^{-0.03t}\right]_0^{40}$

$= 600,000 + \left[-1,000,000e^{-0.04(40)} - \right.$

$\dfrac{100,000}{3}e^{-0.03(40)} - $

$\left.\left(-1,000,000e^{-0.04(0)} - \dfrac{100,000}{3}e^{-0.03(0)}\right)\right]$

$\approx 1,421,397$

The capitalized cost under the given assumptions is $1,421,397.

23. The equation for the consumption of a natural

resource is $\int_0^T P_0 e^{kt}\,dt = \dfrac{P_0}{k}\left(e^{kT} - 1\right)$.

Substituting 101.4 for P_0, 0.026 for k, and 14

$[2020 - 2006 = 14]$ for t, we have

$\int_0^{14} 101.4e^{0.026t} = \dfrac{101.4}{0.026}\left(e^{0.026(14)} - 1\right)$

$= \dfrac{101.4}{0.026}\left(e^{0.364} - 1\right)$

$\approx \dfrac{101.4}{0.026}(0.43907421)$

≈ 1712.4

If demand continues to grow exponentially at 2.6% per year, the world will consume approximately 1712.4 trillion cubic feet of natural gas from 2006 to 2020.

25. The equation for the consumption of a natural

resource is $\int_0^T P_0 e^{kt}\,dt = \dfrac{P_0}{k}\left(e^{kT} - 1\right)$.

Using the information from Exercise 23, we want to find T, such that

$6112 = \dfrac{101.4}{0.026}\left(e^{0.026T} - 1\right)$

We solve for T as follows:

$6112 = 3900\left(e^{0.026T} - 1\right)$

$1.5672 = e^{0.026T} - 1$ Dividing both sides by 3900.

$2.5672 \approx e^{0.026T}$ Adding 1 to both sides.

$\ln(2.5672) \approx \ln\left(e^{0.026T}\right)$ Taking the natural logarithm of each side.

$\ln(2.5672) \approx 0.026T$ Recall that $\ln e^k = k$.

$\dfrac{\ln(2.5672)}{0.026} \approx T$ Dividing both sides by 0.026.

$36.3 \approx T$

Assuming the world consumption of natural gas continues to grow at 2.6% per year, and no new reserves are found, the world reserves of natural gas will be depleted 36 years after 2006, in 2042.

27. We will use the equation

$\int_0^T Pe^{-kt}\,dt = \dfrac{P}{-k}\left(e^{-kT} - 1\right) = \dfrac{P}{k}\left(1 - e^{-kT}\right)$.

Substituting 0.00003 for k, 1 for P, and 20 for T, we have

$\int_0^{20} 1 \cdot e^{-0.00003t}\,dt = \dfrac{1}{0.00003}\left(1 - e^{-0.00003(20)}\right)$

$\approx 33,333.33\left(1 - e^{-0.0006}\right)$

≈ 19.994 Using a calculator.

After 20 years, approximately 19.994 lbs of Plutonium-239 will remain in the atmosphere.

29. a) $P = P_0 e^{kt}$

We will substitute the known information and solve for k.

$34.5 = 30.8e^{k(2010 - 2006)}$

$34.5 = 30.8e^{k \cdot 4}$

$\dfrac{34.5}{30.8} = e^{4k}$

$\ln\left(\dfrac{34.5}{30.8}\right) = \ln\left(e^{4k}\right)$

$\ln\left(\dfrac{34.5}{30.8}\right) = 4k$

$\dfrac{\ln\left(\dfrac{34.5}{30.8}\right)}{4} = k$

$0.02836 \approx k$

The exponential growth rate of demand for oil is 0.0284 or 2.84% per year.

b) Using the growth rate and the initial demand from Part (a) we have the exponential demand function $P = 30.8e^{0.02836t}$.
Substituting 9 $[2015 - 2006 = 9]$ for t, we have:

$$P = 30.8e^{0.02836(9)}$$
$$= 30.8e^{0.25524}$$
$$\approx 30.8(1.29077)$$
$$\approx 39.76$$

In 2015, the world demand for oil will be approximately 39.76 billion barrels.

c) The equation for the consumption of a natural resource is $\int_0^T P_0 e^{kt}\,dt = \dfrac{P_0}{k}\left(e^{kT} - 1\right)$.
We want to find T, such that

$$1293 = \frac{30.8}{0.02836}\left(e^{0.02836T} - 1\right)$$
$$\frac{1293(0.02836)}{30.8} = e^{0.02836T} - 1$$
$$1.1906 \approx e^{0.02836T} - 1$$
$$2.1906 \approx e^{0.02836T}$$
$$\ln(2.1906) \approx \ln e^{0.02836T}$$
$$\ln(2.1906) \approx 0.02836T$$
$$\frac{\ln(2.1906)}{0.02836} \approx T$$
$$27.7 \approx T$$

Assuming the world consumption of oil continues to grow at 2.84% per year, and no new reserves are found, the world reserves of oil will be depleted 27.7 years after 2006, in 2033.

31. $\int_0^T R(t)e^{k(T-t)}\,dt$

Substituting, we have:

$$\int_0^{30} (2000t + 7)e^{0.08(30-t)}\,dt$$
$$= \int_0^{30} 2000te^{0.08(30-t)}\,dt + \int_0^{30} 7e^{0.08(30-t)}\,dt$$
$$= 2000\int_0^{30} te^{0.08(30-t)}\,dt + 7\int_0^{30} e^{0.08(30-t)}\,dt$$

We will use integration by parts to evaluate the first integral.

$$\int te^{0.08(30-t)}\,dt$$

Let

$u = t$ and $dv = e^{0.08(30-t)}\,dt$
Then

$du = dt$ and $v = -\dfrac{1}{0.08}e^{0.08(30-t)}$

Therefore, we have:

$$\int te^{0.08(30-t)}\,dt$$
$$= -\frac{1}{0.08}te^{0.08(30-t)} - \int -\frac{1}{0.08}e^{0.08(30-t)}\,dt$$
$$= -\frac{1}{0.08}te^{0.08(30-t)} - \frac{1}{0.08}\cdot\frac{1}{0.08}e^{0.08(30-t)} + C$$
$$= -12.5te^{0.08(30-t)} - 156.25e^{0.08(30-t)} + C$$

Next, we find the definite integral

$$\int_0^{30} te^{0.08(30-t)}\,dt$$
$$= \left[-12.5te^{0.08(30-t)} - 156.25e^{0.08(30-t)}\right]_0^{30}$$
$$= \left[-12.5(30)e^{0.08(30-30)} - 156.25e^{0.08(30-30)} - \left(-12.5(0)e^{0.08(30-0)} - 156.25e^{0.08(30-0)}\right)\right]$$
$$= \left(-375e^0 - 156.25e^0\right) - \left(0 - 156.25e^{2.4}\right)$$
$$\approx -375 - 156.25 - 0 + 1722.37$$
$$\approx 1191.12$$

Next, we integrate the second part of the original integral. From the process of integration by parts we know:

$$\int e^{0.08(30-t)}\,dt = -\frac{1}{0.08}e^{0.08(30-t)} + C$$
$$= -12.5e^{0.08(30-t)} + C$$

Finding the definite integral, we have:

$$\int_0^{30} e^{0.08(30-t)}\,dt$$
$$= \left[-12.5e^{0.08(30-t)}\right]_0^{30}$$
$$= \left[-12.5e^{0.08(30-30)} - \left(-12.5e^{0.08(30-0)}\right)\right]$$
$$= -12.5e^0 + 12.5e^{2.4}$$
$$\approx -12.5 + 137.79$$
$$\approx 125.29$$

Finally, we have:

$$2000\int_0^{30} te^{0.08(30-t)}\,dt + 7\int_0^{30} e^{0.08(30-t)}\,dt$$
$$\approx 2000(1191.12) + 7(125.29)$$
$$\approx 2,383,117$$

The future value of the continuous money flow is approximately $2,383,117.
Note, if we had done all the calculations before rounding, the result would have been $2,383,119.65.

33. $\int_0^T R(t)e^{-k(T-t)}dt$

$\int_0^{20} te^{-0.08(20-t)}dt$

Use integration by parts.
Let

$u = t$ and $dv = e^{-0.08(20-t)}dt$.
Then,

$du = dt$ and $v = \dfrac{1}{0.08}e^{0.08(20-t)}$

Finding the indefinite integral we have:

$\int te^{-0.08(20-t)}dt$

$= \dfrac{1}{0.08}te^{-0.08(20-t)} - \int \dfrac{1}{0.08}e^{-0.08(20-t)}dt$

$= \dfrac{1}{0.08}te^{-0.08(20-t)} - \dfrac{1}{0.08}\cdot\dfrac{1}{0.08}e^{-0.08(20-t)} + C$

$= 12.5te^{-0.08(20-t)} - 156.25e^{-0.08(20-t)} + C$

Next, find the definite integral.

$\int_0^{20} te^{-0.08(20-t)}dt$

$= \left[12.5te^{-0.08(20-t)} - 156.25e^{-0.08(20-t)}\right]_0^{20}$

$= \left[\left(12.5(20)e^{-0.08(20-20)} - 156.25e^{-0.08(20-20)}\right) - \right.$

$\left. \left(12.5(0)e^{-0.08(20-0)} - 156.25e^{-0.08(20-0)}\right)\right]$

$= \left(250e^0 - 156.25e^0\right) - \left(0 - 156.25e^{-1.6}\right)$

$\approx 250 - 156.25 - 0 + 31.55$

≈ 125.30

The accumulated present value is approximately $125.30.

35. A continuous investment $2500 per year at 6% will yield:

$\int_0^{15} 2500e^{0.06t}dt = \left[\dfrac{2500}{0.06}e^{0.06t}\right]_0^{15}$

$= \dfrac{2500}{0.06}\left(e^{0.06(15)} - e^{0.06(0)}\right)$

$= \dfrac{2500}{0.06}\left(e^{0.9} - 1\right)$

$\approx 60,816.80$

A single initial investment of $20,000 at 6% will yield:

$20,000e^{0.06(15)} = 20,000e^{0.9}$

$\approx 49,192.06$

Therefore, the account in which $2500 per year is continuously invested will have the greatest future value of $60,816.80.

37. [tw]

39.

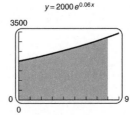

$y = 2000e^{0.06x}$

The area and the future value of the continuous money flow are the same. Both are $20,535.81.

Exercise Set 5.3

1. $\displaystyle\int_2^\infty \dfrac{dx}{x^2}$

$= \lim_{b\to\infty}\int_2^b x^{-2}dx$

$= \lim_{b\to\infty}\left[\dfrac{x^{-1}}{-1}\right]_2^b$

$= \lim_{b\to\infty}\left[-\dfrac{1}{x}\right]_2^b$

$= \lim_{b\to\infty}\left[-\dfrac{1}{b} - \left(-\dfrac{1}{2}\right)\right]$

$= 0 + \dfrac{1}{2} \qquad \left(\text{As } b\to\infty, -\dfrac{1}{b}\to 0\right)$

$= \dfrac{1}{2}$

The limit does exist. The improper integral is convergent.

3. $\displaystyle\int_3^\infty \dfrac{dx}{x}$

$= \lim_{b\to\infty}\int_3^b x^{-1}dx$

$= \lim_{b\to\infty}\left[\ln(x)\right]_3^b$

$= \lim_{b\to\infty}\left[\ln(b) - \ln(3)\right]$

Note that $\ln b$ increases without bound as b increases. Therefore, the limit does not exist. If the limit does not exist, we say the improper integral is divergent.

5. $\displaystyle\int_0^\infty 3e^{-3x}\,dx$

$\displaystyle = \lim_{b\to\infty}\int_0^b 3e^{-3x}\,dx$

$\displaystyle = \lim_{b\to\infty}\left[\frac{3}{-3}e^{-3x}\right]_0^b$

$\displaystyle = \lim_{b\to\infty}\left[-e^{-3x}\right]_0^b$

$\displaystyle = \lim_{b\to\infty}\left[-e^{-3\cdot b}-\left(-e^{-3\cdot 0}\right)\right]$

$\displaystyle = \lim_{b\to\infty}\left[-e^{-3b}+1\right]$

$\displaystyle = \lim_{b\to\infty}\left[1-\frac{1}{e^{3b}}\right]$

$\displaystyle = 1-0 \quad \left[\text{As } b\to\infty, e^{3b}\to\infty, \text{ so } \frac{1}{e^{3b}}\to 0\right]$

$= 1$

The limit does exist. The improper integral is convergent.

7. $\displaystyle\int_1^\infty \frac{dx}{x^3}$

$\displaystyle = \lim_{b\to\infty}\int_1^b x^{-3}\,dx$

$\displaystyle = \lim_{b\to\infty}\left[\frac{x^{-2}}{-2}\right]_1^b$

$\displaystyle = \lim_{b\to\infty}\left[-\frac{1}{2x^2}\right]_1^b$

$\displaystyle = \lim_{b\to\infty}\left[-\frac{1}{2b^2}-\left(-\frac{1}{2(1)^2}\right)\right]$

$\displaystyle = 0+\frac{1}{2} \quad \left(\text{As } b\to\infty, -\frac{1}{2b^2}\to 0\right)$

$\displaystyle = \frac{1}{2}$

The limit does exist. The improper integral is convergent.

9. $\displaystyle\int_0^\infty \frac{dx}{2+x}$

$\displaystyle = \lim_{b\to\infty}\int_0^b \frac{1}{2+x}\,dx$

$\displaystyle = \lim_{b\to\infty}\left[\ln(2+x)\right]_0^b$

$\displaystyle = \lim_{b\to\infty}\left[\ln(2+b)-\ln(2+0)\right]$

$\displaystyle = \lim_{b\to\infty}\left[\ln(2+b)-\ln(2)\right]$

Note that $\ln(2+b)$ increases without bound as b increases. Therefore, the limit does not exist. If the limit does not exist, we say the improper integral is divergent.

11. $\displaystyle\int_2^\infty 4x^{-2}\,dx$

$\displaystyle = \lim_{b\to\infty}\int_2^b 4x^{-2}\,dx$

$\displaystyle = \lim_{b\to\infty}\left[\frac{4x^{-1}}{-1}\right]_2^b$

$\displaystyle = \lim_{b\to\infty}\left[-\frac{4}{x}\right]_2^b$

$\displaystyle = \lim_{b\to\infty}\left[-\frac{4}{b}-\left(-\frac{4}{2}\right)\right]$

$= 0+2 \qquad \left(\text{As } b\to\infty, -\frac{4}{b}\to 0\right)$

$= 2$

The limit does exist. The improper integral is convergent.

13. $\displaystyle\int_0^\infty e^x\,dx$

$\displaystyle = \lim_{b\to\infty}\int_0^b e^x\,dx$

$\displaystyle = \lim_{b\to\infty}\left[e^x\right]_0^b$

$\displaystyle = \lim_{b\to\infty}\left[e^b-\left(e^0\right)\right]$

$\displaystyle = \lim_{b\to\infty}\left[e^b-1\right]$

As $b\to\infty, e^b\to\infty$; thus, the limit does not exist. The improper integral is divergent.

15. $\displaystyle\int_3^\infty x^2\,dx$

$\displaystyle = \lim_{b\to\infty}\int_3^b x^2\,dx$

$\displaystyle = \lim_{b\to\infty}\left[\frac{x^3}{3}\right]_3^b$

$\displaystyle = \lim_{b\to\infty}\left[\frac{(b)^3}{3}-\frac{(3)^3}{3}\right]$

$\displaystyle = \lim_{b\to\infty}\left[\frac{b^3}{3}-9\right]$

As $b\to\infty, \dfrac{b^3}{3}\to\infty$; thus, the limit does not exist. The improper integral is divergent.

17. $\displaystyle\int_0^\infty xe^x\,dx$

$\displaystyle= \lim_{b\to\infty}\int_0^b xe^x\,dx$

$\displaystyle= \lim_{b\to\infty}\left[e^x\left(x-1\right)\right]_0^b$ Using integration by parts.

$\displaystyle= \lim_{b\to\infty}\left[e^b\left(b-1\right)-\left(e^0\left(0-1\right)\right)\right]$

$\displaystyle= \lim_{b\to\infty}\left[e^b\left(b-1\right)+1\right]$

As $b\to\infty, e^b\left(b-1\right)\to\infty$; thus, the limit does not exist. The improper integral is divergent.

19. $\displaystyle\int_0^\infty me^{-mx}\,dx, \qquad m>0$

$\displaystyle= \lim_{b\to\infty}\int_0^b me^{-mx}\,dx$

$\displaystyle= \lim_{b\to\infty}\left[\frac{m}{-m}e^{-mx}\right]_0^b$

$\displaystyle= \lim_{b\to\infty}\left[-e^{-mx}\right]_0^b$

$\displaystyle= \lim_{b\to\infty}\left[-e^{-m\cdot b}-\left(-e^{-m\cdot 0}\right)\right]$

$\displaystyle= \lim_{b\to\infty}\left[-e^{-mb}+1\right]$

$\displaystyle= \lim_{b\to\infty}\left[1-\frac{1}{e^{mb}}\right]$

$\displaystyle= 1-0 \quad \left[\text{As } b\to\infty, e^{mb}\to\infty, \text{so } \tfrac{1}{e^{mb}}\to 0\right]$

$= 1$

The limit does exist. The improper integral is convergent.

21. $\displaystyle\int_\pi^\infty \frac{dt}{t^{1.001}}$

$\displaystyle= \lim_{b\to\infty}\int_\pi^b t^{-1.001}\,dt$

$\displaystyle= \lim_{b\to\infty}\left[\frac{t^{-0.001}}{-0.001}\right]_\pi^b$

$\displaystyle= \lim_{b\to\infty}\left[-\frac{1000}{t^{0.001}}\right]_\pi^b$

$\displaystyle= \lim_{b\to\infty}\left[-\frac{1000}{b^{0.001}}-\left(-\frac{1000}{\left(\pi\right)^{0.001}}\right)\right]$

$\displaystyle= 0+\frac{1000}{\pi^{0.001}} \quad \left(\text{As } b\to\infty, -\tfrac{1000}{b^{0.001}}\to 0\right)$

$\displaystyle= \frac{1000}{\pi^{0.001}}$

≈ 998.86

The limit does exist. The improper integral is convergent.

23. $\displaystyle\int_{-\infty}^\infty t\,dt$

$\displaystyle= \int_{-\infty}^0 t\,dt+\int_0^\infty t\,dt$ Using Definition 2 with $c=0$.

$\displaystyle= \lim_{a\to-\infty}\int_a^0 t\,dt+\lim_{b\to\infty}\int_0^b t\,dt$

$\displaystyle= \lim_{a\to-\infty}\left[\frac{t^2}{2}\right]_a^0+\lim_{b\to\infty}\left[\frac{t^2}{2}\right]_0^b$

$\displaystyle= \lim_{a\to-\infty}\left[\frac{0^2}{2}-\frac{a^2}{2}\right]+\lim_{b\to\infty}\left[\frac{b^2}{2}-\frac{0^2}{2}\right]$

Neither $\displaystyle\lim_{a\to-\infty}\frac{a^2}{2}$ nor $\displaystyle\lim_{b\to\infty}\frac{b^2}{2}$ exists, so the integral is divergent.

25. The area is given by

$\displaystyle\int_2^\infty \frac{1}{x^2}\,dx = \lim_{b\to\infty}\int_2^b x^{-2}\,dx$

$\displaystyle= \lim_{b\to\infty}\left[\frac{x^{-1}}{-1}\right]_2^b$

$\displaystyle= \lim_{b\to\infty}\left[-\frac{1}{x}\right]_2^b$

$\displaystyle= \lim_{b\to\infty}\left[-\frac{1}{b}-\left(-\frac{1}{2}\right)\right]$

$\displaystyle= 0+\frac{1}{2} \qquad \left(\text{As } b\to\infty, -\tfrac{1}{b}\to 0\right)$

$\displaystyle= \frac{1}{2}$

The area of the region is $\dfrac{1}{2}$.

27. The area is given by

$\displaystyle\int_0^\infty 2xe^{-x^2}\,dx$

$\displaystyle= \lim_{b\to\infty}\int_0^b 2xe^{-x^2}\,dx$

$\displaystyle= \lim_{b\to\infty}\left[-e^{-x^2}\right]_0^b \qquad \left[u=-x^2, du=-2x\,dx\right]$

$\displaystyle= \lim_{b\to\infty}\left[-e^{-b^2}-\left(-e^{-0^2}\right)\right]$

$\displaystyle= \lim_{b\to\infty}\left[-e^{-b^2}+1\right]$

$\displaystyle= \lim_{b\to\infty}\left[-\frac{1}{e^{b^2}}+1\right]$

$\displaystyle= -0+1 \qquad \left[\text{As } b\to\infty, -\tfrac{1}{e^{b^2}}\to 0\right]$

$= 1$

The area of the region is 1.

29. From Theorem 2, The accumulated present value is given by

$$\int_0^\infty Pe^{-kt}\,dt = \frac{P}{k}$$

Substituting 3600 for P and 0.07 for k, we have:

$$\int_0^\infty 3600e^{-0.07t}\,dt = \frac{3600}{0.07} \approx 51{,}428.57 \,.$$

The accumulated present value is approximately $51,428.57.

31. Total profit is given by:

$$P(x) = \int_0^\infty 200e^{-0.032x}\,dx$$

$$= \lim_{b \to \infty} \int_0^b 200e^{-0.032x}\,dx$$

$$= \lim_{b \to \infty} \left[\frac{200}{-0.032} e^{-0.032x} \right]_0^b$$

$$= \lim_{b \to \infty} \left[-6250e^{-0.032x} \right]_0^b$$

$$= \lim_{b \to \infty} \left[-6250e^{-0.032b} - \left(-6250e^{-0.032(0)} \right) \right]$$

$$= \lim_{b \to \infty} \left[-\frac{6250}{e^{0.032b}} + 6250 \right]$$

$$= 6250$$

The total profit if it were possible to produce an infinite number of units is $6250.

33. The total cost is given by:

$$C(x) = \int_1^\infty 3600x^{-1.8}\,dx$$

$$= \lim_{b \to \infty} \int_2^b 3600x^{-1.8}\,dx$$

$$= \lim_{b \to \infty} \left[\frac{3600}{-0.8} x^{-0.8} \right]_1^b$$

$$= \lim_{b \to \infty} \left[-4500x^{-0.8} \right]_1^b$$

$$= \lim_{b \to \infty} \left[-\frac{4500}{b^{0.8}} - \left(-\frac{4500}{1^{0.8}} \right) \right]$$

$$= 0 + 4500 \qquad \left(\text{As } b \to \infty, -\frac{4500}{b^{0.8}} \to 0 \right)$$

$$= 4500$$

The total cost would be $4500.

35. From Theorem 2, The accumulated present value is given by

$$\int_0^\infty Pe^{-kt}\,dt = \frac{P}{k}$$

Substituting 5000 for P and 0.08 for k, we have:

$$\int_0^\infty 5000e^{-0.08t}\,dt = \frac{5000}{0.08} \approx 62{,}500 \,.$$

The accumulated present value is $62,500.

37. $c = c_0 + \int_0^\infty m(t)e^{-rt}\,dt$

Substituting 500,000 for c_0, 0.05 for r, and 20,000 for $m(t)$, we have:

$$c = 500{,}000 + \int_0^\infty 20{,}000e^{-0.05t}\,dt$$

$$= 500{,}000 + \lim_{b \to \infty} \int_0^b 20{,}000e^{-0.05t}\,dt$$

$$= 500{,}000 + \lim_{b \to \infty} \left[\frac{20{,}000}{-0.05} e^{-0.05t} \right]_0^b$$

$$= 500{,}000 + \lim_{b \to \infty} \left[-400{,}000 \left(e^{-0.05b} - e^{-0.05(0)} \right) \right]$$

$$= 500{,}000 + \left[-400{,}000(0 - 1) \right]$$

$$= 500{,}000 + 400{,}000$$

$$= 900{,}000$$

The capitalized cost is $900,000.

39. $\int_0^T Pe^{-kt}\,dt = \dfrac{P}{k}\left(1 - e^{-kT}\right)$

As $T \to \infty$, we have:

$$\lim_{T \to \infty} \int_0^T P(t)e^{-kt}\,dt$$

$$= \lim_{T \to \infty} \left[\frac{P}{-k} e^{-kt} \right]_0^T$$

$$= \lim_{T \to \infty} \left[\frac{P}{-k} e^{-kT} - \left(\frac{P}{-k} e^{-k \cdot 0} \right) \right]$$

$$= \lim_{T \to \infty} \left[\frac{P}{k} \left(1 - e^{-k \cdot T} \right) \right]$$

$$= \frac{P}{k}$$

Substituting 0.00003 for k, and 1 for P, we have

$$\frac{P}{k} = \frac{1}{0.00003} \approx 33{,}333\tfrac{1}{3} \,.$$

The limiting value of the radioactive buildup is $33{,}333\tfrac{1}{3}$ pounds.

41. $E = \int_0^a P_0 e^{-kt} dt$

a) Note that 60.1 days is

$$\frac{60.1}{365} \text{ yr} \approx 0.16465753 \text{ yr}.$$

Using the half-life, we find k as follows:

$$\frac{1}{2}P_0 = P_0 e^{-k(0.16465753)}$$

$$\frac{1}{2} = e^{-k(0.16465753)} \qquad \text{Dividing by } P_0$$

$$\ln\left(\frac{1}{2}\right) = \ln\left(e^{-0.16465753k}\right)$$

$$\ln\left(\frac{1}{2}\right) = -0.16465753k$$

$$\frac{\ln\left(\frac{1}{2}\right)}{-0.16465753} = k$$

$$4.20963 \approx k$$

The decay rate is 420.963% per year.

b) The first month is $\frac{1}{12}$ yr.

$$E = \int_0^{1/12} 10 e^{-4.20963t} dt$$

$$= \frac{10}{-4.20963}\left[e^{-4.20963t}\right]_0^{1/12}$$

$$= \frac{10}{-4.20963}\left[e^{-4.20963(1/12)} - e^{-4.20963(0)}\right]$$

$$= \frac{10}{-4.20963}\left[e^{-0.3508025} - 1\right]$$

$$\approx 0.702858$$

In the first month, 0.702858 rems of energy is transmitted.

c) $E = \int_0^\infty 10 e^{-4.20963t} dt$

$$E = \lim_{b \to \infty} \int_0^b 10 e^{-4.20963t} dt$$

$$= \frac{10}{4.20963} \qquad \left[\int_0^\infty P e^{-kt} = \frac{P}{k}\right]$$

$$\approx 2.37551$$

The total amount of energy transmitted is 2.37551 rems.

43. $\int_0^\infty \frac{dx}{x^{2/3}}$

$$= \lim_{b \to \infty} \int_0^b x^{-2/3} dx$$

$$= \lim_{b \to \infty}\left[\frac{x^{1/3}}{1/3}\right]_0^b$$

$$= \lim_{b \to \infty}\left[3x^{1/3}\right]_0^b$$

$$= \lim_{b \to \infty}\left[3b^{1/3} - 3(0)^{1/3}\right]$$

$$= \lim_{b \to \infty}\left[3 \cdot \sqrt[3]{b}\right]$$

As $b \to \infty$, $\sqrt[3]{b} \to \infty$. Therefore, the limit does not exist. The improper integral is divergent.

45. $\int_0^\infty \frac{dx}{(x+1)^{3/2}}$

$$= \lim_{b \to \infty} \int_0^b (x+1)^{-3/2} dx$$

$$= \lim_{b \to \infty}\left[\frac{(x+1)^{-1/2}}{-1/2}\right]_0^b$$

$$= \lim_{b \to \infty}\left[-\frac{2}{\sqrt{x+1}}\right]_0^b$$

$$= \lim_{b \to \infty}\left[-\frac{2}{\sqrt{b+1}} - \left(-\frac{2}{\sqrt{0+1}}\right)\right]$$

$$= \lim_{b \to \infty}\left[-\frac{2}{\sqrt{b+1}} + 2\right]$$

$$= 0 + 2 \qquad \left[\text{As } b \to \infty, -\frac{2}{\sqrt{b+1}} \to 0\right]$$

$$= 2$$

Therefore, the limit exists. The improper integral is convergent.

47. $\displaystyle\int_0^\infty xe^{-x^2}\,dx$

$\displaystyle = \lim_{b\to\infty}\int_0^b xe^{-x^2}\,dx$

$\displaystyle = \lim_{b\to\infty}\int_0^b -\frac{1}{2}\cdot(-2)\,xe^{-x^2}\,dx$ 　　Multiplying by 1.

$\displaystyle = \lim_{b\to\infty}\left[-\frac{1}{2}e^{-x^2}\right]_0^b$ 　　Using substitution where $u=-x^2$ and $du=-2x\,dx$.

$\displaystyle = \lim_{b\to\infty}\left[-\frac{1}{2}e^{-b^2}-\left(-\frac{1}{2}e^{-(0)^2}\right)\right]$

$\displaystyle = \lim_{b\to\infty}\left[-\frac{1}{2e^{b^2}}+\frac{1}{2}\right]$

$\displaystyle = 0+\frac{1}{2}$ 　　$\left[\text{As } b\to\infty,\dfrac{1}{e^{b^2}}\to 0\right]$

$\displaystyle = \frac{1}{2}$

Therefore, the limit exists. The improper integral is convergent.

49. $\displaystyle\int_0^\infty E(t)\,dt$

$\displaystyle = \int_0^\infty te^{-kt}\,dt$

$\displaystyle = \lim_{b\to\infty}\int_0^b te^{-kt}\,dt$

$\displaystyle = \lim_{b\to\infty}\left[\frac{1}{(-k)^2}\cdot e^{-kt}\,(-kt-1)\right]_0^b$

$\displaystyle = \lim_{b\to\infty}\left[-\frac{kt+1}{k^2e^{kt}}\right]_0^b$

$\displaystyle = \lim_{b\to\infty}\left[-\frac{kb+1}{k^2e^{kb}}-\left(-\frac{k(0)+1}{k^2e^{k(0)}}\right)\right]$

$\displaystyle = \lim_{b\to\infty}\left[-\frac{kb+1}{k^2e^{kb}}+\frac{1}{k^2}\right]$

$\displaystyle = \frac{1}{k^2}$ 　　$\left[\text{As } b\to\infty,-\dfrac{kb+1}{k^2e^{kb}}\to 0\right]$

The integral represents the total dose of the drug.

51. 　$\boxed{tw}$

53.

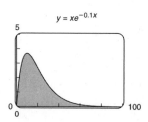
$y=xe^{-0.1x}$

55. Using the fnInt feature on a graphing calculator with a large value for the upper limit, we find

$$\int_1^\infty \frac{6}{5+e^x}\,dx \approx 1.2523.$$

Exercise Set 5.4

1. $\displaystyle f(x)=\frac{1}{4}x,\ \ [1,3]$

$\displaystyle P([1,3])=\int_1^3 f(x)\,dx$

$\displaystyle = \int_1^3 \frac{1}{4}x\,dx$

$\displaystyle = \left[\frac{x^2}{8}\right]_1^3$

$\displaystyle = \frac{(3)^2}{8}-\frac{(1)^2}{8}$

$\displaystyle = \frac{9}{8}-\frac{1}{8}$

$\displaystyle = 1$

3. $\displaystyle f(x)=3,\ \ \left[0,\frac{1}{3}\right]$

$\displaystyle P\left(\left[0,\frac{1}{3}\right]\right)=\int_0^{1/3} f(x)\,dx$

$\displaystyle = \int_0^{1/3} 3\,dx$

$\displaystyle = [3x]_0^{1/3}$

$\displaystyle = 3\left(\frac{1}{3}\right)-3(0)$

$\displaystyle = 1$

5. $f(x) = \dfrac{3}{64}x^2$, $[0,4]$

$$P([0,4]) = \int_0^4 f(x)\,dx$$

$$= \int_0^4 \frac{3}{64}x^2\,dx$$

$$= \left[\frac{x^3}{64}\right]_0^4$$

$$= \frac{(4)^3}{64} - \frac{(0)^3}{64}$$

$$= \frac{64}{64} - 0$$

$$= 1$$

7. $f(x) = \dfrac{1}{x}$, $[1,e]$

$$P([1,e]) = \int_1^e f(x)\,dx$$

$$= \int_1^e \frac{1}{x}\,dx$$

$$= \left[\ln x\right]_1^e$$

$$= \ln(e) - \ln(1)$$

$$= 1 - 0$$

$$= 1$$

9. $f(x) = \dfrac{3}{2}x^2$, $[-1,1]$

$$P([-1,1]) = \int_{-1}^1 f(x)\,dx$$

$$= \int_{-1}^1 \frac{3}{2}x^2\,dx$$

$$= \left[\frac{x^3}{2}\right]_{-1}^1$$

$$= \frac{(1)^3}{2} - \frac{(-1)^3}{2}$$

$$= \frac{1}{2} - \frac{-1}{2}$$

$$= 1$$

11. $f(x) = 3e^{-3x}$, $[0,\infty)$

$$P([0,\infty)) = \int_0^\infty f(x)\,dx$$

$$= \lim_{b \to \infty} \int_0^b 3e^{-3x}\,dx$$

$$= \lim_{b \to \infty} \left[\frac{3}{-3}e^{-3x}\right]_0^b$$

$$= \lim_{b \to \infty} \left[-e^{-3x}\right]_0^b$$

$$= \lim_{b \to \infty} \left[-e^{-3 \cdot b} - \left(-e^{-3(0)}\right)\right]$$

$$= \lim_{b \to \infty} \left[-\frac{1}{e^{3 \cdot b}} + 1\right]$$

$$= 0 + 1 \qquad \left[\text{As } b \to \infty, -\frac{1}{e^{3b}} \to 0\right]$$

$$= 1$$

13. $f(x) = kx$, $[2,5]$

Find k such that $\displaystyle\int_2^5 kx\,dx = 1$

We have

$$\int_2^5 x\,dx = \left[\frac{x^2}{2}\right]_2^5$$

$$= \left[\frac{(5)^2}{2} - \frac{(2)^2}{2}\right]$$

$$= \frac{25}{2} - \frac{4}{2} = \frac{21}{2}$$

Thus $k = \dfrac{1}{\dfrac{21}{2}} = \dfrac{2}{21}$ and the probability density

function is $f(x) = \dfrac{2}{21}x$.

15. $f(x) = kx^2$, $[-1,1]$

Find k such that $\int_{-1}^{1} kx^2 \, dx = 1$

We have

$\int_{-1}^{1} x^2 \, dx = \left[\dfrac{x^3}{3} \right]_{-1}^{1}$

$= \left[\dfrac{(1)^3}{3} - \dfrac{(-1)^3}{3} \right]$

$= \dfrac{1}{3} - \dfrac{-1}{3} = \dfrac{2}{3}$

Thus $k = \dfrac{1}{\frac{2}{3}} = \dfrac{3}{2}$ and the probability density

function is $f(x) = \dfrac{3}{2} x^2$.

17. $f(x) = k$, $[1,7]$

Find k such that $\int_{1}^{7} k \, dx = 1$

We have

$\int_{1}^{7} dx = [x]_{1}^{7}$

$= [7 - 1]$

$= 6$

Thus $k = 6 = \dfrac{1}{6}$ and the probability density

function is $f(x) = \dfrac{1}{6}$.

19. $f(x) = k(2 - x)$, $[0,2]$

Find k such that $\int_{0}^{2} k(2 - x) \, dx = 1$

We have

$\int_{0}^{2} (2 - x) \, dx = \left[2x - \dfrac{x^2}{2} \right]_{0}^{2}$

$= \left[\left(2(2) - \dfrac{(2)^2}{2} \right) - \left(2(0) - \dfrac{0^2}{2} \right) \right]$

$= 4 - \dfrac{4}{2} - 0$

$= 2$

Thus $k = \dfrac{1}{2}$ and the probability density function

is $f(x) = \dfrac{1}{2}(2 - x) = \dfrac{2 - x}{2}$.

21. $f(x) = \dfrac{k}{x}$, $[1,3]$

Find k such that $\int_{1}^{3} \dfrac{k}{x} \, dx = 1$

We have

$\int_{1}^{3} \dfrac{1}{x} \, dx = [\ln(x)]_{1}^{3}$

$= [\ln(3) - \ln(1)]$

$= \ln 3 - 0 = \ln 3$

Thus $k = \dfrac{1}{\ln 3}$ and the probability density

function is $f(x) = \dfrac{1}{\ln 3} \cdot \dfrac{1}{x} = \dfrac{1}{x \ln 3}$.

23. $f(x) = ke^x$, $[0,3]$

Find k such that $\int_{0}^{3} ke^x \, dx = 1$

We have

$\int_{0}^{3} e^x \, dx = [e^x]_{0}^{3}$

$= [e^3 - e^0]$

$= e^3 - 1$

Thus $k = \dfrac{1}{e^3 - 1}$ and the probability density

function is $f(x) = \dfrac{1}{e^3 - 1} \cdot e^x = \dfrac{e^x}{e^3 - 1}$.

25. a) $f(x) = \dfrac{1}{50} x$, for $0 \le x \le 10$

$P(2 \le x \le 6) = \int_{2}^{6} \dfrac{1}{50} x \, dx$

$= \left[\dfrac{1}{50} \cdot \dfrac{x^2}{2} \right]_{2}^{6}$

$= \left[\dfrac{x^2}{100} \right]_{2}^{6}$

$= \dfrac{6^2}{100} - \dfrac{2^2}{100}$

$= \dfrac{36 - 4}{100}$

$= \dfrac{32}{100}$

$= \dfrac{8}{25} = 0.32$

The probability that the dart lands in the

interval $[2,6]$ is $\dfrac{8}{25}$, or 0.32.

b) $\boxed{tw}$ The answer to part (a) is the area under the graph of f over a subinterval from 2 to 6.

27. $f(x) = \dfrac{1}{16}$, for $4 \le x \le 20$

$$P(9 \le x \le 20) = \int_9^{20} \dfrac{1}{16}\,dx$$

$$= \left[\dfrac{1}{16}x\right]_9^{20}$$

$$= \dfrac{1}{16}\cdot 20 - \dfrac{1}{16}\cdot 9$$

$$= \dfrac{20-9}{16}$$

$$= \dfrac{11}{16} = 0.6875$$

The probability that the number selected is in the subinterval $[9,20]$ is $\dfrac{11}{16}$, or 0.6875.

29. From Example 10, we know

$f(x) = ke^{-kx}$, for $0 \le x < \infty$, where $k = \dfrac{1}{a}$ and a is the average distance between successive cars over some period of time.
First, we determine k:

$$k = \dfrac{1}{100} = 0.01 .$$

The probability that the distance between cars is 40 feet or less is calculated as follows:

$$P(0 \le x \le 40) = \int_0^{40} 0.01e^{-0.01x}\,dx$$

$$= \left[\dfrac{0.01}{-0.01}e^{-0.01x}\right]_0^{40}$$

$$= \left[-e^{-0.01x}\right]_0^{40}$$

$$= -e^{-0.01(40)} - \left(-e^{-0.01(0)}\right)$$

$$= -e^{-0.4} + e^0$$

$$= -e^{-0.4} + 1$$

$$\approx -0.670320 + 1$$

$$\approx 0.329680$$

$$\approx 0.3297$$

The probability that the distance between two successive cars, chosen at random, is 40 feet or less is 0.3297.

31. $f(t) = 2e^{-2t}$, for $0 \le t < \infty$

$$P(0 \le t \le 5) = \int_0^5 2e^{-2t}\,dt$$

$$= \left[\dfrac{2}{-2}e^{-2t}\right]_0^5$$

$$= \left[-e^{-2t}\right]_0^5$$

$$= -e^{-2(5)} - \left(-e^{-2(0)}\right)$$

$$= -e^{-10} + e^0$$

$$= -e^{-10} + 1$$

$$\approx -0.0000454 + 1$$

$$\approx 0.9999546$$

$$\approx 0.999955$$

The probability that a phone call will last no more than 5 minutes is 0.999955.

33. $f(t) = ke^{-kt}$, for $0 \le t < \infty$, where $k = \dfrac{1}{a}$ and a is the average amount of time that will pass before a failure occurs.
First, we determine k:

$$k = \dfrac{1}{100} = 0.01 .$$

The probability that a failure will occur in 50 hours or less is

$$P(0 \le x \le 50) = \int_0^{50} 0.01e^{-0.01x}\,dx$$

$$= \left[\dfrac{0.01}{-0.01}e^{-0.01x}\right]_0^{50}$$

$$= \left[-e^{-0.01x}\right]_0^{50}$$

$$= -e^{-0.01(50)} - \left(-e^{-0.01(0)}\right)$$

$$= -e^{-0.5} + e^0$$

$$= -e^{-0.5} + 1$$

$$\approx -0.606531 + 1$$

$$\approx 0.393469$$

$$\approx 0.3935$$

The probability that a failure will occur in 50 hours or less is 0.3935.

35. $f(x) = 8.1305e^{0.074x}$, for $1 \le x \le 85$

   a) Find k such that $\int_1^{85} kf(x)\,dx = 1$

      We have

$$\int_1^{85} 8.1305e^{0.074x}\,dx$$

$$= \left[\frac{8.1305}{0.074} e^{0.074x} \right]_1^{85}$$

$$\approx 109.8716216 \left[e^{0.074(85)} - e^{0.074(1)} \right]$$

$$\approx 59{,}119.34$$

      Thus,

$$k = \frac{1}{59{,}119.34} \approx 0.000017 \ .$$

   b) The probability density function is

$$kf(x) = (0.000017)8.1305e^{0.074x}, \text{ for } [1,85]$$

$$= \left(1.382 \times 10^{-4} \right) e^{0.074x}$$

The probability that a female who died in 2003 was between 25 and 40 years old is:

$$P\big([25,40]\big) = \int_{25}^{40} \left(1.382 \times 10^{-4} \right) e^{0.074x}\,dx$$

$$= \left[\frac{1.382 \times 10^{-4}}{0.074} e^{0.074x} \right]_{25}^{40}$$

$$\approx \left[0.0018676 e^{0.074x} \right]_{25}^{40}$$

$$\approx 0.0018676 \left(e^{0.074(40)} - e^{0.074(25)} \right)$$

$$\approx 0.024$$

The probability that a female who died in 2003 was between 25 and 40 years of age was approximately 0.024.
Note: if we waited until the end of all calculations to round, the probability would be 0.02404.

37. $f(t) = 0.02e^{-0.02t}$, for $0 \le t < \infty$

$$P(0 \le t \le 150) = \int_0^{150} 0.02e^{-0.02t}\,dt$$

$$= \left[\frac{0.02}{-0.02} e^{-0.02t} \right]_0^{150}$$

$$= \left[-e^{-0.02t} \right]_0^{150}$$

$$= -e^{-0.02(150)} - \left(-e^{-0.02(0)} \right)$$

$$= -e^{-3} + e^{0}$$

$$= -e^{-3} + 1$$

$$\approx -0.049787 + 1$$

$$\approx 0.950213$$

The probability that the rat will learn its way through a maze in 150 seconds, or less, is approximately 0.950213.

39. $f(x) = x^3$ is a probability density function over $[0,b]$. Thus,

$$\int_0^b f(x)\,dx = 1$$

$$\int_0^b x^3\,dx = 1$$

$$\left[\frac{x^4}{4} \right]_0^b = 1$$

$$\frac{b^4}{4} - \frac{0^4}{4} = 1$$

$$\frac{b^4}{4} = 1$$

$$b^4 = 1$$

$$b = \sqrt[4]{4}$$

$$b = \sqrt[4]{2^2}$$

$$b = \sqrt{2}$$

$f(x) = x^3$ is a probability density function over $\left[0, \sqrt{2} \right]$.

41. In Exercise 37 we found that
$P(0 \le t \le 150) \approx 0.950213$. Since this is a probability density function, we know

$$\int_0^\infty f(x)\,dx = \int_0^{150} f(x)\,dx + \int_{150}^\infty f(x)\,dx = 1$$

Thus,

$$\int_{150}^\infty f(x)\,dx = 1 - \int_0^{150} f(x)\,dx$$

$$\approx 1 - 0.950213$$

$$= 0.049787$$

Thus, the probability that a rat requires more than 150 seconds to learn its way through the maze is 0.049787.

43 – 54. Check your answers using the solutions to Exercises 1-12.

Exercise Set 5.5

1. $f(x) = \dfrac{1}{4}, \quad [3,7]$

$E(x) = \displaystyle\int_a^b x \cdot f(x)\, dx \qquad$ Expected value of x.

$E(x) = \displaystyle\int_3^7 x \cdot \dfrac{1}{4}\, dx$

$\quad = \dfrac{1}{4}\left[\dfrac{x^2}{2}\right]_3^7$

$\quad = \dfrac{1}{4}\left[\dfrac{7^2}{2} - \dfrac{3^2}{2}\right]$

$\quad = \dfrac{1}{4}\left[\dfrac{49}{2} - \dfrac{9}{2}\right]$

$\quad = \dfrac{1}{4} \cdot \dfrac{40}{2}$

$\quad = 5$

$E(x^2) = \displaystyle\int_a^b x^2 \cdot f(x)\, dx \qquad$ Expected value of x^2.

$E(x^2) = \displaystyle\int_3^7 x^2 \cdot \dfrac{1}{4}\, dx$

$\quad = \dfrac{1}{4}\left[\dfrac{x^3}{3}\right]_3^7$

$\quad = \dfrac{1}{4}\left[\dfrac{7^3}{3} - \dfrac{3^3}{3}\right]$

$\quad = \dfrac{1}{4}\left[\dfrac{343}{3} - \dfrac{27}{3}\right]$

$\quad = \dfrac{1}{4} \cdot \dfrac{316}{3}$

$\quad = \dfrac{79}{3}$

$\mu = E(x) = 5 \qquad$ Mean

$\sigma^2 = E(x^2) - [E(x)]^2$

$\quad = \dfrac{79}{3} - [5]^2 \qquad$ Substituting 79/3 for $E(x^2)$ and 5 for $E(x)$

$\quad = \dfrac{79}{3} - 25$

$\quad = \dfrac{79}{3} - \dfrac{75}{3}$

$\quad = \dfrac{4}{3} \qquad$ Variance

$\sigma = \sqrt{\sigma^2}$

$\quad = \sqrt{\dfrac{4}{3}} = \dfrac{2}{\sqrt{3}} \qquad$ Standard deviation

3. $f(x) = \dfrac{1}{8}x, \quad [0,4]$

$E(x) = \displaystyle\int_a^b x \cdot f(x)\, dx \qquad$ Expected value of x.

$E(x) = \displaystyle\int_0^4 x \cdot \dfrac{1}{8}x\, dx$

$\quad = \displaystyle\int_0^4 \dfrac{1}{8}x^2\, dx$

$\quad = \dfrac{1}{8}\left[\dfrac{x^3}{3}\right]_0^4$

$\quad = \dfrac{1}{8}\left[\dfrac{4^3}{3} - \dfrac{0^3}{3}\right]$

$\quad = \dfrac{1}{8} \cdot \dfrac{64}{3}$

$\quad = \dfrac{8}{3}$

$E(x^2) = \displaystyle\int_a^b x^2 \cdot f(x)\, dx \qquad$ Expected value of x^2.

$E(x^2) = \displaystyle\int_0^4 x^2 \cdot \dfrac{1}{8}x\, dx$

$\quad = \displaystyle\int_0^4 \dfrac{1}{8}x^3\, dx$

$\quad = \dfrac{1}{8}\left[\dfrac{x^4}{4}\right]_0^4$

$\quad = \dfrac{1}{8}\left[\dfrac{4^4}{4} - \dfrac{0^4}{4}\right]$

$\quad = \dfrac{1}{8} \cdot \dfrac{256}{4}$

$\quad = 8$

$\mu = E(x) = \dfrac{8}{3} \qquad$ Mean

$\sigma^2 = E(x^2) - [E(x)]^2$

$\quad = 8 - \left[\dfrac{8}{3}\right]^2 \qquad$ Substituting 8 for $E(x^2)$ and 8/3 for $E(x)$

$\quad = 8 - \dfrac{64}{9}$

$\quad = \dfrac{72}{9} - \dfrac{64}{9}$

$\quad = \dfrac{8}{9} \qquad$ Variance

$\sigma = \sqrt{\sigma^2}$

$= \sqrt{\dfrac{8}{9}} = \dfrac{2\sqrt{2}}{3}$ Standard deviation

5. $f(x) = \dfrac{1}{4}x, \quad [1,3]$

$E(x) = \displaystyle\int_a^b x \cdot f(x)\, dx$ Expected value of x.

$E(x) = \displaystyle\int_1^3 x \cdot \dfrac{1}{4} x\, dx$

$= \displaystyle\int_1^3 \dfrac{x^2}{4}\, dx$

$= \left[\dfrac{x^3}{12} \right]_1^3$

$= \left[\dfrac{3^3}{12} - \dfrac{1^3}{12} \right]$

$= \dfrac{26}{12}$

$= \dfrac{13}{6}$

$E(x^2) = \displaystyle\int_a^b x^2 \cdot f(x)\, dx$ Expected value of x^2.

$E(x^2) = \displaystyle\int_1^3 x^2 \cdot \dfrac{1}{4} x\, dx$

$= \displaystyle\int_1^3 \dfrac{x^3}{4}\, dx$

$= \left[\dfrac{x^4}{16} \right]_1^3$

$= \left[\dfrac{3^4}{16} - \dfrac{1^4}{16} \right]$

$= \dfrac{80}{16}$

$= 5$

$\mu = E(x) = \dfrac{13}{6}$ Mean

$\sigma^2 = E(x^2) - \left[E(x) \right]^2$

$= 5 - \left[\dfrac{13}{6} \right]^2$ Substituting 5 for $E(x^2)$ and 13/6 for $E(x)$

$= 5 - \dfrac{169}{36}$

$= \dfrac{180}{36} - \dfrac{169}{36}$

$= \dfrac{11}{36}$ Variance

$\sigma = \sqrt{\sigma^2}$

$= \sqrt{\dfrac{11}{36}} = \dfrac{\sqrt{11}}{6}$ Standard deviation

7. $f(x) = \dfrac{3}{2}x^2, \quad [-1,1]$

$E(x) = \displaystyle\int_a^b x \cdot f(x)\, dx$ Expected value of x.

$E(x) = \displaystyle\int_{-1}^1 x \cdot \dfrac{3}{2} x^2\, dx$

$= \displaystyle\int_{-1}^1 \dfrac{3}{2} x^3\, dx$

$= \left[\dfrac{3x^4}{8} \right]_{-1}^1$

$= \left[\dfrac{3(1)^4}{8} - \dfrac{3(-1)^4}{8} \right]$

$= \dfrac{3}{8} - \dfrac{3}{8}$

$= 0$

$E(x^2) = \displaystyle\int_a^b x^2 \cdot f(x)\, dx$ Expected value of x^2.

$E(x^2) = \displaystyle\int_{-1}^1 x^2 \cdot \dfrac{3}{2} x^2\, dx$

$= \displaystyle\int_{-1}^1 \dfrac{3}{2} x^4\, dx$

$= \left[\dfrac{3x^5}{10} \right]_{-1}^1$

$= \left[\dfrac{3(1)^5}{10} - \dfrac{3(-1)^5}{10} \right]$

$= \dfrac{3}{10} + \dfrac{3}{10}$

$= \dfrac{6}{10} = \dfrac{3}{5}$

$\mu = E(x) = 0$ Mean

$\sigma^2 = E(x^2) - \left[E(x) \right]^2$

$= \dfrac{3}{5} - [0]^2$ Substituting 3/5 for $E(x^2)$ and 0 for $E(x)$

$= \dfrac{3}{5}$ Variance

$\sigma = \sqrt{\sigma^2}$

$= \sqrt{\dfrac{3}{5}}$ Standard deviation

9. $f(x) = \dfrac{1}{\ln 5} \cdot \dfrac{1}{x}, \quad [1.5, 7.5]$

$E(x) = \displaystyle\int_a^b x \cdot f(x)\, dx \qquad$ Expected value of x.

$E(x) = \displaystyle\int_{1.5}^{7.5} x \cdot \dfrac{1}{\ln 5} \cdot \dfrac{1}{x}\, dx$

$\quad = \displaystyle\int_{1.5}^{7.5} \dfrac{1}{\ln 5}\, dx$

$\quad = \dfrac{1}{\ln 5}\big[x\big]_{1.5}^{7.5}$

$\quad = \dfrac{1}{\ln 5}\big[7.5 - 1.5\big]$

$\quad = \dfrac{6}{\ln 5}$

$E(x^2) = \displaystyle\int_a^b x^2 \cdot f(x)\, dx \qquad$ Expected value of x^2.

$E(x^2) = \displaystyle\int_{1.5}^{7.5} x^2 \cdot \dfrac{1}{\ln 5} \cdot \dfrac{1}{x}\, dx$

$\quad = \displaystyle\int_{1.5}^{7.5} \dfrac{1}{\ln 5} x\, dx$

$\quad = \dfrac{1}{\ln 5}\left[\dfrac{x^2}{2}\right]_{1.5}^{7.5}$

$\quad = \dfrac{1}{\ln 5}\left[\dfrac{(7.5)^2}{2} - \dfrac{(1.5)^2}{2}\right]$

$\quad = \dfrac{1}{\ln 5}\left[\dfrac{56.25}{2} - \dfrac{2.25}{2}\right]$

$\quad = \dfrac{1}{\ln 5}\left[\dfrac{54}{2}\right]$

$\quad = \dfrac{27}{\ln 5}$

$\mu = E(x) = \dfrac{6}{\ln 5} \qquad$ Mean

$\sigma^2 = E(x^2) - \big[E(x)\big]^2$

$\quad = \dfrac{27}{\ln 5} - \left[\dfrac{6}{\ln 5}\right]^2 \qquad$ Substituting $27/\ln 5$ for $E(x^2)$ and $6/\ln 5$ for $E(x)$

$\quad = \dfrac{27}{\ln 5} - \dfrac{36}{(\ln 5)^2}$

$\quad = \dfrac{27\ln 5}{(\ln 5)^2} - \dfrac{36}{(\ln 5)^2}$

$\quad = \dfrac{27\ln 5 - 36}{(\ln 5)^2} \qquad$ Variance

$\sigma = \sqrt{\sigma^2}$

$\quad = \sqrt{\dfrac{27\ln 5 - 36}{(\ln 5)^2}}$

$\quad = \dfrac{\sqrt{27\ln 5 - 36}}{\ln 5} \qquad$ Standard deviation

11.

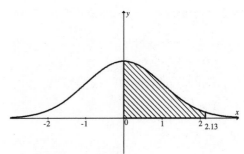

Using Table 2, we have
$P(0 \le x \le 2.13) = 0.4834$

13.

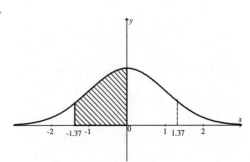

$P(-1.37 \le x \le 0)$
$= P(0 \le x \le 1.37) \qquad$ Symmetry of the graph.
$= 0.4147 \qquad$ Using Table 2.

15. Using Table 2, we have:

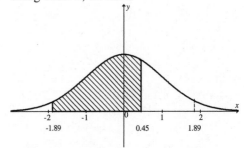

$P(-1.89 \le x \le 0.45)$
$= P(-1.89 \le x \le 0) + P(0 \le x \le 0.45)$
$= P(0 \le x \le 1.89) + P(0 \le x \le 0.45)$
$\qquad\qquad$ Symmetry of the graph
$= 0.4706 + 0.1736 \qquad$ Using Table 2
$= 0.6442$

17. Using Table 2, we have:

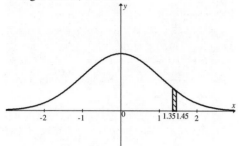

$$P(1.35 \le x \le 1.45)$$
$$= P(0 \le x \le 1.45) - P(0 \le x \le 1.35)$$
$$= 0.4265 - 0.4115$$
$$= 0.0150$$

19. Using Table 2, we have:

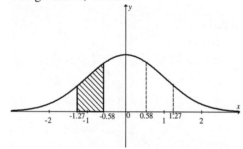

$$P(-1.27 \le x \le -0.58)$$
$$= P(0.58 \le x \le 1.27) \quad \text{Symmetry of the graph.}$$
$$= P(0 \le x \le 1.27) - P(0 \le x \le 0.58)$$
$$= 0.3980 - 0.2190$$
$$= 0.1790$$

21.

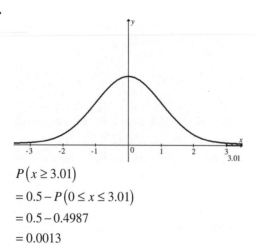

$$P(x \ge 3.01)$$
$$= 0.5 - P(0 \le x \le 3.01)$$
$$= 0.5 - 0.4987$$
$$= 0.0013$$

23. a)

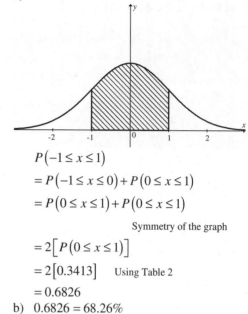

$$P(-1 \le x \le 1)$$
$$= P(-1 \le x \le 0) + P(0 \le x \le 1)$$
$$= P(0 \le x \le 1) + P(0 \le x \le 1)$$
$$\text{Symmetry of the graph}$$
$$= 2\big[P(0 \le x \le 1)\big]$$
$$= 2\big[0.3413\big] \quad \text{Using Table 2}$$
$$= 0.6826$$

b) $0.6826 = 68.26\%$

25. $P(24 \le x \le 30)$

Mean $\mu = 22$

Standard deviation $\sigma = 5$

First, we standardize the numbers 24 and 30.

30 is standardized to $\dfrac{b - \mu}{\sigma} = \dfrac{30 - 22}{5} = \dfrac{8}{5} = 1.6$

24 is standardized to $\dfrac{a - \mu}{\sigma} = \dfrac{24 - 22}{5} = \dfrac{2}{5} = 0.4$

Then,

$$P(24 \le x \le 30)$$
$$= P(0.4 \le z \le 1.6)$$
$$= P(0 \le z \le 1.6) - P(0 \le z \le 0.4)$$
$$= 0.4452 - 0.1554$$
$$= 0.2898$$

27. $P(19 \leq x \leq 25)$

Mean $\mu = 22$

Standard deviation $\sigma = 5$

First, we standardize the numbers 19 and 25.

25 is standardized to $\dfrac{b - \mu}{\sigma} = \dfrac{25 - 22}{5} = \dfrac{3}{5} = 0.6$

19 is standardized to $\dfrac{a - \mu}{\sigma} = \dfrac{19 - 22}{5} = \dfrac{-3}{5} = -0.6$

Then,

$P(19 \leq x \leq 25)$

$= P(-0.6 \leq z \leq 0.6)$

$= P(-0.6 \leq z \leq 0) + P(0 \leq z \leq 0.6)$

$= P(0 \leq z \leq 0.6) + P(0 \leq z \leq 0.6)$

$\qquad\qquad$ Symmetry of the graph

$= 2\big[P(0 \leq z \leq 0.6)\big]$

$= 2[0.2257]$

$= 0.4514$

29 – 45. Check answers using solutions to Exercises 11-28.

47. We first standardize 300 orders. The mean is 250 and the standard deviation is 20, so 300 is standardized to

$\dfrac{b - \mu}{\sigma} = \dfrac{300 - 250}{20} = \dfrac{50}{20} = \dfrac{5}{2} = 2.5$

Therefore,

$P(x \geq 300)$

$= P(z \geq 2.5)$

$= 0.5 - P(z \leq 2.5)$

$= 0.5 - 0.4938 \qquad$ Using Table 2

$= 0.0062$

$= 0.62\%$

The company will have to hire extra help or pay overtime 0.62% of the days.

49. We first standardize 40 seconds. The mean is 38.6 and the standard deviation is 1.729, so 40 is standardized to

$\dfrac{b - \mu}{\sigma} = \dfrac{40 - 38.6}{1.729} \approx 0.81$

Therefore,

$P(x \leq 40)$

$= P(z \leq 0.81)$

$= 0.5 + P(0 \leq z \leq 0.81)$

$= 0.5 + 0.2910 \qquad$ Using Table 2

$= 0.7910$

The probability that the next operation of the robogate will take 40 second, or less, is 0.7910.

51. We first standardize the scores 80 and 89. The mean is 65 and the standard deviation is 20.

89 is standardized to

$\dfrac{b - \mu}{\sigma} = \dfrac{89 - 65}{20} = \dfrac{24}{20} = 1.2$

80 is standardized to

$\dfrac{a - \mu}{\sigma} = \dfrac{80 - 65}{20} = \dfrac{15}{20} = 0.75$

Therefore,

$P(80 \leq x \leq 89)$

$= P(0.75 \leq z \leq 1.2)$

$= P(0 \leq z \leq 1.2) - P(0 \leq z \leq 0.75)$

$= 0.3849 - 0.2734 \qquad$ Using Table 2

$= 0.1115$

The probability of making a B on the biology exam is 0.1115.

53. a) Find z such that $P(x \leq z) = 0.35$

$0.35 = 0.5 - 0.15$ and 0.15 corresponds to $z \approx 0.39$ in Table 2. Then the score that corresponds to the 35th percentile is 0.39 standard deviations less than the mean, or $1020 - 0.39(140) \approx 965$.

b) Find z such that $P(x \leq z) = 0.60$

$0.60 = 0.5 + 0.1$ and 0.1 corresponds to $z \approx 0.25$ in Table 2. Then the score that corresponds to the 60th percentile is 0.25 standard deviations more than the mean, or $1020 + 0.25(140) \approx 1055$.

c) Find z such that $P(x \leq z) = 0.92$

$0.92 = 0.5 + 0.42$ and 0.42 corresponds to $z \approx 1.41$ in Table 2. Then the score that corresponds to the 35th percentile is 1.41 standard deviations less more the mean, or $1020 + 1.41(140) \approx 1217$.

55. $f(x) = \dfrac{1}{b-a}, \quad [a,b]$

$E(x) = \displaystyle\int_a^b x \cdot f(x)\, dx$ Expected value of x.

$E(x) = \displaystyle\int_a^b x \cdot \dfrac{1}{b-a}\, dx$

$\quad = \dfrac{1}{b-a}\left[\dfrac{x^2}{2}\right]_a^b$

$\quad = \dfrac{1}{b-a}\left[\dfrac{b^2}{2} - \dfrac{a^2}{2}\right]$

$\quad = \dfrac{1}{b-a}\left[\dfrac{b^2 - a^2}{2}\right]$

$\quad = \dfrac{1}{b-a} \cdot \dfrac{(b-a)(b+a)}{2}$

$\quad = \dfrac{b+a}{2}$

$E(x^2) = \displaystyle\int_a^b x^2 \cdot f(x)\, dx$ Expected value of x^2.

$E(x^2) = \displaystyle\int_a^b x^2 \cdot \dfrac{1}{b-a}\, dx$

$\quad = \dfrac{1}{b-a}\left[\dfrac{x^3}{3}\right]_a^b$

$\quad = \dfrac{1}{b-a}\left[\dfrac{b^3}{3} - \dfrac{a^3}{3}\right]$

$\quad = \dfrac{1}{b-a}\left[\dfrac{b^3 - a^3}{3}\right]$

$\quad = \dfrac{1}{b-a} \cdot \dfrac{(b-a)(b^2 + ab + a^2)}{3}$

$\quad = \dfrac{b^2 + ab + a^2}{3}$

$\mu = E(x) = \dfrac{b+a}{2}$ Mean

$\sigma^2 = E(x^2) - \left[E(x)\right]^2$

$\quad = \dfrac{b^2 + ab + a^2}{3} - \left[\dfrac{b+a}{2}\right]^2$ Substituting

$\quad = \dfrac{b^2 + ab + a^2}{3} - \left[\dfrac{b^2 + 2ba + a^2}{4}\right]$

$\quad = \dfrac{4b^2 + 4ab + 4a^2}{12} - \dfrac{3b^2 + 6ba + 3a^2}{12}$

$\quad = \dfrac{4b^2 + 4ab + 4a^2 - 3b^2 - 6ba - 3a^2}{12}$

$\quad = \dfrac{b^2 - 2ba + a^2}{12}$

$\quad = \dfrac{(b-a)^2}{12}$ Variance

$\sigma = \sqrt{\sigma^2}$

$\quad = \sqrt{\dfrac{(b-a)^2}{12}}$

$\quad = \dfrac{b-a}{2\sqrt{3}}$ Standard deviation

57. $f(x) = \dfrac{1}{2}x, \quad [0,2]$

$\displaystyle\int_0^m f(x)\, dx = \dfrac{1}{2}$

$\displaystyle\int_0^m \dfrac{1}{2}x\, dx = \dfrac{1}{2}$

$\dfrac{1}{2}\displaystyle\int_0^m x\, dx = \dfrac{1}{2}$

$\displaystyle\int_0^m x\, dx = 1$

$\left[\dfrac{x^2}{2}\right]_0^m = 1$

$\dfrac{m^2}{2} - \dfrac{0^2}{2} = 1$

$\dfrac{m^2}{2} = 1$

$m^2 = 2$

$m = \sqrt{2}$

59. $f(x) = ke^{-kx}$, $[0,\infty)$

$$\int_0^m f(x)\,dx = \frac{1}{2}$$

$$\int_0^m ke^{-kx}\,dx = \frac{1}{2}$$

$$\left[\frac{k}{-k}e^{-kx}\right]_0^m = \frac{1}{2}$$

$$\left[-e^{-kx}\right]_0^m = \frac{1}{2}$$

$$-e^{-k(m)} - \left(-e^{-k(0)}\right) = \frac{1}{2}$$

$$-e^{-k\cdot m} + e^0 = \frac{1}{2}$$

$$-e^{-k\cdot m} + 1 = \frac{1}{2}$$

$$\frac{1}{2} = e^{-k\cdot m}$$

$$\ln\left(\frac{1}{2}\right) = \ln\left(e^{-k\cdot m}\right)$$

$$\ln(1) - \ln(2) = -k\cdot m \qquad [\ln 1 = 0]$$

$$\frac{-\ln 2}{-k} = m$$

$$\frac{\ln 2}{k} = m$$

61. Standardize 8.5.

$$z = \frac{8.5 - \mu}{0.3}$$

We are looking for a value of c for which

$P(z > c) = \dfrac{1}{100} = 0.01$. Now, since we are

guarding against overflow we will have $c \geq 0$.
We have:

$$P(z > c) = P(z \geq 0) - P(0 \leq z \leq c)$$
$$= 0.5 - P(0 \leq z \leq c)$$

Then,

$$P(0 \leq z \leq c) = 0.5 - P(z > c)$$
$$= 0.5 - 0.01$$
$$= 0.4900$$

Looking in Table 2 for the number closest to 0.49, we find that 0.4901 and 0.4898 are the two closest numbers. Since we want to ensure that only 1 cup in 100 will overflow we select the larger value 0.4901. This value corresponds to $c = 2.33$.

We set $z = 2.33$ and solve for μ.

$$2.33 = \frac{8.5 - \mu}{0.3}$$

$$(2.33)(0.3) = 8.5 - \mu$$

$$0.699 = 8.5 - \mu$$

$$\mu + 0.699 = 8.5$$

$$\mu = 8.5 - 0.699$$

$$\mu = 7.801$$

The mean volume should be adjusted to 7.801 oz to ensure that only 1 cup in 100 will overflow.

63. $\boxed{tw}$

Exercise Set 5.6

1. Find the volume of the solid of revolution generated by rotating about the x-axis the region under the graph of
$$y = x$$
from $x = 0$ to $x = 1$

$$V = \int_a^b \pi[f(x)]^2\,dx \quad \text{Volume of a solid of revolution}$$

$$V = \int_0^1 \pi[x]^2\,dx \quad \text{Substituting 0 for } a, 1 \text{ for } b, \text{ and } x \text{ for } f(x).$$

$$V = \int_0^1 \pi x^2\,dx$$

$$= \left[\pi \cdot \frac{x^3}{3}\right]_0^1$$

$$= \frac{\pi}{3}\left[1^3 - 0^3\right]$$

$$= \frac{\pi}{3}[1]$$

$$= \frac{\pi}{3}, \text{ or about } 1.05$$

3. Find the volume of the solid of revolution generated by rotating about the x-axis the region under the graph of

$$y = \sqrt{x}$$

from $x = 1$ to $x = 4$

$V = \int_a^b \pi \left[f(x) \right]^2 dx$ Volume of a solid of revolution

$V = \int_1^4 \pi \left[\sqrt{x} \right]^2 dx$ Substituting 1 for a, 4 for b, and $\sqrt{x}$ for $f(x)$.

$V = \int_1^4 \pi x \, dx$

$= \left[\pi \cdot \dfrac{x^2}{2} \right]_1^4$

$= \dfrac{\pi}{2} \left[4^2 - 1^2 \right]$

$= \dfrac{\pi}{2} [15]$

$= \dfrac{15\pi}{2}$, or about 23.56

5. Find the volume of the solid of revolution generated by rotating about the x-axis the region under the graph of

$$y = e^x$$

from $x = -2$ to $x = 5$

$V = \int_a^b \pi \left[f(x) \right]^2 dx$ Volume of a solid of revolution

$V = \int_{-2}^5 \pi \left[e^x \right]^2 dx$ Substituting -2 for a, 5 for b, and e^x for $f(x)$.

$V = \int_{-2}^5 \pi e^{2x} dx$

$= \left[\pi \cdot \dfrac{1}{2} e^{2x} \right]_{-2}^5$

$= \dfrac{\pi}{2} \left[e^{2(5)} - e^{2(-2)} \right]$

$= \dfrac{\pi}{2} \left[e^{10} - e^{-4} \right]$, or about 34,599.06

7. Find the volume of the solid of revolution generated by rotating about the x-axis the region under the graph of

$$y = \dfrac{1}{x}$$

from $x = 1$ to $x = 3$

$V = \int_a^b \pi \left[f(x) \right]^2 dx$ Volume of a solid of revolution

$V = \int_1^3 \pi \left[\dfrac{1}{x} \right]^2 dx$ Substituting 1 for a, 3 for b, and $1/x$ for $f(x)$.

$V = \int_1^3 \pi \cdot \dfrac{1}{x^2} dx$

$V = \int_1^3 \pi x^{-2} dx$

$= \left[\pi \cdot \dfrac{x^{-1}}{-1} \right]_1^3$

$= -\pi \left[\dfrac{1}{x} \right]_1^3$

$= -\pi \left[\dfrac{1}{3} - \dfrac{1}{1} \right]$

$= -\pi \left[-\dfrac{2}{3} \right]$

$= \dfrac{2\pi}{3}$, or about 2.09

9. Find the volume of the solid of revolution generated by rotating about the x-axis the region under the graph of

$$y = \dfrac{2}{\sqrt{x}}$$

from $x = 4$ to $x = 9$

$V = \int_a^b \pi \left[f(x) \right]^2 dx$ Volume of a solid of revolution

$V = \int_4^9 \pi \left[\dfrac{2}{\sqrt{x}} \right]^2 dx$ Substituting 4 for a, 9 for b, and $2/\sqrt{x}$ for $f(x)$.

$V = \int_4^9 \pi \cdot \dfrac{4}{x} dx$

$V = 4\pi \int_4^9 \dfrac{1}{x} dx$

$= 4\pi \left[\ln x \right]_4^9$

$= 4\pi \left[\ln 9 - \ln 4 \right]$

$= 4\pi \ln \left(\dfrac{9}{4} \right)$, or about 10.19

11. Find the volume of the solid of revolution generated by rotating about the x-axis the region under the graph of

$$y = 4$$

from $x = 1$ to $x = 3$

$$V = \int_a^b \pi \left[f(x) \right]^2 dx \qquad \text{Volume of a solid of revolution}$$

$$V = \int_1^3 \pi \left[4 \right]^2 dx \qquad \text{Substituting 1 for } a, 3 \text{ for } b, \text{ and 4 for } f(x).$$

$$V = \int_1^3 16\pi dx$$

$$= 16\pi \left[x \right]_1^3$$

$$= 16\pi \left[3 - 1 \right]$$

$$= 16\pi \left[2 \right]$$

$$= 32\pi, \text{ or about } 100.53$$

13. Find the volume of the solid of revolution generated by rotating about the x-axis the region under the graph of

$$y = x^2$$

from $x = 0$ to $x = 2$

$$V = \int_a^b \pi \left[f(x) \right]^2 dx \qquad \text{Volume of a solid of revolution}$$

$$V = \int_0^2 \pi \left[x^2 \right]^2 dx \qquad \text{Substituting 0 for } a, 2 \text{ for } b, \text{ and } x^2 \text{ for } f(x).$$

$$V = \int_0^2 \pi x^4 dx$$

$$= \left[\pi \cdot \frac{x^5}{5} \right]_0^2$$

$$= \frac{\pi}{5} \left[2^5 - 0^5 \right]$$

$$= \frac{\pi}{5} \left[32 \right]$$

$$= \frac{32\pi}{5}, \text{ or about } 20.11$$

15. Find the volume of the solid of revolution generated by rotating about the x-axis the region under the graph of

$$y = \sqrt{1 + x}$$

from $x = 2$ to $x = 10$

$$V = \int_a^b \pi \left[f(x) \right]^2 dx \qquad \text{Volume of a solid of revolution}$$

$$V = \int_2^{10} \pi \left[\sqrt{1 + x} \right]^2 dx \qquad \text{Substituting 2 for } a, 10 \text{ for } b, \text{ and } \sqrt{1+x} \text{ for } f(x).$$

$$V = \int_2^{10} \pi (1 + x) dx$$

$$= \pi \left[\frac{(1 + x)^2}{2} \right]_2^{10}$$

$$= \frac{\pi}{2} \left[(1 + 10)^2 - (1 + 2)^2 \right]$$

$$= \frac{\pi}{2} \left[(11)^2 - (3)^2 \right]$$

$$= \frac{\pi}{2} \left[121 - 9 \right]$$

$$= \frac{\pi}{2} \cdot 112$$

$$= 56\pi, \text{ or about } 175.93$$

17. Find the volume of the solid of revolution generated by rotating about the x-axis the region under the graph of

$$y = \sqrt{4 - x^2}$$

from $x = -2$ to $x = 2$

$$V = \int_a^b \pi \left[f(x) \right]^2 dx \qquad \text{Volume of a solid of revolution}$$

$$V = \int_{-2}^2 \pi \left[\sqrt{4 - x^2} \right]^2 dx \qquad \text{Substituting } -2 \text{ for } a, 2 \text{ for } b, \text{ and } \sqrt{4-x^2} \text{ for } f(x).$$

$$V = \int_{-2}^2 \pi (4 - x^2) dx$$

$$= \pi \left[4x - \frac{x^3}{3} \right]_{-2}^2$$

$$= \pi \left[\left(4(2) - \frac{(2)^3}{3} \right) - \left(4(-2) - \frac{(-2)^3}{3} \right) \right]$$

$$= \pi \left[\left(8 - \frac{8}{3} \right) - \left(-8 + \frac{8}{3} \right) \right]$$

$$= \pi \left[\frac{16}{3} - \left(-\frac{16}{3} \right) \right]$$

$$= \pi \left(\frac{32}{3} \right)$$

$$= \frac{32\pi}{3}, \text{ or about } 33.51$$

19. Graphing the equations, we have

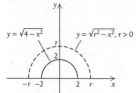

The graphs are semicircles. Their rotation about the x-axis creates spheres of radius 2 and radius r respectively. In Exercise 18, by finding the volume of the solid of revolution created by rotating $y = \sqrt{r^2 - x^2}$, we actually derived the general formula for finding the volume of a sphere with radius r.

21. Find the volume of the solid of revolution generated by rotating about the x-axis the region under the graph of

$$y = \sqrt{xe^{-x}}$$

from $x = 1$ to $x = 2$.

$$V = \int_a^b \pi \left[f(x) \right]^2 dx \qquad \text{Volume of a solid of revolution}$$

$$V = \int_1^2 \pi \left[\sqrt{xe^{-x}} \right]^2 dx \qquad \text{Substituting 1 for } a, \text{ 2 for } b, \text{ and } \sqrt{xe^{-x}} \text{ for } f(x).$$

$$V = \int_1^2 \pi xe^{-x} dx$$

$$= \pi \left[\frac{1}{(-1)^2} e^{-x} (-x - 1) \right]_1^2 \qquad \text{Using formula 6}$$

$$= \pi \left[\frac{-x-1}{e^x} \right]_1^2$$

$$= \pi \left[\left(\frac{-2-1}{e^2} \right) - \left(\frac{-1-1}{e^1} \right) \right]$$

$$= \pi \left[\left(\frac{-3}{e^2} \right) - \left(\frac{-2}{e} \right) \right]$$

$$= \pi \left[\frac{-3}{e^2} + \frac{2}{e} \right]$$

$$= \pi \left(\frac{-3 + 2e}{e^2} \right), \text{ or about } 1.04$$

23. Using the fnInt feature on a calculator with successively larger values of the upper limit, we find that the integral diverges.

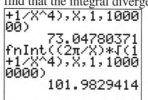

Therefore, the surface area of Gabriel's horn does not exist.

Exercise Set 5.7

1. Solve: $y' = 5x^4$

$$y = \int 5x^4 dx$$

$$= 5 \left(\frac{x^5}{5} \right) + C$$

$$= x^5 + C$$

This solution is called a general solution because taking all values of C gives all the solutions. Taking specific values of C gives particular solutions. The following are particular solutions of $y' = 5x^4$. Answers may vary depending on the choice of C.

$$y = x^5$$

$$y = x^5 - 1$$

$$y = x^5 + \pi$$

3. Solve: $y' = e^{2x} + x$

$$y = \int \left(e^{2x} + x \right) dx$$

$$= \frac{1}{2} e^{2x} + \frac{1}{2} x^2 + C$$

This solution is called a general solution because taking all values of C gives all the solutions. Taking specific values of C gives particular solutions. The following are particular solutions of $y' = e^{2x} + x$. Answers may vary depending on the choice of C.

$$y = \frac{1}{2} e^{2x} + \frac{1}{2} x^2$$

$$y = \frac{1}{2} e^{2x} + \frac{1}{2} x^2 - 3$$

$$y = \frac{1}{2} e^{2x} + \frac{1}{2} x^2 + 3$$

5. Solve: $y' = \dfrac{8}{x} - x^2 + x^5$

$$y = \int\left(\frac{8}{x} - x^2 + x^5\right)dx$$

$$= 8\ln x - \frac{1}{3}x^3 + \frac{1}{6}x^6 + C$$

This solution is called a general solution because taking all values of C gives all the solutions. Taking specific values of C gives particular solutions. The following are particular solutions of $y' = \dfrac{8}{x} - x^2 + x^5$. Answers may vary depending on the choice of C.

$$y = 8\ln x - \frac{1}{3}x^3 + \frac{1}{6}x^6$$

$$y = 8\ln x - \frac{1}{3}x^3 + \frac{1}{6}x^6 + 5$$

$$y = 8\ln x - \frac{1}{3}x^3 + \frac{1}{6}x^6 - 17$$

7. Solve $y' = x^2 + 2x - 3$, given that $y = 4$ when $x = 0$.

First, find the general solution.

$$y = \int y'\,dx$$

$$= \int\left(x^2 + 2x - 3\right)dx$$

$$y = \frac{1}{3}x^3 + x^2 - 3x + C$$

We substitute $y = 4$ and $x = 0$ into the general solution to find C.

$$4 = \frac{1}{3}(0)^3 + (0)^2 - 3(0) + C$$

$$4 = C$$

Thus the particular solution determined by the given condition is

$$y = \frac{1}{3}x^3 + x^2 - 3x + 4.$$

9. Solve $f'(x) = x^{\frac{2}{3}} - x$, given $f(1) = -6$.

First, find the general solution.

$$f(x) = \int f'(x)\,dx$$

$$= \int\left(x^{\frac{2}{3}} - x\right)dx$$

$$f(x) = \frac{3}{5}x^{\frac{5}{3}} - \frac{1}{2}x^2 + C$$

We substitute -6 for $f(x)$ and 1 for x into the general solution to find C.

$$-6 = \frac{3}{5}(1)^{\frac{5}{3}} - \frac{1}{2}(1)^2 + C$$

$$-6 = \frac{3}{5} - \frac{1}{2} + C$$

$$-6 - \frac{3}{5} + \frac{1}{2} = C$$

$$\frac{-60}{10} - \frac{6}{10} + \frac{5}{10} = C$$

$$-\frac{61}{10} = C$$

Thus the particular solution determined by the given condition is

$$f(x) = \frac{3}{5}x^{\frac{5}{3}} - \frac{1}{2}x^2 - \frac{61}{10}.$$

11. Show that $y = x\ln x + 3x - 2$ is a solution to

$$y'' - \frac{1}{x} = 0.$$

First find y' and y''.

$$y = x\ln x + 3x - 2$$

$$y' = x\cdot\frac{1}{x} + 1\cdot\ln x + 3$$

$$= \ln x + 4$$

$$y'' = \frac{1}{x}$$

Then we substitute into the differential equation, as follows

$$y'' - \frac{1}{x} = 0$$

$$\begin{array}{c|c} \dfrac{1}{x} - \dfrac{1}{x} & 0 \\ \hline 0 & 0 \end{array}$$

Since a true equation $0 = 0$ results, we know that $y = x\ln x + 3x - 2$ is a solution of the differential equation.

13. Show that $y = e^x + 3xe^x$ is a solution to $y'' - 2y' + y = 0$.

First find y' and y''.

$$y = e^x + 3xe^x$$

$$y' = e^x + 3(xe^x + e^x)$$

$$= e^x + 3xe^x + 3e^x$$

$$= 4e^x + 3xe^x$$

$$y'' = 4e^x + 3(xe^x + e^x)$$

$$= 4e^x + 3xe^x + 3e^x$$

$$= 7e^x + 3xe^x$$

Then we substitute into the differential equation, as follows

$$y'' - 2y' + y = 0$$

$(7e^x + 3xe^x) - 2(4e^x + 3xe^x) + (e^x + 3xe^x)$	0
$7e^x + 3xe^x - 8e^x - 6xe^x + e^x + 3xe^x$	0
$(7 - 8 + 1)e^x + (3 - 6 + 3)xe^x$	0
$0 + 0$	0
0	0

Since a true equation $0 = 0$ results, we know that $y = e^x + 3xe^x$ is a solution of the differential equation.

15. Solve $\dfrac{dy}{dx} = 4x^3 y$

First, we separate the variables, as follows:

$$dy = 4x^3 y\, dx \qquad \text{Multiplying by } dx.$$

$$\frac{dy}{y} = 4x^3\, dx \qquad \text{Dividing by } y.$$

We then integrate both sides:

$$\int \frac{dy}{y} = \int 4x^3\, dx$$

$$\ln|y| = x^4 + C$$

$$|y| = e^{x^4 + C} \qquad \text{Definition of logarithms.}$$

$$y = \pm e^{x^4 + C}$$

$$y = \pm e^{x^4} \cdot e^C \qquad \text{Properties of exponents.}$$

$$y = C_1 e^{x^4} \qquad \left[C_1 = \pm e^C \right]$$

Thus, the general solution to the differential equation is $y = C_1 e^{x^4}$, where $C_1 = \pm e^C$.

17. Solve $3y^2 \dfrac{dy}{dx} = 8x$

First, we separate the variables, as follows:

$$3y^2\, dy = 8x\, dx \qquad \text{Multiplying by } dx.$$

We then integrate both sides:

$$\int 3y^2\, dy = \int 8x\, dx$$

$$y^3 = 4x^2 + C$$

$$y = \sqrt[3]{4x^2 + C} \qquad \begin{array}{l}\text{Taking the cubed root}\\ \text{of both sides.}\end{array}$$

Thus, the general solution to the differential equation is $y = \sqrt[3]{4x^2 + C}$.

19. Solve $\dfrac{dy}{dx} = \dfrac{2x}{y}$

First, we separate the variables, as follows:

$$dy = \frac{2x}{y}\, dx \qquad \text{Multiplying by } dx.$$

$$y\, dy = 2x\, dx \qquad \text{Multiplying by } y.$$

We then integrate both sides:

$$\int y\, dy = \int 2x\, dx$$

$$\frac{1}{2} y^2 = x^2 + C$$

$$y^2 = 2x^2 + 2C$$

$$y = \pm\sqrt{2x^2 + C_1}, \qquad \left[C_1 = 2C \right]$$

Thus, the general solutions to the differential equation are

$$y = \sqrt{2x^2 + C_1} \text{ and } y = -\sqrt{2x^2 + C_1}, \text{ where}$$

$$C_1 = 2C.$$

21. Solve $\dfrac{dy}{dx} = \dfrac{6}{y}$

First, we separate the variables, as follows:

$$dy = \frac{6}{y}\, dx \qquad \text{Multiplying by } dx.$$

$$y\, dy = 6\, dx \qquad \text{Multiplying by } y.$$

We then integrate both sides:

$$\int y\, dy = \int 6\, dx$$

$$\frac{1}{2} y^2 = 6x + C$$

$$y^2 = 12x + 2C$$

$$y = \pm\sqrt{12x + C_1}, \qquad \left[C_1 = 2C \right]$$

Thus, the general solutions to the differential equation are

$$y = \sqrt{12x + C_1} \text{ and } y = -\sqrt{12x + C_1}, \text{ where}$$

$$C_1 = 2C.$$

23. Solve $y' = 3x + xy;\ y = 5$, when $x = 0$.

First, we separate the variables, as follows:

$\dfrac{dy}{dx} = 3x + xy \qquad$ Replacing y' with $\dfrac{dy}{dx}$.

$\dfrac{dy}{dx} = x(3 + y) \qquad$ Factoring.

$dy = x(3 + y)\,dx \qquad$ Multiplying by dx.

$\dfrac{dy}{3 + y} = x\,dx \qquad$ Dividing by $(3 + y)$.

We then integrate both sides:

$\displaystyle\int \dfrac{dy}{3 + y} = \int x\,dx$

$\ln(3 + y) = \dfrac{1}{2}x^2 + C \qquad [3 + y > 0]$

$3 + y = e^{\frac{x^2}{2} + C}$

$3 + y = e^{\frac{x^2}{2}} \cdot e^C$

$y = e^{\frac{x^2}{2}} \cdot e^C - 3$

$y = C_1 e^{\frac{x^2}{2}} - 3 \qquad \left[C_1 = e^C\right]$

Thus, the general solution to the differential

equation is $y = C_1 e^{\frac{x^2}{2}} - 3$ where $C_1 = e^C$.

We substitute 0 for x and 5 for y to find C_1.

$5 = C_1 e^{\frac{0^2}{2}} - 3$

$8 = C_1 e^0$

$8 = C_1 \cdot 1$

$8 = C_1$

Therefore, the particular solution is

$y = 8e^{\frac{x^2}{2}} - 3$.

25. Solve $y' = 5y^{-2};\ y = 3$, when $x = 2$.

First, we separate the variables, as follows:

$\dfrac{dy}{dx} = 5y^{-2} \qquad$ Replacing y' with $\dfrac{dy}{dx}$.

$dy = 5y^{-2}\,dx \qquad$ Multiplying by dx.

$y^2\,dy = 5\,dx \qquad$ Multiplying by y^2.

We then integrate both sides:

$\displaystyle\int y^2\,dy = \int 5\,dx$

$\dfrac{1}{3}y^3 = 5x + C$

$y^3 = 15x + 3C$

$y = \sqrt[3]{15x + C_1} \qquad [C_1 = 3C]$

Thus, the general solution to the differential

equation is $y = \sqrt[3]{15x + C_1}$ where $C_1 = 3C$.

We substitute 2 for x and 3 for y to find C_1.

$3 = \sqrt[3]{15(2) + C_1}$

$3 = \sqrt[3]{30 + C_1}$

$27 = 30 + C_1$

$-3 = C_1$

Therefore, the particular solution is

$y = \sqrt[3]{15x - 3}$.

27. Solve $\dfrac{dy}{dx} = 3y$

First, we separate the variables, as follows:

$dy = 3y\,dx \qquad$ Multiplying by dx.

$\dfrac{dy}{y} = 3\,dx \qquad$ Dividing by y.

We then integrate both sides:

$\displaystyle\int \dfrac{dy}{y} = \int 3\,dx$

$\ln|y| = 3x + C$

$|y| = e^{3x + C}$

$y = \pm e^{3x + C}$

$y = \pm e^{3x} \cdot e^C$

$y = C_1 e^{3x} \qquad \left[C_1 = \pm e^C\right]$

Thus, the general solution to the differential

equation is $y = C_1 e^{3x}$, where $C_1 = \pm e^C$.

29. Solve $\dfrac{dP}{dt} = 2P$

First, we separate the variables, as follows:

$dP = 2P\,dt \qquad$ Multiplying by dt.

$\dfrac{dP}{P} = 2\,dt \qquad$ Dividing by P.

We then integrate both sides:

$\displaystyle\int \dfrac{dP}{P} = \int 2\,dt$

$\ln|P| = 2t + C$

$|P| = e^{2t + C}$

$P = \pm e^{2t + C}$

$P = \pm e^{2t} \cdot e^C$

$P = C_1 e^{2t} \qquad \left[C_1 = \pm e^C\right]$

Thus, the general solution to the differential

equation is $P = C_1 e^{2t}$, where $C_1 = \pm e^C$.

31. Solve $f'(x) = \dfrac{1}{x} - 4x + \sqrt{x}$, given that

$f(1) = \dfrac{23}{3}$.

Integrating, we have:

$f(x) = \int f'(x)\,dx$

$\quad = \int \left(\dfrac{1}{x} - 4x + \sqrt{x} \right) dx$

$\quad = \ln x - 2x^2 + \dfrac{2}{3} x^{\frac{3}{2}} + C$

We substitute 1 for x and $\dfrac{23}{3}$ for $f(1)$ to find C.

$\dfrac{23}{3} = \ln(1) - 2(1)^2 + \dfrac{2}{3}(1)^{\frac{3}{2}} + C$

$\dfrac{23}{3} = 0 - 2 + \dfrac{2}{3} + C$

$\dfrac{23}{3} = -\dfrac{6}{3} + \dfrac{2}{3} + C$

$\dfrac{23}{3} = -\dfrac{4}{3} + C$

$\dfrac{23}{3} + \dfrac{4}{3} = C$

$\dfrac{27}{3} = C$

$9 = C$

Therefore, the particular solution is

$f(x) = \ln x - 2x^2 + \dfrac{2}{3} x^{\frac{3}{2}} + 9.$

33. $C'(x) = 2.6 - 0.02x$

$C(x) = \int C'(x)\,dx$

$C(x) = \int (2.6 - 0.02x)\,dx$

$\quad = 2.6x - 0.01x^2 + K$

We use K as the constant of integration so we do not confuse it with total cost.
Next, we substitute to find K.

$C(0) = 120$

$2.6(0) - 0.01(0)^2 + K = 120$

$K = 120$

Thus, the total cost function is

$C(x) = 2.6x - 0.01x^2 + 120.$

The average cost function is given by:

$A(x) = \dfrac{C(x)}{x}$

$\quad = \dfrac{2.6x - 0.01x^2 + 120}{x}$

$\quad = 2.6 - 0.01x + \dfrac{120}{x}.$

35. $\dfrac{dP}{dC} = \dfrac{-200}{(C+3)^{\frac{3}{2}}}$

a) Integrating, we have

$P(C) = \int \dfrac{-200}{(C+3)^{\frac{3}{2}}}\,dC$

$\quad = -200 \int (C+3)^{-\frac{3}{2}}\,dC$

$\quad = -200 \left[\dfrac{(C+3)^{-\frac{1}{2}}}{-\frac{1}{2}} + K \right]$

$\quad = \dfrac{400}{(C+3)^{\frac{1}{2}}} + K_1 \qquad [K_1 = -200K]$

$\quad = \dfrac{400}{\sqrt{C+3}} + K_1$

We substitute 10 for P and 61 for C to find K_1.

$10 = \dfrac{400}{\sqrt{61+3}} + K_1$

$10 = \dfrac{400}{8} + K_1$

$10 = 50 + K_1$

$-40 = K_1$

Thus, the particular solution, given the conditions is $P(C) = \dfrac{400}{\sqrt{C+3}} - 40.$

b) The firm will break even when $P = 0$.
Substituting, we have:

$0 = \dfrac{400}{\sqrt{C+3}} - 40$

$40 = \dfrac{400}{\sqrt{C+3}}$

$40\sqrt{C+3} = 400$

$\sqrt{C+3} = 10$

$C + 3 = 100$

$C = 97$

When total cost is \$97, the firm will break even.

37. a) $\dfrac{dR}{dS} = \dfrac{k}{S+1}$

First, we separate the variables, as follows:

$dR = \dfrac{k}{S+1}\,dS$ Multiplying by dS.

We then integrate both sides:

$\displaystyle\int dR = \int \dfrac{k}{S+1}\,dS$

$R = k\ln(S+1) + C$ $[S+1>0]$

The general solution is $R = k\ln(S+1) + C$.

b) We substitute the initial conditions to find C.

$R = k\ln(S+1) + C$

$0 = k\ln(0+1) + C$

$0 = k\ln(1) + C$

$0 = k\cdot 0 + C$

$0 = C$

Thus the particular solution given the initial conditions is $R = k\ln(S+1)$.

c) $\boxed{\text{tw}}$

39. $E(x) = \dfrac{x}{200-x}$; $q = 190$ when $x = 10$

Elasticity is given by

$E(x) = -\dfrac{x}{q}\cdot\dfrac{dq}{dx}$

Therefore,

$-\dfrac{x}{q}\cdot\dfrac{dq}{dx} = \dfrac{x}{200-x}.$

Solving the differential equation by separation of variables, we have:

$-\dfrac{x}{q}\cdot\dfrac{dq}{dx} = \dfrac{x}{200-x}$

$\dfrac{dq}{q} = -\dfrac{1}{200-x}\,dx$

$\displaystyle\int \dfrac{dq}{q} = -\int \dfrac{1}{200-x}\,dx$

$\ln q = \ln(200-x) + C$ $\left[\begin{smallmatrix} q>0,\\ 200-x>0 \end{smallmatrix}\right]$

$q = e^{\ln(200-x)+C}$

$q = e^{\ln(200-x)}\cdot e^{C}$

$q = C_1(200-x)$ $\left[C_1 = e^{C}\right]$

Substitute to find C_1.

$190 = C(200-10)$

$190 = C(190)$

$1 = C$

Therefore, the demand function is

$q = 200 - x.$

41. $E(x) = n$; for some constant n and all $x > 0$.

Elasticity is given by

$E(x) = -\dfrac{x}{q}\cdot\dfrac{dq}{dx}$

Therefore,

$-\dfrac{x}{q}\cdot\dfrac{dq}{dx} = n.$

Solving the differential equation by separation of variables, we have:

$-\dfrac{x}{q}\cdot\dfrac{dq}{dx} = n$

$\dfrac{dq}{q} = -\dfrac{n}{x}\,dx$

$\displaystyle\int \dfrac{dq}{q} = -\int \dfrac{n}{x}\,dx$

$\ln q = -n\ln(x) + C$ $[q>0, x>0]$

$\ln q = \ln x^{-n} + C$

$q = e^{\ln(x^{-n})+C}$

$q = e^{\ln x^{-n}}\cdot e^{C}$

$q = C_1 e^{\ln x^{-n}}$ $\left[C_1 = e^{C}\right]$

$q = C_1\cdot x^{-n}$

$q = \dfrac{C_1}{x^{n}}$

Therefore, the demand function is $q = \dfrac{C_1}{x^{n}}.$

43. $\dfrac{dR}{dS} = k \cdot \dfrac{R}{S}$

First, we separate the variables

$\dfrac{dR}{R} = k \cdot \dfrac{dS}{S}$ Multiplying by $\dfrac{dS}{R}$.

Then we integrate both sides.

$\displaystyle \int \dfrac{dR}{R} = \int k \cdot \dfrac{dS}{S}$

$\ln R = k \ln S + C$ $[S, R > 0]$

$\ln R = \ln S^k + C$

$R = e^{\ln S^k + C}$

$R = e^{\ln S^k} \cdot e^C$

$R = C_1 e^{\ln S^k}$ $\left[C_1 = e^C \right]$

$R = C_1 \cdot S^k$

Thus the solution to the differential equation is
$R = C_1 \cdot S^k$ where $C_1 = e^C$.

45. Solve $e^{-1/x} \cdot \dfrac{dy}{dx} = x^{-2} \cdot y^2$

First, we separate the variables, as follows

$\dfrac{dy}{y^2} = \dfrac{e^{1/x}}{x^2} dx.$

We then integrate both sides:

$\displaystyle \int \dfrac{dy}{y^2} = \int x^{-2} e^{1/x} dx$

$-\dfrac{1}{y} = -e^{1/x} + C \quad \left[u = x^{-1}, du = -x^{-2} \right]$

$\dfrac{1}{y} = e^{1/x} - C$

$\dfrac{1}{y} = e^{1/x} - C$

$1 = \left(e^{1/x} - C \right) y$

$\dfrac{1}{e^{1/x} - C_1} = y$

Thus, the general solution to the differential

equation is $y = \dfrac{1}{e^{1/x} + C_1}$, where $C_1 = -C$.

47. $\boxed{tw}$

Chapter 6

Functions of Several Variables

Exercise Set 6.1

1. $f(x,y) = x^2 - 3xy$

$f(0,-2) = (0)^2 - 3(0)(-2)$ Substituting 0 for x and -2 for y.

$= 0 - 0$

$= 0$

$f(2,3) = (2)^2 - 3(2)(3)$ Substituting 2 for x and 3 for y.

$= 4 - 18$

$= -14$

$f(10,-5) = (10)^2 - 3(10)(-5)$ Substituting 10 for x and -5 for y.

$= 100 + 150$

$= 250$

3. $f(x,y) = 3^x + 7xy$

$f(0,-2) = 3^0 + 7(0)(-2)$ Substituting 0 for x and -2 for y.

$= 1 + 0$

$= 1$

$f(-2,1) = 3^{-2} + 7(-2)(1)$ Substituting -2 for x and 1 for y.

$= \frac{1}{9} - 14$

$= -\frac{125}{9}$

$f(2,1) = 3^2 + 7(2)(1)$ Substituting 2 for x and 1 for y.

$= 9 + 14$

$= 23$

5. $f(x,y) = \ln x + y^3$

$f(e,2) = \ln e + (2)^3$ Substituting e for x and 2 for y.

$= 1 + 8$

$= 9$

$f(e^2,4) = \ln e^2 + (4)^3$ Substituting e^2 for x and 4 for y.

$= 2 + 64$

$= 66$

$f(e^3,4) = \ln e^3 + (5)^3$ Substituting e^3 for x and 5 for y.

$= 3 + 125$

$= 128$

7. $f(x,y,z) = x^2 - y^2 + z^2$

We substitute -1 for x, 2 for y, and 3 for z.

$f(-1,2,3) = (-1)^2 - (2)^2 + (3)^2$

$= 1 - 4 + 9$

$= 6$

We substitute 2 for x, -1 for y, and 3 for z.

$f(2,-1,3) = (2)^2 - (-1)^2 + (3)^2$

$= 4 - 1 + 9$

$= 12$

9. $R(P,E) = \dfrac{P}{E}$

Substituting 32.03 for P, and 1.25 for E, gives

$R(32.03,1.25) = \dfrac{32.03}{1.25}$

≈ 25.624

$\approx 25.62.$

The price-earnings ration for Hewlett-Packard was 25.62.

11. From Example 3 we have $C_2 = \left(\dfrac{V_2}{V_1}\right)^{0.6} C_1$.

Where C_1 is the cost of the original piece of equipment, V_1 is the capacity of the original piece of equipment, and V_2 is the capacity of the new piece of equipment.

We substitute 100,000 for C_1, 80,000 for V_1 and 160,000 for V_2.

$C_2 = \left(\dfrac{160,000}{80,000}\right)^{0.6} (100,000)$

$= (2)^{0.6} (100,000)$

$= 151,571.6567$

We estimate the cost of the new tank to be $151,571.66.

13. $S(a,d,V) = \dfrac{aV}{0.51d^2}$

Substituting 100 for d, 1,600,000 for V, and 0.78 for a, we have:

$$S(0.78,100,1,600,000) = \dfrac{(0.78)(1,600,000)}{0.51(100)^2}$$

$$= \dfrac{1,248,000}{5100}$$

$$\approx 244.70588$$

The approximate wind speed 100 ft from the center of the tornado is 244.7 miles per hour.

15. $S(h,w) = 0.024265h^{0.3964}w^{0.5378}$

Substituting 165 for h and 80 for w, we have:

$$S(165,80) = 0.024265(165)^{0.3964}(80)^{0.5378}$$

$$\approx 0.024265(7.56851)(10.55557)$$

$$\approx 1.93852$$

The approximate surface area for the person is 1.939 square meters.

17. Using the formula $S(a,d,V) = \dfrac{aV}{0.51d^2}$, we remember from problem 13, that $a = 0.78$, $V = 1,600,000$. We substitute 200 for S, and solve for d.

$$200 = \dfrac{(0.78)(1,600,000)}{0.51 \cdot d^2}$$

$$200 = \dfrac{1,248,000}{0.51 \cdot d^2}$$

$$102d^2 = 1,248,000$$

$$d^2 = \dfrac{1,248,000}{102}$$

$$d^2 \approx 12,235.29$$

$$d \approx \pm\sqrt{12,235.29} \quad \text{Taking the square root of both sides.}$$

$$d \approx 110.6132 \quad \text{d Must be positive.}$$

The measurement was taken approximately 110.6 feet from the center of the tornado.

19. $\boxed{tw}$

21. $W(v,T) =$

$$91.4 - \dfrac{\left(10.45 + 6.68\sqrt{v} - 0.447v\right)(457 - 5T)}{110}$$

$W(25,30)$

$$= 91.4 - \dfrac{\left(10.45 + 6.68\sqrt{25} - 0.447 \cdot 25\right)(457 - 5 \cdot 30)}{110}$$

$$= 91.4 - \dfrac{\left(10.45 + 33.4 - 11.175\right)(307)}{110}$$

$$\approx 91.4 - \dfrac{10,031.225}{110}$$

$$\approx 91.4 - 91.2$$

$$\approx 0.2$$

The wind chill, rounded to the nearest degree is $0°F$.

23. $W(v,T) =$

$$91.4 - \dfrac{\left(10.45 + 6.68\sqrt{v} - 0.447v\right)(457 - 5T)}{110}$$

$W(40,20)$

$$= 91.4 - \dfrac{\left(10.45 + 6.68\sqrt{40} - 0.447 \cdot 40\right)(457 - 5 \cdot 20)}{110}$$

$$\approx 91.4 - \dfrac{\left(10.45 + 42.248 - 17.88\right)(357)}{110}$$

$$\approx 91.4 - \dfrac{(34.818)(357)}{110}$$

$$\approx 91.4 - \dfrac{12,430.026}{110}$$

$$\approx 91.4 - 113.0$$

$$\approx -21.6$$

The wind chill, rounded to the nearest degree is $-22°F$.

25.

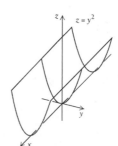

27.

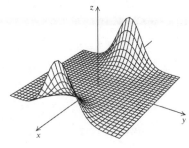

$$z = (x^4 - 16x^2)e^{-y^2}$$

29.

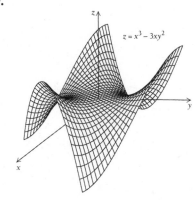

$$z = x^3 - 3xy^2$$

Exercise Set 6.2

1. $z = 2x - 3y$

Find $\dfrac{\partial z}{\partial x}$.

$z = 2\underline{x} - 3y$ The variable is underlined; y is treated as a constant.

$\dfrac{\partial z}{\partial x} = 2$

Find $\dfrac{\partial z}{\partial y}$.

$z = 2x - 3\underline{y}$ The variable is underlined; x is treated as a constant.

$\dfrac{\partial z}{\partial y} = -3$

Find $\left.\dfrac{\partial z}{\partial x}\right|_{(-2,-3)}$

$\dfrac{\partial z}{\partial x} = 2$ The partial derivative is constant for all values of x and y. Therefore:

$\left.\dfrac{\partial z}{\partial x}\right|_{(-2,-3)} = 2$

Find $\left.\dfrac{\partial z}{\partial y}\right|_{(0,-5)}$

$\dfrac{\partial z}{\partial x} = -3$ The partial derivative is constant for all values of x and y. Therefore:

$\left.\dfrac{\partial z}{\partial y}\right|_{(0,-5)} = -3$

3. $z = 3x^2 - 2xy + y$

Find $\dfrac{\partial z}{\partial x}$.

$z = 3\underline{x}^2 - 2\underline{x}y + y$ The variable is underlined; y is treated as a constant.

$\dfrac{\partial z}{\partial x} = 6x - 2y$

Find $\dfrac{\partial z}{\partial y}$.

$z = 3x^2 - 2x\underline{y} + \underline{y}$ The variable is underlined; x is treated as a constant.

$\dfrac{\partial z}{\partial x} = -2x + 1$

Find $\left.\dfrac{\partial z}{\partial x}\right|_{(-2,-3)}$

$\dfrac{\partial z}{\partial x} = 6x - 2y$

$\left.\dfrac{\partial z}{\partial x}\right|_{(-2,-3)} = 6(-2) - 2(-3)$ Substituting -2 for x and -3 for y.

$= -12 + 6$

$= -6$

Find $\left.\dfrac{\partial z}{\partial y}\right|_{(0,-5)}$

$\dfrac{\partial z}{\partial y} = -2x + 1$

$\left.\dfrac{\partial z}{\partial y}\right|_{(0,-5)} = -2(0) + 1$ Substituting 0 for x.

$= 1$

5. $f(x, y) = 2x - 5xy$

Find f_x.

$f(x, y) = 2\underline{x} - 5\underline{x}y$ The variable is underlined; y is treated as a constant.

$f_x = 2 - 5y$

Find f_y.

$f(x, y) = 2x - 5x\underline{y}$ The variable is underlined; x is treated as a constant.

$f_y = -5x$

Find $f_x(-2,4)$.

$$f_x = 2 - 5y$$

$$f_x(-2,4) = 2 - 5(4) \qquad \text{Substituting 4 for } y.$$

$$= 2 - 20$$

$$= -18$$

Find $f_y(4,-3)$.

$$f_y = -5x$$

$$f_y(4,-3) = -5(4) \quad \text{Substituting 4 for } x.$$

$$= -20$$

7. $f(x,y) = \sqrt{x^2 + y^2} = \left(x^2 + y^2\right)^{\frac{1}{2}}$

Find f_x.

$$f(x,y) = \left(\underline{x}^2 + y^2\right)^{\frac{1}{2}} \qquad \begin{array}{l}\text{The variable is underlined; } y \\ \text{is treated as a constant.}\end{array}$$

$$f_x = \frac{1}{2}\left(x^2 + y^2\right)^{-\frac{1}{2}} \cdot 2x$$

$$= x\left(x^2 + y^2\right)^{-\frac{1}{2}}, \text{ or } \frac{x}{\sqrt{x^2 + y^2}}$$

Find f_y.

$$f(x,y) = \left(x^2 + \underline{y}^2\right)^{\frac{1}{2}} \qquad \begin{array}{l}\text{The variable is underlined; } x \\ \text{is treated as a constant.}\end{array}$$

$$f_y = \frac{1}{2}\left(x^2 + y^2\right)^{-\frac{1}{2}} \cdot 2y$$

$$= y\left(x^2 + y^2\right)^{-\frac{1}{2}}, \text{ or } \frac{y}{\sqrt{x^2 + y^2}}$$

Find $f_x(-2,1)$.

$$f_x = \frac{x}{\sqrt{x^2 + y^2}}$$

$$f_x(-2,1) = \frac{(-2)}{\sqrt{(-2)^2 + (1)^2}} \qquad \begin{array}{l}\text{Substituting } -2 \text{ for} \\ x \text{ and 1 for } y.\end{array}$$

$$= \frac{-2}{\sqrt{4+1}}$$

$$= \frac{-2}{\sqrt{5}}$$

Find $f_y(-3,-2)$.

$$f_y = \frac{y}{\sqrt{x^2 + y^2}}$$

$$f_y(-3,-2) = \frac{(-2)}{\sqrt{(-3)^2 + (-2)^2}} \qquad \begin{array}{l}\text{Substituting } -3 \text{ for} \\ x \text{ and } -2 \text{ for } y.\end{array}$$

$$= \frac{-2}{\sqrt{9+4}}$$

$$= \frac{-2}{\sqrt{13}}$$

9. $f(x,y) = e^{2x-y}$

Find f_x.

$$f(x,y) = e^{2\underline{x}-y} \qquad \begin{array}{l}\text{The variable is underlined; } y \\ \text{is treated as a constant.}\end{array}$$

$$f_x = e^{2x-y} \cdot (2)$$

$$= 2e^{2x-y}$$

Find f_y.

$$f(x,y) = e^{2x-\underline{y}} \qquad \begin{array}{l}\text{The variable is underlined; } x \\ \text{is treated as a constant.}\end{array}$$

$$f_y = e^{2x-y} \cdot (-1)$$

$$= -e^{2x-y}$$

11. $f(x,y) = e^{xy}$

Find f_x.

$$f(x,y) = e^{\underline{x}y} \qquad \begin{array}{l}\text{The variable is underlined; } y \\ \text{is treated as a constant.}\end{array}$$

$$f_x = e^{xy} \cdot (y)$$

$$= ye^{xy}$$

Find f_y.

$$f(x,y) = e^{x\underline{y}} \qquad \begin{array}{l}\text{The variable is underlined; } x \\ \text{is treated as a constant.}\end{array}$$

$$f_y = e^{xy} \cdot (x)$$

$$= xe^{xy}$$

13. $f(x,y) = y \ln(x + 2y)$

Find f_x.

$$f(x,y) = y \ln(\underline{x} + 2y) \qquad \begin{array}{l}\text{The variable is} \\ \text{underlined.}\end{array}$$

$$f_x = y \cdot \frac{1}{x + 2y} \cdot 1$$

$$= \frac{y}{x + 2y}$$

Find f_y.

$$f(x,y) = \underline{y}\ln(x+2\underline{y}) \quad \text{The variable is underlined.}$$

$$f_y = y \cdot \frac{1}{x+2y} \cdot 2 + 1 \cdot \ln(x+2y)$$

$$= \frac{2y}{x+2y} + \ln(x+2y)$$

15. $f(x,y) = x\ln(xy)$

Find f_x.

$$f(x,y) = \underline{x}\ln(\underline{x}y) \quad \text{The variable is underlined.}$$

$$f_x = x \cdot \left(\frac{1}{xy} \cdot y\right) + 1 \cdot \ln(xy)$$

$$= 1 + \ln(xy)$$

Find f_y.

$$f(x,y) = x\ln(x\underline{y}) \quad \text{The variable is underlined.}$$

$$f_y = x \cdot \left(\frac{1}{xy} \cdot x\right)$$

$$= \frac{x}{y}$$

17. $f(x,y) = \dfrac{x}{y} - \dfrac{y}{3x}$

Find f_x.

$$f(x,y) = \frac{1}{y}\underline{x} - \frac{y}{3} \cdot \underline{x}^{-1} \quad \text{The variable is underlined.}$$

$$f_x = \frac{1}{y} - \frac{y}{3}\left(-1x^{-2}\right)$$

$$= \frac{1}{y} + \frac{y}{3x^2}$$

Find f_y.

$$f(x,y) = x\underline{y}^{-1} - \frac{1}{3x} \cdot \underline{y} \quad \text{The variable is underlined.}$$

$$f_y = x\left(-1y^{-2}\right) - \frac{1}{3x} \cdot 1$$

$$= -\frac{x}{y^2} - \frac{1}{3x}$$

19. $f(x,y) = 3(2x+y-5)^2$

Find f_x.

$$f(x,y) = 3(2\underline{x}+y-5)^2 \quad \text{The variable is underlined.}$$

$$f_x = 3\big[2(2x+y-5)\cdot 2\big]$$

$$= 12(2x+y-5)$$

Find f_y.

$$f(x,y) = 3(2x+\underline{y}-5)^2 \quad \text{The variable is underlined.}$$

$$f_y = 3\big[2(2x+y-5)\cdot 1\big]$$

$$= 6(2x+y-5)$$

21. $f(b,m) =$
$$m^3 + 4m^2b - b^2 + (2m+b-5)^2 + (3m+b-6)^2$$

Find $\dfrac{\partial f}{\partial b}$.

$f(b,m) =$
$$m^3 + 4m^2\underline{b} - \underline{b}^2 + (2m+\underline{b}-5)^2 + (3m+\underline{b}-6)^2$$
The variable is underlined.

$$\frac{\partial f}{\partial b} = 4m^2 - 2b + 2(2m+b-5)\cdot 1 +$$

$$2(3m+b-6)\cdot 1$$

$$= 4m^2 - 2b + 4m + 2b - 10 + 6m + 2b - 12$$

$$= 4m^2 + 10m + 2b - 22$$

Find $\dfrac{\partial f}{\partial m}$.

$f(b,m) =$
$$\underline{m}^3 + 4\underline{m}^2b - b^2 + (2\underline{m}+b-5)^2 + (3\underline{m}+b-6)^2$$
The variable is underlined.

$$\frac{\partial f}{\partial m} = 3m^2 + 8mb + 2(2m+b-5)\cdot 2 +$$

$$2(3m+b-6)\cdot 3$$

$$= 3m^2 + 8mb + 8m + 4b - 20 +$$

$$18m + 6b - 36$$

$$= 3m^2 + 8mb + 26m + 10b - 56$$

23. $f(x,y,\lambda) = 5xy - \lambda(2x + y - 8)$

Find f_x.

$f(x,y,\lambda) = 5\underline{x}y - \lambda(2\underline{x} + y - 8)$ The variable is underlined.

$f_x = 5y - \lambda \cdot 2$

$= 5y - 2\lambda$

Find f_y.

$f(x,y,\lambda) = 5x\underline{y} - \lambda(2x + \underline{y} - 8)$ The variable is underlined.

$f_y = 5x - \lambda \cdot 1$

$= 5x - \lambda$

Find f_λ.

$f(x,y,\lambda) = 5xy - \underline{\lambda}(2x + y - 8)$ The variable is underlined.

$f_\lambda = -1 \cdot (2x + y - 8)$

$= -(2x + y - 8)$

25. $f(x,y,\lambda) = x^2 + y^2 - \lambda(10x + 2y - 4)$

Find f_x.

$f(x,y,\lambda) = \underline{x}^2 + y^2 - \lambda(10\underline{x} + 2y - 4)$

The variable is underlined.

$f_x = 2x - \lambda \cdot 10$

$= 2x - 10\lambda$

Find f_y.

$f(x,y,\lambda) = x^2 + \underline{y}^2 - \lambda(10x + 2\underline{y} - 4)$

The variable is underlined.

$f_y = 2y - \lambda \cdot 2$

$= 2y - 2\lambda$

Find f_λ.

$f(x,y,\lambda) = x^2 + y^2 - \underline{\lambda}(10x + 2y - 4)$

The variable is underlined.

$f_\lambda = -1(10x + 2y - 4)$

$= -(10x + 2y - 4)$

27. $f(x,y) = 5xy$

First, we find the partial derivatives.

We find f_x first.

$f(x,y) = 5\underline{x}y$ The variable is underlined.

$f_x = 5y$

Then we find f_y.

$f(x,y) = 5x\underline{y}$ The variable is underlined.

$f_y = 5x$

We find f_{xx} by taking the partial derivative with respect to x of f_x.

$f_{xx} = \dfrac{\partial}{\partial x}(f_x) = \dfrac{\partial}{\partial x}(5y) = 0$

We find f_{xy} by taking the partial derivative with respect to y of f_x.

$f_{xy} = \dfrac{\partial}{\partial y}(f_x) = \dfrac{\partial}{\partial y}(5y) = 5$

We find f_{yx} by taking the partial derivative with respect to x of f_y.

$f_{yx} = \dfrac{\partial}{\partial x}(f_y) = \dfrac{\partial}{\partial x}(5x) = 5$

We find f_{yy} by taking the partial derivative with respect to y of f_y.

$f_{yy} = \dfrac{\partial}{\partial y}(f_y) = \dfrac{\partial}{\partial y}(5x) = 0$

29. $f(x,y) = 7xy^2 + 5xy - 2y$

First, we find the partial derivatives.

We find f_x first.

$f(x,y) = 7\underline{x}y^2 + 5\underline{x}y - 2y$ The variable is underlined.

$f_x = 7y^2 + 5y$

Then we find f_y.

$f(x,y) = 7x\underline{y}^2 + 5x\underline{y} - 2\underline{y}$ The variable is underlined.

$f_y = 14xy - 5x - 2$

We find f_{xx} by taking the partial derivative with respect to x of f_x.

$f_{xx} = \dfrac{\partial}{\partial x}(f_x)$

$= \dfrac{\partial}{\partial x}(7y^2 + 5y)$ y is treated as a constant.

$= 0$

We find f_{xy} by taking the partial derivative with respect to y of f_x.

$f_{xy} = \dfrac{\partial}{\partial y}(f_x)$

$= \dfrac{\partial}{\partial y}(7\underline{y}^2 + 5\underline{y})$ The variable is underlined.

$= 14y + 5$

We find f_{yx} by taking the partial derivative with respect to x of f_y.

$$f_{yx} = \frac{\partial}{\partial x}\left(f_y\right)$$

$$= \frac{\partial}{\partial x}\left(14\underline{x}y + 5\underline{x} - 2\right) \qquad \text{The variable is underlined.}$$

$$= 14y + 5$$

We find f_{yy} by taking the partial derivative with respect to y of f_y.

$$f_{yy} = \frac{\partial}{\partial y}\left(f_y\right)$$

$$= \frac{\partial}{\partial y}\left(14x\underline{y} + 5x - 2\right) \qquad \text{The variable is underlined.}$$

$$= 14x$$

31. $f(x,y) = x^5y^4 + x^3y^2$

First, we find the partial derivatives.
We find f_x first.

$$f(x,y) = \underline{x}^5y^4 + \underline{x}^3y^2 \qquad \text{The variable is underlined.}$$

$$f_x = 5x^4y^4 + 3x^2y^2$$

Then we find f_y.

$$f(x,y) = x^5\underline{y}^4 + x^3\underline{y}^2 \qquad \text{The variable is underlined.}$$

$$f_y = 4x^5y^3 + 2x^3y$$

We find f_{xx} by taking the partial derivative with respect to x of f_x.

$$f_{xx} = \frac{\partial}{\partial x}\left(f_x\right)$$

$$= \frac{\partial}{\partial x}\left(5\underline{x}^4y^4 + 3\underline{x}^2y^2\right) \qquad \text{The variable is underlined.}$$

$$= 20x^3y^4 + 6xy^2$$

We find f_{xy} by taking the partial derivative with respect to y of f_x.

$$f_{xy} = \frac{\partial}{\partial y}\left(f_x\right)$$

$$= \frac{\partial}{\partial y}\left(5x^4\underline{y}^4 + 3x^2\underline{y}^2\right) \qquad \text{The variable is underlined.}$$

$$= 20x^4y^3 + 6x^2y$$

We find f_{yx} by taking the partial derivative with respect to x of f_y.

$$f_{yx} = \frac{\partial}{\partial x}\left(f_y\right)$$

$$= \frac{\partial}{\partial x}\left(4x^5y^3 + 2x^3y\right) \qquad \text{The variable is underlined.}$$

$$= 20x^4y^3 + 6x^2y$$

We find f_{yy} by taking the partial derivative with respect to y of f_y.

$$f_{yy} = \frac{\partial}{\partial y}\left(f_y\right)$$

$$= \frac{\partial}{\partial y}\left(4x^5y^3 + 2x^3y\right) \qquad \text{The variable is underlined.}$$

$$= 12x^5y^2 + 2x^3$$

33. $f(x,y) = 2x - 3y$

First, we find the partial derivatives.
We find f_x first.

$$f(x,y) = 2\underline{x} - 3y \qquad \text{The variable is underlined.}$$

$$f_x = 2$$

Then we find f_y.

$$f(x,y) = 2x - 3\underline{y} \qquad \text{The variable is underlined.}$$

$$f_y = -3$$

We find f_{xx} by taking the partial derivative with respect to x of f_x.

$$f_{xx} = \frac{\partial}{\partial x}\left(f_x\right) = \frac{\partial}{\partial x}(2) = 0$$

We find f_{xy} by taking the partial derivative with respect to y of f_x.

$$f_{xy} = \frac{\partial}{\partial y}\left(f_x\right) = \frac{\partial}{\partial y}(2) = 0$$

We find f_{yx} by taking the partial derivative with respect to x of f_y.

$$f_{yx} = \frac{\partial}{\partial x}\left(f_y\right) = \frac{\partial}{\partial x}(-3) = 0$$

We find f_{yy} by taking the partial derivative with respect to y of f_y.

$$f_{yy} = \frac{\partial}{\partial y}\left(f_y\right) = \frac{\partial}{\partial y}(-3) = 0$$

35. $f(x,y) = e^{2xy}$

First, we find the partial derivatives.
We find f_x first.

$f(x,y) = e^{2\underline{x}y}$ The variable is underlined.

$\quad f_x = 2ye^{2xy}$

Then we find f_y.

$f(x,y) = e^{2x\underline{y}}$ The variable is underlined.

$\quad f_y = 2xe^{2xy}$

We find f_{xx} by taking the partial derivative with respect to x of f_x.

$f_{xx} = \dfrac{\partial}{\partial x}(f_x)$

$\quad = \dfrac{\partial}{\partial x}\left(2ye^{2\underline{x}y}\right)$ The variable is underlined.

$\quad = 2y\left(2ye^{2xy}\right)$

$\quad = 4y^2 e^{2xy}$

We find f_{xy} by taking the partial derivative with respect to y of f_x.

$f_{xy} = \dfrac{\partial}{\partial y}(f_x)$

$\quad = \dfrac{\partial}{\partial y}\left(2ye^{2x\underline{y}}\right)$ The variable is underlined.

$\quad = 2y\left(2xe^{2xy}\right) + 2\left(e^{2xy}\right)$

$\quad = 4xye^{2xy} + 2e^{2xy}$

We find f_{yx} by taking the partial derivative with respect to x of f_y.

$f_{yx} = \dfrac{\partial}{\partial x}(f_y)$

$\quad = \dfrac{\partial}{\partial x}\left(2\underline{x}e^{2\underline{x}y}\right)$ The variable is underlined.

$\quad = 2x\left(2ye^{2xy}\right) + 2\left(e^{2xy}\right)$

$\quad = 4xye^{2xy} + 2e^{2xy}$

We find f_{yy} by taking the partial derivative with respect to y of f_y.

$f_{yy} = \dfrac{\partial}{\partial y}(f_y)$

$\quad = \dfrac{\partial}{\partial y}\left(2xe^{2x\underline{y}}\right)$ The variable is underlined.

$\quad = 2x\left(2xe^{2xy}\right)$

$\quad = 4x^2 e^{2xy}$

37. $f(x,y) = x + e^y$

First, we find the partial derivatives.
We find f_x first.

$f(x,y) = \underline{x} + e^y$ The variable is underlined.

$\quad f_x = 1$

Then we find f_y.

$f(x,y) = x + e^{\underline{y}}$ The variable is underlined.

$\quad f_y = e^y$

We find f_{xx} by taking the partial derivative with respect to x of f_x.

$f_{xx} = \dfrac{\partial}{\partial x}(f_x) = \dfrac{\partial}{\partial x}(1) = 0$

We find f_{xy} by taking the partial derivative with respect to y of f_x.

$f_{xy} = \dfrac{\partial}{\partial y}(f_x) = \dfrac{\partial}{\partial y}(1) = 0$

We find f_{yx} by taking the partial derivative with respect to x of f_y.

$f_{yx} = \dfrac{\partial}{\partial x}(f_y)$

$\quad = \dfrac{\partial}{\partial x}(e^y)$ e^y is treated as a constant.

$\quad = 0$

We find f_{yy} by taking the partial derivative with respect to y of f_y.

$f_{yy} = \dfrac{\partial}{\partial y}(f_y)$

$\quad = \dfrac{\partial}{\partial y}(e^{\underline{y}})$ The variable is underlined.

$\quad = e^y$

39. $f(x,y) = y \ln x$

First, we find the partial derivatives.
We find f_x first.

$f(x,y) = y \ln \underline{x}$ The variable is underlined.

$\quad f_x = y \cdot \dfrac{1}{x} = \dfrac{y}{x}$

Then we find f_y.

$f(x,y) = \underline{y} \ln x$ The variable is underlined.

$\quad f_y = 1 \cdot \ln x = \ln x$

We find f_{xx} by taking the partial derivative with respect to x of f_x.

$$f_{xx} = \frac{\partial}{\partial x}(f_x)$$

$$= \frac{\partial}{\partial x}\left(\frac{y}{\underline{x}}\right) = \frac{\partial}{\partial x}\left(y\underline{x}^{-1}\right) \quad \text{The variable is underlined.}$$

$$= -yx^{-2}$$

$$= \frac{-y}{x^2}$$

We find f_{xy} by taking the partial derivative with respect to y of f_x.

$$f_{xy} = \frac{\partial}{\partial y}(f_x)$$

$$= \frac{\partial}{\partial y}\left(\underline{y} \cdot \frac{1}{x}\right) \quad \text{The variable is underlined.}$$

$$= \frac{1}{x}$$

We find f_{yx} by taking the partial derivative with respect to x of f_y.

$$f_{yx} = \frac{\partial}{\partial x}(f_y)$$

$$= \frac{\partial}{\partial x}(\ln \underline{x}) \quad \text{The variable is underlined.}$$

$$= \frac{1}{x}$$

We find f_{yy} by taking the partial derivative with respect to y of f_y.

$$f_{yy} = \frac{\partial}{\partial y}(f_y)$$

$$= \frac{\partial}{\partial y}(\ln x) \quad \ln x \text{ is treated as a constant.}$$

$$= 0$$

41. $p(x,y) = 2400x^{2/5}y^{3/5}$

a) Substituting 32 for x and 1024 for y, we have:

$$p(32,1024) = 2400(32)^{2/5}(1024)^{3/5}$$

$$= 2400(4)(64)$$

$$= 614,400.$$

Using 32 units of labor and 1024 units of capital, Lincolnville Sporting goods will produce 614,400 units.

b) We find the marginal productivity of labor by taking the partial derivative with respect to x.

$$\frac{\partial p}{\partial x} = \frac{\partial}{\partial x}\left(2400\underline{x}^{2/5}y^{3/5}\right) \quad \text{The variable is underlined.}$$

$$= 2400\left(\frac{2}{5}x^{-3/5}\right)y^{3/5}$$

$$= 960x^{-3/5}y^{3/5}$$

$$= 960\left(\frac{y}{x}\right)^{3/5}$$

We find the marginal productivity of capital by taking the partial derivative with respect to y.

$$\frac{\partial p}{\partial y} = \frac{\partial}{\partial y}\left(2400x^{2/5}\underline{y}^{3/5}\right) \quad \text{The variable is underlined.}$$

$$= 2400x^{2/5}\left(\frac{3}{5}y^{-2/5}\right)$$

$$= 1440x^{2/5}y^{-2/5}$$

$$= 1440\left(\frac{x}{y}\right)^{2/5}$$

c) Substituting 32 for x and 1024 for y into $\frac{\partial p}{\partial x}$, we have:

$$\left.\frac{\partial p}{\partial x}\right|_{(32,1024)} = 960\left(\frac{1024}{32}\right)^{3/5}$$

$$= 960(32)^{3/5}$$

$$= 960(8)$$

$$= 7680$$

The marginal productivity of labor when 32 units of labor and 1024 units of capital are currently being used is 7680 units per unit labor.

Substituting 32 for x and 1024 for y into $\dfrac{\partial p}{\partial y}$, we have:

$$\left.\frac{\partial p}{\partial y}\right|_{(32,1024)} = 1440\left(\frac{32}{1024}\right)^{2/5}$$

$$= 1440\left(\frac{1}{32}\right)^{2/5}$$

$$= 1440\left(\tfrac{1}{4}\right)$$

$$= 360$$

The marginal productivity of capital when 32 units of labor and 1024 units of capital are currently being used is 360 units per unit capital.

d) $\boxed{tw}$

43. $P(w,r,s,t) = 0.007955 w^{-0.638} r^{1.038} s^{0.873} t^{2.468}$

a) We substitute 20 for w, 70 for r, 400,000 for s, and 8 for t.

$P(20,70,400,000,8)$

$= 0.007955(20)^{-0.638}(70)^{1.038}(400,000)^{0.873}(8)^{2.468}$

$\approx 1,274,146$

The nursing home's annual profit is approximately \$1,274,146.

b) Taking the partial derivative with respect to each variable, we have:

$$\frac{\partial P}{\partial w} = 0.007955\left(-0.638 w^{-1.638}\right) r^{1.038} s^{0.873} t^{2.468}$$

$$= -0.005075 w^{-1.638} r^{1.038} s^{0.873} t^{2.468}$$

$$\frac{\partial P}{\partial r} = 0.007955 w^{-0.638}\left(1.038 r^{1-1.038}\right) s^{0.873} t^{2.468}$$

$$= 0.008257 w^{-0.638} r^{0.038} s^{0.873} t^{2.468}$$

$$\frac{\partial P}{\partial s} = 0.007955 w^{-0.638} r^{1.038}\left(0.873 s^{-0.127}\right) t^{2.468}$$

$$= 0.006945 w^{-0.638} r^{1.038} s^{-0.127} t^{2.468}$$

$$\frac{\partial P}{\partial t} = 0.007955 w^{-0.638} r^{1.038} s^{0.873}\left(2.468 t^{1.468}\right)$$

$$= 0.019633 w^{-0.638} r^{1.038} s^{0.873} t^{1.468}$$

c) $\boxed{tw}$

45. $T_h = 1.98T - 1.09(1-H)(T-58) - 56.9$

Substituting 85 for T, and 60%=0.60 for H, we have:

$T_h = 1.98(85) - 1.09(1-0.6)(85-58) - 56.9$

$= 168.3 - 1.09(0.4)(27) - 56.9$

$= 168.3 - 11.772 - 56.9$

$= 99.628$

≈ 99.6

The temperature-humidity index is about $99.6°F$.

47. $T_h = 1.98T - 1.09(1-H)(T-58) - 56.9$

Substituting 90 for T, and 100%=1.0 for H, we have:

$T_h = 1.98(90) - 1.09(1-1.0)(90-58) - 56.9$

$= 178.2 - 1.09(0)(32) - 56.9$

$= 178.2 - 0 - 56.9$

$= 121.3$

The temperature-humidity index is $121.3°F$.

49. $\boxed{tw}$

51. $S = \dfrac{\sqrt{hw}}{60} = \dfrac{(hw)^{1/2}}{60} = \dfrac{h^{1/2}w^{1/2}}{60}$

a) $\dfrac{\partial S}{\partial h} = \dfrac{1}{60}\left(\dfrac{1}{2}h^{-1/2}w^{1/2}\right)$

$= \dfrac{1}{120}\left(\dfrac{w^{1/2}}{h^{1/2}}\right)$

$= \dfrac{\sqrt{w}}{120\sqrt{h}}$

b) $\dfrac{\partial S}{\partial w} = \dfrac{1}{60}h^{1/2}\left(\dfrac{1}{2}w^{-1/2}\right)$

$= \dfrac{1}{120}\left(\dfrac{h^{1/2}}{w^{1/2}}\right)$

$= \dfrac{\sqrt{h}}{120\sqrt{w}}$

c) $\Delta S \approx \dfrac{\partial S}{\partial w}\Delta w$

$$\Delta S \approx \left[\dfrac{\sqrt{h}}{120\sqrt{w}}\right]\Delta w$$

$$\approx \dfrac{\sqrt{170}}{120\sqrt{80}}(-2)$$

$$\approx -0.0243$$

The change in the surface area is approximately $-0.0243 m^2$.

53. $E = 206.835 - 0.846w - 1.015s$

Substituting 146 for w and 5 for s, we have:

$E = 206.835 - 0.846(146) - 1.015(5)$

$\quad = 206.835 - 123.516 - 5.075$

$\quad = 78.244$

The reading ease is 78.244.

55. $E = 206.835 - 0.846w - 1.015s$

$$\dfrac{\partial E}{\partial w} = -0.846$$

57. $f(x,t) = \dfrac{x^2 + t^2}{x^2 - t^2}$

Find f_x.

$$f(x,t) = \dfrac{\underline{x}^2 + t^2}{\underline{x}^2 - t^2} \qquad \text{The variable is underlined.}$$

$$f_x = \dfrac{(x^2 - t^2)(2x) - (x^2 + t^2)(2x)}{(x^2 - t^2)^2}$$

$$= \dfrac{2x^3 - 2xt^2 - 2x^3 - 2xt^2}{(x^2 - t^2)^2}$$

$$= \dfrac{-4xt^2}{(x^2 - t^2)^2}$$

Find f_t.

$$f(x,t) = \dfrac{x^2 + \underline{t}^2}{x^2 - \underline{t}^2} \qquad \text{The variable is underlined.}$$

$$f_t = \dfrac{(x^2 - t^2)(2t) - (x^2 + t^2)(-2t)}{(x^2 - t^2)^2}$$

$$= \dfrac{2x^2t - 2t^3 + 2x^2t + 2t^3}{(x^2 - t^2)^2}$$

$$= \dfrac{4x^2t}{(x^2 - t^2)^2}$$

59. $f(x,t) = \dfrac{2\sqrt{x} - 2\sqrt{t}}{1 + 2\sqrt{t}} = \dfrac{2x^{\frac{1}{2}} - 2t^{\frac{1}{2}}}{1 + 2t^{\frac{1}{2}}}$

Find f_x.

$$f(x,t) = \dfrac{2\underline{x}^{\frac{1}{2}} - 2t^{\frac{1}{2}}}{1 + 2t^{\frac{1}{2}}}$$

$$= \dfrac{2x^{\frac{1}{2}}}{1 + 2t^{\frac{1}{2}}} - \dfrac{2t^{\frac{1}{2}}}{1 + 2t^{\frac{1}{2}}} \qquad \text{The variable is underlined.}$$

$$f_x = \dfrac{2\left(\dfrac{1}{2}x^{-\frac{1}{2}}\right)}{1 + 2t^{\frac{1}{2}}} - 0$$

$$= \dfrac{x^{-\frac{1}{2}}}{1 + 2t^{\frac{1}{2}}}$$

$$= \dfrac{1}{x^{\frac{1}{2}}\left(1 + 2t^{\frac{1}{2}}\right)}$$

$$= \dfrac{1}{\sqrt{x}\left(1 + 2\sqrt{t}\right)}$$

Find f_t.

$$f(x,t) = \dfrac{2x^{\frac{1}{2}} - 2\underline{t}^{\frac{1}{2}}}{1 + 2\underline{t}^{\frac{1}{2}}} \qquad \text{The variable is underlined.}$$

$$f_t = \dfrac{\left(1 + 2t^{\frac{1}{2}}\right)\left(-1t^{-\frac{1}{2}}\right) - \left(2x^{\frac{1}{2}} - 2t^{\frac{1}{2}}\right)\left(t^{-\frac{1}{2}}\right)}{\left(1 + 2t^{\frac{1}{2}}\right)^2}$$

$$= \dfrac{-t^{-\frac{1}{2}} - 2t^0 - 2x^{\frac{1}{2}}t^{-\frac{1}{2}} + 2t^0}{\left(1 + 2t^{\frac{1}{2}}\right)^2}$$

$$= \dfrac{t^{-\frac{1}{2}}\left(-1 - 2x^{\frac{1}{2}}\right)}{\left(1 + 2t^{\frac{1}{2}}\right)^2}$$

$$= \dfrac{-1 - 2\sqrt{x}}{\sqrt{t}\left(1 + 2\sqrt{t}\right)^2}$$

61. $f(x,t) = 6x^{2/3} - 8x^{1/4}t^{1/2} - 12x^{-1/2}t^{3/2}$

Find f_x.

$f(x,t) = 6\underline{x}^{2/3} - 8\underline{x}^{1/4}t^{1/2} - 12\underline{x}^{-1/2}t^{3/2}$

$$\text{The variable is underlined.}$$

$$f_x = 6\left(\frac{2}{3}x^{-1/3}\right) - 8\left(\frac{1}{4}x^{-3/4}\right)t^{1/2} -$$

$$12\left(\frac{-1}{2}x^{-3/2}\right)t^{3/2}$$

$$= 4x^{-1/3} - 2x^{-3/4}t^{1/2} + 6x^{-3/2}t^{3/2}$$

Find f_t.

$$f(x,t) = 6x^{2/3} - 8x^{1/4}\underline{t}^{1/2} - 12x^{-1/2}\underline{t}^{3/2}$$

$$\text{The variable is underlined.}$$

$$f_t = -8x^{1/4}\left(\frac{1}{2}t^{-1/2}\right) - 12x^{-1/2}\left(\frac{3}{2}t^{1/2}\right)$$

$$= -4x^{1/4}t^{-1/2} - 18x^{-1/2}t^{1/2}$$

63. $f(x,y) = \dfrac{x}{y^2} - \dfrac{y}{x^2} = xy^{-2} - yx^{-2}$

First, we find the partial derivatives.
We find f_x first.

$f(x,y) = \underline{x}y^{-2} - y\underline{x}^{-2}$ The variable is underlined.

$$f_x = y^{-2} + 2yx^{-3}$$

Then we find f_y.

$f(x,y) = x\underline{y}^{-2} - \underline{y}x^{-2}$ The variable is underlined.

$$f_y = -2xy^{-3} - x^{-2}$$

We find f_{xx} by taking the partial derivative with respect to x of f_x.

$$f_{xx} = \frac{\partial}{\partial x}(f_x)$$

$$= \frac{\partial}{\partial x}\left(y^{-2} + 2yx^{-3}\right)$$

$$= -6yx^{-4}$$

$$= \frac{-6y}{x^4}$$

We find f_{xy} by taking the partial derivative with respect to y of f_x.

$$f_{xy} = \frac{\partial}{\partial y}(f_x)$$

$$= \frac{\partial}{\partial y}\left(y^{-2} + 2yx^{-3}\right)$$

$$= -2y^{-3} + 2x^{-3}$$

$$= -\frac{2}{y^3} + \frac{2}{x^3}$$

We find f_{yx} by taking the partial derivative with respect to x of f_y.

$$f_{yx} = \frac{\partial}{\partial x}(f_y)$$

$$= \frac{\partial}{\partial x}\left(-2xy^{-3} - x^{-2}\right)$$

$$= -2y^{-3} + 2x^{-3}$$

$$= -\frac{2}{y^3} + \frac{2}{x^3}$$

We find f_{yy} by taking the partial derivative with respect to y of f_y.

$$f_{yy} = \frac{\partial}{\partial y}(f_y)$$

$$= \frac{\partial}{\partial y}\left(-2xy^{-3} - x^{-2}\right)$$

$$= 6xy^{-4}$$

$$= \frac{6x}{y^4}$$

65. tw

67. $f(x,y) = \ln\left(x^2 + y^2\right)$

We need to find the second-order partial derivatives.
First, we find f_x.

$$f(x,y) = \ln\left(\underline{x}^2 + y^2\right)$$

$$f_x = \frac{1}{x^2 + y^2} \cdot 2x = \frac{2x}{x^2 + y^2}$$

Then we find f_{xx}.

$$f_{xx} = \frac{\partial}{\partial x}(f_x)$$

$$= \frac{\partial}{\partial x}\left(\frac{2\underline{x}}{x^2 + y^2}\right)$$

$$= \frac{(x^2 + y^2)(2) - (2x)(2x)}{(x^2 + y^2)^2}$$

$$= \frac{-2x^2 + 2y^2}{(x^2 + y^2)^2}$$

Now we find f_y.

$$f(x, y) = \ln(x^2 + \underline{y}^2)$$

$$f_x = \frac{1}{x^2 + y^2} \cdot 2y = \frac{2y}{x^2 + y^2}$$

Then we find f_{yy}.

$$f_{yy} = \frac{\partial}{\partial y}(f_y)$$

$$= \frac{\partial}{\partial y}\left(\frac{2\underline{y}}{x^2 + \underline{y}^2}\right)$$

$$= \frac{(x^2 + y^2)(2) - (2y)(2y)}{(x^2 + y^2)^2}$$

$$= \frac{2x^2 - 2y^2}{(x^2 + y^2)^2}$$

Therefore,

$$\frac{\partial^2 f}{\partial x^2} + \frac{\partial^2 f}{\partial y^2} = 0$$

$$\begin{array}{c|c} \dfrac{-2x^2 + 2y^2}{(x^2 + y^2)^2} + \dfrac{2x^2 - 2y^2}{(x^2 + y^2)^2} & 0 \\[3mm] \dfrac{0}{(x^2 + y^2)^2} & 0 \\[3mm] 0 & 0 \end{array}$$

Thus, f is a solution to $\dfrac{\partial^2 f}{\partial x^2} + \dfrac{\partial^2 f}{\partial y^2} = 0$.

69.

$$f(x, y) = \begin{cases} \dfrac{xy(x^2 - y^2)}{x^2 + y^2}, & \text{for } (x, y) \neq (0, 0), \\ 0, & \text{for } (x, y) = (0, 0). \end{cases}$$

a) Find $f_x(0, y)$.

$$\lim_{h \to 0} \frac{f(h, y) - f(0, y)}{h}$$

$$= \lim_{h \to 0} \frac{\dfrac{hy(h^2 - y^2)}{h^2 + y^2} - \dfrac{0 \cdot y(0^2 - y^2)}{0^2 + y^2}}{h}$$

$$= \lim_{h \to 0} \frac{hy(h^2 - y^2)}{h(h^2 + y^2)}$$

$$= \lim_{h \to 0} \frac{y(h^2 - y^2)}{(h^2 + y^2)}$$

$$= \frac{y(-y^2)}{y^2}$$

$$= -y$$

Thus, $f_x(0, y) = -y$.

b) Find $f_y(x, 0)$.

$$\lim_{h \to 0} \frac{f(x, h) - f(x, 0)}{h}$$

$$= \lim_{h \to 0} \frac{\dfrac{xh(x^2 - h^2)}{x^2 + h^2} - \dfrac{x \cdot 0(x^2 - 0^2)}{x^2 + 0^2}}{h}$$

$$= \lim_{h \to 0} \frac{xh(x^2 - h^2)}{h(x^2 + h^2)}$$

$$= \lim_{h \to 0} \frac{x(x^2 - h^2)}{(x^2 + h^2)}$$

$$= \frac{x(x^2)}{x^2}$$

$$= x$$

Thus, $f_y(x, 0) = x$.

c) Find $f_{yx}(0, 0)$

$$\lim_{h \to 0} \frac{f_y(h, 0) - f_y(0, 0)}{h}$$

$$= \lim_{h \to 0} \frac{h - 0}{h} \qquad \text{Substituting } f_y(x, 0) = x.$$

$$= \lim_{h \to 0} 1$$

$$= 1$$

Find $f_{xy}(0,0)$

$$\lim_{h \to 0} \frac{f_x(0,h) - f_x(0,0)}{h}$$

$$= \lim_{h \to 0} \frac{-h - (0)}{h} \quad \text{Substituting } f_x(0,y) = -y.$$

$$= \lim_{h \to 0} (-1)$$

$$= -1$$

Thus, $f_{yx}(0,0) \neq f_{xy}(0,0)$. The mixed partials are not equal at $(0,0)$

Exercise Set 6.3

1. $f(x,y) = x^2 + xy + y^2 - y$

Find f_x.

$f(x,y) = \underline{x}^2 + x\underline{y} + y^2 - y,$ The variable is underlined.

$\quad f_x = 2x + y.$

Find f_y.

$f(x,y) = x^2 + x\underline{y} + \underline{y}^2 - \underline{y},$ The variable is underlined.

$\quad f_y = x + 2y - 1.$

Find f_{xx} and f_{xy}.

$f_x = 2\underline{x} + y, \quad f_x = 2x + \underline{y},$

$f_{xx} = 2. \quad\quad f_{xy} = 1.$

Find f_{yy}.

$f_y = x + 2\underline{y} - 1,$

$f_{yy} = 2.$

Solve the system of equations

$f_x = 0$ and $f_y = 0$:

$\quad 2x + y = 0, \quad\quad (1)$

$x + 2y - 1 = 0. \quad\quad (2)$

Solving Eq. (1) for y, we get $y = -2x$.

Substituting $-2x$ for y in Eq. (2) and solving, we get

$x + 2(-2x) - 1 = 0$

$\quad -3x - 1 = 0$

$\quad\quad -3x = 1$

$\quad\quad\quad x = -\frac{1}{3}.$

To find y when $x = -\frac{1}{3}$, we substitute $-\frac{1}{3}$ for x in either Eq. (1) or Eq. (2). We use Eq. (1):

$$2\left(-\frac{1}{3}\right) + y = 0$$

$$-\frac{2}{3} + y = 0$$

$$y = \frac{2}{3}.$$

Thus, $\left(-\frac{1}{3}, \frac{2}{3}\right)$ is our candidate for a maximum or minimum.

We have to check to see if $f\left(-\frac{1}{3}, \frac{2}{3}\right)$ is a maximum or minimum:

$$D = f_{xx}(a,b) \cdot f_{yy}(a,b) - \left[f_{xy}(a,b)\right]^2$$

$$D = f_{xx}\left(-\frac{1}{3}, \frac{2}{3}\right) \cdot f_{yy}\left(-\frac{1}{3}, \frac{2}{3}\right) -$$

$$\left[f_{xy}\left(-\frac{1}{3}, \frac{2}{3}\right)\right]^2$$

$D = 2 \cdot 2 - 1^2$ For all values of x and y, $f_{xx} = 2, f_{yy} = 2,$ and $f_{xy} = 1.$

$D = 3.$

Thus, $D = 3$ and $f_{xx} = 2$. Since $D > 0$ and

$$f_{xx}\left(-\frac{1}{3}, \frac{2}{3}\right) = 2 > 0, \text{ it follows that } f \text{ has a}$$

relative minimum at $\left(-\frac{1}{3}, \frac{2}{3}\right)$. The minimum is found as follows:

$f(x,y) = x^2 + xy + y^2 - y$

$$f\left(-\frac{1}{3}, \frac{2}{3}\right) = \left(-\frac{1}{3}\right)^2 + \left(-\frac{1}{3}\right)\left(\frac{2}{3}\right) + \left(\frac{2}{3}\right)^2 - \left(\frac{2}{3}\right)$$

$$= \frac{1}{9} - \frac{2}{9} + \frac{4}{9} - \frac{2}{3}$$

$$= \frac{1}{9} - \frac{2}{9} + \frac{4}{9} - \frac{6}{9}$$

$$= -\frac{3}{9}$$

$$= -\frac{1}{3}.$$

The relative minimum value of f is $-\frac{1}{3}$

at $\left(-\frac{1}{3}, \frac{2}{3}\right)$.

3. $f(x,y) = 2xy - x^3 - y^2$

Find f_x.

$f(x,y) = 2x\underline{y} - \underline{x}^3 - y^2,$ The variable is underlined.

$\quad f_x = 2y - 3x^2.$

Find f_y.

$f(x,y) = 2x\underline{y} - x^3 - \underline{y}^2,$ The variable is underlined.

$\quad f_y = 2x - 2y.$

Find f_{xx} and f_{xy}.

$f_x = 2y - 3\underline{x}^2, \quad f_x = 2\underline{y} - 3x^2,$

$f_{xx} = -6x. \qquad f_{xy} = 2.$

Find f_{yy}.

$f_y = 2x - 2\underline{y},$

$f_{yy} = -2.$

Solve the system of equations

$f_x = 0$ and $f_y = 0$:

$2y - 3x^2 = 0, \qquad (1)$

$2x - 2y = 0. \qquad (2)$

Solving Eq. (1) for $2y$, we get $2y = 3x^2$.

Substituting $3x^2$ for $2y$ in Eq. (2) and solving,

we get

$2x - 3x^2 = 0$

$x(2 - 3x) = 0$

$x = 0 \quad$ or $\quad 2 - 3x = 0$

$x = 0 \quad$ or $\quad -3x = -2$

$x = 0 \quad$ or $\quad x = \dfrac{2}{3}$

To find y when $x = 0$, we substitute 0 for x in
either Eq. (1) or Eq. (2). We use Eq. (1):

$2y - 3(0)^2 = 0$

$\qquad 2y = 0$

$\qquad y = 0.$

Thus, $(0,0)$ is one critical point, and $f(0,0)$ is
a candidate for a maximum or minimum value.
To find the other critical point we substitute
$\dfrac{2}{3}$ for x in either Eq. (1) or Eq. (2). We use
Eq. (2):

$2\left(\dfrac{2}{3}\right) - 2y = 0$

$\dfrac{4}{3} - 2y = 0$

$\qquad -2y = -\dfrac{4}{3}$

$\qquad y = \dfrac{2}{3}.$

Thus, $\left(\dfrac{2}{3}, \dfrac{2}{3}\right)$ is the other critical point, and

$f\left(\dfrac{2}{3}, \dfrac{2}{3}\right)$ is another candidate for maximum or

minimum value.

We must check both $(0,0)$ and $\left(\dfrac{2}{3}, \dfrac{2}{3}\right)$ to see

whether they yield maximum or minimum
values.

For $(0,0)$

$D = f_{xx}(0,0) \cdot f_{yy}(0,0) - \left[f_{xy}(0,0)\right]^2$

$D = 0 \cdot (-2) - 2^2 \quad \begin{bmatrix} f_{xx}(0,0) = -6 \cdot 0 = 0 \\ f_{yy}(0,0) = -2 \\ f_{xy}(0,0) = 2 \end{bmatrix}$

$D = -4.$

Since $D < 0$, it follows that $f(0,0)$ is neither a
maximum nor a minimum, but a saddle point.

For $\left(\dfrac{2}{3}, \dfrac{2}{3}\right)$

$D = f_{xx}\left(\dfrac{2}{3}, \dfrac{2}{3}\right) \cdot f_{yy}\left(\dfrac{2}{3}, \dfrac{2}{3}\right) - \left[f_{xy}\left(\dfrac{2}{3}, \dfrac{2}{3}\right)\right]^2$

$D = (-4) \cdot (-2) - 2^2 \quad \begin{bmatrix} f_{xx}\left(\frac{2}{3}, \frac{2}{3}\right) = -6 \cdot \frac{2}{3} = -4 \\ f_{yy}\left(\frac{2}{3}, \frac{2}{3}\right) = -2 \\ f_{xy}\left(\frac{2}{3}, \frac{2}{3}\right) = 2 \end{bmatrix}$

$D = 8 - 4$

$D = 4.$

Thus, $D = 4$ and $f_{xx}\left(\dfrac{2}{3}, \dfrac{2}{3}\right) = -4$. Since

$D > 0$ and $f_{xx}\left(\dfrac{2}{3}, \dfrac{2}{3}\right) < 0$, it follows that f has a

relative maximum at $\left(\dfrac{2}{3}, \dfrac{2}{3}\right)$.

The maximum is found as follows:

$$f(x,y) = 2xy - x^3 - y^2$$

$$f\left(\frac{2}{3}, \frac{2}{3}\right) = 2\left(\frac{2}{3}\right)\left(\frac{2}{3}\right) - \left(\frac{2}{3}\right)^3 - \left(\frac{2}{3}\right)^2$$

$$= \frac{8}{9} - \frac{8}{27} - \frac{4}{9}$$

$$= \frac{4}{9} - \frac{8}{27}$$

$$= \frac{12}{27} - \frac{8}{27}$$

$$= \frac{4}{27}.$$

The relative maximum value of f is $\dfrac{4}{27}$ at $\left(\dfrac{2}{3}, \dfrac{2}{3}\right)$.

5. $f(x,y) = x^3 + y^3 - 3xy$

Find f_x.

$f(x,y) = \underline{x}^3 + y^3 - 3\underline{x}y,$ The variable is underlined.

$\qquad f_x = 3x^2 - 3y.$

Find f_y.

$f(x,y) = x^3 + \underline{y}^3 - 3x\underline{y},$ The variable is underlined.

$\qquad f_y = 3y^2 - 3x.$

Find f_{xx} and f_{xy}.

$f_x = 3\underline{x}^2 - 3y, \quad f_x = 3x^2 - 3\underline{y},$

$f_{xx} = 6x. \qquad f_{xy} = -3.$

Find f_{yy}.

$f_y = 3\underline{y}^2 - 3x,$

$f_{yy} = 6y.$

Solve the system of equations
$f_x = 0$ and $f_y = 0$:

$$3x^2 - 3y = 0, \qquad (1)$$

$$3y^2 - 3x = 0. \qquad (2)$$

We multiply each equation by $\dfrac{1}{3}$.

$$x^2 - y = 0, \qquad (1)$$

$$y^2 - x = 0. \qquad (2)$$

Solving Eq. (1) for y, we get $y = x^2$.

Substituting x^2 for y in Eq. (2) and solving, we get

$$\left(x^2\right)^2 - x = 0$$

$$x^4 - x = 0$$

$$x\left(x^3 - 1\right) = 0$$

$x = 0 \quad$ or $\quad x^3 - 1 = 0$

$x = 0 \quad$ or $\quad x^3 = 1$

$x = 0 \quad$ or $\quad x = 1$

To find y when $x = 0$, we substitute 0 for x in either Eq. (1) or Eq. (2). We use Eq. (2):

$$y^2 - (0) = 0$$

$$y^2 = 0$$

$$y = 0.$$

Thus, $(0,0)$ is one critical point.

To find the other critical point we substitute 1 for x in either Eq. (1) or Eq. (2). We use Eq. (2):

$$y^2 - 1 = 0$$

$$y^2 = 1$$

$$y = \pm 1.$$

However, we notice that when $(1,-1)$ is not a solution to Eq. (1) $1^2 - (-1) = 2 \neq 0$.

Thus, $(1,1)$ is the other critical point.

We must check both $(0,0)$ and $(1,1)$ to see whether they yield maximum or minimum values.

For $(0,0)$

$$D = f_{xx}(0,0) \cdot f_{yy}(0,0) - \left[f_{xy}(0,0)\right]^2$$

$$D = 0 \cdot (0) - (-3)^2 \qquad \begin{bmatrix} f_{xx}(0,0) = 6 \cdot 0 = 0 \\ f_{yy}(0,0) = 6 \cdot 0 = 0 \\ f_{xy}(0,0) = -3 \end{bmatrix}$$

$$D = -9.$$

Since $D < 0$, it follows that $f(0,0)$ is neither a maximum nor a minimum, but a saddle point.

For $(1,1)$

$$D = f_{xx}(1,1) \cdot f_{yy}(1,1) - \left[f_{xy}(1,1) \right]^2$$

$$D = 6 \cdot (6) - (-3)^2 \qquad \begin{bmatrix} f_{xx}(1,1) = 6 \cdot 1 = 6 \\ f_{yy}(1,1) = 6 \cdot (1) = 6 \\ f_{xy}(1,1) = -3 \end{bmatrix}$$

$$D = 36 - 9$$

$$D = 27.$$

Thus, $D = 27$ and $f_{xx}(1,1) = 6$. Since $D > 0$ and $f_{xx}(1,1) > 0$, it follows that f has a relative minimum at $(1,1)$. The minimum is found as follows:

$$f(x,y) = x^3 + y^3 - 3xy$$

$$f(1,1) = 1^3 + 1^3 - 3(1)(1)$$

$$= 1 + 1 - 3$$

$$= -1.$$

The relative minimum value of f is -1 at $(1,1)$.

7. $f(x,y) = x^2 + y^2 - 2x + 4y - 2$

Find f_x.

$$f(x,y) = \underline{x}^2 + y^2 - 2\underline{x} + 4y - 2,$$

$$f_x = 2x - 2.$$

Find f_y.

$$f(x,y) = x^2 + \underline{y}^2 - 2x + 4\underline{y} - 2,$$

$$f_y = 2y + 4.$$

Find f_{xx} and f_{xy}.

$$f_x = 2\underline{x} - 2, \qquad f_x = 2x - 2,$$

$$f_{xx} = 2. \qquad f_{xy} = 0.$$

Find f_{yy}.

$$f_y = 2\underline{y} + 4,$$

$$f_{yy} = 2.$$

Solve the system of equations $f_x = 0$ and $f_y = 0$:

$$2x - 2 = 0, \quad 2y + 4 = 0,$$

$$2x = 2, \qquad 2y = -4,$$

$$x = 1. \qquad y = -2.$$

The only critical point is $(1,-2)$.

We must check $(1,-2)$ to see whether it yields a maximum or minimum value.

For $(1,-2)$

$$D = f_{xx}(1,-2) \cdot f_{yy}(1,-2) - \left[f_{xy}(1,-2) \right]^2$$

$$D = 2 \cdot (2) - (0)^2 \qquad \begin{bmatrix} f_{xx}(1,-2) = 2 \\ f_{yy}(1,-2) = 2 \\ f_{xy}(1,-2) = 0 \end{bmatrix}$$

$$D = 4.$$

Thus, $D = 4$ and $f_{xx}(1,-2) = 2$. Since $D > 0$ and $f_{xx}(1,-2) > 0$, it follows that f has a relative minimum at $(1,-2)$. The minimum is found as follows:

$$f(x,y) = x^2 + y^2 - 2x + 4y - 2$$

$$f(1,-2) = 1^2 + (-2)^2 - 2(1) + 4(-2) - 2$$

$$= 1 + 4 - 2 - 8 - 2$$

$$= -7.$$

The relative minimum value of f is -7 at $(1,-2)$.

9. $f(x,y) = x^2 + y^2 + 2x - 4y$

Find f_x.

$$f(x,y) = \underline{x}^2 + y^2 + 2\underline{x} - 4y,$$

$$f_x = 2x + 2.$$

Find f_y.

$$f(x,y) = x^2 + \underline{y}^2 + 2x - 4\underline{y},$$

$$f_y = 2y - 4.$$

Find f_{xx} and f_{xy}.

$$f_x = 2\underline{x} + 2, \qquad f_x = 2x + 2,$$

$$f_{xx} = 2. \qquad f_{xy} = 0.$$

Find f_{yy}.

$$f_y = 2\underline{y} - 4,$$

$$f_{yy} = 2.$$

Solve the system of equations $f_x = 0$ and $f_y = 0$:

$$2x + 2 = 0, \quad 2y - 4 = 0,$$

$$2x = -2, \qquad 2y = 4,$$

$$x = -1. \qquad y = 2.$$

The only critical point is $(-1,2)$.

We must check $(-1,2)$ to see whether it yields a maximum or minimum value.

For $(-1,2)$, we have:

$$D = f_{xx}(-1,2) \cdot f_{yy}(-1,2) - \left[f_{xy}(-1,2)\right]^2$$

$$D = 2 \cdot (2) - (0)^2 \qquad \begin{bmatrix} f_{xx}(-1,2) = 2 \\ f_{yy}(-1,2) = 2 \\ f_{xy}(-1,2) = 0 \end{bmatrix}$$

$D = 4.$

Thus, $D = 4$ and $f_{xx}(-1,2) = 2.$ Since $D > 0$ and $f_{xx}(-1,2) > 0$, it follows that f has a relative minimum at $(-1,2)$. The minimum is found as follows:

$$f(x,y) = x^2 + y^2 + 2x - 4y$$

$$f(-1,2) = (-1)^2 + (2)^2 + 2(-1) - 4(2)$$

$$= 1 + 4 - 2 - 8$$

$$= -5.$$

The relative minimum value of f is -5 at $(-1,2)$.

11. $f(x,y) = 4x^2 - y^2$

Find f_x.

$$f(x,y) = 4\underline{x}^2 - y^2,$$

$$f_x = 8x.$$

Find f_y.

$$f(x,y) = 4x^2 - \underline{y}^2,$$

$$f_y = -2y.$$

Find f_{xx} and f_{xy}.

$$f_x = 8\underline{x}, \quad f_x = 8x,$$

$$f_{xx} = 8. \quad f_{xy} = 0.$$

Find f_{yy}.

$$f_y = -2\underline{y},$$

$$f_{yy} = -2.$$

Solve the system of equations $f_x = 0$ and $f_y = 0$:

$$8x = 0, \qquad -2y = 0,$$

$$x = 0. \qquad y = 0.$$

The only critical point is $(0,0)$.

We must check $(0,0)$ to see whether it yields a maximum or minimum value.

For $(0,0)$

$$D = f_{xx}(0,0) \cdot f_{yy}(0,0) - \left[f_{xy}(0,0)\right]^2$$

$$D = 8 \cdot (-2) - (0)^2 \qquad \begin{bmatrix} f_{xx}(0,0) = 8 \\ f_{yy}(0,0) = -2 \\ f_{xy}(0,0) = 0 \end{bmatrix}$$

$D = -16.$

Since $D < 0$, it follows that $f(0,0)$ is neither a maximum nor a minimum, but a saddle point.

13. $f(x,y) = e^{x^2+y^2+1}$

Find f_x.

$$f(x,y) = e^{x^2+y^2+1},$$

$$f_x = 2xe^{x^2+y^2+1}.$$

Find f_y.

$$f(x,y) = e^{x^2+\underline{y}^2+1},$$

$$f_y = 2ye^{x^2+y^2+1}.$$

Find f_{xx}.

$$f_x = 2\underline{x}e^{x^2+y^2+1}$$

$$f_{xx} = 2x\left(2xe^{x^2+y^2+1}\right) + 2e^{x^2+y^2+1}$$

$$= 4x^2e^{x^2+y^2+1} + 2e^{x^2+y^2+1}.$$

Find f_{xy}.

$$f_x = 2xe^{x^2+\underline{y}^2+1}$$

$$f_{xy} = 2x\left(2ye^{x^2+y^2+1}\right)$$

$$= 4xye^{x^2+y^2+1}.$$

Find f_{yy}.

$$f_y = 2\underline{y}e^{x^2+y^2+1}$$

$$f_{yy} = 2y\left(2ye^{x^2+y^2+1}\right) + 2e^{x^2+y^2+1}$$

$$= 4y^2e^{x^2+y^2+1} + 2e^{x^2+y^2+1}.$$

Solve the system of equations $f_x = 0$ and $f_y = 0$:

$$2xe^{x^2+y^2+1} = 0, \qquad 2ye^{x^2+y^2+1} = 0,$$

$$x = 0. \qquad y = 0.$$

The only critical point is $(0,0)$.

We must check $(0,0)$ to see whether it yields a maximum or minimum value.

For $(0,0)$

$$D = f_{xx}(0,0) \cdot f_{yy}(0,0) - \left[f_{xy}(0,0) \right]^2$$

$$D = 2e \cdot (2e) - (0)^2 \qquad \begin{bmatrix} f_{xx}(0,0) = 2e \\ f_{yy}(0,0) = 2e \\ f_{xy}(0,0) = 0 \end{bmatrix}$$

$$D = 4e^2.$$

Thus, $D > 0$ and $f_{xx}(0,0) = 2e > 0$, it follows that f has a relative minimum at $(0,0)$. The minimum is found as follows:

$$f(x,y) = e^{x^2 + y^2 + 1}$$

$$f(0,0) = e^{0^2 + 0^2 + 1}$$

$$= e.$$

The relative minimum value of f is e at $(0,0)$.

15. $R(x,y) = 17x + 21y$

$C(x,y) = 4x^2 - 4xy + 2y^2 - 11x + 25y - 3$

Total profit, $P(x,y)$ is given by

$P(x,y)$

$= R(x,y) - C(x,y)$

$= (17x + 21y) - (4x^2 - 4xy + 2y^2 - 11x + 25y - 3)$

$= -4x^2 + 4xy - 2y^2 + 28x - 4y + 3$

Find P_x.

$P(x,y) = -4\underline{x}^2 + 4\underline{x}y - 2y^2 + 28\underline{x} - 4y + 3$

$\qquad P_x = -8x + 4y + 28$

Find P_y.

$P(x,y) = -4x^2 + 4x\underline{y} - 2\underline{y}^2 + 28x - 4\underline{y} + 3$

$\qquad P_y = 4x - 4y - 4$

Find P_{xx} and P_{xy}.

$P_x = -8\underline{x} + 4y + 28 \qquad P_x = -8x + 4\underline{y} + 28$

$P_{xx} = -8. \qquad\qquad P_{xy} = 4.$

Find P_{yy}.

$P_y = 4x - 4\underline{y} - 4$

$P_{yy} = -4.$

Solve the system of equations
$P_x = 0$ and $P_y = 0$:

$-8x + 4y + 28 = 0, \qquad (1)$

$4x - 4y - 4 = 0. \qquad (2)$

Adding these equations, we get:

$-4x + 24 = 0.$

Then,

$-4x = -24$

$x = 6.$

To find y when $x = 6$, we substitute 6 for x into either Eq. (1) or Eq. (2). We use Eq. (1):

$-8(6) + 4y + 28 = 0$

$4y - 20 = 0$

$4y = 20$

$y = 5.$

Thus, $(6,5)$ is the only critical point, and $P(6,5)$ is a candidate for a maximum or minimum value.

We must check to see whether $P(6,5)$ is a maximum or minimum value:

$$D = P_{xx}(6,5) \cdot P_{yy}(6,5) - \left[P_{xy}(6,5) \right]^2$$

$$= (-8)(-4) - 4^2 \qquad \begin{bmatrix} P_{xx}(6,5) = -8 \\ P_{yy}(6,5) = -4 \\ P_{xy}(6,5) = 4 \end{bmatrix}$$

$$= 32 - 16$$

$$= 16.$$

Thus, $D = 16$ and $P_{xx}(6,5) = -8$. Since $D > 0$ and $P_{xx}(6,5) < 0$, it follows that P has a relative maximum at $(6,5)$. Thus, to maximize profit, the company must produce and sell 6 thousand of the $17 sunglasses and 5 thousand of the $21 sunglasses.

17. $P(a,p) = 2ap + 80p - 15p^2 - \dfrac{1}{10}a^2 p - 80$

Find P_a.

$P(a,p) = 2\underline{a}p + 80p - 15p^2 - \dfrac{1}{10}\underline{a}^2 p - 80,$

$\qquad P_a = 2p - \dfrac{1}{5}ap.$

Find P_p.

$P(a,p) = 2a\underline{p} + 80\underline{p} - 15\underline{p}^2 - \dfrac{1}{10}a^2\underline{p} - 80,$

$\qquad P_p = 2a + 80 - 30p - \dfrac{1}{10}a^2.$

Find P_{aa} and P_{ap}.

$$P_a = 2p - \frac{1}{5}ap \qquad P_a = 2\underline{p} - \frac{1}{5}a\underline{p}$$

$$P_{aa} = -\frac{1}{5}p. \qquad P_{ap} = 2 - \frac{1}{5}a.$$

Find P_{pp}.

$$P_p = 2a + 80 - 30\underline{p} - \frac{1}{10}a^2$$

$$P_{pp} = -30.$$

Solve the system of equations
$P_a = 0$ and $P_p = 0$:

$$2p - \frac{1}{5}ap = 0, \qquad (1)$$

$$2a + 80 - 30p - \frac{1}{10}a^2 = 0. \qquad (2)$$

Solving Eq. (1) for a, we get $a = 10$.

Substituting 10 for a in Eq. (2) and solving for p we get

$$2(10) + 80 - 30p - \frac{1}{10}(10)^2 = 0$$

$$20 + 80 - 10 = 30p$$

$$90 = 30p$$

$$3 = p.$$

Thus, $(10, 3)$ is the only critical point, and $P(10, 3)$ is a candidate for a maximum or minimum value.

We must check to see whether $P(10, 3)$ is a maximum or minimum value:

$$D = P_{aa}(10, 3) \cdot P_{pp}(10, 3) - \left[P_{ap}(10, 3) \right]^2$$

$$= \left(-\frac{3}{5} \right)(-30) - 0^2$$

$$\begin{bmatrix} P_{aa}(10, 3) = -\frac{1}{5} \cdot 3 = \frac{-3}{5} \\ P_{pp}(10, 3) = -30 \\ P_{ap}(10, 3) = 2 - \frac{1}{5} \cdot 10 = 0 \end{bmatrix}$$

$$= 18.$$

Since $D > 0$ and $P_{aa}(10, 3) = -\frac{3}{5} < 0$, it follows

that P has a relative maximum at $(10, 3)$. Thus, to maximize profit, the company must spend 10 million dollars on advertising and charge $3 per item. The maximum profit is found as follows:

$$P(10, 3) = 2(10)(3) + 80(3) - 15(3)^2 - \frac{1}{10}(10)^2(3) - 80$$

$$= 60 + 240 - 135 - 30 - 80$$

$$= 55.$$

The maximum profit is $55 million.

19. Sketch a drawing of the container.

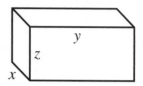

Let x, y and z represent the dimensions of the container as shown in the drawing.

$$V = x \cdot y \cdot z$$

$$320 = x \cdot y \cdot z \qquad \left[V = 320 \text{ft}^3 \right]$$

$$\frac{320}{x \cdot y} = z$$

Now we can express the cost as a function of two variables x and y. The area of the bottom is xy ft^2, so the cost of the bottom is $5xy$, two of the sides have area

$$xz \text{ ft}^2, \text{ or } x\left(\frac{320}{xy} \right) = \frac{320}{y} \text{ each. The area of each}$$

of the remaining two sides is

$$yz, \text{ or } y\left(\frac{320}{xy} \right) = \frac{320}{x}. \text{ Then, the total area of}$$

all four sides is

$$2\left(\frac{320}{y} + \frac{320}{x} \right) = \frac{640}{y} + \frac{640}{x}, \text{ and the cost of the}$$

four sides is $4\left(\frac{640}{y} + \frac{640}{x} \right)$, or $\frac{2560}{y} + \frac{2560}{x}$.

Now we can write the total cost function.

$$\underset{\text{cost}}{\text{Total}} = \underset{\text{bottom}}{\text{Cost of}} + \underset{\text{sides}}{\text{Cost of}}$$

$$C(x, y) = 5xy + \left(\frac{2560}{y} + \frac{2560}{x} \right).$$

Now, we find a minimum for $C(x, y)$

1. Find $C_x, C_y, C_{xx}, C_{yy},$ and C_{xy} :

$$C_x = 5y - \frac{2560}{x^2}, \qquad C_y = 5x - \frac{2560}{y^2},$$

$$C_{xx} = \frac{5120}{x^3}; \qquad C_{yy} = \frac{5120}{y^3};$$

$$C_{xy} = 5.$$

2. Solve the system of equations
$C_x = 0$ and $C_y = 0$:

$$5y - \frac{2560}{x^2} = 0, \qquad (1)$$

$$5x - \frac{2560}{y^2} = 0. \qquad (2)$$

Solving Eq. (1) for y:

$$5y - \frac{2560}{x^2} = 0$$

$$5y = \frac{2560}{x^2}$$

$$y = \frac{512}{x^2}$$

Substitute $\dfrac{512}{x^2}$ for y into Eq. (2) and solve
for x:

$$5x - \frac{2560}{\left(\dfrac{512}{x^2}\right)^2} = 0$$

$$5x - \frac{2560}{\dfrac{262,144}{x^4}} = 0$$

$$5x - \frac{2560x^4}{262,144} = 0$$

$$5x - \frac{5x^4}{512} = 0$$

$$2560x - 5x^4 = 0 \qquad \text{Multiplying by 512.}$$

$$5x\left(512 - x^3\right) = 0$$

$$5x = 0 \quad \text{or} \quad 512 - x^3 = 0$$

$$x = 0 \quad \text{or} \qquad x^3 = 512$$

$$x = 0 \quad \text{or} \qquad x = 8.$$

Since none of the dimensions can be 0, only
$x = 8$ has meaning in this application.

Substitute 8 for x into Eq. (1) to find y:

$$5y - \frac{2560}{(8)^2} = 0$$

$$5y - \frac{2560}{64} = 0$$

$$5y - 40 = 0$$

$$5y = 40$$

$$y = 8.$$

Thus, $(8,8)$ is the only critical point, and
$C(8,8)$ is a candidate for a maximum or
minimum value.

3. We must check to see whether $C(8,8)$ is a
maximum or minimum value:

$$D = C_{xx}(8,8) \cdot C_{yy}(8,8) - \left[C_{xy}(8,8)\right]^2$$

$$= \left(\frac{5120}{8^3}\right)\left(\frac{5120}{8^3}\right) - 5^2 \quad \text{Using step 1 above}$$

$$= \frac{5120}{512} \cdot \frac{5120}{512} - 25$$

$$= 10 \cdot 10 - 25$$

$$= 100 - 25$$

$$= 75.$$

4. Since $D > 0$ and $C_{xx}(8,8) = 10 > 0$, it
follows that C has a relative minimum
at $(8,8)$. Thus, to minimize cost, the
dimensions of the bottom of the container
should be 8 ft. by 8 ft. The height of the
container should be $\dfrac{320}{8 \cdot 8}$, or 5 ft.

21. a) $q_1 = 64 - 4p_1 - 2p_2 \qquad (1)$

$q_2 = 56 - 2p_1 - 4p_2 \qquad (2)$

$R(p_1, p_2)$

$= p_1 q_1 + p_2 q_2$

$= p_1(64 - 4p_1 - 2p_2) +$

$\quad p_2(56 - 2p_1 - 4p_2)$

$= 64p_1 - 4p_1^2 - 2p_1 p_2 +$

$\quad 56p_2 - 2p_1 p_2 - 4p_2^2$

$= 64p_1 - 4p_1^2 - 4p_1 p_2 + 56p_2 - 4p_2^2.$

b) We now find the values of p_1 and p_2 to maximize total revenue.

$$R_{p_1} = 64 - 8p_1 - 4p_2,$$

$$R_{p_2} = -4p_1 + 56 - 8p_2,$$

$$R_{p_1 p_1} = -8,$$

$$R_{p_2 p_2} = -8,$$

$$R_{p_1 p_2} = -4.$$

Solve the system of equations $R_{p_1} = 0$ and $R_{p_2} = 0$:

$$64 - 8p_1 - 4p_2 = 0$$

$$-4p_1 + 56 - 8p_2 = 0$$

The solution to this system is $p_1 = 6$ and $p_2 = 4$.

We check to see if $R(6,4)$ is a maximum or a minimum value.

$$D = R_{p_1 p_1}(6,4) \cdot R_{p_2 p_2}(6,4) -$$

$$\left[R_{p_1 p_2}(6,4) \right]^2$$

$$= (-8)(-8) - (-4)^2$$

$$= 64 - 16$$

$$= 48.$$

Since $D > 0$ and $R_{p_1 p_1}(6,4) = -8 < 0$, it follows that R has a relative maximum at $(6,4)$. Thus, in order to maximize revenue, p_1 must be $6 \cdot 10 = \$60$ and p_2 must be $4 \cdot 10 = \$40$.

c) We substitute 6 for p_1 and 4 for p_2 into the demand equations to find q_1 and q_2.

$$q_1 = 64 - 4p_1 - 2p_2$$

$$q_1 = 64 - 4(6) - 2(4)$$

$$= 64 - 24 - 8$$

$$= 32$$

$$q_2 = 56 - 2p_1 - 4p_2$$

$$q_2 = 56 - 2(6) - 4(4)$$

$$= 56 - 12 - 16$$

$$= 28$$

32 hundred units of q_1 will be demanded and 28 hundred units of q_2 will be demanded.

d) To maximize revenue 3200 units of the \$60 calculator and 2800 units of the \$40 calculator must be produced and sold. The maximum revenue is found as follows:

$$R = 60 \cdot 3200 + 40 \cdot 2800$$

$$= 192,000 + 112,000$$

$$= 304,000.$$

The maximum revenue is \$304,000.

23. $f(x,y) = e^x + e^y - e^{x+y}$

1. Find f_x, f_y, f_{xx}, f_{yy}, and f_{xy}:

$$f_x = e^x - e^{x+y}$$

$$f_y = e^y - e^{x+y}$$

$$f_{xx} = e^x - e^{x+y}$$

$$f_{yy} = e^y - e^{x+y}$$

$$f_{xy} = -e^{x+y}$$

2. Solve the system of equations $f_x = 0$ and $f_y = 0$:

$$e^x - e^{x+y} = 0 \qquad (1)$$

$$e^y - e^{x+y} = 0 \qquad (2)$$

We can solve the first equation for y:

$$e^x - e^{x+y} = 0$$

$$e^x = e^{x+y}$$

$$x = x + y$$

$$y = 0.$$

We can solve the second equation for x:

$$e^y - e^{x+y} = 0$$

$$e^y = e^{x+y}$$

$$y = x + y$$

$$x = 0.$$

Thus, $(0,0)$ is a critical point, and $f(0,0)$ is a candidate for a maximum or minimum.

3. We must check to see if $f(0,0)$ is a maximum or minimum value:

$$D = f_{xx}(0,0) \cdot f_{yy}(0,0) - \left[f_{xy}(0,0) \right]^2$$

$$D = 0 \cdot 0 - (-1)^2$$

$$D = -1.$$

4. Since $D < 0$, it follows that $f(0,0)$ is neither a maximum nor a minimum, but a saddle point.

25. $f(x,y) = 2y^2 + x^2 - x^2 y$

1. Find f_x, f_y, f_{xx}, f_{yy}, and f_{xy}:

$$f_x = 2x - 2xy \qquad f_y = 4y - x^2$$

$$f_{xx} = 2 - 2y; \qquad f_{yy} = 4;$$

$$f_{xy} = -2x.$$

2. Solve the system of equations
$f_x = 0$ and $f_y = 0$:

$$2x - 2xy = 0, \qquad (1)$$

$$4y - x^2 = 0. \qquad (2)$$

Solving Eq. (2) for y, we get

$$4y = x^2$$

$$y = \frac{x^2}{4}$$

Substituting $\frac{x^2}{4}$ for y in Eq. (1) and solving, we get

$$2x - 2x \cdot \frac{x^2}{4} = 0$$

$$2x - \frac{x^3}{2} = 0$$

$$4x - x^3 = 0$$

$$x(4 - x^2) = 0$$

$$x = 0 \quad \text{or} \quad 4 - x^2 = 0$$

$$x = 0 \quad \text{or} \quad x^2 = 4$$

$$x = 0 \quad \text{or} \quad x = \pm 2$$

When $x = 0$, $y = \dfrac{0^2}{4} = 0$.

When $x = 2$, $y = \dfrac{2^2}{4} = 1$.

When $x = -2$, $y = \dfrac{(-2)^2}{4} = 1$.

The critical points are
$(0,0), (2,1),$ and $(-2,1)$

3. We must check all the critical points to determine whether they yield maximum or minimum values.
For $(0,0)$

$$D = f_{xx}(0,0) \cdot f_{yy}(0,0) - \left[f_{xy}(0,0) \right]^2$$

$$D = (2) \cdot (4) - 0^2 \qquad \begin{bmatrix} f_{xx}(0,0) = 2 \\ f_{yy}(0,0) = 4 \\ f_{xy}(0,0) = 0 \end{bmatrix}$$

$$D = 8.$$

Since $D > 0$ and $f_{xx}(0,0) = 2 > 0$, it follows that f has a relative minimum at $(0,0)$. The minimum is found as follows:

$$f(x,y) = 2y^2 + x^2 - x^2 y$$

$$f(0,0) = 2 \cdot 0^2 + 0^2 - 0^2 \cdot 0 = 0$$

The relative minimum value of f is 0 at $(0,0)$.

For $(2,1)$

$$D = f_{xx}(2,1) \cdot f_{yy}(2,1) - \left[f_{xy}(2,1) \right]^2$$

$$D = (0) \cdot (4) - (-4)^2 \qquad \begin{bmatrix} f_{xx}(2,1) = 0 \\ f_{yy}(2,1) = 4 \\ f_{xy}(2,1) = -4 \end{bmatrix}$$

$$D = -16.$$

For $(-2,1)$

$$D = f_{xx}(-2,1) \cdot f_{yy}(-2,1) - \left[f_{xy}(-2,1) \right]^2$$

$$D = (0) \cdot (4) - (4)^2 \qquad \begin{bmatrix} f_{xx}(-2,1) = 0 \\ f_{yy}(-2,1) = 4 \\ f_{xy}(-2,1) = 4 \end{bmatrix}$$

$$D = -16.$$

Since $D < 0$ for both $(2,1)$ and $(-2,1)$, it follows that f has neither a maximum nor a minimum, but a saddle point at both of these points. Therefore, the only relative extrema of f is a relative minimum of 0 occurring at $(0,0)$.

27. $\boxed{tw}$

29. $f(x,y)$ has a relative minimum of -5 at $(0,0)$.

31. $f(x,y)$ has no relative extrema.

Exercise Set 6.4

1. Find the regression line for the data set:

x	1	2	4	5
y	1	3	3	4

The data points are $(1,1),(2,3),(4,3),$ and $(5,4)$.

The points on the regression line are $(1,y_1),(2,y_2),(4,y_3),$ and $(5,y_4)$.

The y-deviations are
$$y_1 - 1,\ y_2 - 3,\ y_3 - 3,\ y_4 - 4.$$

We want to minimize
$$S = (y_1 - 1)^2 + (y_2 - 3)^2 + (y_3 - 3)^2 + (y_4 - 4)^2$$

Where:
$$y_1 = m \cdot 1 + b$$
$$y_2 = m \cdot 2 + b$$
$$y_3 = m \cdot 4 + b$$
$$y_4 = m \cdot 5 + b$$

Substituting we get:
$$S = (m + b - 1)^2 + (2m + b - 3)^2 + (4m + b - 3)^2 + (5m + b - 4)^2$$

In order to minimize S, we need to find the first partial derivatives.
$$\frac{\partial S}{\partial b} = 2(m + b - 1) + 2(2m + b - 3) + 2(4m + b - 3) + 2(5m + b - 4)$$
$$= 2m + 2b - 2 + 4m + 2b - 6 + 8m + 2b - 6 + 10m + 2b - 8$$
$$= 24m + 8b - 22$$

$$\frac{\partial S}{\partial m} = 2(m + b - 1) \cdot 1 + 2(2m + b - 3) \cdot 2 + 2(4m + b - 3) \cdot 4 + 2(5m + b - 4) \cdot 5$$
$$= 2m + 2b - 2 + 8m + 4b - 12 + 32m + 8b - 24 + 50m + 10b - 40$$
$$= 92m + 24b - 78$$

We set these derivatives equal to 0 and solve the resulting system.
$$24m + 8b - 22 = 0$$
$$92m + 24b - 78 = 0$$

The solution to this system is $b = 0.95$, $m = 0.6$.

We use the D-test to verify that $S(0.6, 0.95)$ is a relative minimum.

We first find the second-order partial derivatives.
$$S_{bb} = 8, S_{bm} = 24, S_{mm} = 92$$
$$D = S_{bb}(0.6, 0.95) \cdot S_{mm}(0.6, 0.95) - \left[S_{bm}(0.6, 0.95) \right]^2$$
$$D = 8 \cdot 92 - [24]^2$$
$$= 160$$

Since $D > 0$ and $S_{bb}(0.6, 0.95) = 8 > 0$, S has a relative minimum at $(0.6, 0.95)$. The regression line is $y = 0.6x + 0.95$.

3. Find the regression line for the data set:

x	1	2	3	5
y	0	1	3	4

The data points are $(1,0),(2,1),(3,3),$ and $(5,4)$.

The points on the regression line are $(1,y_1),(2,y_2),(3,y_3),$ and $(5,y_4)$.

The y-deviations are
$$y_1 - 0,\ y_2 - 1,\ y_3 - 3,\ \text{and } y_4 - 4.$$

We want to minimize
$$S = (y_1 - 0)^2 + (y_2 - 1)^2 + (y_3 - 3)^2 + (y_4 - 4)^2$$

Where:
$$y_1 = m \cdot 1 + b$$
$$y_2 = m \cdot 2 + b$$
$$y_3 = m \cdot 3 + b$$
$$y_4 = m \cdot 5 + b$$

Substituting we get:
$$S = (m + b)^2 + (2m + b - 1)^2 + (3m + b - 3)^2 + (5m + b - 4)^2$$

In order to minimize S, we need to find the first partial derivatives.
$$\frac{\partial S}{\partial b} = 2(m + b) + 2(2m + b - 1) + 2(3m + b - 3) + 2(5m + b - 4)$$
$$= 2m + 2b + 4m + 2b - 2 + 6m + 2b - 6 + 10m + 2b - 8$$
$$= 22m + 8b - 16$$

$\dfrac{\partial S}{\partial m} = 2(m+b)\cdot 1 + 2(2m+b-1)\cdot 2 +$

$\qquad 2(3m+b-3)\cdot 3 + 2(5m+b-4)\cdot 5$

$\qquad = 2m + 2b + 8m + 4b - 4 +$

$\qquad\qquad 18m + 6b - 18 + 50m + 10b - 40$

$\qquad = 78m + 22b - 62$

We set these derivatives equal to 0 and solve the resulting system.

$22m + 8b - 16 = 0$

$78m + 22b - 62 = 0$

The solution to this system is $b = -\dfrac{29}{35}$, $m = \dfrac{36}{35}$

We use the D-test to verify that $S\left(-\frac{29}{35}, \frac{36}{35}\right)$ is a relative minimum.

We first find the second-order partial derivatives.

$S_{bb} = 8, S_{bm} = 22, S_{mm} = 78$

$D = S_{bb}\left(-\frac{29}{35}, \frac{36}{35}\right)\cdot S_{mm}\left(-\frac{29}{35}, \frac{36}{35}\right) -$

$\qquad\qquad \left[S_{bm}\left(-\frac{29}{35}, \frac{36}{35}\right)\right]^2$

$D = 8\cdot 78 - [22]^2$

$\quad = 140$

Since $D > 0$ and $S_{bb}\left(-\frac{29}{35}, \frac{36}{35}\right) = 8 > 0$, S has a relative minimum at $\left(-\frac{29}{35}, \frac{36}{35}\right)$. The regression line is $y = \dfrac{36}{35}x - \dfrac{29}{35}$.

5. a) The data points are
    $(0, 3.10), (1, 3.35), (10, 3.80), (11, 4.25),$
    $(16, 4.75),$ and $(17, 5.15).$
    The points on the regression line are
    $(0, y_1), (1, y_2), (10, y_3), (11, y_4),$
    $(16, y_5),$ and $(17, y_6).$
    The y-deviations are
    $y_1 - 3.10,\ y_2 - 3.35,\ y_3 - 3.80,$
    $y_4 - 4.25, y_5 - 4.75,$ and $y_6 - 5.15.$
    We want to minimize

    $S = (y_1 - 3.10)^2 + (y_2 - 3.35)^2 + (y_3 - 3.80)^2$

    $\quad + (y_4 - 4.25)^2 + (y_5 - 4.75)^2 + (y_6 - 5.15)^2$

Where:

$y_1 = m\cdot 0 + b$

$y_2 = m\cdot 1 + b$

$y_3 = m\cdot 10 + b$

$y_4 = m\cdot 11 + b$

$y_5 = m\cdot 16 + b$

$y_6 = m\cdot 17 + b$

Substituting we get:

$S = (b - 3.10)^2 + (m + b - 3.35)^2 +$

$\qquad (10m + b - 3.80)^2 + (11m + b - 4.25)^2 +$

$\qquad (16m + b - 4.75)^2 + (17m + b - 5.15)^2$

In order to minimize S, we need to find the first partial derivatives.

$\dfrac{\partial S}{\partial b} = 2(b - 3.10) + 2(m + b - 3.35) +$

$\qquad 2(10m + b - 3.80) + 2(11m + b - 4.25) +$

$\qquad 2(16m + b - 4.75) + 2(17m + b - 5.15)$

$\qquad = 2b - 6.2 + 2m + 2b - 6.7 + 20m + 2b - 7.6 +$

$\qquad\quad 22m + 2b - 8.5 + 32m + 2b - 9.5 + 34m +$

$\qquad\quad 2b - 10.3$

$\qquad = 110m + 12b - 48.8$

$\dfrac{\partial S}{\partial m}$

$\qquad = 2(m + b - 3.35)\cdot 1 + 2(10m + b - 3.80)\cdot 10 +$

$\qquad\quad 2(11m + b - 4.25)\cdot 11 + 2(16m + b - 4.75)\cdot 16 +$

$\qquad\quad 2(17m + b - 5.15)\cdot 17$

$\qquad = 2m + 2b - 6.7 + 200m + 20b - 76 + 242m +$

$\qquad\quad 22b - 93.5 + 512m + 32b - 152 + 578m +$

$\qquad\quad 34b - 175.1$

$\qquad = 1534m + 110b - 503.3$

We set these derivatives equal to 0 and solve the resulting system.

$\qquad 110m + 12b - 48.8 = 0$

$1534m + 110b - 503.3 = 0$

The solution to this system is

$b = 3.090710209$

$m = 0.1064679772$

We use the D-test to verify that $S(b, m)$ is a relative minimum.

We first find the second-order partial derivatives.

$S_{bb} = 12, S_{bm} = 110, S_{mm} = 1534$

$D = S_{bb} \cdot S_{mm} - [S_{bm}]^2$

$D = 12 \cdot 1534 - [110]^2$

$\quad = 6308$

Since $D > 0$ and $S_{bb} = 12 > 0$, S has a relative minimum at $(3.090710209, 0.1064679772)$. The regression line is
$y = 0.1064679772x + 3.090710209$.

b) In 2010, $x = 2010 - 1980 = 30$

$y = 0.1064679772(30) + 3.090710209$

$\quad \approx 6.28$

The minimum wage will be about \$6.28 in 2010.

In 2015, $x = 2015 - 1980 = 35$

$y = 0.1064679772(35) + 3.090710209$

$\quad \approx 6.82$

The minimum wage will be about \$6.82 in 2015.

7. a) The data points are
$(1950, 71.1), (1960, 73.1), (1970, 74.7),$
$(1980, 77.4), (1990, 78.8), (2000, 79.5),$
and $(2003, 80.1)$.

The points on the regression line are
$(1950, y_1), (1960, y_2), (1970, y_3),$
$(1980, y_4), (1990, y_5), (2000, y_6),$
and $(2003, y_7)$.

The y-deviations are
$y_1 - 71.1, \; y_2 - 73.1, \; y_3 - 74.7,$
$y_4 - 77.4, y_5 - 78.8, y_6 - 79.5,$
and $y_7 - 80.1$.

We want to minimize
$S = (y_1 - 71.1)^2 + (y_2 - 73.1)^2 +$
$\quad (y_3 - 74.7)^2 + (y_4 - 77.4)^2 +$
$\quad (y_5 - 78.8)^2 + (y_6 - 79.5)^2 +$
$\quad (y_7 - 80.1)^2$

Where:

$y_1 = m \cdot 1950 + b$

$y_2 = m \cdot 1960 + b$

$y_3 = m \cdot 1970 + b$

$y_4 = m \cdot 1980 + b$

$y_5 = m \cdot 1990 + b$

$y_6 = m \cdot 2000 + b$

$y_7 = m \cdot 2003 + b$

Substituting we get:

$S = (1950m + b - 71.1)^2 + (1960m + b - 73.1)^2 +$
$\quad (1970m + b - 74.7)^2 + (1980m + b - 77.4)^2 +$
$\quad (1990m + b - 78.8)^2 + (2000m + b - 79.5)^2 +$
$\quad (2003m + b - 80.1)^2$

In order to minimize S, we need to find the first partial derivatives.

$\dfrac{\partial S}{\partial b}$

$= 2(1950m + b - 71.1) + 2(1960m + b - 73.1) +$
$2(1970m + b - 74.7) + 2(1980m + b - 77.4) +$
$2(1990m + b - 78.8) + 2(2000m + b - 79.5) +$
$2(2003m + b - 80.1)$

$= 3900m + 2b - 142.2 + 3920m + 2b -$
$146.2 + 3940m + 2b - 149.4 + 3960m +$
$2b - 154.8 + 3980m + 2b - 157.6 +$
$4000m + 2b - 159 + 4006m + 2b - 160.2$

$= 27706m + 14b - 1069.4$

$\dfrac{\partial S}{\partial m} = 2(1950m + b - 71.1) \cdot 1950 +$

$\quad 2(1960m + b - 73.1) \cdot 1960 +$

$\quad 2(1970m + b - 74.7) \cdot 1970 +$

$\quad 2(1980m + b - 77.4) \cdot 1980 +$

$\quad 2(1990m + b - 78.8) \cdot 1990 +$

$\quad 2(2000m + b - 79.5) \cdot 2000 +$

$\quad 2(2003m + b - 80.1) \cdot 2003$

$\quad = 7,605,000m + 3900b - 277,290 +$

$\quad 7,683,200m + 3920b - 286,552 +$

$\quad 7,761,800m + 3940b - 294,318 +$

$\quad 7,840,800m + 3960b - 306,504 +$

$\quad 7,920,200m + 3980b - 313,624 +$

$\quad 8,000,000m + 4000b - 318,000 +$

$\quad 8,024,018m + 4006b - 320,880.6$

$\quad = 54,835,018m + 27,706b - 2,117,168.6$

We set these derivatives equal to 0 and solve the resulting system.
$$27,706m + 14b - 1069.4 = 0$$
$$54,835,018m + 27,706b - 2,117,168.6 = 0$$
The solution to this system is
$$b = -261.0738233$$
$$m = 0.1705202312$$
We use the D-test to verify that $S(b,m)$ is a relative minimum.
We first find the second-order partial derivatives.
$$S_{bb} = 14, S_{bm} = 27,706, S_{mm} = 54,835,018$$
$$D = S_{bb} \cdot S_{mm} - [S_{bm}]^2$$
$$D = 14 \cdot 54,835,018 - [27,706]^2$$
$$= 67,816$$

Since $D > 0$ and $S_{bb} = 14 > 0$, S has a relative minimum at
$(-261.0738233, 0.1705202312)$. The regression line is
$y = 0.1705202312x - 261.0738233$.

b) In 2010,
$$y = 0.1705202312(2010) - 261.0738233$$
$$\approx 81.7.$$
In 2010, the average life expectancy of women will be about 81.7 years.
In 2015,
$$y = 0.1705202312(2015) - 261.0738233$$
$$\approx 82.5.$$
In 2015, the average life expectancy of women will be about 82.5 years.

9. a) The data points are
$(70,75), (60,62),$ and $(85,89).$
The points on the regression line are
$(70, y_1), (60, y_2),$ and $(85, y_3).$
The y-deviations are
$y_1 - 75,$ $y_2 - 62,$ and $y_3 - 89.$
We want to minimize
$$S = (y_1 - 75)^2 + (y_2 - 62)^2 + (y_3 - 89)^2$$
Where:
$$y_1 = m \cdot 70 + b$$
$$y_2 = m \cdot 60 + b$$
$$y_3 = m \cdot 85 + b$$

Substituting we get:
$$S = (70m + b - 75)^2 + (60m + b - 62)^2 +$$
$$(85m + b - 89)^2$$
In order to minimize S, we need to find the first partial derivatives.
$$\frac{\partial S}{\partial b}$$
$$= 2(70m + b - 75) + 2(60m + b - 62) +$$
$$2(85m + b - 89)$$
$$= 140m + 2b - 150 + 120m + 2b - 124 +$$
$$170m + 2b - 178$$
$$= 430m + 6b - 452$$
$$\frac{\partial S}{\partial m}$$
$$= 2(70m + b - 75) \cdot 70 +$$
$$2(60m + b - 62) \cdot 60 +$$
$$2(85m + b - 89) \cdot 85$$
$$= 9800m + 140b - 10,500 + 7200m + 120b -$$
$$7440 + 14,450m + 170b - 15,130$$
$$= 31,450m + 430b - 33,070$$
We set these derivatives equal to 0 and solve the resulting system.
$$430m + 6b - 452 = 0$$
$$31,450m + 430b - 33,070 = 0$$
The solution to this system is
$$b = -1.236842105$$
$$m = 1.068421053$$
We use the D-test to verify that $S(b,m)$ is a relative minimum.
We first find the second-order partial derivatives.
$$S_{bb} = 6, S_{bm} = 430, S_{mm} = 31,450$$
$$D = S_{bb} \cdot S_{mm} - [S_{bm}]^2$$
$$D = 6 \cdot 31,450 - [430]^2$$
$$= 3800$$
Since $D > 0$ and $S_{bb} = 6 > 0$, S has a relative minimum at
$(-1.236842105, 1.068421053)$.
The regression line is
$y = 1.068421053x - 1.236842105$

b) $x = 81$
$$y = 1.068421053(81) - 1.236842105$$
$$\approx 85.$$
A student who scores 81% on the midterm will score about 85% on the final.

11. $\boxed{tw}$

13. a) Converting the times to decimal notation and using the STAT package on a calculator, we get the regression equation
$y = -0.0059379586x + 15.57191398$.

   b) We predict that the world record in 2010 will be about 3.636617 minutes or $3{:}38.2$. We predict that the world record in 2015 will be about 3.60693 minutes or $3{:}36.4$.

   c) According to the regression model, we would predict the world record in 1999 to be 3.7019 minutes or $3{:}42.1$. This is about a second faster than the actual world record.

Exercise Set 6.5

1. Find the maximum value of
$$f(x,y) = xy$$
subject to the constraint
$3x + y = 10$.
We first express $3x + y = 10$ as $3x + y - 10 = 0$.
We form the new function F, given by:
$$F(x,y,\lambda) = xy - \lambda(3x + y - 10).$$
We find the first partial derivatives:
$$F(x,y,\lambda) = x\underline{y} - \lambda(3\underline{x} + y - 10)$$
$$F_x = y - 3\lambda,$$
$$F(x,y,\lambda) = x\underline{y} - \lambda(3x + \underline{y} - 10)$$
$$F_y = x - \lambda,$$
$$F(x,y,\lambda) = xy - \underline{\lambda}(3x + y - 10)$$
$$F_\lambda = -(3x + y - 10).$$
We set each derivative equal to 0 and solve the resulting system:
$$y - 3\lambda = 0 \qquad (1)$$
$$x - \lambda = 0 \qquad (2)$$
$$3x + y - 10 = 0 \qquad (3) \begin{bmatrix} {\scriptstyle -(3x+y-10)=0,\ \text{or}} \\ {\scriptstyle 3x+y-10=0} \end{bmatrix}$$
Solving Eq. (2) for λ, we get:
$$\lambda = x.$$
Substituting into Eq. (1) for λ, we get:
$$y - 3(x) = 0, \text{ or } y = 3x. \qquad (4)$$

Substituting $3x$ for y in Eq. (3), we get:
$$3x + 3x - 10 = 0$$
$$6x = 10$$
$$x = \frac{10}{6} = \frac{5}{3}$$
Then, using Eq. (4), we have:
$$y = 3\left(\frac{5}{3}\right) = 5.$$
The maximum value of f subject to the
constraint occurs at $\left(\dfrac{5}{3}, 5\right)$ and is
$$f\left(\frac{5}{3}, 5\right) = \frac{5}{3} \cdot 5 = \frac{25}{3}.$$

3. Find the maximum value of
$$f(x,y) = 4 - x^2 - y^2$$
subject to the constraint
$x + 2y = 10$.
We first express $x + 2y = 10$ as $x + 2y - 10 = 0$.
We form the new function F, given by:
$$F(x,y,\lambda) = 4 - x^2 - y^2 - \lambda(x + 2y - 10).$$
We find the first partial derivatives:
$$F(x,y,\lambda) = 4 - \underline{x}^2 - y^2 - \lambda(\underline{x} + 2y - 10)$$
$$F_x = -2x - \lambda,$$
$$F(x,y,\lambda) = 4 - x^2 - \underline{y}^2 - \lambda(x + 2\underline{y} - 10)$$
$$F_y = -2y - 2\lambda,$$
$$F(x,y,\lambda) = 4 - x^2 - y^2 - \underline{\lambda}(x + 2y - 10)$$
$$F_\lambda = -(x + 2y - 10).$$
We set each derivative equal to 0 and solve the resulting system:
$$-2x - \lambda = 0 \qquad (1)$$
$$-2y - 2\lambda = 0 \qquad (2)$$
$$x + 2y - 10 = 0 \qquad (3)\begin{bmatrix} {\scriptstyle -(x+2y-10)=0,\ \text{or}} \\ {\scriptstyle x+2y-10=0} \end{bmatrix}$$
Solving Eq. (1) for λ, we get:
$$\lambda = -2x.$$
Substituting into Eq. (2) for λ, we get:
$$-2y - 2(-2x) = 0, \text{ or } y = 2x. \qquad (4)$$
Substituting $2x$ for y in Eq. (3), we get:
$$x + 2(2x) - 10 = 0$$
$$5x = 10$$
$$x = 2.$$

Then, using Eq. (4), we have:

$y = 2(2) = 4.$

The maximum value of f subject to the constraint occurs at $(2,4)$ and is

$f(2,4) = 4 - (2)^2 - (4)^2$

$= 4 - 4 - 16$

$= -16.$

5. Find the minimum value of
    $f(x,y) = x^2 + y^2$
    subject to the constraint
    $2x + y = 10$.
    We first express $2x + y = 10$ as $2x + y - 10 = 0$.
    We form the new function F, given by:
    $F(x,y,\lambda) = x^2 + y^2 - \lambda(2x + y - 10).$
    We find the first partial derivatives:
    $F(x,y,\lambda) = \underline{x}^2 + y^2 - \lambda(2\underline{x} + y - 10).$

    $F_x = 2x - 2\lambda,$

    $F(x,y,\lambda) = x^2 + \underline{y}^2 - \lambda(2x + \underline{y} - 10).$

    $F_y = 2y - \lambda,$

    $F(x,y,\lambda) = x^2 + y^2 - \underline{\lambda}(2x + y - 10)$

    $F_\lambda = -(2x + y - 10).$

    We set each derivative equal to 0 and solve the resulting system:

    $2x - 2\lambda = 0 \qquad (1)$

    $2y - \lambda = 0 \qquad (2)$

    $2x + y - 10 = 0 \qquad (3)\begin{bmatrix} -(2x+y-10)=0,\ \text{or} \\ 2x+y-10=0 \end{bmatrix}$

    Solving Eq. (2) for λ, we get:

    $\lambda = 2y.$

    Substituting into Eq. (1) for λ, we get:

    $2x - 2(2y) = 0$, or $x = 2y.$ \qquad (4)

    Substituting $2y$ for x in Eq. (3), we get:

    $2(2y) + y - 10 = 0$

    $5y = 10$

    $y = 2.$

    Then, using Eq. (4), we have:

    $x = 2(2) = 4.$

The minimum value of f subject to the constraint occurs at $(4,2)$ and is

$f(4,2) = 4^2 + 2^2$

$= 16 + 4$

$= 20.$

7. Find the minimum value of
    $f(x,y) = 2y^2 - 6x^2$
    subject to the constraint
    $2x + y = 4$.
    We first express $2x + y = 4$ as $2x + y - 4 = 0$.
    We form the new function F, given by:
    $F(x,y,\lambda) = 2y^2 - 6x^2 - \lambda(2x + y - 4).$
    We find the first partial derivatives:
    $F(x,y,\lambda) = 2y^2 - 6\underline{x}^2 - \lambda(2\underline{x} + y - 4)$

    $F_x = -12x - 2\lambda,$

    $F(x,y,\lambda) = 2\underline{y}^2 - 6x^2 - \lambda(2x + \underline{y} - 4)$

    $F_y = 4y - \lambda,$

    $F(x,y,\lambda) = 2y^2 - 6x^2 - \underline{\lambda}(2x + y - 4)$

    $F_\lambda = -(2x + y - 4).$

    We set each derivative equal to 0 and solve the resulting system:

    $-12x - 2\lambda = 0 \qquad (1)$

    $4y - \lambda = 0 \qquad (2)$

    $2x + y - 4 = 0 \qquad (3)\begin{bmatrix} -(2x+y-4)=0,\ \text{or} \\ 2x+y-4=0 \end{bmatrix}$

    Solving Eq. (2) for λ, we get:

    $\lambda = 4y.$

    Substituting into Eq. (1) for λ, we get:

    $-12x - 2(4y) = 0$

    $8y = -12x$

    $y = -\dfrac{3}{2}x.$ \qquad (4)

    Substituting $-\dfrac{3}{2}x$ for y in Eq. (3), we get:

    $2x + \left(-\dfrac{3}{2}x\right) - 4 = 0$

    $\dfrac{1}{2}x = 4$

    $x = 8.$

    Then, using Eq. (4), we have:

    $y = -\dfrac{3}{2}(8) = -12.$

The minimum value of f subject to the constraint occurs at $(8, -12)$ and is

$$f(8, -12) = 2(-12)^2 - 6(8)^2$$
$$= 2(144) - 6(64)$$
$$= 288 - 384$$
$$= -96.$$

9. Find the minimum value of
$$f(x, y, z) = x^2 + y^2 + z^2$$
subject to the constraint
$y + 2x - z = 3$.
We first express $y + 2x - z = 3$
as $y + 2x - z - 3 = 0$.
We form the new function F, given by:
$$F(x, y, z, \lambda)$$
$$= x^2 + y^2 + z^2 - \lambda(y + 2x - z - 3)$$
We find the first partial derivatives:
$$F_x = 2x - 2\lambda,$$
$$F_y = 2y - \lambda,$$
$$F_z = 2z + \lambda,$$
$$F_\lambda = -(y + 2x - z - 3).$$
We set each derivative equal to 0 and solve the resulting system:

$$2x - 2\lambda = 0 \qquad (1)$$
$$2y - \lambda = 0 \qquad (2)$$
$$2z + \lambda = 0 \qquad (3)$$
$$y + 2x - z - 3 = 0 \qquad (4)\begin{bmatrix} -(y+2x-z-3)=0, \text{ or} \\ y+2x-z-3=0 \end{bmatrix}$$

Solving Eq. (1) for x, we get:

$x = \lambda.$

Solving Eq. (2) for y, we get:

$$y = \frac{1}{2}\lambda.$$

Solving Eq. (3) for z, we get:

$$z = -\frac{1}{2}\lambda.$$

Substituting λ for x, $\frac{1}{2}\lambda$ for y, and $-\frac{1}{2}\lambda$ for z
into Eq. (4), we get:

$$y + 2x - z - 3 = 0$$
$$\frac{1}{2}\lambda + 2\lambda - \left(-\frac{1}{2}\lambda\right) - 3 = 0$$
$$3\lambda = 3$$
$$\lambda = 1.$$

Then,
$x = \lambda = 1$
$$y = \frac{1}{2}\lambda = \frac{1}{2}$$
$$z = -\frac{1}{2}\lambda = -\frac{1}{2}$$
The minimum value of f subject to the

constraint occurs at $\left(1, \frac{1}{2}, -\frac{1}{2}\right)$ and is

$$f\left(1, \frac{1}{2}, -\frac{1}{2}\right) = 1^2 + \left(\frac{1}{2}\right)^2 + \left(-\frac{1}{2}\right)^2$$
$$= 1 + \frac{1}{4} + \frac{1}{4}$$
$$= \frac{3}{2}.$$

11. Find the maximum value of
$$f(x, y) = xy \qquad (\text{Product is } x \cdot y)$$
subject to the constraint
$$x + y = 50. \qquad (\text{Sum is 50.}).$$
We first express $x + y = 50$ as $x + y - 50 = 0$.
We form the new function F, given by:
$$F(x, y, \lambda) = xy - \lambda(x + y - 50).$$
We find the first partial derivatives:
$$F_x = y - \lambda,$$
$$F_y = x - \lambda,$$
$$F_\lambda = -(x + y - 50).$$
We set each derivative equal to 0 and solve the resulting system:

$$y - \lambda = 0 \qquad (1)$$
$$x - \lambda = 0 \qquad (2)$$
$$x + y - 50 = 0 \qquad (3)\begin{bmatrix} -(x+y-50)=0, \text{ or} \\ x+y-50=0 \end{bmatrix}$$

Solving Eq. (2) for λ, we get:

$\lambda = x.$

Substituting into Eq. (1) for λ, we get:

$$y - (x) = 0, \text{ or } y = x. \qquad (4)$$

Substituting x for y in Eq. (3), we get:

$$x + x - 50 = 0$$
$$2x = 50$$
$$x = 25.$$
Then, using Eq. (4), we have:
$y = 25.$

The maximum value of f subject to the constraint occurs at $(25, 25)$. Thus, the two numbers whose sum is 50 that have the maximum product are 25 and 25.

13. Find the minimum value of

$$f(x, y) = xy \qquad (\text{Product is } x \cdot y)$$

subject to the constraint

$$x - y = 6. \qquad (\text{Difference is 6.}).$$

We first express $x - y = 6$ as $x - y - 6 = 0$.

We form the new function F, given by:

$$F(x, y, \lambda) = xy - \lambda(x - y - 6).$$

We find the first partial derivatives:

$$F_x = y - \lambda,$$

$$F_y = x + \lambda,$$

$$F_\lambda = -(x - y - 6).$$

We set each derivative equal to 0 and solve the resulting system:

$$y - \lambda = 0 \qquad (1)$$

$$x + \lambda = 0 \qquad (2)$$

$$x - y - 6 = 0 \qquad (3) \left[\begin{smallmatrix} -(x-y-6)=0, \text{ or} \\ x-y-6=0 \end{smallmatrix}\right]$$

Solving Eq. (1) for λ, we get:

$$\lambda = y.$$

Substituting into Eq. (1) for λ, we get:

$$x + (y) = 0, \text{ or } y = -x. \qquad (4)$$

Substituting $-x$ for y in Eq. (3), we get:

$$x - (-x) - 6 = 0$$

$$2x = 6$$

$$x = 3.$$

Then, using Eq. (4), we have:

$$y = -3.$$

The minimum value of f subject to the constraint occurs at $(3, -3)$. Thus, the two numbers whose difference is 6 that have the minimum product are 3 and -3.

15. Find the minimum value of

$$f(x, y, z) = (x - 1)^2 + (y - 1)^2 + (z - 1)^2$$

subject to the constraint

$$x + 2y + 3z = 13.$$

We first express $x + 2y + 3z = 13$

as $x + 2y + 3z - 13 = 0$.

We form the new function F, given by:

$$F(x, y, z, \lambda)$$

$$= (x - 1)^2 + (y - 1)^2 + (z - 1)^2 -$$

$$\lambda(x + 2y + 3z - 13)$$

We find the first partial derivatives:

$$F_x = 2(x - 1) - \lambda,$$

$$F_y = 2(y - 1) - 2\lambda,$$

$$F_z = 2(z - 1) - 3\lambda,$$

$$F_\lambda = -(x + 2y + 3z - 13).$$

We set each derivative equal to 0 and solve the resulting system:

$$2(x - 1) - \lambda = 0 \qquad (1)$$

$$2(y - 1) - 2\lambda = 0 \qquad (2)$$

$$2(z - 1) - 3\lambda = 0 \qquad (3)$$

$$x + 2y + 3z - 13 = 0 \qquad (4) \left[\begin{smallmatrix} -(x+2y+3z-13)=0, \text{ or} \\ x+2y+3z-13=0 \end{smallmatrix}\right]$$

Solving Eq. (1) for x, we get:

$$2x - 2 - \lambda = 0$$

$$2x = 2 + \lambda$$

$$x = 1 + \frac{1}{2}\lambda.$$

Solving Eq. (2) for y, we get:

$$2y - 2 - 2\lambda = 0$$

$$2y = 2 + 2\lambda$$

$$y = 1 + \lambda.$$

Solving Eq. (3) for z, we get:

$$2z - 2 - 3\lambda = 0$$

$$2z = 2 + 3\lambda$$

$$z = 1 + \frac{3}{2}\lambda.$$

Substituting $1 + \frac{1}{2}\lambda$ for x, $1 + \lambda$ for y, and

$1 + \frac{3}{2}\lambda$ for z into Eq. (4), we get:

$$x + 2y + 3z - 13 = 0$$

$$\left(1 + \frac{1}{2}\lambda\right) + 2(1 + \lambda) + 3\left(1 + \frac{3}{2}\lambda\right) - 13 = 0$$

$$1 + \frac{1}{2}\lambda + 2 + 2\lambda + 3 + \frac{9}{2}\lambda = 13$$

$$6 + 7\lambda = 13$$

$$7\lambda = 7$$

$$\lambda = 1.$$

Then,

$$x = 1 + \frac{1}{2}\lambda = 1 + \frac{1}{2} \cdot \frac{3}{2}$$

$$y = 1 + \lambda = 1 + 1 = 2$$

$$z = 1 + \frac{3}{2}\lambda = 1 + \frac{3}{2} = \frac{5}{2}.$$

The minimum value of f subject to the

constraint occurs at $\left(\frac{3}{2}, 2, \frac{5}{2}\right)$.

17. The area of the page is given by $A = xy$ and the perimeter of the page is given by $P = 2x + 2y$. See the figure.

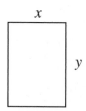

We want to maximize the area

$A = xy$

Subject to the constraint

$2x + 2y = 39$.

We first express $2x + 2y = 39$ as

$2x + 2y - 39 = 0$.

We form the new function F, given by:

$F(x, y, \lambda) = xy - \lambda(2x + 2y - 39)$.

We find the first partial derivatives:

$F_x = y - 2\lambda$,

$F_y = x - 2\lambda$,

$F_\lambda = -(2x + 2y - 39)$.

We set each derivative equal to 0 and solve the resulting system:

$$y - 2\lambda = 0 \qquad (1)$$

$$x - 2\lambda = 0 \qquad (2)$$

$$2x + 2y - 39 = 0 \qquad (3) \begin{bmatrix} -(2x+2y-39)=0, \text{ or} \\ 2x+2y-39=0 \end{bmatrix}$$

From Eqs. (1) and (2) we see:

$y = 2\lambda = x$.

Substituting x for y in Eq. (3), we get:

$2x + 2x - 39 = 0$

$4x = 39$

$x = \frac{39}{4} = 9\frac{3}{4}$.

Then,

$y = x = 9\frac{3}{4}$.

The maximum area subject to the constraint occurs at $\left(9\frac{3}{4}, 9\frac{3}{4}\right)$. The maximum area is

$A = 9\frac{3}{4} \cdot 9\frac{3}{4} = 95\frac{1}{16}$ in^2. The area of the standard

$8\frac{1}{2} \times 11$ paper is not the maximum area of paper that has a perimeter of 39 in.

19. We want to minimize the function s given by

$s(h, r) = 2\pi rh + 2\pi r^2$

subject to the volume constraint

$\pi r^2 h = 27$, or $\pi r^2 h - 27 = 0$.

We form the new function S given by

$S(h, r, \lambda) = 2\pi rh + 2\pi r^2 - \lambda(\pi r^2 h - 27)$.

We find the first partial derivatives.

$S_h = 2\pi r - \lambda\pi r^2$,

$S_r = 2\pi h + 4\pi r - 2\lambda\pi rh$,

$S_\lambda = -(\pi r^2 h - 27)$.

We set these derivatives equal to 0 and solve the resulting system.

$$2\pi r - \lambda\pi r^2 = 0 \quad (1)$$

$$2\pi h + 4\pi r - 2\lambda\pi rh = 0 \quad (2)$$

$$\pi r^2 h - 27 = 0. \quad (3) \begin{bmatrix} -(\pi r^2 h - 27)=0, \text{ or} \\ \pi r^2 h - 27 = 0 \end{bmatrix}$$

We solve Eq. (1) for r:

$\pi r(2 - \lambda r) = 0$

$\pi r = 0 \quad$ or $\quad 2 - \lambda r = 0$

$r = 0 \quad$ or $\quad r = \frac{2}{\lambda}$

Note, $r = 0$ can not be a solution to the original

problem, so we continue by substituting $\frac{2}{\lambda}$ for r

in Eq. (2).

$$2\pi h + 4\pi\left(\frac{2}{\lambda}\right) - 2\lambda\pi\left(\frac{2}{\lambda}\right)h = 0$$

$$2\pi h + \frac{8\pi}{\lambda} - 4\pi h = 0$$

$$\frac{8\pi}{\lambda} - 2\pi h = 0$$

$$h = \frac{4}{\lambda}$$

Since $h = \dfrac{4}{\lambda}$ and $r = \dfrac{2}{\lambda}$, it follows that $h = 2r$.

substituting $2r$ for h in Eq. (3) yields:

$$\pi r^2 (2r) - 27 = 0$$
$$2\pi r^3 = 27$$
$$r^3 = \frac{27}{2\pi}$$
$$r = \sqrt[3]{\frac{27}{2\pi}} \approx 1.6$$

So when $r \approx 1.6$ ft and $h = 2(1.6) \approx 3.2$ ft, the surface area of the oil drum is a minimum. The minimum area is about

$$2\pi(1.6)(3.2) + 2\pi(1.6)^2 \approx 48.3 \text{ ft}^2.$$

(Answers will vary due to rounding differences.)

21. We want maximize
$$S(L,M) = ML - L^2$$
subject to the constraint
$M + L = 90$.
We first express $M + L = 90$ as $M + L - 90 = 0$.
We form the new function F, given by:
$$F(x,y,\lambda) = ML - L^2 - \lambda(M + L - 90).$$
We find the first partial derivatives:
$$F_L = M - 2L - \lambda,$$
$$F_M = L - \lambda,$$
$$F_\lambda = -(M + L - 90).$$
We set each derivative equal to 0 and solve the resulting system:

$M - 2L - \lambda = 0$	(1)
$L - \lambda = 0$	(2)
$M + L - 90 = 0$	$(3)\ \left[\begin{smallmatrix}-(M+L-90)=0,\ \text{or}\\ M+L-90=0\end{smallmatrix}\right]$

Solving Eq. (2) for λ, we get:
$$\lambda = L.$$
Substituting into Eq. (1) for λ, we get:
$$M - 2L - L = 0$$
$$M - 3L = 0$$
$$M = 3L. \qquad (4)$$
Substituting $3L$ for M in Eq. (3), we get:
$$3L + L - 90 = 0$$
$$4L = 90$$
$$L = 22.5.$$

Then, using Eq. (4), we have:
$$M = 3(22.5) = 67.5.$$
The maximum value of S subject to the constraint occurs at $(22.5, 67.5)$ and is

$$S(22.5, 67.5) = (67.5)(22.5) - (22.5)^2$$
$$= 1518.75 - 506.25$$
$$= 1012.5$$

23. a) The area of the floor is xy.
The cost of the floor is $4xy$.
The area of the walls is $2xz + 2yz$.
The cost of the walls is $3(2xz + 2yz)$.
The area of the ceiling is xy.
The cost of the ceiling is $3xy$.
Therefore, the total cost function is
$$C(x,y,z) = 4xy + 3(2xz + 2yz) + 3xy$$
$$= 7xy + 6xz + 6yz.$$

b) We want to minimize the value of
$$C(x,y,z) = 7xy + 6xz + 6yz$$
subject to the constraint of
$$x \cdot y \cdot z = 252{,}000 \qquad (\text{Volume} = l \cdot w \cdot h)$$
We first express $x \cdot y \cdot z = 252{,}000$
as $x \cdot y \cdot z - 252{,}000 = 0$.
We form the new function F, given by:
$$F(x,y,z,\lambda)$$
$$= 7xy + 6xz + 6yz - \lambda(x \cdot y \cdot z - 252{,}000)$$
We find the first partial derivatives:
$$F_x = 7y + 6z - \lambda yz,$$
$$F_y = 7x + 6z - \lambda xz,$$
$$F_z = 6x + 6y - \lambda xy,$$
$$F_\lambda = -(x \cdot y \cdot z - 252{,}000).$$
We set each derivative equal to 0 and solve the resulting system:

$7y + 6z - \lambda yz = 0$	(1)
$7x + 6z - \lambda xz = 0$	(2)
$6x + 6y - \lambda xy = 0$	(3)
$xyz - 252{,}000 = 0$	$(4)\ \left[\begin{smallmatrix}-(x\cdot y\cdot z-252{,}000)=0,\ \text{or}\\ x\cdot y\cdot z-252{,}000=0\end{smallmatrix}\right]$

Solving Eq. (2) for x and Eq. (1) for y, we get:
$$x = \frac{6z}{\lambda z - 7} \quad \text{and} \quad y = \frac{6z}{\lambda z - 7}.$$
Thus, $x = y$.

Substituting x for y we get the following system:

$7x + 6z - \lambda xz = 0$

$6x + 6x - \lambda xx = 0$

$xxz - 252,000 = 0$

Which simplifies to:

$7x + 6z - \lambda xz = 0 \qquad (5)$

$12x - \lambda x^2 = 0 \qquad (6)$

$x^2 z - 252,000 = 0 \qquad (7)$

Solving Eq. (6) for x, we get

$12x - \lambda x^2 = 0$

$x(12 - \lambda x) = 0$

$x = 0 \quad$ or $\quad 12 - \lambda x = 0$

$x = 0 \quad$ or $\qquad x = \dfrac{12}{\lambda}$

We only consider $x = \dfrac{12}{\lambda}$ since x cannot be

0 in the original problem. We continue by

substituting $\dfrac{12}{\lambda}$ for x into Eq. (7) and solving

for z.

$\left(\dfrac{12}{\lambda}\right)^2 z - 252,000 = 0$

$\qquad \dfrac{144}{\lambda^2} \cdot z = 252,000$

$\qquad z = \dfrac{252,000}{144} \lambda^2$

$\qquad z = 1750 \lambda^2$

Next we substitute $\dfrac{12}{\lambda}$ for x and $1750\lambda^2$ for

z in Eq. (5) and solve for λ.

$7\left(\dfrac{12}{\lambda}\right) + 6 \cdot 1750\lambda^2 - \lambda\left(\dfrac{12}{\lambda}\right)1750\lambda^2 = 0$

$\dfrac{84}{\lambda} + 10,500\lambda^2 - 21,000\lambda^2 = 0$

$\qquad 10,500\lambda^2 = \dfrac{84}{\lambda}$

$\qquad\qquad \lambda^3 = \dfrac{84}{10,500}$

$\qquad\qquad \lambda^3 = \dfrac{1}{125}$

$\qquad\qquad \lambda = \dfrac{1}{5}$

Thus,

$x = \dfrac{12}{\lambda} = \dfrac{12}{\frac{1}{5}} = 12 \cdot \dfrac{5}{1} = 60$

$y = \dfrac{12}{\lambda} = \dfrac{12}{\frac{1}{5}} = 12 \cdot \dfrac{5}{1} = 60$

$z = 1750\lambda^2 = 1750\left(\dfrac{1}{5}\right)^2 = 70$

The minimum total cost subject to the constraint occurs when the dimensions are 60 ft by 60ft by 70 ft. The minimum cost is found as follows:

$C(60,60,70) = 7 \cdot 60 \cdot 60 + 6 \cdot 60 \cdot 70 + 6 \cdot 60 \cdot 70$

$\qquad\qquad = 25,200 + 25,200 + 25,200$

$\qquad\qquad = 75,600.$

The minimum total cost of the building is $75,600.

25. $C(x,y) = C(x) + C(y)$

$C(x,y) = 10 + \dfrac{x^2}{6} + 200 + \dfrac{y^3}{9}$

$\qquad\quad = 210 + \dfrac{x^2}{6} + \dfrac{y^3}{9}$

We need to minimize

$C(x,y) = 210 + \dfrac{x^2}{6} + \dfrac{y^3}{9}$

subject to the constraint

$x + y = 10,100.$

We first express $x + y = 10,100$ as $x + y - 10,100 = 0.$.

We form the new function F, given by:

$F(x,y,\lambda)$

$= 210 + \dfrac{x^2}{6} + \dfrac{y^3}{9} - \lambda(x + y - 10,100).$

We find the first partial derivatives:

$F_x = \dfrac{x}{3} - \lambda,$

$F_y = \dfrac{1}{3}y^2 - \lambda,$

$F_\lambda = -(x + y - 10,100).$

We set each derivative equal to 0 and solve the resulting system:

$$\frac{x}{3} - \lambda = 0 \quad (1)$$

$$\frac{1}{3}y^2 - \lambda = 0 \quad (2)$$

$$x + y - 10{,}100 = 0 \quad (3) \begin{bmatrix} -(x+y-10{,}100)=0, \text{ or} \\ x+y-10{,}100=0 \end{bmatrix}$$

From Eq. (1) and Eq. (2) we see:

$$x = 3\lambda = y^2.$$

Thus, $x = y^2$.

Substituting y^2 for x in Eq. (3), we get:

$$y^2 + y - 10{,}100 = 0$$

$$(y+101)(y-100) = 0$$

$$y + 101 = 0 \quad \text{or} \quad y - 100 = 0$$

$$y = -101 \quad \text{or} \quad y = 100.$$

Since y cannot be -101 in the original problem, we only consider $y = 100$. If $y = 100$, then

$x = 100^2 = 10{,}000$. To minimize total costs, 10,000 units should be made on machine A and 100 units should be made on machine B.

27. Find the minimum value of

$$f(x,y) = 2x^2 + y^2 + 2xy + 3x + 2y$$

subject to the constraint

$$y^2 = x + 1.$$

We first express $y^2 = x+1$ as $y^2 - x - 1 = 0$.

We form the new function F, given by:

$$F(x,y,\lambda)$$
$$= 2x^2 + y^2 + 2xy + 3x + 2y - \lambda(y^2 - x - 1).$$

We find the first partial derivatives:

$$F_x = 4x + 2y + 3 + \lambda,$$

$$F_y = 2y + 2x + 2 - 2\lambda y,$$

$$F_\lambda = -(y^2 - x - 1).$$

We set each derivative equal to 0 and solve the resulting system:

$$4x + 2y + 3 + \lambda = 0 \quad (1)$$

$$2y + 2x + 2 - 2\lambda y = 0 \quad (2)$$

$$y^2 - x - 1 = 0 \quad (3) \begin{bmatrix} -(y^2-x-1)=0, \text{ or} \\ y^2-x-1=0 \end{bmatrix}$$

Solving Eq. (1) for λ, we get:

$$4x + 2y + 3 + \lambda = 0$$

$$\lambda = -4x - 2y - 3. \quad (4)$$

Solving Eq. (2) for λ, we get:

$$2y + 2x + 2 - 2\lambda y = 0$$

$$2\lambda y = 2y + 2x + 2$$

$$\lambda = \frac{2y + 2x + 2}{2y}$$

$$\lambda = \frac{y + x + 1}{y}. \quad (5)$$

Setting Eq. (4) equal to Eq. (5) and solving for x we have:

$$-4x - 2y - 3 = \frac{y + x + 1}{y}$$

$$-4xy - 2y^2 - 3y = y + x + 1$$

$$-2y^2 - 3y - y - 1 = x + 4xy$$

$$-2y^2 - 4y - 1 = x(1 + 4y)$$

$$\frac{-2y^2 - 4y - 1}{1 + 4y} = x. \quad (6)$$

Solving Eq. (3) for x we have:

$$y^2 - x - 1 = 0$$

$$y^2 - 1 = x \quad (7)$$

Substituting Eq. (7) into Eq. (6), we have:

$$\frac{-2y^2 - 4y - 1}{1 + 4y} = y^2 - 1$$

$$-2y^2 - 4y - 1 = (y^2 - 1)(1 + 4y)$$

$$-2y^2 - 4y - 1 = y^2 + 4y^3 - 1 - 4y$$

$$0 = 4y^3 + 3y^2$$

$$0 = y^2(4y + 3)$$

$$y^2 = 0 \quad \text{or} \quad 4y + 3 = 0$$

$$y = 0 \quad \text{or} \quad y = -\frac{3}{4}$$

Using equation (7) When $y = 0$,

$$x = (0)^2 - 1$$

$$= -1$$

$$f(-1,0) = 2(-1)^2 + (0)^2 + 2(-1)(0) +$$
$$3(-1) + 2(0)$$

$$= 2 - 3$$

$$= -1.$$

Using Eq. (7), when $y = -\dfrac{3}{4}$,

$$x = \left(\dfrac{-3}{4}\right)^2 - 1$$

$$= \dfrac{9}{16} - 1$$

$$= -\dfrac{7}{16}.$$

$$f\left(-\dfrac{7}{16}, -\dfrac{3}{4}\right) = 2\left(-\dfrac{7}{16}\right)^2 + \left(-\dfrac{3}{4}\right)^2 +$$

$$2\left(-\dfrac{7}{16}\right)\left(-\dfrac{3}{4}\right) + 3\left(-\dfrac{7}{16}\right) +$$

$$2\left(-\dfrac{3}{4}\right)$$

$$= \dfrac{49}{128} + \dfrac{9}{16} + \dfrac{21}{32} - \dfrac{21}{16} - \dfrac{3}{2}$$

$$= -\dfrac{155}{128}.$$

The minimum value of f subject to the constraint occurs at $\left(-\dfrac{7}{16}, -\dfrac{3}{4}\right)$ and is

$$f\left(-\dfrac{7}{16}, -\dfrac{3}{4}\right) = -\dfrac{155}{128}.$$

29. Find the maximum value of
$$f(x, y, z) = x^2 y^2 z^2$$
subject to the constraint
$$x^2 + y^2 + z^2 = 2.$$
We first express $x^2 + y^2 + z^2 = 2$
as $x^2 + y^2 + z^2 - 2 = 0$.
We form the new function F, given by:
$$F(x, y, z, \lambda)$$
$$= x^2 y^2 z^2 - \lambda(x^2 + y^2 + z^2 - 2)$$
We find the first partial derivatives:
$$F_x = 2xy^2 z^2 - 2\lambda x$$
$$F_y = 2x^2 yz^2 - 2\lambda y,$$
$$F_z = 2x^2 y^2 z - 2\lambda z,$$
$$F_\lambda = -(x^2 + y^2 + z^2 - 2).$$

We set each derivative equal to 0 and solve the resulting system:

$$2xy^2 z^2 - 2\lambda x = 0$$
$$2x^2 yz^2 - 2\lambda y = 0$$
$$2x^2 y^2 z - 2\lambda z = 0$$
$$x^2 + y^2 + z^2 - 2 = 0 \quad \left[\begin{array}{c} -(x^2+y^2+z^2-2)=0,\ \text{or} \\ x^2+y^2+z^2-2=0 \end{array}\right]$$

Rewriting the system we get:

$$x(2y^2 z^2 - 2\lambda) = 0 \quad (1)$$
$$y(2x^2 z^2 - 2\lambda) = 0 \quad (2)$$
$$z(2x^2 y^2 - 2\lambda) = 0 \quad (3)$$
$$x^2 + y^2 + z^2 - 2 = 0 \quad (4) \left[\begin{array}{c} -(x^2+y^2+z^2-2)=0,\ \text{or} \\ x^2+y^2+z^2-2=0 \end{array}\right]$$

Note that for
$x = 0, y = 0,$ or $z = 0$, $f(x, y, z) = 0$. For all
values of $x, y,$ and $z \neq 0$, $f(x, y, z) > 0$. Thus the
maximum value of f cannot occur when any or
all of the variables is 0. Thus we will only
consider nonzero values of $x, y,$ and z.
Using the Principle of Zero Products, we get:

From Eq. (1) From Eq. (2) From Eq. (3)
$y^2 z^2 - \lambda = 0$ $x^2 z^2 - \lambda = 0$ $x^2 y^2 - \lambda = 0$
$y^2 z^2 = \lambda$ $x^2 z^2 = \lambda$ $x^2 y^2 = \lambda$

Thus, $y^2 z^2 = x^2 z^2 = x^2 y^2$ and $x^2 = y^2 = z^2$.
Substituting x^2 for y^2 and z^2 in Eq. (4), we
have:
$$x^2 + x^2 + x^2 - 2 = 0$$
$$3x^2 = 2$$
$$x^2 = \dfrac{2}{3}$$
$$x = \pm\sqrt{\dfrac{2}{3}}$$

Since $x^2 = y^2 = z^2$ it follows that

$$y^2 = \dfrac{2}{3} \quad \text{and} \quad z^2 = \dfrac{2}{3}$$

$$y = \pm\sqrt{\dfrac{2}{3}} \quad \text{and} \quad z = \pm\sqrt{\dfrac{2}{3}}.$$

For $\left(\pm\sqrt{\dfrac{2}{3}}, \pm\sqrt{\dfrac{2}{3}}, \pm\sqrt{\dfrac{2}{3}} \right)$:

$$f(x,y,z) = \left(\pm\sqrt{\dfrac{2}{3}} \right)^2 \left(\pm\sqrt{\dfrac{2}{3}} \right)^2 \left(\pm\sqrt{\dfrac{2}{3}} \right)^2$$

$$= \dfrac{2}{3} \cdot \dfrac{2}{3} \cdot \dfrac{2}{3}$$

$$= \dfrac{8}{27}$$

Thus $f(x,y,z)$ has a maximum value of $\dfrac{8}{27}$ at

$\left(\pm\sqrt{\dfrac{2}{3}}, \pm\sqrt{\dfrac{2}{3}}, \pm\sqrt{\dfrac{2}{3}} \right)$.

31. Find the maximum value of

$f(x,y,z,t) = x + y + z + t$

subject to the constraint

$x^2 + y^2 + z^2 + t^2 = 1$.

We first express $x^2 + y^2 + z^2 + t^2 = 1$

as $x^2 + y^2 + z^2 + t^2 - 1 = 0$.

We form the new function F, given by:

$F(x,y,z,t,\lambda)$

$= x + y + z + t - \lambda\left(x^2 + y^2 + z^2 + t^2 - 1 \right)$

We find the first partial derivatives:

$F_x = 1 - 2\lambda x,$

$F_y = 1 - 2\lambda y,$

$F_z = 1 - 2\lambda z,$

$F_t = 1 - 2\lambda t,$

$F_\lambda = -\left(x^2 + y^2 + z^2 + t^2 - 1 \right).$

We set each derivative equal to 0 and solve the resulting system:

$$1 - 2\lambda x = 0 \quad (1)$$

$$1 - 2\lambda y = 0 \quad (2)$$

$$1 - 2\lambda z = 0 \quad (3)$$

$$1 - 2\lambda t = 0 \quad (4)$$

$$x^2 + y^2 + z^2 + t^2 - 1 = 0 \quad (5)$$

$$\left[\begin{array}{l} -\left(x^2 + y^2 + z^2 + t^2 - 1 \right) = 0, \text{ or} \\ x^2 + y^2 + z^2 + t^2 - 1 = 0 \end{array} \right]$$

From Eq. (1), Eq. (2), Eq. (3), and Eq. (4),

we see:

$$\dfrac{1}{2\lambda} = x = y = z = t.$$

Substituting x for y, z, and t in Eq. (5), we have:

$$x^2 + (x)^2 + (x)^2 + (x)^2 - 1 = 0$$

$$4x^2 = 1$$

$$x^2 = \dfrac{1}{4}$$

$$x = \pm\dfrac{1}{2}$$

Since $x = y = z = t$ it follows that

When $x = \dfrac{1}{2}$

$y = \dfrac{1}{2}$ and $z = \dfrac{1}{2}$ and $t = \dfrac{1}{2}$.

When $x = -\dfrac{1}{2}$

$y = -\dfrac{1}{2}$ and $z = -\dfrac{1}{2}$ and $t = -\dfrac{1}{2}$.

However, it is clear looking at the function that the point that will yield a maximum value is:

$\left(\dfrac{1}{2}, \dfrac{1}{2}, \dfrac{1}{2}, \dfrac{1}{2} \right)$.

The maximum value is found as follows:

$$f\left(\dfrac{1}{2}, \dfrac{1}{2}, \dfrac{1}{2}, \dfrac{1}{2} \right) = \dfrac{1}{2} + \dfrac{1}{2} + \dfrac{1}{2} + \dfrac{1}{2} = 2.$$

33. We want to maximize

$p(x,y)$

subject to the constraint.

$B = c_1 x + c_2 y.$

We first express $B = c_1 x + c_2 y$ as

$c_1 x + c_2 y - B = 0$

Then we form the new function P, given by:

$P(x,y,\lambda) = p(x,y) - \lambda\left(c_1 x + c_2 y - B \right).$

We find the first partial derivatives.

$P_x = p_x - \lambda c_1$

$P_y = p_y - \lambda c_2.$

We set these derivatives equal to 0 and solve for λ.

$$p_x - \lambda c_1 = 0 \qquad\qquad p_y - \lambda c_2 = 0$$

$$p_x = \lambda c_1 \qquad\qquad\quad p_y = \lambda c_2$$

$$\dfrac{p_x}{c_1} = \lambda \qquad\qquad\quad \dfrac{p_y}{c_2} = \lambda$$

Thus, $\lambda = \dfrac{p_x}{c_1} = \dfrac{p_y}{c_2}.$

35. $\boxed{tw}$

37 – 43. Left to the student.

Exercise Set 6.6

1. $\displaystyle\int_0^3\int_0^1 2y\,dx\,dy$

$= \displaystyle\int_0^3\left(\int_0^1 2y\,dx\right)dy$

We first evaluate the inside x-integral, treating y as a constant:

$\displaystyle\int_0^1 2y\,dx = 2y\,[x]_0^1$

$= 2y\,[1-0]$

$= 2y.$

Then we evaluate the outside y-integral:

$\displaystyle\int_0^3\left(\int_0^1 2y\,dx\right)dy = \int_0^3 2y\,dy \qquad \left(\int_0^1 2y\,dx = 2y\right)$

$= \left[y^2\right]_0^3$

$= 3^2 - 0^2$

$= 9.$

3. $\displaystyle\int_{-1}^3\int_1^2 x^2 y\,dy\,dx$

$= \displaystyle\int_{-1}^3\left(\int_1^2 x^2 y\,dy\right)dx$

We first evaluate the inside y-integral, treating x as a constant:

$\displaystyle\int_1^2 x^2 y\,dy = x^2\left[\frac{1}{2}y^2\right]_1^2$

$= x^2\left[\frac{1}{2}(2)^2 - \frac{1}{2}(1)^2\right]$

$= x^2\left[2 - \frac{1}{2}\right]$

$= \frac{3}{2}x^2.$

Then we evaluate the outside x-integral:

$\displaystyle\int_{-1}^3\left(\int_1^2 x^2 y\,dy\right)dx = \int_{-1}^3\frac{3}{2}x^2\,dx \qquad \left(\int_1^2 x^2 y\,dy = \frac{3}{2}x^2\right)$

$= \left[\frac{1}{2}x^3\right]_{-1}^3$

$= \frac{1}{2}\left[3^3 - (-1)^3\right]$

$= \frac{1}{2}\left[27 - (-1)\right]$

$= \frac{1}{2}[28]$

$= 14.$

5. $\displaystyle\int_0^5\int_{-2}^{-1}(3x+y)\,dx\,dy$

$= \displaystyle\int_0^5\left(\int_{-2}^{-1}(3x+y)\,dx\right)dy$

We first evaluate the inside x-integral, treating y as a constant:

$\displaystyle\int_{-2}^{-1}(3x+y)\,dx$

$= \left[\frac{3}{2}x^2 + yx\right]_{-2}^{-1}$

$= \left[\frac{3}{2}(-1)^2 + y(-1)\right] - \left[\frac{3}{2}(-2)^2 + y(-2)\right]$

$= \left[\frac{3}{2} - y\right] - \left[\frac{3}{2}\cdot 4 - 2y\right]$

$= y - \frac{9}{2}.$

Then we evaluate the outside y-integral:

$\displaystyle\int_0^5\left(\int_{-2}^{-1}(3x+y)\,dx\right)dy$

$= \displaystyle\int_0^5\left(y - \frac{9}{2}\right)dy \qquad \left(\int_{-2}^{-1}(3x+y)\,dx = y - \frac{9}{2}\right)$

$= \left[\frac{1}{2}y^2 - \frac{9}{2}y\right]_0^5$

$= \left[\frac{1}{2}(5)^2 - \frac{9}{2}(5)\right] - \left[\frac{1}{2}(0)^2 - \frac{9}{2}(0)\right]$

$= \frac{25}{2} - \frac{45}{2} - 0$

$= \frac{-20}{2}$

$= -10.$

7. $\int_{-1}^{1}\int_{x}^{1}xy\,dy\,dx$

$= \int_{-1}^{1}\left(\int_{x}^{1}xy\,dy\right)dx$

We first evaluate the inside y-integral, treating x as a constant:

$\int_{x}^{1}xy\,dy = x\left[\frac{1}{2}y^{2}\right]_{x}^{1}$

$= x\left[\frac{1}{2}(1)^{2} - \frac{1}{2}(x)^{2}\right]$

$= x\left[\frac{1}{2} - \frac{1}{2}x^{2}\right]$

$= \frac{1}{2}x - \frac{1}{2}x^{3}.$

Then we evaluate the outside x-integral:

$\int_{-1}^{1}\left(\int_{x}^{1}xy\,dy\right)dx$

$= \int_{-1}^{1}\left(\frac{1}{2}x - \frac{1}{2}x^{3}\right)dx \quad \left(\int_{x}^{1}xy\,dy = \frac{1}{2}x - \frac{1}{2}x^{3}\right)$

$= \left[\frac{1}{4}x^{2} - \frac{1}{8}x^{4}\right]_{-1}^{1}$

$= \left[\frac{1}{4}(1)^{2} - \frac{1}{8}(1)^{4}\right] - \left[\frac{1}{4}(-1)^{2} - \frac{1}{8}(-1)^{4}\right]$

$= \left[\frac{1}{4} - \frac{1}{8}\right] - \left[\frac{1}{4} - \frac{1}{8}\right]$

$= 0.$

9. $\int_{0}^{1}\int_{x^{2}}^{x}(x+y)\,dy\,dx$

$= \int_{0}^{1}\left(\int_{x^{2}}^{x}(x+y)\,dy\right)dx$

We first evaluate the inside y-integral, treating x as a constant:

$\int_{x^{2}}^{x}(x+y)\,dy$

$= \left[xy + \frac{1}{2}y^{2}\right]_{x^{2}}^{x}$

$= \left[x(x) + \frac{1}{2}(x)^{2}\right] - \left[x(x^{2}) + \frac{1}{2}(x^{2})^{2}\right]$

$= \left[x^{2} + \frac{1}{2}x^{2}\right] - \left[x^{3} + \frac{1}{2}x^{4}\right]$

$= -\frac{1}{2}x^{4} - x^{3} + \frac{3}{2}x^{2}.$

Then we evaluate the outside x-integral:

$\int_{0}^{1}\left(\int_{x^{2}}^{x}(x+y)\,dy\right)dx$

$= \int_{0}^{1}\left(-\frac{1}{2}x^{4} - x^{3} + \frac{3}{2}x^{2}\right)dx$

$\left(\int_{x^{2}}^{x}(x+y)\,dy = -\frac{1}{2}x^{4} - x^{3} + \frac{3}{2}x^{2}\right)$

$= \left[-\frac{1}{10}x^{5} - \frac{1}{4}x^{4} + \frac{1}{2}x^{3}\right]_{0}^{1}$

$= \left[-\frac{1}{10}(1)^{5} - \frac{1}{4}(1)^{4} + \frac{1}{2}(1)^{3}\right] -$

$\qquad \left[-\frac{1}{10}(0)^{5} - \frac{1}{4}(0)^{4} + \frac{1}{2}(0)^{3}\right]$

$= \left[-\frac{1}{10} - \frac{1}{4} + \frac{1}{2}\right] - [0]$

$= \frac{3}{20}.$

11. $\int_{0}^{1}\int_{1}^{e^{x}}\frac{1}{y}\,dy\,dx$

$= \int_{0}^{1}\left(\int_{1}^{e^{x}}\frac{1}{y}\,dy\right)dx$

We first evaluate the inside y-integral, treating x as a constant:

$\int_{1}^{e^{x}}\frac{1}{y}\,dy = \left[\ln y\right]_{1}^{e^{x}}$

$= \ln e^{x} - \ln 1$

$= x.$

Then we evaluate the outside x-integral:

$\int_{0}^{1}\left(\int_{1}^{e^{x}}\frac{1}{y}\,dy\right)dx$

$= \int_{0}^{1}x\,dx \qquad \left(\int_{1}^{e^{x}}\frac{1}{y}\,dy = x\right)$

$= \left[\frac{1}{2}x^{2}\right]_{0}^{1}$

$= \left[\frac{1}{2}(1)^{2} - \frac{1}{2}(0)^{2}\right]$

$= \frac{1}{2}.$

13. $\int_0^2 \int_0^x (x+y^2)\, dy\, dx$

$= \int_0^2 \left(\int_0^x (x+y^2)\, dy \right) dx$

We first evaluate the inside y-integral, treating x as a constant:

$\int_0^x (x+y^2)\, dy$

$= \left[xy + \frac{1}{3}y^3 \right]_0^x$

$= \left[x(x) + \frac{1}{3}(x)^3 \right] - \left[x(0) + \frac{1}{3}(0)^3 \right]$

$= \left[x^2 + \frac{1}{3}x^3 \right] - [0]$

$= \frac{1}{3}x^3 + x^2$

Then we evaluate the outside x-integral:

$\int_0^2 \left(\int_0^x (x+y^2)\, dy \right) dx$

$= \int_0^2 \left(\frac{1}{3}x^3 + x^2 \right) dx$

$\qquad \left(\int_0^x (x+y^2)\, dy = \frac{1}{3}x^3 + x^2 \right)$

$= \left[\frac{1}{12}x^4 + \frac{1}{3}x^3 \right]_0^2$

$= \left[\frac{1}{12}(2)^4 + \frac{1}{3}(2)^3 \right] - \left[\frac{1}{12}(0)^4 + \frac{1}{3}(0)^3 \right]$

$= \left[\frac{16}{12} + \frac{8}{3} \right] - [0]$

$= \frac{4}{3} + \frac{8}{3}$

$= 4.$

15. $\int_0^1 \int_0^{1-x^2} (1-y-x^2)\, dy\, dx$

$= \int_0^1 \left(\int_0^{1-x^2} (1-y-x^2)\, dy \right) dx$

We first evaluate the inside y-integral, treating x as a constant:

$\int_0^{1-x^2} (1-y-x^2)\, dy$

$= \left[y - \frac{1}{2}y^2 - x^2 y \right]_0^{1-x^2}$

$= \left[(1-x^2) - \frac{1}{2}(1-x^2)^2 - x^2(1-x^2) \right] -$

$\qquad\qquad \left[(0) - \frac{1}{2}(0)^2 - x^2(0) \right]$

$= 1 - x^2 - \frac{1}{2}(1 - 2x^2 + x^4) - x^2 + x^4$

$= 1 - x^2 - \frac{1}{2} + x^2 - \frac{1}{2}x^4 - x^2 + x^4$

$= \frac{1}{2}x^4 - x^2 + \frac{1}{2}$

Then we evaluate the outside x-integral:

$= \int_0^1 \left(\int_0^{1-x^2} (1-y-x^2)\, dy \right) dx$

$= \int_0^1 \left(\frac{1}{2}x^4 - x^2 + \frac{1}{2} \right) dx$

$= \left[\frac{1}{10}x^5 - \frac{1}{3}x^3 + \frac{1}{2}x \right]_0^1$

$= \left[\frac{1}{10}(1)^5 - \frac{1}{3}(1)^3 + \frac{1}{2}(1) \right] -$

$\qquad\qquad \left[\frac{1}{10}(0)^5 - \frac{1}{3}(0)^3 + \frac{1}{2}(0) \right]$

$= \frac{1}{10} - \frac{1}{3} + \frac{1}{2}$

$= \frac{4}{15}.$

The volume of the solid is $\frac{4}{15}$ units^3.

17. $f(x,y) = x^2 + \frac{1}{3}xy$

$0 \le x \le 1$

$0 \le y \le 2$

Find

$\int_0^2 \int_0^1 f(x,y)\, dx\, dy$

$= \int_0^2 \left(\int_0^1 \left(x^2 + \frac{1}{3}xy \right) dx \right) dy$

We first evaluate the inside x-integral, treating y as a constant:

$$\int_0^1 \left(x^2 + \frac{1}{3}xy \right)dx$$

$$= \left[\frac{1}{3}x^3 + \frac{1}{6}x^2 y \right]_0^1$$

$$= \left[\frac{1}{3}(1)^3 + \frac{1}{6}(1)^2 y \right] - \left[\frac{1}{3}(0)^3 + \frac{1}{6}(0)^2 y \right]$$

$$= \frac{1}{3} + \frac{1}{6}y.$$

Then we evaluate the outside y-integral:

$$\int_0^2 \left(\int_0^1 \left(x^2 + \frac{1}{3}xy \right)dx \right)dy$$

$$= \int_0^2 \left(\frac{1}{3} + \frac{1}{6}y \right)dy$$

$$= \left[\frac{1}{3}y + \frac{1}{12}y^2 \right]_0^2$$

$$= \left[\frac{1}{3}(2) + \frac{1}{12}(2)^2 \right] - \left[\frac{1}{3}(0) + \frac{1}{12}(0)^2 \right]$$

$$= \frac{2}{3} + \frac{4}{12}$$

$$= 1.$$

19. $f(x,y) = x^2 - 3x + \frac{1}{3}xy - \frac{1}{3}y + 2$

$1 \le x \le 2$

$3 \le y \le 5$

Find

$$\int_3^4 \int_1^2 f(x,y)\,dx\,dy$$

$$= \int_3^4 \left(\int_1^2 \left(x^2 - 3x + \frac{1}{3}xy - \frac{1}{3}y + 2 \right)dx \right)dy$$

We first evaluate the inside x-integral, treating y as a constant:

$$\int_1^2 \left(x^2 - 3x + \frac{1}{3}xy - \frac{1}{3}y + 2 \right)dx$$

$$= \left[\frac{1}{3}x^3 - \frac{3}{2}x^2 + \frac{1}{6}x^2 y - \frac{1}{3}xy + 2x \right]_1^2$$

$$= \left[\frac{1}{3}(2)^3 - \frac{3}{2}(2)^2 + \frac{1}{6}(2)^2 y - \frac{1}{3}(2)y + 2(2) \right] -$$

$$\left[\frac{1}{3}(1)^3 - \frac{3}{2}(1)^2 + \frac{1}{6}(1)^2 y - \frac{1}{3}(1)y + 2(1) \right]$$

$$= \left[\frac{8}{3} - 6 + \frac{2}{3}y - \frac{2}{3}y + 4 \right] -$$

$$\left[\frac{1}{3} - \frac{3}{2} + \frac{1}{6}y - \frac{1}{3}y + 2 \right]$$

$$= \frac{2}{3} - \left[\frac{5}{6} - \frac{1}{6}y \right]$$

$$= \frac{1}{6}y - \frac{1}{6}.$$

Then we evaluate the outside y-integral:

$$= \int_3^4 \left(\int_1^2 \left(x^2 - 3x + \frac{1}{3}xy - \frac{1}{3}y + 2 \right)dx \right)dy$$

$$= \int_3^4 \left(\frac{1}{6}y - \frac{1}{6} \right)dy$$

$$= \left[\frac{1}{12}y^2 - \frac{1}{6}y \right]_3^4$$

$$= \left[\frac{1}{12}(4)^2 - \frac{1}{6}(4) \right] - \left[\frac{1}{12}(3)^2 - \frac{1}{6}(3) \right]$$

$$= \left[\frac{4}{3} - \frac{2}{3} \right] - \left[\frac{3}{4} - \frac{1}{2} \right]$$

$$= \frac{2}{3} - \frac{1}{4}$$

$$= \frac{5}{12}.$$

21. $\int_0^1 \int_1^3 \int_{-1}^2 (2x + 3y - z)\, dx\, dy\, dz$

$= \int_0^1 \int_1^3 \left(\int_{-1}^2 (2x + 3y - z)\, dx \right) dy\, dz$

We first evaluate the inside x-integral, treating y and z as constants:

$\int_{-1}^2 (2x + 3y - z)\, dx$

$= \left[x^2 + 3yx - zx \right]_{-1}^2$

$= \left[(2)^2 + 3y(2) - z(2) \right] -$

$\qquad \left[(-1)^2 + 3y(-1) - z(-1) \right]$

$= \left[4 + 6y - 2z \right] - \left[1 - 3y + z \right]$

$= 3 + 9y - 3z$

Then we evaluate the middle y-integral, treating z as a constant:

$\int_1^3 \left(\int_{-1}^2 (2x + 3y - z)\, dx \right) dy$

$= \int_1^3 (3 + 9y - 3z)\, dy$

$= \left[3y + \frac{9}{2}y^2 - 3zy \right]_1^3$

$= \left[3(3) + \frac{9}{2}(3)^2 - 3z(3) \right] -$

$\qquad \left[3(1) + \frac{9}{2}(1)^2 - 3z(1) \right]$

$= \left[9 + \frac{81}{2} - 9z \right] - \left[3 + \frac{9}{2} - 3z \right]$

$= 42 - 6z.$

Finally, we evaluate the outside z integral:

$\int_0^1 \left(\int_1^3 \left(\int_{-1}^2 (2x + 3y - z)\, dx \right) dy \right) dz$

$= \int_0^1 (42 - 6z)\, dz$

$= \left[42z - 3z^2 \right]_0^1$

$= \left[42(1) - 3(1)^2 \right] - \left[42(0) - 3(0)^2 \right]$

$= 42 - 3$

$= 39.$

23. $\int_0^1 \int_0^{1-x} \int_0^{2-x} (xyz)\, dz\, dy\, dx$

$= \int_0^1 \int_0^{1-x} \left(\int_0^{2-x} (xyz)\, dz \right) dy\, dx$

We first evaluate the inside z-integral, treating x and y as constants:

$\int_0^{2-x} (xyz)\, dz$

$= \left[\frac{1}{2} xyz^2 \right]_0^{2-x}$

$= \left[\frac{1}{2} xy(2 - x)^2 \right] - \left[\frac{1}{2} xy(0)^2 \right]$

$= \left[\frac{1}{2} xy(2 - x)^2 \right] - [0]$

$= \frac{1}{2} x(2 - x)^2 y$

Then we evaluate the middle y-integral, treating x as a constant:

$\int_0^{1-x} \left(\int_0^{2-x} (xyz)\, dz \right) dy$

$= \int_0^{1-x} \left(\frac{1}{2} x(2 - x)^2 y \right) dy$

$= \frac{1}{2} x(2 - x)^2 \left[\frac{1}{2} y^2 \right]_0^{1-x}$

$= \frac{1}{2} x(2 - x)^2 \left[\frac{1}{2}(1 - x)^2 - \frac{1}{2}(0)^2 \right]$

$= \frac{1}{4} x(2 - x)^2 \left[(1 - x)^2 \right]$

$= \frac{1}{4} x(4 - 4x + x^2) \left[1 - 2x + x^2 \right]$

$= \left[x - x^2 + \frac{1}{4} x^3 \right] \left[1 - 2x + x^2 \right]$

$= x - 2x^2 + x^3 - x^2 + 2x^3 - x^4 +$

$\qquad \frac{1}{4} x^3 - \frac{1}{2} x^4 + \frac{1}{4} x^5$

$= \frac{1}{4} x^5 - \frac{3}{2} x^4 + \frac{13}{4} x^3 - 3x^2 + x$

Finally, we evaluate the outside x integral:

$$= \int_0^1 \int_0^{1-x} \left(\int_0^{2-x} (xyz)\,dz \right) dy\,dx$$

$$= \int_0^1 \left(\frac{1}{4}x^5 - \frac{3}{2}x^4 + \frac{13}{4}x^3 - 3x^2 + x \right) dx$$

$$= \left[\frac{1}{24}x^6 - \frac{3}{10}x^5 + \frac{13}{16}x^4 - x^3 + \frac{1}{2}x^2 \right]_0^1$$

$$= \left[\frac{1}{24}(1)^6 - \frac{3}{10}(1)^5 + \frac{13}{16}(1)^4 - (1)^3 + \frac{1}{2}(1)^2 \right] -$$

$$\qquad \left[\frac{1}{24}(0)^6 - \frac{3}{10}(0)^5 + \frac{13}{16}(0)^4 - (0)^3 + \frac{1}{2}(0)^2 \right]$$

$$= \left(\frac{1}{24} - \frac{3}{10} + \frac{13}{16} - 1 + \frac{1}{2} \right) - (0)$$

$$= \frac{10}{240} - \frac{72}{240} + \frac{195}{240} - \frac{240}{240} + \frac{120}{240}$$

$$= \frac{13}{240}.$$

25. $\boxed{tw}$

27. Left to the student.